Overland flow

Hydraulics and erosion mechanics

EDITED BY

Anthony J. Parsons & Athol D. Abrahams

UCL
PRESS

First published in 1992 by UCL Press

UCL Press Limited
University College London
Gower Street
London WC1E 6BT

The name of University College London (UCL) is a registered trade mark used by UCL Press with the consent of the owner.

ISBN: 1-85728-006-7 HB

A CIP catalogue record for this book is available from the British Library.

This book was typeset by Anthony J. Parsons.

Typeset in Times Roman.
Printed and bound by
Biddles Ltd, King's Lynn and Guildford, England.

Contents

Preface

Overland flow on hillslopes is of two types. Horton overland flow occurs where rainfall intensity exceeds the surface infiltration rate, whereas saturation overland flow takes place where water is forced to the surface or remains on it as a result of saturation of the soil beneath. Saturation overland flow is usually confined to footslopes marginal to stream channels, to topographic hollows and to areas of thin soil cover. In contrast, Horton overland flow most often occurs where sparse vegetation permits surface sealing or where bedrock outcrops limit infiltration. Horton overland flow is thus widespread in arid and semi-arid landscapes and on agricultural lands. For this reason studies of overland flow have tended to be conducted in these settings.

Research on overland flow has largely been conducted by geomorphologists and agricultural engineers/hydrologists. Because geomorphologists have generally worked in arid and semi-arid regions while agricultural engineers/hydrologists have focused upon agricultural lands, the research has proceeded more or less separately in each discipline. This separation has been exacerbated by the fact that most geomorphologists conducting this type of research have been located in Europe, whereas most agricultural engineers/hydrologists have been situated in the USA.

Much of the research of both geomorphologists and agricultural engineers/hydrologists is driven by a need to understand the causes and effects of soil erosion. During the past 15 years, models used to predict soil erosion have changed fundamentally from empirical "black-box" models to process-based "white-box" models. The integrity of process-based models depends upon an understanding of the hydraulics and erosion mechanics of overland flow. There have been a variety of methodologies employed to study this topic: field experiments, laboratory experiments and mathematical modelling. These methodologies are found in both geomorphology and agricultural engineering/hydrology. Although there has been much interaction among practitioners of the different methodologies within each discipline, there has been hardly any at all between the disciplines. The need for interaction across disciplinary boundaries is widely acknowledged in both disciplines, but hitherto no suitable forum has existed for this purpose.

To meet this need, we convened a workshop whose specific goals were to promote interaction between modellers, field workers and laboratory experimentalists both within and between the disciplines of geomorphology and agricultural engineering/hydrology. This workshop was held at the University of Keele, 8–11 July 1991, and its proceedings are contained in this volume. In addition to the papers presented at the workshop, two papers by authors who were unable to attend are included in the book. Substantial workshop time was devoted to discussions; the main themes that arose are summarized at the end of this preface.

Book structure and summary

Although no classification of any research activity can be clear-cut, we may broadly divide the topic of overland flow into the study of flow hydraulics, on the one hand, and of sediment entrainment and transport, on the other. Within these major divisions it is possible further to subdivide research according to whether it is based primarily upon field or laboratory experiments or upon modelling. The arrangement of this book reflects this classification.

The first nine chapters are concerned primarily with the nature of overland flow – that is, its hydraulics. The Abrahams et al. report on field experiments and associated laboratory work aimed at identifying the components and controls of resistance to interrill overland flow. Using the Darcy–Weisbach friction factor f to characterize resistance to flow, these authors show that at Froude numbers less than 0.5 form resistance is dominant and typically accounts for about 95% of f. At Froude numbers between 0.5 and 2.0 wave resistance becomes dominant. Both form resistance and wave resistance are due to irregularities in the ground surface. For gravel-covered surfaces on semi-arid hillslopes, Abrahams et al. produce predictive equations for f on gravel-covered surfaces of semi-arid hillslopes based upon gravel size and concentration and argue that, although on agricultural lands the sources of roughness will differ, resistance to flow will still be dominated by form and wave resistance. Gilley et al. continue the theme of analyzing resistance to overland flow through Darcy– Weisbach's f. Rather than presenting f as a function of such elements as form resistance, grain resistance, etc., these authors identify the physical contributors to resistance. So, for example, the roughness coefficient for rills on cropland is presented as the sum of roughness coefficients due to rills, to gravel and cobble materials, to surface residue and to plants. For each contributing factor an equation to predict it is developed.

At a larger scale, issues that are important in understanding overland flow are those of sources and routing. Yair examines the hydrometeorological conditions required for the initiation of overland flow on various geomorphic units within the Sede Boquer experimental watershed, Israel, and from this examination identifies the relative contribution of each of these units to storm channel flow. Not only are the runoff-producing properties of the various geomorphic units important in understanding storm channel flow, but so, too, are their spatial associations. Yair shows that rocky, upper valley-side slopes frequently produce runoff, but that this runoff is absorbed by lower colluvial slopes, thus generating no channel flow. In contrast, overland flow in rocky headwater areas that lead directly into the channel is not infiltrated. Thus it is these areas that are most significant in generating channel runoff in this arid watershed.

In Chapter 4 attention shifts to modelling. Lima addresses the issue of overland-flow generation on a surface that is simultaneously infiltrating, a topic which has generally been oversimplified (Stephenson & Meadows 1986, 15). Lima's approach recognizes that both processes may be described by continuity equations and that these equations must be solved simultaneously because their boundary conditions are interdependent. Aside from the particular details of the model he presents, Lima makes the important point that, inasmuch as all models are simplifications, a primary consideration in the choice of a model is the determination of the physical processes that are important in a particular situation. One element of this determination already alluded to is that of spatial scale. For both logistical and conceptual reasons, studies that focus on understanding processes are generally undertaken on a very small scale (tens of square metres, at most), whereas application of that understanding, whether for predictions of soil erosion on agricultural land or for conceptualizing long-term landform evolution, generally needs to be at a much larger scale (hundreds or thousands of square metres). In her distributed model for dynamic overland-flow hydraulics, Scoging attempts to tackle this issue.

Scoging recognizes that overland-flow hydraulics depend upon how runoff is generated at a point and upon how that runoff is routed. In her cell-based, distributed model, runoff generation is determined for individual cells, based upon fitting the parameters of an infiltration equation to some measurable ground-surface property (e.g. stone cover) of the cells. Runoff is routed out of the cells according to their slope vector. The hydraulics of flow are determined from the Darcy–Weisbach flow equation. As with parameters of the infiltration equation, the Darcy–Weisbach friction factor is defined by relating it to a measurable ground-surface property of the cells (cf. Abrahams et al. and Gilley et al.). The ability of this model to predict overland-flow hydraulics at a larger scale is examined by Scoging et al. who apply it to an 18 m by 35 m runoff plot at Walnut Gulch Experimental Watershed, Arizona. The model achieves significant success in correctly predicting the observed spatial distribution of overland-flow hydraulics. More important, perhaps, than its successes, are the implications of the problems encountered in applying the model, and the differences between its predictions and the empirical data. The former highlight the problems of using data obtained from small-plot experiments to make predictions at a bigger scale; the latter demonstrate the questions that arise, and hence the progress that can be made, by close linkage of models and empirical studies.

Whether or not obtaining the "right" answer is a good test of models of overland flow is a question that is raised by Grayson & Moore. Applying a flexible, distributed parameter, physically based model to a catchment near Wagga Wagga, New South Wales, the authors parameterize the model to simulate hydrographs for a series of recorded rainfall events. By varying surface roughness (Manning's n) and considering saturation overland flow, Horton overland flow, rill flow and ephemeral gullies in a series of simulations, a variety of runoff-producing areas is simulated. An inappropriate model is shown to give results similar to those from a model that incorporates the processes known to be occurring in the catchment. Although the values of Manning's n that are used are all within the range suggested as valid for surfaces of the type found in this catchment, the authors point to the fact that, because there is no physical basis for representing surface flow on natural surfaces as broad sheet flow, neither can there be a physical basis for such estimated flow characteristics. The authors conclude that the apparent conceptual sophistication of distributed-parameter models masks the frequently invalid or questionable algorithms on which they are based, and draws into question our ability to model overland flow in a deterministic manner.

Deterministic complexity is examined further by Phillips from the point of view that some of that complexity may, in part, be independent of environmental variability and be inherent in the system dynamics – that is, chaotic. Phillips presents a theoretical analysis which shows that saturation overland flow is stable and cannot be expected to display chaotic behaviour, whereas Horton overland flow may be unstable and, hence, may do so. Examples drawn from field studies suggest that chaos is not commonplace in surface runoff but that the possibility of chaos needs to inform decisions about appropriate approaches to modelling. Specifically, the presence of chaos limits deterministic prediction and hence the utility of increasingly sophisticated deterministic modelling. For practical purposes, prediction must rely more on stochastic approaches.

Such a stochastic approach characterizes the model presented by Baird et al. These authors argue that models which assume a uniform water depth for overland flow and average roughness of the slope cannot be realistically applied to complex slope topography or the evolution of three-dimensional topography. Likewise, the assumption of a rill–interrill dichotomy, common in many models, negates the possibility of rill formation and growth during landform evolution. To overcome these limitations the authors develop a reticular flow model in which it is assumed that water arriving from an upslope reach is distributed on a topographically complex surface from the deepest point upwards. This model allows flow concentrations to change in size, shape and number, both over time and down slope.

For the second half of the book, attention shifts towards the entrainment and transportation capabilities of overland flow. Because overland flow is conceptually no different from very shallow channel flow and because transport equations have been developed for rivers, it has been tempting simply to apply these transport equations to overland flow. Both Guy et al. and Govers examine the difficulties of this approach. As Govers points out, existing formulae are empirical and their use in overland flow is outside the domain for which they were developed. In consequence, they are of limited utility in overland flow studies. There are two solutions to this problem; either to develop more physically based transport equations or to develop empirical equations for use with overland flow. In the short term the latter solution is likely to offer the greater prospect of success. For transport inception Guy et al. and Poesen develop equations to predict the critical discharge for transport inception that performs better than the Shields criterion. Guy et al. also demonstrate the importance of rainfall impact on transport capacity of overland flow, showing that it can account for 85% of total transport capacity where the flow is very shallow (i.e. less than half the mean raindrop diameter).

It is, of course, impossible to separate, in any real sense, the generation and routing of overland flow from its erosive effects. Furthermore, since the former determine the latter, any physical modelling of soil erosion by overland flow must begin with overland-flow generation and routing. However, research ranges from that which is driven by a desire to develop predictive equations (Poesen and Simanton

& Renard) to that which is more concerned with conceptual understanding of the mechanics of erosion (Parsons & Abrahams). The two goals ought to be convergent, but the studies presented here indicate that, at the present time, focus on one limits the attention given to the other. Poesen reports on a series of experiments on soils of different textures and differing rock-fragment content to determine their susceptibility to sealing, runoff coefficients and sediment yield. He is able to present a series of empirical equations for predicting sediment yield from such soil parameters as rock fragment cover, shear strength and particle size. Likewise, Simanton & Renard's focus leads to a series of empirical relationships between erosion rate and such variables as time, plot treatment and vegetation. In contrast, Parsons & Abrahams present empirical results which challenge commonly held beliefs but do not yield predictive equations. Their studies focus on the relative importance of grain and form shear-stress in explaining the paradoxically low flow detachment by interrill overland flow, the inter-relationships of gradient, rock-fragment size and runoff coefficient, and the effects of flow routing on competence of overland flow to remove particles detached by raindrop impact.

It is the desire to make physically based predictions that underlies the US Department of Agriculture Water Erosion Prediction Project (WEPP) model (Lane et al.). The contrast in approach between WEPP and the Universal Soil Loss Equation (USLE) is evident from a comparison of the chapters by Simanon & Renard and Lane et al. Inevitably, questions similar to those posed by Grayson & Moore must arise. It is early days yet for the WEPP model, but tests of its predictions, of the type undertaken by Grayson & Moore for overland-flow generation, need to be made.

The link between models of overland-flow generation and routing and those of sediment yield is considered further by Lea who presents an aspect-driven kinematic routing algorithm and considers its utility for sediment routing in a river basin. The issues of delivery ratio, akin to those considered by Yair for overland flow, become critical at this scale.

In the final chapter, which parallels that of Baird et al. in terms of its perspective, Schmidt draws our attention to the fact that physically based models are not solely of use in predicting short-term soil erosion, but need to be employed by geomorphologists seeking to understand long-term landform evolution. In a series of simulations, Schmidt shows predicted landform change under such a model. Of interest to geomorphologists and agricultural engineers alike are the results of his sensitivity analysis. Although Schmidt's concern in these simulations is with dependence of landform change on conditions of hillslope geometry and surface condition rather than with processes, his analysis is, in effect, the converse of that of Grayson & Moore. Landform evolution and sediment yield may not be strongly dependent upon the types of process active, so that the "right" answers, in a predictive sense, do not depend upon having the right model.

Discussion themes

Although most of the discussion at the workshop was specific to particular papers, there were a number of recurring themes of a broader nature. These themes reflect current ideas on the nature and future direction of research on overland flow. For that reason, they are included here. Because these themes represent questions about the purpose and future direction of research on overland flow, they are presented below under headings that take the form of questions.

What are the purposes of mathematical models?
Mathematical models represent a formalization of our conceptualization of reality. Such models may have one of two objectives: to make predictions for practical applications, or to yield inferences about the operation of physical processes. Agricultural engineers/hydrologists have been primarily concerned with the former, while geomorphologists have concentrated on the latter. Thus agricultural engineers/hydrologists have developed the kinematic wave model for overland flow (Woolhiser & Liggett 1967) and predictive soil erosion models such as CREAMS, KINEROS and WEPP. On the other hand, geomorphologists have proposed models for long-term landform evolution (Kirkby 1971), saturation overland flow (Kirkby & Chorley 1967, Bevan & Kirkby 1979) and soil erosion (Kirkby 1980, Schmidt 1991).

How useful are soil erosion models developed for agricultural lands for geomorphic studies of "natural" landscapes?
To a large extent models developed by geomorphologists have tended to lag behind those developed by agricultural engineers/hydrologists. For example, geomorphologists continue to model hillslope evolution assuming transport-limited erosion (e.g. Band 1990), whereas agricultural engineers/hydrologists have long abandoned this notion. There is a need to apply the more sophisticated models developed by agricultural engineers/hydrologists to hillslope evolution. However, there are a number of problems with using soil erosion models in this way. First, soil erosion models are "event" models which fail to take into account the effects of variations in antecedent conditions (Simanton & Renard, this volume, Govers 1991). Secondly, geomorphic studies need to take into account the climatic and tectonic histories of landscapes and the topographic settings of hillslopes within the landscape as a whole (Carter & Chorley 1961, Soons & Rainer 1968, Arnett 1971, Frostick & Reid 1982, Yair, this volume). Thirdly, the amount of sediment storage, particularly along stream channels, is underestimated by soil erosion models (Walling 1990). Fourthly, it is doubtful whether models of soil erosion can usefully be applied for periods beyond the order of a few decades on agricultural land and a few centuries on rangeland, whereas geomorphologists are interested in landscape evolution over longer time spans. Given the limitations that these problems impose, it is not clear that modern soil erosion models can be used to investigate long-term landscape evolution. However, they may have a rôle in identifying the sensitivity and character of landscape change over secular time spans.

Do modern process-based soil erosion models yield better predictions than better-established empirically derived soil erosion models?
Empirical models, such as the Revised Universal Soil Loss Equation (RUSLE), are at the limit of their predictive capabilities, and they lack the potential to incorporate new understanding of the erosion mechanics of overland flow. Process-based models, such as that generated by the Water Erosion Prediction Project (WEPP), do not appear to perform significantly better at the present time than the former models. However, they do have the potential for incorporating new knowledge, and will presumably provide better predictions in the future.

What is wrong with current practice in model validation and how might it be improved?
In lumped empirical models, the only available test of the model is the degree to which its output agrees with reality. Unfortunately, this mode of model testing has often been transferred to process-based models. Because it is relatively easy to obtain the right answer for the wrong reasons (Grayson & Moore, this volume), by itself this mode of testing is not sufficient to validate process-based models, which aim to obtain the right answers for the right reasons! It is essential in process-based models to test the process components of the model as well as the final output (Scoging et al., this volume). Further, it is necessary to examine the sensitivity of the model output to error in each of the process components. Finally, parameter optimization should be avoided, as it often leads to model errors being hidden in the optimized parameter values. Estimates of parameter values should be based on independent field or laboratory data, so that model errors can be identified.

How can field and laboratory experiments be better integrated with modelling?
Field and laboratory experiments which improve physical understanding of overland flow should precede and guide model design. To be most effective, the results of such experiments need to be presented in a such a manner that they are readily assimilated into mathematical models. Effectively, this means that experiments need to produce equations that link measurable parameters to desired model output quantities. Field and laboratory experiments may also be employed to parameterize models. Where such experiments identify diversity in the real world, such diversity needs to be expressed in terms of statistical distributions of parameters that can be utilized in stochastic models (e.g. Abrahams et al. 1989, Baird et al., this volume). Some field and laboratory experiments should have the specific goal of model testing. Field testing is particularly important where surface material properties are a significant determinant of the behaviour of the system being modelled. Laboratory testing is especially useful for isolating multiple system controls and testing the sensitivity of the model to each of them. However, given the scale limitations that often constrain laboratory and field experiments, such experiments may be able to test models only over the lower end of their range of applications.

What are the problems of using data collected at the small-plot scale in runoff/soil erosion models at field/hillslope scales?
Small plots are incapable of capturing either across-slope variation or downslope changes in overland flow. First, they fail to identify the full range of infiltration values. Secondly, they do not sample the range of overland-flow depths and velocities either in interrill areas or in rills. Thirdly, they fail to capture systematic downslope changes in flow concentration and its distribution between rill and interrill flow. These deficiencies have serious ramifications for modelling overland-flow hydraulics and, hence, sediment transport. Finally, simple extrapolation of sediment yields from small plots fails totally to take account of sediment storage at the foot of hillslopes and thus leads to overestimation erosion rates at the hillslope scale.

Can we identify the key variables that control overland flow?
In all probablility this is easier in arid and semi-arid areas than it is on agricultural lands. Yair (1990, this volume) showed the importance of rock:soil ratio and the presence of colluvial footslopes as controls of runoff, and of biological activity as a determinant of sediment yield. In contrast, many and varied physical and chemical properties of soils have been shown to affect runoff (Musgrave & Holtan 1964) and sediment yields from agricultural lands (Gilley et al. 1992).

References

Abrahams, A. D., A. J. Parsons, S.-H. Luk 1989. Distribution of depth of overland flow on desert hillslopes and its implications for modeling soil erosion. *Journal of Hydrology* **106**, 177–84.

Arnett, R. R. 1970. Slope form and geomorphological processes: an Australian example. In *Slopes: form and process*, 81–92. Institute of British Geographers special publication no. 3. London: Institute of British Geographers.

Band, L. E. 1990. Grain size catenas amd hillslope evolution. *Catena Supplement* **17**, 167–76.

Bevan, K. J. & M. J. Kirkby 1979. A physically-based variable contributing area model of basin hydrology. *Hydrological Sciences Bulletin* **24**, 43–69.

Carter, C. A. & R. J. Chorley 1961. Early slope development in an expanding stream system. *Geological Magazine* **98**, 117–30.

Gilley, J. E., D. C. Kincaid, W. J. Elliot, J. M. Laflan 1992. Sediment delivery on rill and interrill areas. *Journal of Hydrology* (in press).

Govers, G. 1991. Time-dependency of runoff velocity and erosion: the effect of the initial soil moisture profile. *Earth Surface Processes and Landforms* **16**, 713–29.

Frostick, L. E. & I. Reid 1982. Talluvial processes, mass wasting and slope evolution in arid environments. *Zeitschrift für Geomorphologie Supplementband* **44**, 53–67.

Kirkby, M. J. 1971. Hillslope process-response models based on the continuity equation. In *Slopes: form and process*, 15–30. Institute of British Geographers special publication no. 3. London: Institute of British Geographers.

Kirkby, M. J. 1980. Modelling water erosion processes. In *Soil erosion*, Kirkby, M. J. & R. P. C. Morgan (eds.), 183–216. Chichester, England: John Wiley.

Kirkby, M. J. & R. J. Chorley 1967. Throughflow, overland flow and erosion. *Bulletin of the International Association for Scientific Hydrology* **12**, 5–21.

Musgrave, G. W. & H. N. Holtan 1964. Infiltration. In *Handbook of applied hydrology*, V. T. Chow (ed.), 12-1–12-30. New York: McGraw–Hill.

Schmidt, J. 1991. A mathematical model to simulate rainfall erosion. *Catena Supplement* **19**, 101–9.

Soons J. M. & J. N. Rainer 1968. Micro-climate and erosion processes in the Southern Alps, New Zealand. *Geografiska Annaler,* Series A **50**, 1–15.

Stephenson D. & M. E. Meadows 1986. *Kinematic hydrology and modelling.* Amsterdam: Elsevier.

Walling, D. A. 1990. Linking the field to the river: sediment delivery from agricultural land. In *Soil erosion on agricultural land*, J. Boardman, I. D. L. Forster, J. A. Dearing (eds), 130–52. Chichester, England: John Wiley

Woolhiser D. A. & J. A. Liggett 1967. Unsteady, one-dimensional flow over a plane – the rising hydrograph. *Water Resources Research* **3**, 753–71.

Yair, A. 1990. Runoff generation in a sandy area – the Nizzana Sands, western Negev, Israel. *Earth Surface Processes and Landforms* **15**, 597–609.

Acknowledgements

Keeping to schedule in a multi-authored book depends largely on the willingness of the contributors to meet deadlines. We would like to express our thanks to the contributors to this book for their prompt and helpful responses throughout. We owe gratitude to John Wainwright, particulary for his help in formatting the tables. Roger Simanton kindly provided the photograph used on the cover.

Contributors

Athol D. Abrahams *Department of Geography, State University of New York at Buffalo, Buffalo, New York 14260, USA*

Andrew J. Baird *Dept of Geography and Geology, Cheltenham and Gloucester College of Higher Education, Cheltenham GL50 3PP, England*

W. T. Dickinson *School of Engineering, University of Guelph, Guelph, Ontario, Canada*

Dennis C. Flanagan *USDA–ARS, Purdue University, West Lafayette, Indiana 47907, USA*

G. R. Foster *Aridland Watershed Management Research Unit, USDA–ARS, 2000 E. Allen Road, Tucson, Arizona 85719, USA*

John E. Gilley *USDA–ARS, University of Nebraska, Lincoln, Nebraska 68583-0729, USA*

Gerard Govers *Laboratory for Experimental Geomorphology, Catholic University of Leuven, Redingenstraat 16 bis, B-3000 Leuven, Belgium*

Rodger B. Grayson *Centre for Environmental Applied Hydrology, Department of Civil and Agricultural Engineering, University of Melbourne, Parkville, Victoria 3052, Australia*

B. T. Guy *TRITON Environmental Consultants Ltd., Richmond, BC, Canada*

Paul J. Hirsch *Department of Geography, State University of New York at Buffalo, Buffalo, New York 14260, USA*

Eugene R. Kottwitz *Department of Biological Systems Engineering, University of Nebraska, Lincoln, Nebraska 68583–0729, USA*

J. M. Laflen *Southwest Watershed Research, USDA–ARS, 2000 E. Allen Road, Tucson, Arizona 85719, USA*

L. J. Lane *Southwest Watershed Research, USDA–ARS, 2000 E. Allen Road, Tucson, Arizona 85719, USA*

Nicholas J. Lea *Hydraulics Research, Wallingford, Oxfordshire, OX10 8BA, England*

J. L. M. P. de Lima *Department of Civil Engineering, Faculty of Science and Technology, University of Coimbra, 3049 Coimbra Codex, Portugal*

Ian D. Moore *Centre for Resource and Environmental Studies, The Australian National University, Canberra, ACT 2601, Australia*

M. A. Nearing *Southwest Watershed Research, USDA–ARS, 2000 E. Allen Road, Tucson, Arizona 85719, USA*

M. H. Nichols *Southwest Watershed Research, USDA–ARS, 2000 E. Allen Road, Tucson, Arizona 85719, USA*

Anthony J. Parsons *Department of Geography, University of Keele, Keele, Staffordshire ST5 5BG, England*

Jonathan D. Phillips *Department of Geography, East Carolina University, Greenville, North Carolina 27858-4353, USA*

J. W. A. Poesen *Laboratory for Experimental Geomorphology, Catholic University of Leuven, Redingenstraat 16 bis, B-3000 Leuven, Belgium*

K. G. Renard *Southwest Watershed Research, USDA–ARS, 2000 E. Allen Road, Tucson, Arizona 85719, USA*

R. P. Rudra *School of Engineering, University of Guelph, Guelph, Ontario, Canada*

Jürgen Schmidt *Sächsisches Landesamt für Umwelt und Geologie, Halsbrücker Str. 31a, D(O) - 9200 Freiberg, Germany*

H. M. Scoging *Department of Geography, London School of Economics, Houghton Street, London WC2A 2AE, England*

J. R. Simanton *Southwest Watershed Research, USDA–ARS, 2000 E. Allen Road, Tucson, Arizona 85719, USA*

John B. Thornes *Department of Geography, King's College London, Strand, London WC2R 2LS, England*

Glenn P. Watts *National Rivers Authority, Manley House, Exeter, England*

Mark A. Weltz *Southwest Watershed Research, USDA–ARS, 2000 E. Allen Road, Tucson, Arizona 85719, USA*

Aaron Yair *Institute of Earth Sciences, The Hebrew University of Jerusalem, Jerusalem 91904, Israel*

1

Field and laboratory studies of resistance to interrill overland flow on semi-arid hillslopes, southern Arizona

Athol D. Abrahams, Anthony J. Parsons, Paul J. Hirsch

Abstract

Resistance to overland flow is divided into grain resistance, form resistance and wave resistance. The conventional relation between Darcy–Weisbach friction factor f and Reynolds number Re for shallow flow over a plane bed applies only where f is due entirely to grain resistance, which is manifestly not the case on gravel-covered semi-arid hillslopes. Field experiments at Walnut Gulch Experimental Watershed, Tombstone, Arizona, indicate that f-Re relations are convex-upward, positively sloping and negatively sloping. These shapes are attributed to the progressive inundation of the roughness elements. A second set of field experiments reveals that 50.1% of the variance in f can be explained by percentage gravel cover. Quantitative estimates of the relative magnitudes of grain resistance f_g and form resistance f_f for the latter experiments disclose that f_g is typically about 5% and f_f 95% of f.

Wave resistance f_w becomes significant when the Froude number F lies between 0.5 and 2. Flume experiments conducted on a plane bed with cylindrical roughness elements protruding through the flow suggest that f_w is insensitive to Re but very sensitive to concentration of roughness elements (i.e. percentage cover) C. f_w increases rapidly with C and becomes larger than f_g and f_f once C exceeds 10%. At C values of 20%, f_w is about 70% of f. Despite their limitations, the flume experiments leave little doubt that f_w is a major contributor to f on gravel-covered semi-arid hillslopes when $0.5 < F < 2$. It is concluded that resistance to interrill overland flow is determined largely by surface form through form resistance and wave resistance.

Introduction

On semi-arid hillslopes virtually all runoff occurs in the form of overland flow which is generated when rainfall intensity exceeds the surface infiltration rate. Overland flow may be conveniently divided into rill and interrill flow, of which only the latter is examined here. Interrill overland flow generally appears as a sheet of water with threads of deeper, faster flow diverging and converging around surface protuberances, rocks and vegetation. As a result of these diverging and converging threads, flow depth and velocity may vary markedly over short distances, giving rise to changes in the state of flow. Thus over a small area the flow may be wholly laminar, wholly turbulent, wholly transitional or consist of patches of any of these three flow states. Flow resistance may be estimated by the dimensionless Darcy–Weisbach friction factor

$$f = \frac{8gds}{v^2} \tag{1.1}$$

where g is the acceleration of gravity, d the mean depth of flow, s the energy slope, and v the mean flow velocity.

Total resistance to overland flow, f, may be divided into grain resistance, form resistance, wave resistance and rain resistance. Grain resistance is imparted by soil particles and microaggregates that protrude into the flow less than about 10 times the thickness, δ, of the viscous sublayer (Yen 1965). Traditionally, grain resistance has been viewed as the kinetic energy dissipated by the flow in overcoming the no-slip condition at the boundary. But it also includes energy dissipated in separation eddies generated by the particles and microaggregates that protrude above the viscous sublayer. Form resistance is exerted by microtopographic protuberances, stones and vegetation that protrude more than 10δ into the flow or give rise to spatially varying flow cross sections or changing flow directions. Energy is dissipated in separation eddies, secondary circulation and locally increased shear. Wave resistance is the energy loss due to deformation of the free water surface or, put another way, the energy dissipated in maintaining an uneven water surface. Wave resistance includes spill resistance, defined as the sudden forced reduction in flow velocity due to an abrupt diverging of the streamlines (Leopold et al. 1960). Rapid decelerations in overland flow occur at the base of local steepenings in the ground surface and where flow emerges from between large roughness elements. Finally, rain resistance is due to velocity retardation as flow momentum is transferred to accelerate the raindrop mass to the velocity of the flow. Laboratory studies have shown that for laminar flow on gentle slopes rain resistance may attain 20% of f (Savat 1977), but generally it is a much smaller proportion. For transitional and turbulent flow this proportion is even smaller (Yoon & Wenzel 1971, Shen & Li 1973). Field studies have indicated that on semi-arid hillslopes, rain resistance is typically several orders of magnitude less than f (Dunne & Dietrich 1980). Because rain resistance appears to be relatively insignificant on such hillslopes, this paper will focus on grain, form and wave resistance.

Resistance to flow generally varies with the rate of flow, which is represented by the dimensionless Reynolds number

$$Re = \frac{4vd}{\nu} \tag{1.2}$$

where ν is the kinematic fluid viscosity. Laboratory experiments and theoretical analyses since the 1930s have established that the relation between f and Re for shallow flow is a power relation whose slope coefficient (i.e. exponent) depends on the state of flow. The relation has a slope of -1.0 where the flow is laminar and a slope of approximately -0.25 where it is turbulent (Fig. 1.1). Virtually all models of hillslope runoff have employed this relation between f and Re (or surrogates thereof). However, this relation applies only to plane beds where the resistance to flow is due entirely to grain resistance, whereas the surfaces of semi-arid hillslopes are rarely, if ever, planar, and the anastomosing pattern of overland flow around microtopographic protuberances, rocks and vegetation attests to the importance of form and wave resistance. If form and wave resistance are important, their influence might be expected to be reflected in the shape of the f–Re relation, which should be quite different from that in Figure 1.1.

Emmett (1970) investigated the relation between f and Re for semi-arid hillslopes in Wyoming. However, his analysis was concerned with the downslope relation between these variables rather than the at-a-section relation, which is of concern here. Dunne & Dietrich (1980) graphed f against Re for runoff plots on savanna hillslopes in Kenya. But these authors lumped together both at-a-section and downslope data on their graphs, thereby obscuring the actual shapes of the at-a-section relations. Moreover, they forced linear f–Re relations with slopes of -1.0 through the lumped data on the assumption that the flow was laminar. Roels (1984) computed the f–Re relations for 11 runoff plots with rilled and prerilled surfaces in the Ardeche drainage basin, France. Although Re ranged from 75 to 14,000, none of the relations exhibited a break of slope corresponding to transitional flow. Indeed, all the relations were linear with negative slopes ranging from -0.07 to -0.90. Although Roels did not recognize the significance of his findings, they represent the first empirical evidence that on rough natural hillslope surfaces where form and wave resistance dominate, the f–Re relation departs from the conventional relation for shallow flow over a plane bed.

This chapter argues that resistance to overland flow on gravel-covered, semi-arid hillslopes is primarily a function of surface form, which acts through form

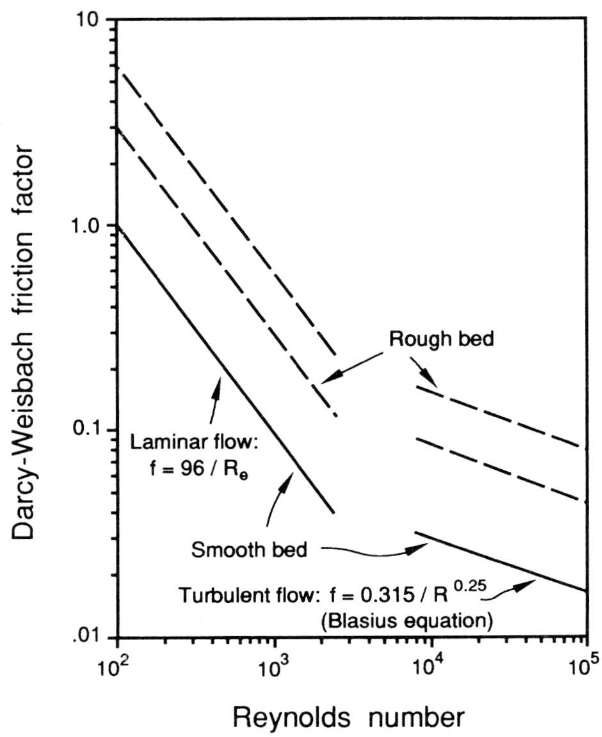

Figure 1.1 Graph of Darcy–Weisbach friction factor against Reynolds number, showing the relations for smooth and rough plane beds.

resistance and wave resistance. The first part of the chapter describes two sets of field experiments which demonstrate the role of surface form and the importance of form resistance where Froude numbers are relatively low. The second part of the chapter summarizes a set of laboratory experiments which show that at higher Froude numbers, wave resistance may dominate resistance to overland flow.

Field site

Both sets of field experiments were conducted at Walnut Gulch Experimental Watershed, Tombstone, Arizona (31°43′N, 110°41′W). The watershed has a warm, semi-arid climate with a mean annual precipitation of 288 mm and a mean monthly temperature range of 8°–27°C (Osborn 1983). Toward the end of the 19th century, the vegetation over much of the watershed underwent a transition from a grass- to a shrub-dominated community. A similar transition has been recorded over many parts of the northern Sonoran Desert (Glendening

1952, Hastings & Turner 1965). As the shrubs grew and the understorey grasses thinned, the A-horizon of the pre-existing soil was selectively eroded in the intershrub areas by a combination of rainsplash and overland flow, leaving behind a gravel lag. Under many of the larger shrubs the A-horizon has been preserved. In addition, under virtually all shrubs with moderate to dense canopies, sand-sized particles have accumulated as a result of differential splash – that is, the transport by rainsplash of more sediment from intershrub to shrub areas than in the reverse direction, due to protection by the shrub canopy of the ground surface beneath it from the action of raindrops. The result is a mosaic of gravel-covered intershrub areas interspersed with shrub-protected areas of fines, forming microtopographic highs (Parsons et al. 1992).

The experiments were all performed on the interrill portions of shrub-covered piedmont hillslopes underlain by well-cemented, coarse Quaternary alluvium. The shrub community is dominated by *Larrea tridentata*, *Mortonia scabrella*, *Dasylirion wheeleri*, *Acacia constricta*, *Rhus microphylla*, *Nolina microcarpa*, and *Yucca elata*, with a ground layer principally of *Zinnia pumila* and *Dyssodia acerosa*. The vegetation typically covers less than half the ground surface, and almost all the shrubs stand atop mounds. These mounds contain a higher proportion of sand and larger quantities of organic matter than adjacent gravel-covered intershrub areas. They are also subject to greater amounts of digging and burrowing by animals, notably rabbits and pack rats. These factors, along with the effect of the shrub canopy itself in dissipating the kinetic energy of raindrops and thereby inhibiting surface sealing, cause infiltration rates to be much higher beneath shrubs than between them (Abrahams & Parsons 1991a). As a result, overland flow is generated preferentially in the gravel-covered intershrub areas and travels down slope in well defined threads that are largely confined to these areas that form swales between shrub-protected mounds.

The friction factor–Reynolds number relation

The first set of experiments (Abrahams et al. 1986) was conducted on six plots 1.8 m wide and 5.5 m long that ranged in gradient from 6° to 33°. These plots were all located in gravel-covered intershrub areas. Although gravel mantled the plot surfaces, clipped plant stems occupied as much as 10% of their area, and the steeper plots had quite irregular surfaces. Overland flow was simulated by trickling water onto each plot at six different rates. The outflow rate for each run was measured volumetrically at the lower end of the plot. Three cross sections, C1, C2 and C3, were established on each plot at 1.25, 3.25 and 5.25 m from the upper end of the plot. Discharges, Q, at these cross sections were computed by assuming that infiltration and evaporation losses were uniformly distributed over the plot. Once equilibrium discharge was obtained, flow depth was measured with a thin scale at intervals of 5 cm across each cross section, and these measurements were averaged to yield mean flow depth, d. The mean velocity, v, was calculated by dividing Q by wd, where w is the width of the cross section. The energy slope was approximated by the local ground slope, kinematic viscosity, ν, was estimated from water temperature, and f and *Re* were calculated using Equations 1.1 & 1.2.

Four of the 18 cross sections were discarded because they were affected by erosion of a plunge pool, formation of an organic dam, or scouring by debris flows. For the remaining 14 cross sections, f was plotted against *Re* on log–log axes (Fig. 1.2). The relation between f and *Re* for each cross section was then determined algebraically from the corresponding d–Q relation. Because the general form of the d–Q relation is log-quadratic, so is the general form of the f–*Re* relation. The computed f–*Re* relations for the 14 cross sections are displayed in Figure 1.2. Six of these relations are convex-upward, three are positively sloping, and five negatively sloping. The range of slope coefficients for the negatively sloping relations is -0.078 to -0.893, which is almost identical to that reported by Roels (1984). These shapes are attributed to the progressive inundation of the roughness elements (i.e. gravel, plant stems and microtopographic protuberances). So long as these elements are emergent from the flow, f increases with *Re* as the upstream wetted projected area of the elements increases. However, once the elements become submerged, f decreases as *Re* increases, and the ability of the elements to retard the flow progressively decreases.

These field experiments show that the conventional f–*Re* relation for shallow flow over a plane bed does not apply to the semi-arid hillslopes under study. Furthermore, they suggest that the shape of the relevant f–*Re* relation is not a function of the state of flow, as is the case for the conventional relation for a plane bed, but of surface form. The same is likely to be true of other semi-arid hillslopes with gravel-covered surfaces. Consequently, such hillslopes might be expected to have f–*Re* relations similar in shape to those in Figure 1.2, and the conventional relation should not be employed in runoff models for such hillslopes.

Multivariate relations for friction factor

There is, however, a more general problem with the use of f–*Re* relations in runoff models for gravel-covered semi-arid hillslopes. If f is largely determined by surface form, as the evidence suggests, f might be expected to be predicted by one or more measures of surface form better than by *Re*.

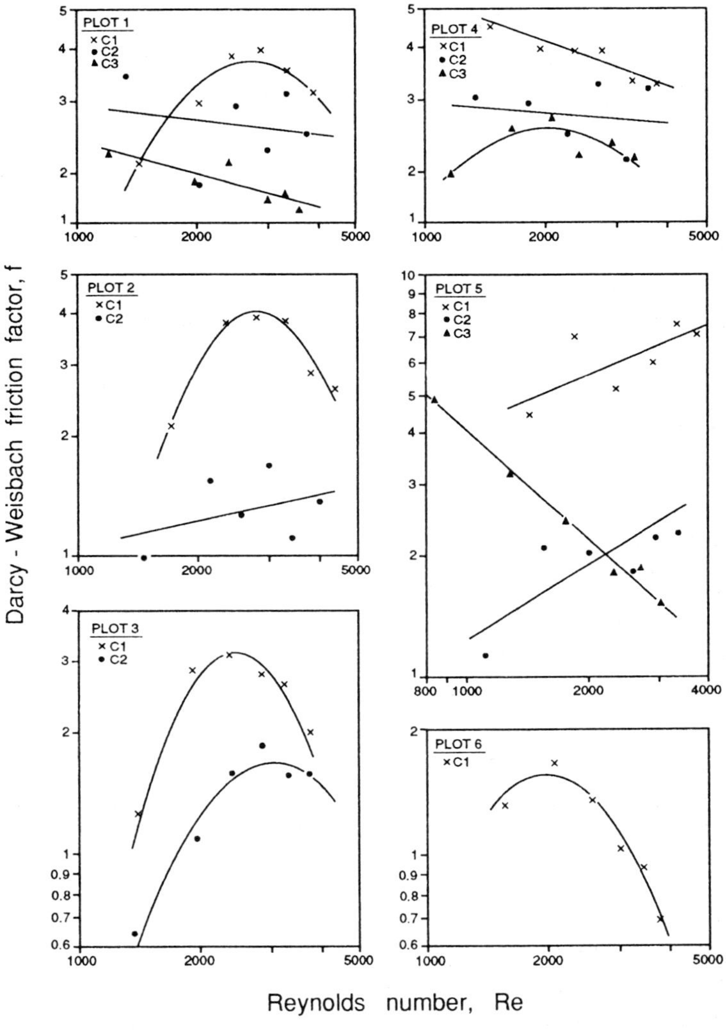

Figure 1.2 Graphs of Darcy–Weisbach friction factor against Reynolds number for 14 cross sections (C1, C2 and C3) on six runoff plots at Walnut Gulch Experimental Watershed, Arizona.

Indeed, as Figure 1.2 illustrates, owing to the dependence of *f* on surface form, the shape of the *f–Re* relation changes markedly over short distances across a hillslope as surface form changes. As a result, there is unlikely to be any single *f–Re* relation that applies to an entire hillslope. This suggests that *f–Re* relations are of limited utility in hillslope runoff models for gravel-covered hillslopes.

A possible solution to this problem is to replace the *f–Re* relation with a multivariate relation that includes surface form variables. Abrahams et al. (1986) attempted to relate observed values of *f* to microrelief and surface sediment characteristics for the six runoff plots investigated in the above experiments. However, no significant correlations were found. The reasons for this were not clear, but they were probably related to the wide range of plot gradients and the fact that surface gravel, microtopographic protuberances, and plant stems all contributed to f to varying degrees on the different plots. Accordingly, a second set of field experiments was conducted on eight plots also sited in intershrub areas at Walnut Gulch (Abrahams & Parsons 1991b). These plots, which were 0.61 m wide and 1.5 m long, had a limited range of gradients (1.33° and 4.86°), no plants and imperceptible microtopography; only the surface gravel protruded into the overland flow.

Water was trickled onto the upper end of each plot at a variety of rates, and outflow from the plot was measured volumetrically. A 0.61 m by 0.61 m test section was established on the lower half of each plot. Discharge, Q, at the middle of the test section was computed from the plot inflow and outflow rates assuming that infiltration and evaporation losses were uniform over the plot. Once equilibrium runoff was achieved, flow depth was measured using a millimetre scale at 55 points arranged in a grid over the test section. The 55 depth measurements so recorded, excluding those where the flow depth was zero (Abrahams & Parsons 1990), were averaged to obtain the mean flow depth, d. The mean velocity was then calculated by dividing Q by wd, where w is the plot width reduced in proportion to the number of points where the flow depth was zero. The kinematic viscosity of the flow was estimated from water temperature, the energy slope, s, was assumed to be equal to the ground slope, and f and Re were computed from Equations 1.1 & 1.2. The surface form variables included in the analysis were mean gravel size D_g and percentage gravel cover, $\%G$. Both variables were determined from a grid sample of 56 to 70 particles from the test section. D_g was obtained by measuring the length (in millimetres) of the intermediate axis of each gravel-sized particle and calculating the arithmetic mean, whereas $\%G$ was set equal to the percentage of gravel-sized particles in the grid sample. Altogether, 73 experiments were conducted on the eight runoff plots.

Stepwise multiple regression analyses for the 73 experiments yielded the equations

$$\log f = -5.960 - 0.306 \log Re + 3.481 \log \%G + 0.998 \log D_g \qquad (1.3)$$

with $r^2 = 0.61$ and

$$\log f = 0.699 - 1.091 \log Re + 3.151 \log \%G + 1.31 \log R_s \qquad (1.4)$$

with $r^2 = 0.81$. In Equation 1.4, R_s is a measure of the roughness size defined by $D_g d/A_b$, where A_b is the bed area over which D_g and d were measured (i.e. the area of the test section). The product $D_g d$ represents the wetted upstream projected area of the average-sized gravel particle, whereas the division by A_b makes R_s dimensionless (which D_g in Eq. 1.3 is not). Although the improvement in r^2 from Equation 1.3 to 1.4 may be largely spurious (as d appears in both f and R_s), it does not alter the fact that Equation 1.4 affords an excellent empirical model for predicting the value of f on the gravel-covered hillslopes under study.

Equations 1.3 & 1.4 have much higher r^2 values than the corresponding *f*–*Re* relation ($r^2 = 0.032$: Table 1.1) and therefore might be expected to predict

Table 1.1 Coefficients of determination between f and predictor variables in equations 1.3 and 1.4.

Predictor variable	Coefficient of determination
Re	0.032
%G	0.50*
D_g	0.0029
R_s	0.024

* Statistically significant at the 0.05 level

hillslope hydrographs more accurately. Moreover, of the independent variables in the equations, %*G* is by far the single best predictor of f, explaining 50.1% of the variance (Table 1.1, Fig. 1.3). These results confirm the important effect of surface form on resistance to overland flow on gravel-covered semi-arid hillslopes and suggest that, wherever practical, multivariate relations that include a measure of gravel cover should be used in place of simple *f*–*Re* relations in runoff models for such hillslopes.

Grain and form resistance

In the preceding field experiments, the Froude number, F, never exceeds 0.57 and in 90 per cent of the experiments it is less than 0.50. These data signify that wave resistance is relatively unimportant in these experiments (see below) and, hence, that resistance to flow must consist almost entirely of grain and form resistance. The above regression analyses imply that form resistance is

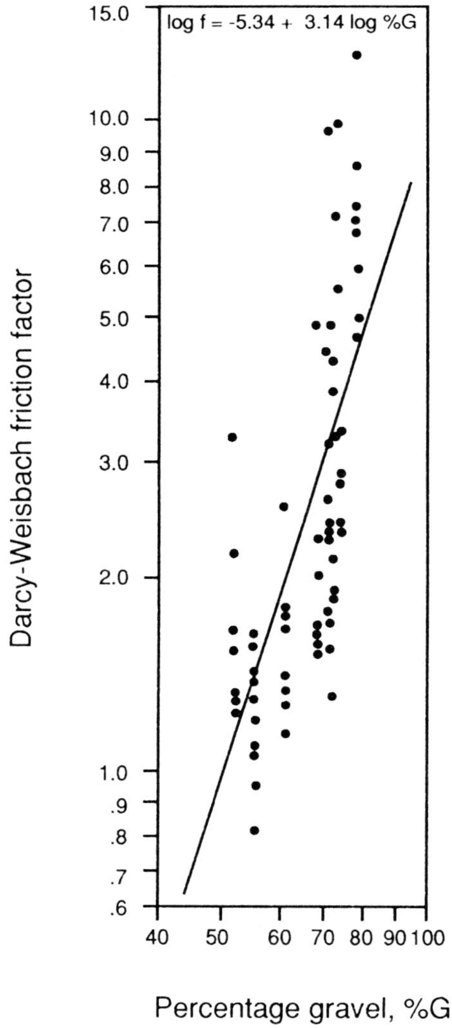

Figure 1.3 Graph of Darcy–Weisbach friction factor against percentage gravel for 73 experiments on eight runoff plots at Walnut Gulch Experimental Watershed, Arizona.

much greater than grain resistance, but they do not permit the relative magnitudes of these resistance components to be calculated. However, a procedure for calculating these relative magnitudes was proposed by Govers & Rauws (1986) and verified in the laboratory by Rauws (1988). This procedure is based on the assumption that grain resistance, f_g, and form resistance, f_f, are additive (Einstein & Banks 1950, Einstein & Barbarossa 1952) and that there are no other sources of hydraulic resistance – that is,

$$f = f_g + f_f \ . \qquad (1.5)$$

Meyer-Peter & Müller (1948) divided the energy slope into one component to overcome grain resistance and another to overcome form resistance. But Govers & Rauws (1986) pointed out that in the case of overland flow it is not the energy slope but the flow depth that is increased when form resistance is added to a surface. Consequently,

$$f = f_g + f_f = \frac{8gd_g s}{v^2} + \frac{8gd_f s}{v^2} \qquad (1.6)$$

where d_g and d_f are the components of flow depth due to grain resistance and form resistance, respectively. Govers & Rauws (1986) and Rauws (1988) then assumed that the grain resistance, f_g, on an irregular bed is equal to the total resistance, *f,* on a plane bed covered with the same-sized grains (Einstein & Barbarossa 1952), and they employed Savat's (1980) algorithm to calculate *f* for the plane bed. Inasmuch as Rauws (1988) demonstrated that Savat's program yields realistic values of f_g for irregular beds in the laboratory, it seems likely that this procedure will also work well for irregular beds in the field. Accordingly, f_g was estimated for the 73 experiments on the eight plots using Savat's program and measured values of *s*, *dv* (unit discharge) and temperature. Bed roughness for every experiment was set equal to 205 μm, which is the median size of the fine material between the gravel on the plots. Grain resistance, f_g, is graphed against *Re* for the eight plots in Figure 1.4. Savat's algorithm ensures that the slope of the f_g–*Re* relation is −1.0 where *Re* < 2440 and less negative than −1.0 where *Re* > 2440. *Re* = 2440 was considered by Savat to mark the boundary between laminar and turbulent flow. Also in Figure 1.4, grain resistance, expressed as a percentage of total resistance ($\%f_g$), is graphed against *Re* for the eight plots. For plots 1 to 5 there is no significant relation between $\%f_g$ and *Re*, whereas for plots 6 to 8 the relations are negatively sloping. The slope of the $\%f_g$–*Re* relation for each plot depends on the relative slopes of the *f*–*Re* and f_g–*Re* relations. As might be expected, the slopes of the *f*–*Re* relations are similar to those of the f_g–*Re* relations for plots 1 to 5 and gentler than those for the latter relations for plots 6 to 8.

Perhaps of greater significance, however, are the absolute values of $\%f_g$. A relative frequency distribution of the computed values of $\%f_g$ for the 73 experiments is presented in Figure 1.5. The mode of this distribution occurs at 4.55% and the median at 4.53%. Moreover, almost 77% of the values are smaller than 6%. Thus on the gravel-covered interrill portions of the semi-arid hillslopes under study, grain resistance is usually a very small proportion of the total resistance to overland flow. For convenience, we might say that $\%f_g$ is typically about 5%. At first glance, this proportion may seem to be unrealistically small. However, when one considers the large size of the gravel roughness elements relative to the small depth of overland flow, the proportion does not seem out of line with the other values of $\%f_g$ reported in the literature for flow through channels with less rough boundaries. For example, Petit (1989) obtained $\%f_g$ values of 31 to 46% for flow over a gravel bed with pools and riffles; Vanoni & Hwang (1967) recorded values of 11.1 to 29.4% for flow over a sand bed with ripple and dune bedforms; and Foster et al. (1984) computed values of 8.6 to 38.5% for flow through a rill. Thus $\%f_g = 5\%$ is a plausible result for overland flow over a gravel-covered surface. According to Equation 1.5, form resistance therefore accounts for 95% of the resistance to flow and clearly dominates on the gravel-covered surfaces under study.

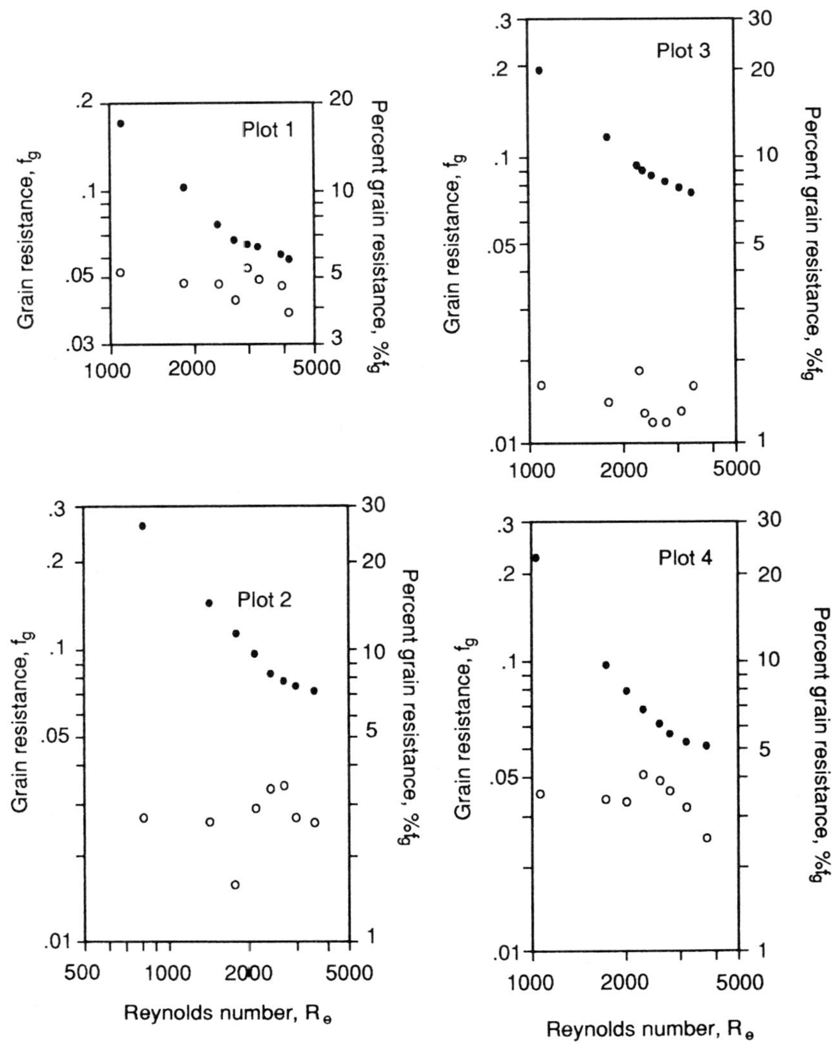

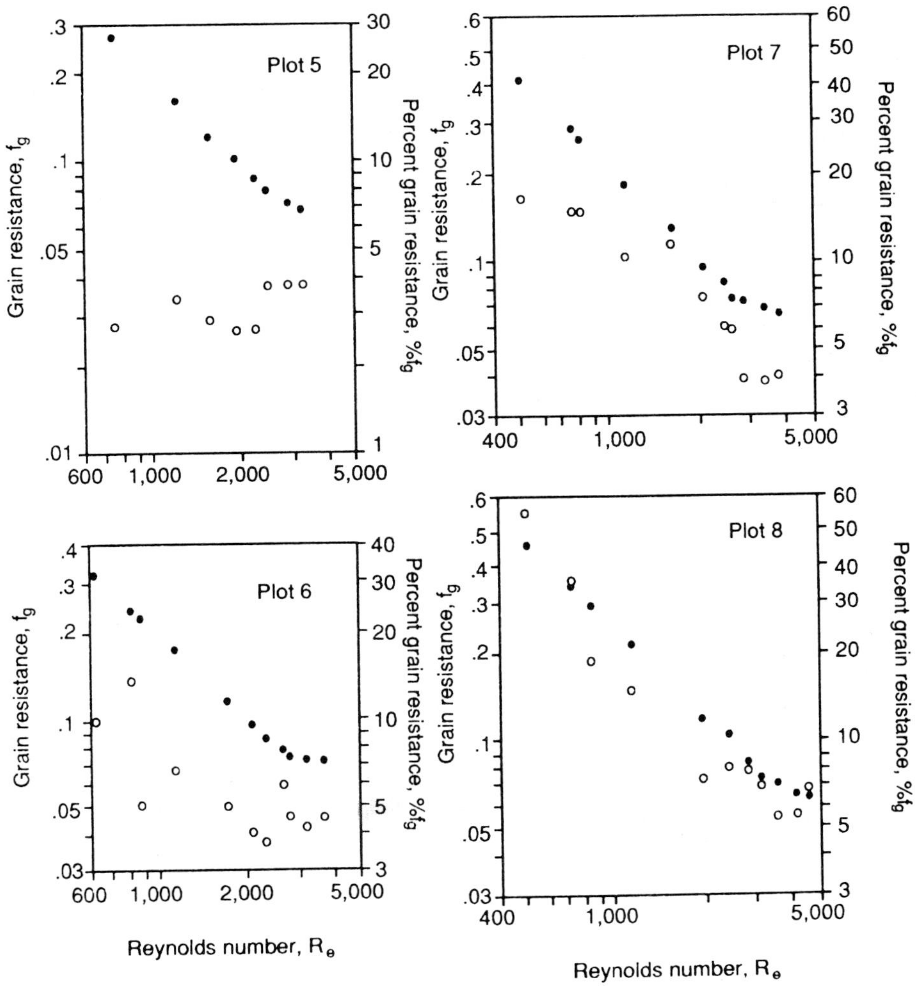

Figure 1.4 Graphs of grain resistance (closed circles) and percentage grain resistance (open circles) against Reynolds number for eight runoff plots, Walnut Gulch Experimental Watershed, Arizona.

Wave resistance

The foregoing analysis is based on the assumption that Equation 1.5 is an accurate characterization of the resistance to overland flow. Disregarding rain resistance, which is assumed to be small, this assumption is probably an acceptable approximation to reality so long as F remains below about 0.5, as it does in the preceding experiments. However, field experiments suggest that F values in excess of 0.5 are not uncommon in interrill overland flow, especially in threads of flow. In such threads, therefore, wave resistance, f_w, may be expected to be significant.

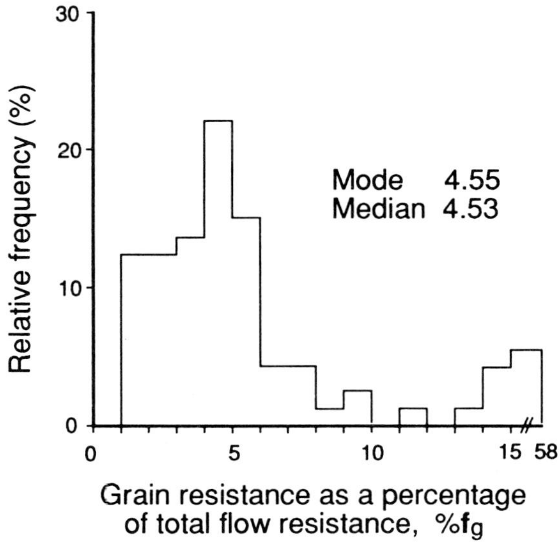

Figure 1.5 Relative frequency distribution of grain resistance expressed as a percentage of total flow resistance for 73 experiments on eight runoff plots, Walnut Gulch Experimental Watershed, Arizona.

In a flume study of the effect of cylindrical piers on open-channel flow, Hsieh (1964) found that, although the relation between f_w and F depends on pier concentration, in general f_w is small for F less than 0.3 to 0.5, rises steeply to a peak at an F value of 0.5 to 0.6, then decreases less precipitously to become small again where F exceeds 2 (Fig. 1.6). Yen (1965) reanalysed the flume data of Herbich & Shulits (1964), who used concrete cubes as roughness elements, and obtained a similar relationship. Flammer et al. (1970) evaluated the resistance to flow of a hemisphere and found that wave resistance was pronounced where the relative depth (i.e. flow depth divided by hemisphere height) was less than 1.6 and $F < 1.5$. Wave resistance peaked at $F = 0.5$ and was up to 10 times greater than surface and form resistance combined. Although these laboratory studies employed cylinders, cubes and hemispheres, it seems very likely that their findings relate in a general way to overland flow over gravel-covered semi-arid hillslopes. The implication of these studies is that where $0.5 < F < 2$, wave resistance makes an important contribution to flow resistance. The crucial question in the present context of semi-arid hillslopes is how important is this contribution relative to those of the other resistance components.

To investigate this question, Hirsch (1992) performed a set of flume experiments in which overland flow across a gravel-covered semi-arid hillslope is mimicked by shallow flow over a plane bed with cylindrical roughness elements. The results of the experiments were analysed with the aid of a mathematical model that permits the total flow resistance to be partitioned into

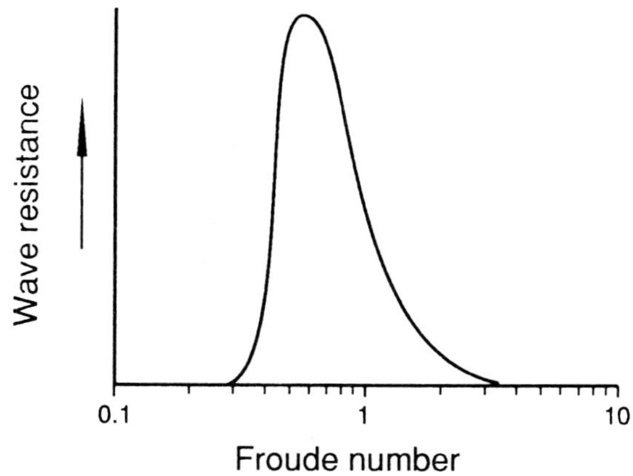

Figure 1.6 Generalized relation between wave resistance and Froude number (after Hsieh 1964, Flammer et al. 1970).

its surface, form and wave components. The variation in the relative magnitudes of these components with variation of Reynolds number and roughness concentration (i.e. percentage cover) was then examined.

Mathematical model

The model assumes that

$$f = function(f_g, f_f, f_w) \qquad (1.7)$$

where the symbol f_g is retained for surface resistance. Inasmuch as f_g and f_f are generally assumed to be additive, Equation 1.7 may be rewritten as

$$f = function\ [(f_g + f_f), f_w)]\ . \qquad (1.8)$$

A mathematical expression may now be developed for $(f_g + f_f)$ in steady uniform flow with an undisturbed water surface (i.e. $f_w = 0$). In this type of flow

$$\Sigma F_x = 0 \qquad (1.9)$$

where ΣF_x is the sum of the forces in the direction of flow. Expanding Equation 1.9 yields

$$\gamma ALs \ - \ \Sigma D_i \ - \ \tau_g PL \ = \ 0 \qquad (1.10)$$

driving force	form resistance	surface resistance

where γ = the specific weight of the fluid, A is the cross-sectional area of the flow, L is the flow length, D_i is the drag force on the i-th roughness element, τ_g is the (grain) shear stress on the boundary, and P is the wetted perimeter. Substituting standard expressions for D_i and τ_g, Hirsch (1992) obtained

$$f = f_g + \frac{4C_d \Sigma A_i}{A_b} \qquad (1.11)$$

total resistance   surface resistance   form resistance

where C_d is the drag coefficient, A_i is the wetted cross-sectional area of the i-th element, and A_b is the area of the flume bed. It is well established that

$$f_g = kRe^x . \qquad (1.12)$$

Thus if the values of k, x, and C_d can be determined experimentally, then (f_g + f_f) can be calculated using Equation 1.11 and subtracted from measured values of f to obtain f_w.

Flume experiments

Flume experiments were therefore conducted in which shallow flows were introduced onto a plane bed of aluminum approximately 5 m long and 0.5 m wide. During each series of experiments Re varied between approximately 1,000 and 25,000 and F between 0.5 and 2.8. In other words, the flows were both transitional and turbulent and both subcritical and supercritical. Because the value of F decreased as roughness concentration, C, increased, Figure 1.6 suggests that f_w should increase with C.

The first series of experiments was performed on a plane bed with no obstacles and yielded the relation

$$f_g = 3.19Re^{-0.45} \qquad (1.13)$$

indicating that k = 3.19 and x = −0.45. During subsequent series of experiments, regular patterns of lead cylinders were placed on the bed with values of C ranging from 2% to 24%. For all experiments, C_d was set equal to 1.2, the average value for an array of cylinders obtained experimentally by Petryk (1969, 43, 118). Separate experiments with two sizes of cylinder demonstrated that the results were independent of cylinder size.

As expected, the calculated values of (f_g + f_f) for each series of experiments were consistently smaller than the computed values of f. Moreover, the slope of the $(f_g + f_f)$–*Re* relation for each series was always the same as that of the f–*Re* relation, with the difference measured along the ordinate axis between the two relations (i.e. the difference between the intercepts) being directly proportional to C. These findings strongly suggest that the difference between f and $(f_g + f_f)$ is in fact equal to f_w. The difference was therefore expressed in terms of C. Because f_w is a constant proportion of f for a given C, the complete model has the form

$$\log f = \log\left[3.19Re^{-0.45} + \frac{4.8\Sigma A_i}{A_b}\right] + 2.80C \,. \qquad (1.14)$$

total resistance	surface resistance	form resistance	wave resistance

Figure 1.7 shows the model fitted to two sets of flume data.

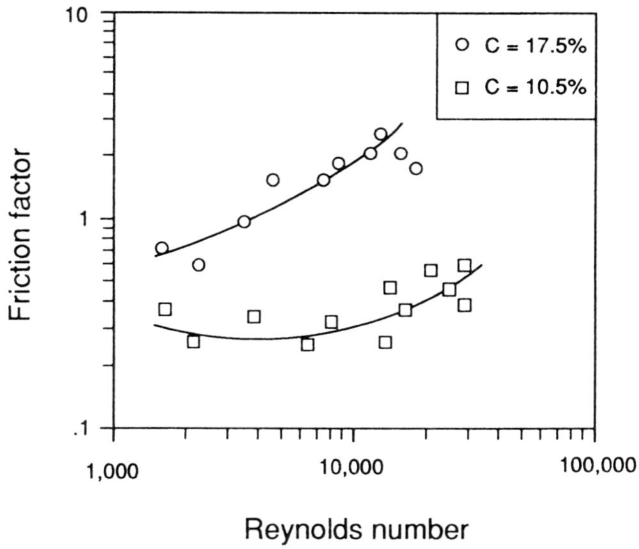

Figure 1.7 Graph of Darcy–Weisbach friction factor against Reynolds number showing the fit of the model (Eq. 1.14) to the flume data for two concentrations of cylinders: C = 10.5% and C = 17.5%.

Using the model, it is possible to examine the variation in the relative magnitudes of f_g, f_f and f_w as *Re* and C vary. The percentage contributions of f_g, f_f and f_w to f are denoted by $\%f_g$, $\%f_f$ and $\%f_w$, respectively. Figure 1.8 shows that where $C = 2\%$, $\%f_w = 12\%$ and resistance to flow is dominated by f_g at low *Re* values but is similar to f_f at *Re* values in the order of 20,000. In contrast, where $C = 17.5\%$, $\%f_w = 68\%$ and dominates f. Figure 1.9 reveals that at $Re = 2000$, $\%f_f$ varies conservatively with C and remains relatively small. This leaves $\%f_g$ and $\%f_w$ to more or less complement one

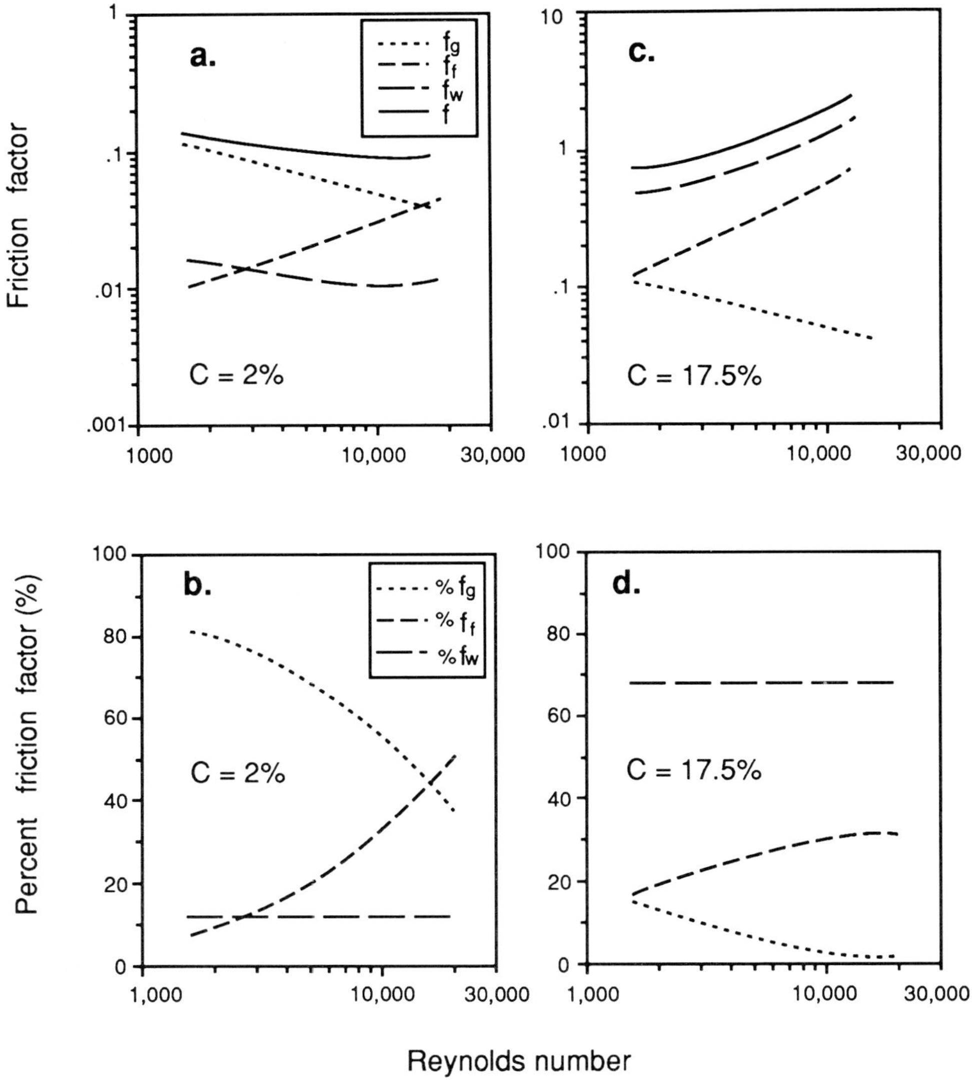

Figure 1.8 Graphs of Darcy–Weisbach friction factor against Reynolds number for two concentrations of cylinders: C = 2% (a, b) and C = 17.5% (c, d).

another, with $\%f_g$ decreasing and $\%f_w$ increasing as C increases: f_g is dominant where $C < 8\%$ and f_w dominant where $C > 8\%$. At Re = 12,000, $\%f_f$ behaves less conservatively and actually exceeds both $\%f_g$ and $\%f_w$ where $2\% < C < 10\%$. The two most salient points that emerge from these graphs is that over the measured ranges of Re and C, $\%f_w$ is very sensitive to C but insensitive to Re. f_w contributes more than either f_g or f_f to f where $C > 10\%$, and the contribution by f_w exceeds 70% where $C > 20\%$.

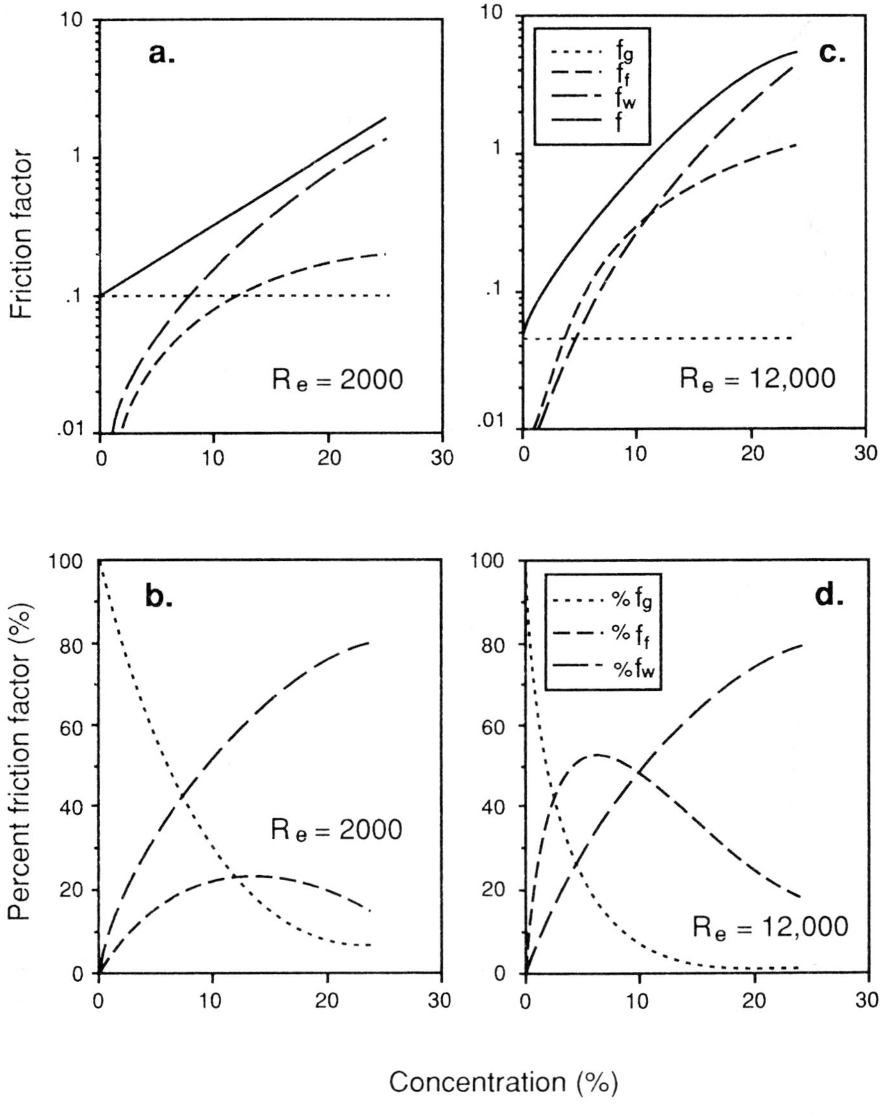

Figure 1.9 Graphs of Darcy–Weisbach friction factor and percentage friction factor against concentration of cylinders for two Reynolds numbers: *Re* = 2000 (a, b) and *Re* = 12,000 (c, d).

At the highest *Re* values during the flume experiments, the cylindrical roughness elements were overtopped by the flow, and f began to decline. Equation 1.14 does not apply to these flows. A model comparable to this equation could be derived in a similar way, but too few data were collected during the present experiments to justify its development here. Nevertheless, a qualitative appraisal of the behaviour of f_g, f_f and f_w is possible. Once the roughness elements are overtopped, f_g might be expected to continue to decline as *Re* increases, as indicated by the f_g term in Equation 1.14. f_f should remain constant as *Re* increases because the upstream wetted projected area of the roughness elements ΣA_i remains constant. Finally, f_w might be expected to decline as *Re* increases because the elements become submerged under increasing depths of water and so disturb the water surface less (Flammer et al. 1970).

Discussion

The two main findings of the present flume experiments are: (a) that where $0.5 < F < 2.8$, wave resistance makes a significant contribution to total flow resistance; and (b) for concentrations in excess of about 10%, wave resistance exceeds the other resistance components. The principal limitations of the present experiments in terms of their applicability to gravel-covered semi-arid hillslopes are: (a) that a plane bed covered with cylinders does not closely resemble a gravel-covered hillslope surface; and (b) that the experiments were confined to roughness concentrations of $\leq$ 24%, whereas semi-arid hillslopes typically have gravel covers in excess of 60%. Inasmuch as an analytical expression for f_f could be developed only for regular-shaped roughness elements (for which the form drag coefficient is known) resting on a plane bed, there was no way of avoiding limitation (a). Limitation (b) is regrettable, as there is no certainty that the expression for f_w derived here for $C \leq 24\%$ can be extended to values of *C* in excess of 60%. Intuitively, it seems unlikely that it can. But, by the same token, it is difficult to imagine that f_w will not dominate the other resistance components at these high *C* values. Thus the significance of the present flume experiments is that they strongly suggest that wave resistance not only contributes to resistance to overland flow on gravel-covered semi-arid hillslopes but that where $0.5 < F < 2$ its contribution is probably larger than that of any other component.

Conclusion

The field and laboratory experiments described in this chapter suggest that resistance to overland flow in interrill parts of semi-arid hillslopes is largely controlled by surface form, which gives rise to form and wave resistance. Form resistance is likely to dominate at Froude numbers less than about 0.5 and wave resistance at higher Froude numbers. Consequently, resistance equations for use in process-based mathematical models of interrill overland flow should include surface-form variables such as percentage gravel cover. Although this recommendation represents a break with the traditional practice of employing resistance equations which simply relate *f* to *Re* (or surrogates thereof), it does not necessarily pose daunting problems, as surface form variables can be measured at any time and applied to all runoff events. For example, in the case of the interrill area sampled in the second set of field experiments, total flow resistance can be well predicted by a simple relation between f and percentage gravel cover (Fig. 1.3).

Although this chapter has been concerned with gravel-covered semi-arid hillslopes in southern Arizona, it seems likely that the above conclusions will apply equally well to all irregular hillslope surfaces, whether natural or disturbed by agriculture. In such settings, however, surface-form variables other than percentage gravel cover, such as microtopography, litter cover or density of plant stems, may be better predictors of resistance to interrill overland flow.

Acknowledgements

We acknowledge the assistance and co-operation of the USDA/ARS Aridland Watershed Management Research Unit, Tucson, in conducting the field experiments at Walnut Gulch. We thank the many friends and colleagues who helped with these experiments, especially Shiu-hung Luk, who collaborated with us in our early work. The flume experiments were performed in the Hydraulics Laboratory, Department of Civil Engineering, SUNY-Buffalo. The research was supported by NATO grant RG. 85/0066 and NSF grants SES-8614255 and SES-8812587.

References

Abrahams, A. D. & A. J. Parsons 1990. Determining the mean depth of overland flow in field studies of flow hydraulics. *Water Resources Research* **27**, 501–3.

Abrahams, A. D. & A. J. Parsons 1991a. Relation between infiltration and stone cover on a semiarid hillslope, southern Arizona. *Journal of Hydrology* **122**, 49–59.

Abrahams, A. D. & A. J. Parsons 1991b. Resistance to overland flow on desert pavement and its implications for sediment transport modeling. *Water Resources Research* **27**, 1827–36.

Abrahams, A. D., A. J. Parsons, S.-H. Luk 1986. Resistance to overland flow on desert hillslopes. *Journal of Hydrology* **88**, 343–63.

Dunne, T. & W. E. Dietrich 1980. Experimental study of Horton overland flow on tropical hillslopes. 2. Hydraulic characteristics and hillslope hydrographs. *Zeitschrift für Geomorphologie Supplementband* **35**, 60–80.

Einstein, H. A. & R. B. Banks 1950. Fluid resistance of composite roughness. *Transactions of the American Geophysical Union* **31**, 603–10.

Einstein, H. A. & N. L. Barbarossa 1952. River channel roughness. *Transactions of the*

American Society of Civil Engineers **117**, 1121–32.

Emmett, W. W. 1970. *The hydraulics of overland flow on hillslopes*. US Geological Survey Professional Paper 662-A. Washington DC: US Government Printing Office.

Flammer, G. H., J. P. Tullis, E. S. Mason 1970. Free surface, velocity gradient flow past hemisphere. *Journal of the Hydraulics Division, Proceedings of the American Society of Civil Engineers* **96**, 1485–502.

Foster, G. R., L. F. Huggins, L. D. Meyer 1984. A laboratory study of rill hydraulics: II. Shear stress relationships. *Transactions of the American Society of Agricultural Engineers* **27**, 797–804.

Glendening, G. E. 1952. Some quantitative data on the increase of mesquite and cactus on a desert grassland range in southern Arizona. *Ecology* **33**, 319–28.

Govers, G. & G. Rauws 1986. Transporting capacity of overland flow on plane and on irregular beds. *Earth Surface Processes and Landforms* **11**, 515–24.

Hastings, J. R. & R. M. Turner 1965. *The changing mile*. Tucson: University of Arizona Press.

Herbich, J. B. & S. Shulits 1964. Large-scale roughness in open-channel flow. *Journal of the Hydraulics Division, Proceedings of the American Society of Civil Engineers* **90**, HY6 203–30.

Hirsch, P. J. 1992. Hydraulic resistance to overland flow on semiarid hillslopes: a physical simulation. Unpublished Ph.D. dissertation. Department of Geography, State University of New York at Buffalo, Buffalo, New York.

Hsieh, T. 1964. Resistance of cylindrical piers in open-channel flow. *Journal of the Hydraulics Division, Proceedings of the American Society of Civil Engineers* **90**, HY1 161–73.

Leopold, L. B., R. A. Bagnold, M. G. Wolman, L. M. Brush, Jr 1960. *Flow resistance in sinuous or irregular channels*. US Geological Survey Professional Paper 282-D. Washington DC: US Government Printing Office.

Meyer-Peter, E. & R. Müller 1948. Formulas for bed load transport. *Proceedings of the Second Congress of the International Association for Hydraulic Research*, 39–64.

Osborn, H. B. 1983. *Precipitation characteristics affecting hydrological response of Southwestern rangelands*. Agricultural Reviews and Manuals, Western Series, No. 34, Oakland, Calif., US Department of Agriculture.

Parsons, A. J., A. D. Abrahams, J. R. Simanton 1992. Microtopography and soil-surface materials on semi-arid piedmont hillslopes, southern Arizona. *Journal of Arid Environments* **22**, 107–15.

Petit, F. 1989. The evaluation of grain shear stress from experiments in a pebble-bedded flume. *Earth Surface Processes and Landforms* **14**, 499–508.

Petryck, S. 1969. *Drag on cylinders in open channel flow. Unpublished* Ph.D. dissertation. Department of Hydraulic Engineering, Colorado State University, Fort Collins, Colorado.

Rauws, G. 1988. Laboratory experiments on resistance to overland flow due to composite roughness. *Journal of Hydrology* **103**, 37–52.

Roels, J. M. 1984. Flow resistance in concentrated overland flow on rough slope surfaces. *Earth Surface Processes and Landforms* **9**, 541–51.

Savat, J. 1977. The hydraulics of sheet flow on a smooth surface and the effect of simulated rainfall. *Earth Surface Processes* **2**, 125–40.

Savat, J. 1980. Resistance to flow in rough supercritical sheet flow. *Earth Surface Processes* **5**, 103–22.

Shen, H. W. & R.-M. Li 1973. Rainfall effect on sheet flow over smooth surface. *Journal of the Hydraulics Division, Proceedings of the American Society of Civil Engineers* **99**, 771–92.

Vanoni, V. A. & L. S. Hwang 1967. Relation between bed forms and friction in streams. *Journal of the Hydraulics Division, Proceedings of the American Society of Civil Engineers* **93**, HY5 121–44.

Yen, B.-C. 1965. Discussion of "Large-scale roughness in open-channel flow." *Journal of the Hydraulics Division, Proceedings of the American Society of Civil Engineers* **91**, 257–62.

Yoon, Y. & H. G. Wenzel 1971. Mechanics of sheet flow under simulated rainfall. *Journal of the Hydraulics Division, Proceedings of the American Society of Civil Engineers* **97**, 1367–86.

2

Darcy–Weisbach roughness coefficients for overland flow

John E. Gilley, Dennis C. Flanagan, Eugene R. Kottwitz, Mark A. Weltz

Abstract

Analysis of surface runoff on upland areas requires identification of hydraulic roughness coefficients. Procedures were identified for estimating total hydraulic roughness for rill and interrill areas on both croplands and rangelands. Equations were presented to determine roughness coefficients for: (a) rills; (b) gravel and cobble materials; (c) surface residue on croplands; (d) interrill areas; (e) plants on cropland areas; and (f) rangeland areas. Experimental procedures used to determine the roughness coefficients were summarized. Finally, future needs for estimating roughness coefficients were identified.

Introduction

Analysis of surface runoff on upland areas requires identification of hydraulic roughness coefficients. Roughness coefficients are used in the calculation of flow velocity and the routing of runoff hydrographs. Understanding and properly modelling upland flow hydraulics is also essential in developing process-based erosion models.

Meyer-Peter & Müller (1948) suggested that the Darcy–Weisbach roughness coefficient for open channels be composed of two components, f_g and f_b, which denote roughness coefficients associated with grain roughness and bed-form roughness, respectively. It has been assumed that f_g and f_b are additive, with the total roughness coefficient, f, representing their sums or

$$f = f_g + f_b. \tag{2.1}$$

Several subsequent investigators utilized this concept, including Einstein & Barbarossa (1952), Engelund (1966), Alam & Kennedy (1969) and Lovera & Kennedy (1969).

Shen & Li (1973) used the concept of additive roughness for use with overland flow. They assumed the total roughness coefficient under rainfall conditions to be the sum of f_{wo}, the roughness coefficient without rainfall, and f_{ra}, the added roughness coefficient due to rainfall, or

$$f = f_{wo} + f_{ra}. \tag{2.2}$$

Regression analysis was performed by Shen & Li (1973) to identify empirical equations for estimating f_{ra}. Rainfall was found to influence total hydraulic resistance significantly primarily on smooth surfaces with small discharge rates. For most overland-flow conditions, rainfall would be expected to have a minimal effect on total hydraulic resistance.

Laboratory measurements of roughness coefficients on surfaces covered with sand or gravel were made by Woo & Brater (1961), Emmett (1970), Phelps (1975) and Savat (1980). Similar tests were performed on natural landscapes by Dunne & Dietrich (1980), Roels (1984), Abrahams et al. (1986) and Parsons et al. (1990). In these studies, a significant correlation was established between Reynolds number and roughness coefficient.

Roughness coefficients were also significantly influenced by flow depth. For flow depths less than the height of the roughness elements, roughness coefficients increase with greater Reynolds number. Once roughness elements are submerged, their ability to retard overland flow is reduced as flow depth becomes larger. As a result, the roughness coefficient usually decreases with an increasing Reynolds number.

A comprehensive review of previous studies involving evaluation of roughness coefficients on agricultural and natural areas was provided by Engman (1986). Hydraulic roughness coefficients were developed from runoff data originally collected for erosion studies on experimental plots. Roughness coefficients were presented in a tabular format with a description of various surfaces and land uses.

Equations for estimating total hydraulic resistance on cropland and rangeland areas are presented below. Procedures are identified for estimating roughness coefficients caused by several factors. Roughness coefficients computed for these individual factors can be added to obtain total hydraulic resistance for a particular site.

Hydraulic equations

Overland-flow hydraulics

The Darcy–Weisbach equation is frequently used to model hydraulic characteristics of overland flow. Under uniform flow conditions, the Darcy–Weisbach roughness coefficient, f, is given as

$$f = \frac{8gRs}{V^2} \tag{2.3}$$

where g is acceleration due to gravity, s is average slope, V is mean flow velocity, and hydraulic radius, R, is defined as

$$R = \frac{A}{P} \tag{2.4}$$

where A is cross-sectional flow area and P is wetted perimeter (Chow 1959). For a rectangular flow geometry

$$R = \frac{by}{b+2y} \tag{2.5}$$

where b is flow width and y is flow depth. For overland flow conditions where b is much greater than y, hydraulic radius can be assumed to be approximately equal to flow depth.

The continuity equation is defined as

$$Q = VA \tag{2.6}$$

where Q is flow rate. For a rectangular channel, water depth is given as

$$y = \frac{Q}{Vb}. \tag{2.7}$$

Reynolds number, Re, which is used to express the ratio of inertial forces to viscous forces is given as

$$Re = \frac{VR}{\nu} \tag{2.8}$$

where ν is kinematic viscosity. Kinematic viscosity can be determined directly from water temperature.

Investigation of the correlation between roughness coefficient and Reynolds number requires the determination of shallow flow depths existing under field conditions. Since it may be difficult to identify the soil–water interface for eroding situations, direct measurement of flow depth may not be possible.

Thus, it may be necessary to determine water depth indirectly using Equation 2.7. Water depth can be substituted into Equation 2.5 to calculate hydraulic radius. Finally, roughness coefficient and Reynolds number values can be obtained from Equations 2.3 & 2.8, respectively.

Roughness coefficient equations

The total roughness coefficient for rills on croplands, f_r , can be represented as

$$f_r = f_{sr} + f_{rk} + f_{cr} + f_{st} \tag{2.9}$$

where f_{sr} is the roughness coefficient for rills, f_{rk} is the roughness coefficient for gravel and cobble materials, f_{cr} is the roughness coefficient for surface residue on croplands, and f_{st} is the roughness coefficient for plants on cropland areas. For rills on rangelands, the total roughness coefficient, f_{rr}, is given as

$$f_{rr} = f_{sr} + f_{rk} + f_{lt} + f_{pb} \tag{2.10}$$

where f_{lt} is the roughness coefficient for litter and organic residue on rangelands, and f_{pb} is the roughness coefficient for plants on rangeland areas. It can be seen from Equations 2.9 & 2.10 that two of the factors contributing to hydraulic roughness of rills are the same on cropland and rangeland areas.

The total roughness coefficient for interrill cropland areas, f_i, can be represented as

$$f_i = f_{si} + f_{rk} + f_{cr} + f_{st} \tag{2.11}$$

where f_{si} is the roughness coefficient for interrill areas. For interrill rangeland areas, the total roughness coefficient, f_{ir} , is given as

$$f_{ir} = f_{si} + f_{rk} + f_{lt} + f_{pb}. \tag{2.12}$$

Again, several of the same factors contribute to hydraulic resistance on interrill areas for both croplands and rangelands.

Equations 2.9–2.12 each contain four factors which may contribute to hydraulic roughness. Some of these factors may not be present at a given location. Even if a particular component is represented, its contribution to total hydraulic resistance may be minimal.

Roughness coefficients for rills

Experimental procedures

A field study was conducted by Gilley et al. (1990) at 11 sites located throughout the eastern United States to measure hydraulic characteristics of rills. The location, slope and particle size analysis of soils at the study sites are shown in Table 2.1. These soils were selected to cover a broad range of physical, chemical, biological and mineralogical properties. These properties resulted from diverse soil-forming factors acting through time, including climate, parent material, vegetation, biological activity and topography. Each soil is considered to be of regional or national importance.

Table 2.1 Location, slope and particle size analysis of selected soils used to measure roughness coefficients for rills.

Soil	Location		Slope	Particle size analysis % by weight		
	County	State	%	Sand	Silt	Clay
Caribou	Aroostook	Maine	6.4	47.0	40.3	12.7
Cecil	Oconee	Georgia	6.2	64.6	15.6	19.8
Collamer	Tompkins	New York	8.2	7.0	78.0	15.0
Gaston	Rown	North Carolina	5.9	35.5	25.4	39.1
Grenada	Panola	Mississippi	6.7	2.0	77.8	20.2
Lewisburg	Whitley	Indiana	9.6	38.5	32.2	29.3
Manor	Howard	Maryland	9.8	43.6	30.7	25.7
Mexico	Boone	Missouri	3.8	5.3	68.7	26.0
Miami	Montgomery	Indiana	6.4	4.2	72.7	23.1
Miamian	Montgomery	Ohio	8.8	30.6	44.1	25.3
Tifton	Worth	Georgia	5.5	86.4	10.8	2.8

The study areas were located on uniform slopes having relatively homogeneous soil characteristics. Either corn or small grains had been planted the previous year. Preparing the study areas for testing required removing all surface residue and then moldboard ploughing 3–12 months before the tests were conducted. After ploughing, the sites were disked lightly and maintained free of vegetation either by tillage or by application of herbicide. The study areas were disked immediately preceding testing. Two plots, 3.7 m across the slope by 10.7 m long, were established at each site using sheet metal borders. The plots were raked by hand prior to testing to provide a uniform surface.

A portable rainfall simulator designed by Swanson (1965) was used to apply rainfall at an intensity of approximately 57 mm h^{-1}. The first rainfall application (initial run) of 1 h duration occurred at existing soil-water conditions. A second rainfall simulation run (wet run) was conducted approximately 24 h later, again for a duration of 1 h. A final, very wet rainfall application was applied within one hour after completion of the wet run.

After steady-state conditions had become established during the very wet rainfall application, inflow was added at the top of each plot to simulate greater slope lengths. Flow addition for each of four inflow increments occurred only after steady-state runoff conditions for the previous inflow increment had become established and selected hydraulic measurements had been made. A trough extending across the bottom of each plot gathered runoff, which was continuously measured using an HS flume with stage recorder. Steady-state runoff conditions were determined using the stage recorder and HS flume. A thermometer was used to measure water temperature, and flow width was determined using a ruler.

To determine rill discharge, a bromide solution of known concentration was continuously injected into each rill at a constant rate (Replogle et al. 1966). Runoff samples containing the diluted bromide solution were collected at the point where each rill discharged into the collection trough. Samples of approximately 800 ml were obtained using polyethylene bags. The concentrations of diluted bromide in each of these samples were determined later using an ion analyser. From measurements of the bromide injection rate and concentration, and diluted concentration, rill discharge rate was determined.

Mean flow velocity in each rill was measured using a fluorometer (Hubbard et al. 1982). A slug of dye was injected into the rill and the time required for the concentration peak to travel a known distance to a downstream point was identified. A time-concentration curve resulted from continuous pumping of runoff from the rill through the fluorometer flow cell. Due to the symmetrical shape of the dye concentration curve, the velocity associated with the peak concentration was assumed to equal mean flow velocity. Mean flow velocity was obtained by dividing travel distance by time of travel.

Roughness coefficient equations

The regression equations shown in Table 2.2 relate roughness coefficients calculated using Equation 2.3 to Reynolds number values obtained from

Table 2.2 Regression equations for roughness coefficients for rills versus Reynolds number.

Soil	Regression coefficients* a	b	Coefficient of determination r^2
Caribou	4.99×10^3	−1.120	0.825
Cecil	9.72×10^2	−0.874	0.702
Collamer	1.14×10^2	−0.670	0.678
Gaston	2.57×10^2	−0.767	0.702
Grenada	3.41×10^2	−0.695	0.601
Lewisburg	8.75×10^2	−0.889	0.614
Manor	6.01×10^3	−1.120	0.879
Mexico	5.27×10^5	−1.850	0.860
Miami	1.51×10^2	−0.621	0.816
Tifton	2.36×10^4	−1.240	0.731
All soils combined**	1.35×10^3	−0.934	0.655

* Regression coefficients a and b are used in the equation

$$f_{sr} = a\,(Re)^b$$

where f_{sr} is the roughness coefficient for rills and Re is Reynolds number.

** For the "All soils combined" analysis, Darcy–Weisbach roughness coefficients ranged from 0.17 to 8.0 while Reynolds number varied from approximately 300 to 10,000.

Equation 2.8. Regression coefficients are reported for each of the individual soils and for all soils combined. The fluorometer that was used to measure flow velocity was not functioning properly during most of the run on the Miamian soil. As a result, information from this site was omitted from Table 2.2.

Analyses of all soils combined provided roughness coefficient values ranging from 0.17 to 8.0, while Reynolds number varied from 300 to 10,000. Results from the all soils combined analysis can be used to estimate roughness coefficients for rills from the equation

$$f_{sr} = \frac{1350}{Re^{0.934}} \qquad (2.13)$$

Roughness coefficients for gravel and cobble materials

Experimental procedures

Gilley et al. (1992) performed a laboratory study to measure roughness coefficients for gravel and cobble materials. The diameters of gravel and cobble materials used in the investigation are shown in Table 2.3. The gravel material, varying in size from 0.25 to 12.70 cm, was removed from a rangeland site near Tombstone, Arizona. Cobble material, with dimensions of 12.70–25.40 cm, was obtained near Lincoln, Nebraska.

Table 2.3 Diameter, surface cover and shape factor for gravel and cobble surfaces.

Diameter cm	Surface cover %	Shape factor*
0.25 - 1.27	6, 15, 37, 66, 90	0.51
1.27 - 2.54	7, 13, 32, 61, 90	0.52
2.54 - 3.81	4, 16, 32, 56, 80	0.49
3.81 - 12.70	6, 17, 33, 61, 89	0.47
12.70 - 25.40	9, 13, 24, 61, 83	0.52

* Shape factor, *SF*, is given as (Guy 1969)

$$SF = c \,/\, (ab)^{0.5}$$

where a = longest axis, b = intermediate axis and c = shortest axis

Shape factors (Guy 1969) determined from measurements on 10 samples from each of the size classes are shown in Table 2.3. Shape factors provide a relative estimate of the physical configuration of gravel and cobble material. Little variation in shape factor was found between size classes. For natural sediments with much smaller diameters, a shape factor of 0.7 is typical (Guy 1969).

Gravel and cobble materials were glued in a random orientation onto a section of reinforced fibreglass sheeting located within a flume. Surface cover values for each of the size classes are shown in Table 2.3. The percentage of surface cover was obtained using a photographic grid procedure (Laflen et al. 1978). Gravel and cobble materials on the fibreglass sheets were photographed using 35 mm colour slide film. The slides were projected onto a screen on which a grid had been superimposed. The number of grid intersections over gravel and cobble material was determined visually from the projected slides, and surface cover was then calculated.

The flume, which was 0.91 m wide, 7.31 m long and 0.279 m deep, was maintained at a slope of 1.35%. Water was supplied to the flume using a constant-head tank. Two replicate tests were run at selected flow rates. Flow rate was determined immediately before and after each test to confirm steady-state conditions. Water temperature was measured following flow rate determinations.

Reynolds number values varied from approximately 500 to 16,000. Uniform flow conditions were difficult to maintain on the gravel- and cobble-covered surfaces for Reynolds numbers less than approximately 500. For Reynolds numbers greater than 16,000, little variation in roughness coefficient values was found.

Once steady-state runoff conditions had become established, line sources of fluorescent dye were simultaneously injected across the flume at downslope distances of 0.91 and 7.01 m. A fluorometer was used to determine time of travel of the dye concentration peaks. Mean flow velocity was calculated by dividing the distance between the two line sources of dye (6.10 m) by the difference in travel time of the two dye concentration peaks. For each test sequence, three measurements of flow velocity were made.

Roughness coefficients for the fibreglass sheets supporting the gravel and cobble materials were also identified. The experimental procedures used to measure roughness coefficients for the fibreglass sheets with and without gravel and cobble material were identical. Roughness coefficients induced by the bare fibreglass sheets at a given Reynolds number were subtracted from measurements obtained with gravel and cobble material to determine hydraulic resistance caused by the gravel and cobble material alone.

Tests were also conducted to measure total hydraulic roughness for three distributions of size classes. The purpose of these tests was to validate the addition of roughness coefficients for individual size classes to obtain total hydraulic roughness. The percentage of surface cover contributed by each size class for each distribution is shown in Table 2.4.

Roughness coefficient equations

Darcy–Weisbach roughness coefficients at varying Reynolds number for gravel material with dimensions of 2.54–3.81 cm are shown in Figure 2.1. The trends

Table 2.4 Percent cover for selected size classes used in validation test series for gravel and cobble surfaces.

Diameter	Percent cover in test series		
cm	1	2	3
0.25 - 1.27	21	3	16
1.27 - 2.54	31	11	9
2.54 - 3.81	14	18	38
3.81 - 12.70	13	28	11
12.70 - 25.40	9	30	15
Total cover	88	90	89

presented in Figure 2.1 are characteristic of all but the largest size class of gravel and cobble material. For the experimental results shown in Figure 2.1 with surface covers of 56 and 80%, water depth was usually greater than the height of the gravel material. As a result, Darcy–Weisbach roughness coefficients consistently decreased as Reynolds number became larger. In contrast, water depths at lower Reynolds numbers for the test runs with surface cover values of 4, 16 and 32% were typically less than the height of the gravel material. As a result, roughness coefficients initially increased with Reynolds number. Once flow depth exceeded roughness element height, roughness coefficients became smaller as Reynolds number increased.

Water depths were usually smaller than the height of the roughness elements for cobble materials having a diameter of 12.70–25.40 cm. As a result, Darcy–Weisbach roughness coefficients generally increased with Reynolds number (Fig. 2.2). However, the surfaces with 61 and 83% cover showed a substantial reduction in roughness coefficient values at the highest Reynolds number, where flow depth exceeded the height of many of the roughness elements.

Regression equations that relate roughness coefficients for gravel and cobble materials to percentage cover and Reynolds number are shown in Table 2.5. Regression relations are presented for five selected size classes with dimensions ranging from 0.25 to 25.40 cm. Use of the regression equations shown in Table 2.5 requires information on the percentage of the ground surface covered with gravel and cobble materials. If ground cover percentages are available for the separate size classes shown in Table 2.5, then the individual regression equations can be used. If only total ground cover is known, the friction coefficient for gravel and cobble materials can be estimated using a generalized

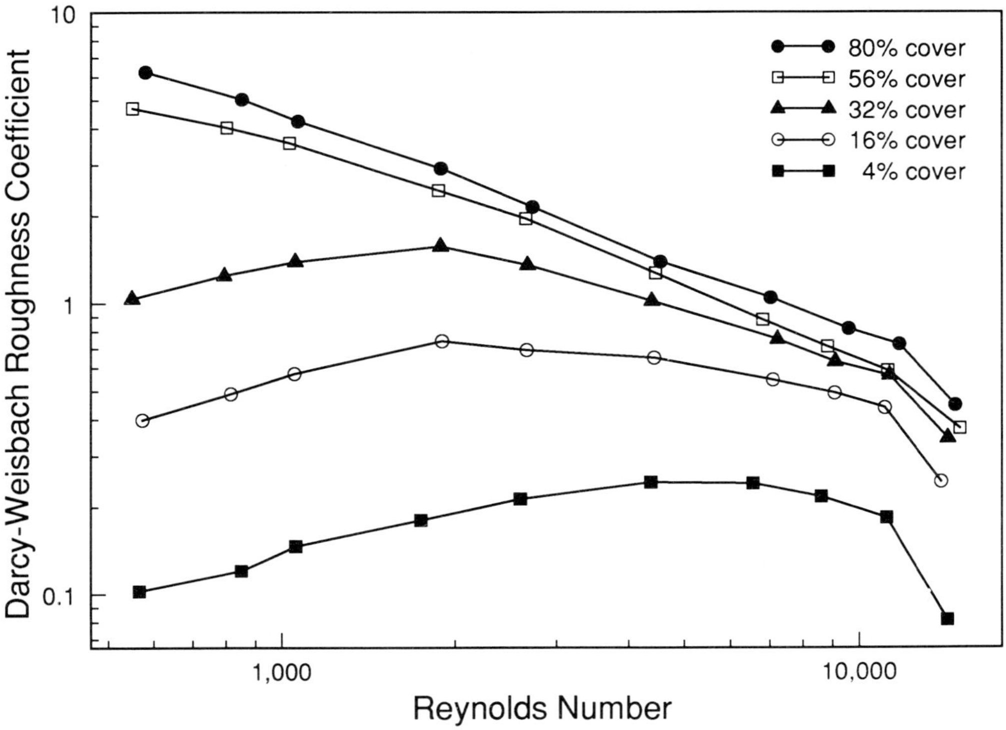

Figure 2.1 Darcy–Weisbach roughness coefficients vs. Reynolds number for gravel material with a diameter of 2.54 to 3.81 cm.

regression equation. Data for gravel and cobble materials having a diameter range of 0.25–12.70 cm were combined to obtain

$$f_{rk} = \frac{2.16(\textit{percentage cover})^{0.953}}{Re^{0.550}}. \tag{2.14}$$

Equation 2.14 was derived using roughness coefficient values ranging from 0.05 to 7.8.

Information on the size distribution of surface material obtained on the basis of mass may be more readily available and easier to obtain. Gilley et al. (1992) made measurements of the mass of gravel or cobble material corresponding to a given surface cover. This data was used to develop regression equations for relating surface cover for a given size class to gravel or cobble mass.

Laboratory data collected on the surfaces described in Table 2.4, which contained multiple size classes, were used to test the reliability of the regression equations. Roughness coefficients were first determined for each size class using information presented in Table 2.5. Roughness contributions for each of

Table 2.5 Regression equations for roughness coefficients for gravel and cobble materials vs. percentage cover and Reynolds number.

Diameter	Regression coefficients*			Coefficient of determination
cm	a	b	c	r^2
0.25 - 1.27	1.68×10^{1}	5.78×10^{-1}	7.09×10^{-1}	0.985
1.27 - 2.54	1.18×10^{1}	6.78×10^{-1}	6.67×10^{-1}	0.945
2.54 - 3.81	1.91	1.19	6.28×10^{-1}	0.943
3.81 - 12.70	1.11×10^{-1}	1.61	4.68×10^{-1}	0.944
12.70 - 25.40	1.25×10^{-5}	1.63	-5.68×10^{-1}	0.944
0.25 - 12.70**	2.16	9.53×10^{-1}	5.50×10^{-1}	0.672

* Regression coefficients a, b and c are used in the equation

$$f_{rk} = a\,(\text{percentage cover})^{b} \,/\, (Re)^{c}$$

where f_{rk} is the roughness coefficient for gravel and cobble materials and Re is Reynolds number.

** Data for gravel and cobble surfaces having a diameter range of 0.25–12.70 cm were combined to obtain a generalized regression equation. Darcy–Weisbach roughness coefficients for this generalized equation ranged from 0.05 to 7.8 while Reynolds number varied from approximately 500 to 16,000.

the five size classes were then added to find total hydraulic resistance for the given test series. Hydraulic roughness coefficients were determined for each Reynolds number value used in the laboratory tests.

Predicted versus measured roughness coefficients are presented in Figure 2.3. Close agreement between predicted and measured values was found for each test series. Linear regression analysis of predicted versus measured roughness coefficients yielded an r^2 value of 0.983. Thus, reliable estimates of roughness coefficients for gravel and cobble materials were obtained by adding the roughness contributions of individual size classes.

Roughness coefficients for surface residue on croplands

Experimental procedures

A laboratory study was conducted by Gilley et al. (1991) to identify roughness coefficients for selected crop residue materials. The types of residue used in the investigation included corn, cotton, peanut, pine needles, sorghum, soybeans, sunflower and wheat. Needles produced by ponderosa pine were

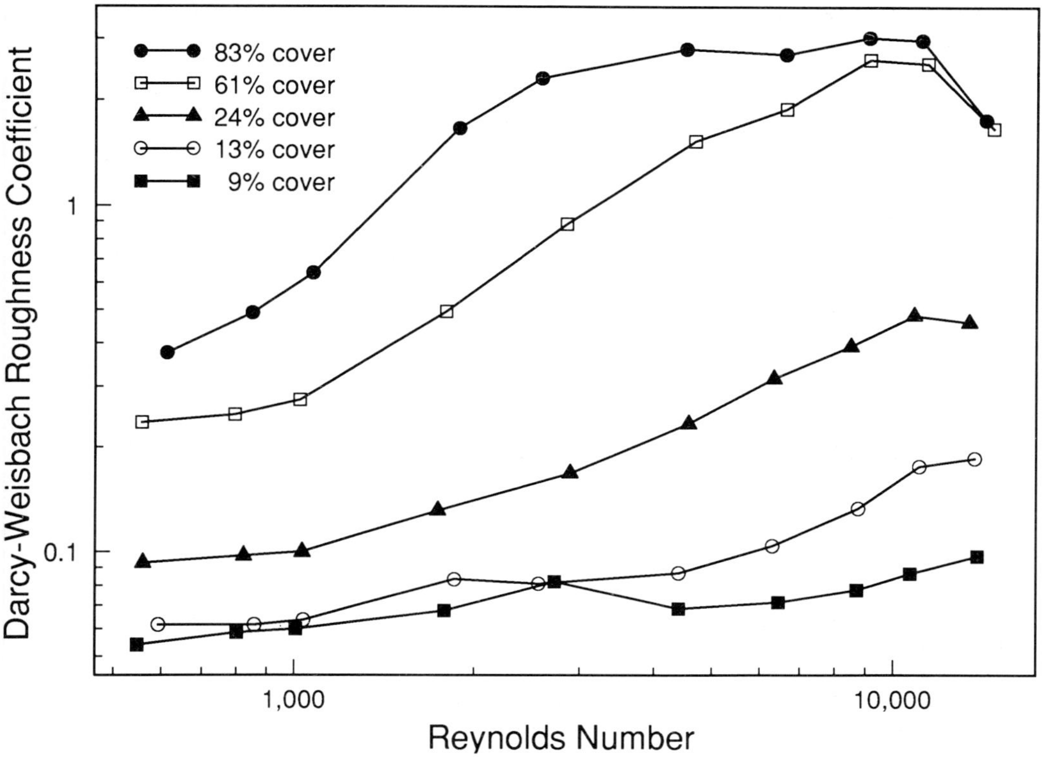

Figure 2.2 Darcy–Weisbach roughness coefficients vs. Reynolds number for cobble material with a diameter of 12.7–25.4 cm.

used to obtain an estimate of roughness coefficients on forested areas. After the residue materials had been removed from the field, they were placed in an oven and dried. For each residue type, 10 separate residue elements were randomly selected for measurement of residue dimensions. Mean residue diameters and lengths are shown in Table 2.6.

A measured mass of residue material was glued in a random orientation onto a section of reinforced fibreglass sheeting. For each residue type, five residue rates were selected. All of the residue materials, except pine needles and wheat, were applied at rates equivalent to 2, 4, 6, 8 and 10 t ha^{-1}. Rates equivalent to 0.75, 2, 4, 6 and 8 t ha^{-1} were used for pine needles, while wheat straw was applied at rates equivalent to 0.25, 0.50, 1, 2 and 4 t ha^{-1}. Since pine needle and wheat residue elements had smaller diameters than the other residue materials, they furnished greater surface cover at a given residue rate.

The percentage of surface cover provided at a given residue rate (Table 2.6) was obtained using the photographic grid procedure (Laflen et al. 1978) described previously. Testing procedures used to measure roughness coefficients for crop residues were similar to those used for gravel and cobble

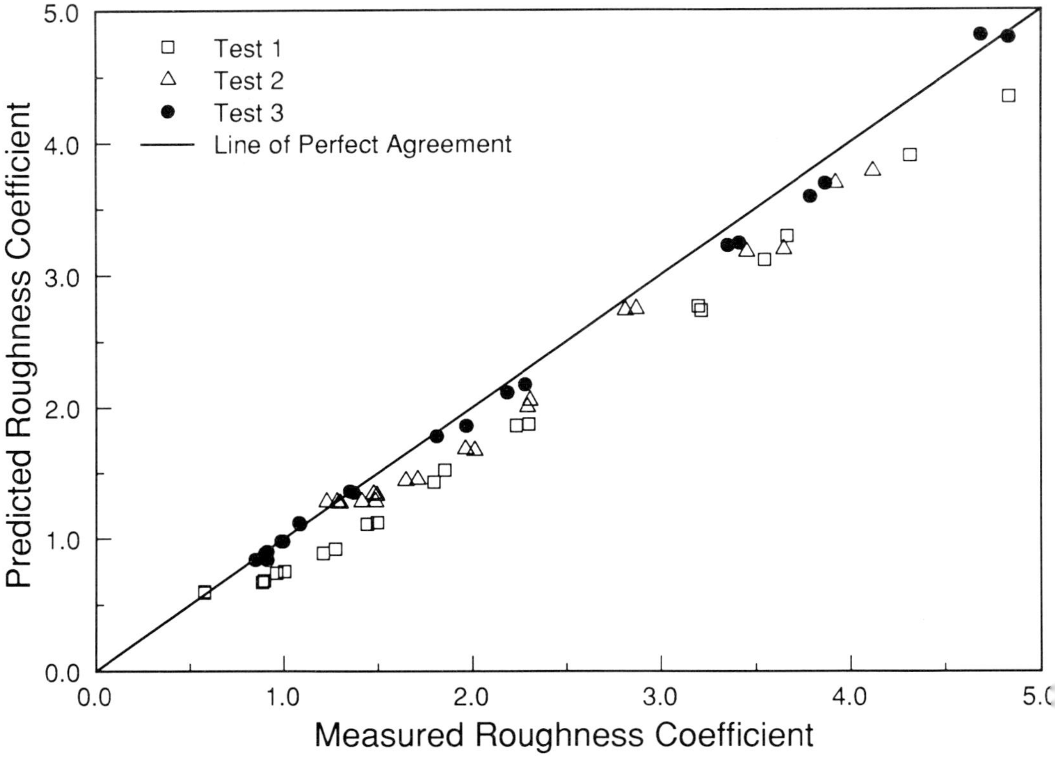

Figure 2.3 Predicted vs. measured Darcy–Weisbach roughness coefficients for surfaces containing gravel and cobble materials.

materials. Results reported here may be used for Reynolds number values ranging from approximately 500 to 16,000.

Roughness coefficient equations

Darcy–Weisbach roughness coefficients at varying Reynolds numbers for selected rates of wheat residue are shown in Figure 2.4. The trends presented in Figure 2.4 are characteristic not only of wheat residue but also of the other vegetative materials used in this investigation. Data presented in Figure 2.4 indicates that for a given residue rate, the Darcy–Weisbach friction factor usually decreased as Reynolds number increased.

The regression coefficients presented in Table 2.7 can be used to relate roughness coefficients for crop residue materials to percentage residue cover and Reynolds number. Regression coefficients are reported for selected residue types and for all residue types combined. Results for the all residue types combined analysis can be used to estimate the roughness coefficient for residue materials not used in this investigation using the relation

Table 2.6 Diameter, length, residue rate and surface cover of crop residue materials.

Residue type	Mean diameter cm	Mean length cm	Residue rate t ha^{-1}	Surface cover %
Corn	1.87	42.9	2 - 10	25 - 81
Cotton	0.73	36.2	2 - 10	12 - 50
Peanut	0.36	20.2	2 - 10	17 - 84
Pine needles	0.12	12.6	0.75 - 8	30 - 93
Sorghum	1.59	35.7	2 - 10	22 - 91
Soybeans	0.40	13.1	2 - 10	32 - 93
Sunflower	1.93	42.2	2 - 10	15 - 63
Wheat	0.30	19.4	0.25 - 4	26 - 99

$$f_{cr} = \frac{0.127(\mathit{percentage\ cover})^{1.55}}{Re^{0.388}} . \qquad (2.15)$$

Roughness coefficient values varying from 0.17 to 18.7 were used in the derivation of Equation 2.15.

Information on the rate of residue present at a particular site may be more readily available than surface cover data. Regression equations relating roughness coefficients to residue rate and Reynolds number were presented by Gilley et al. (1991). Procedures for estimating surface cover from values of residue rate were also identified for the selected residue materials.

Roughness coefficients on interrill areas

Experimental procedures

Field tests to determine roughness coefficients for interrill areas were conducted by Gilley and Finkner (1991) at the University of Nebraska Rogers Memorial Farm located in Lancaster County, approximately 18 km east of Lincoln, Nebraska. The Sharpsburg silty clay loam at the site (fine, montmorillonitic, mesic Typic Argiudolls) formed on loess under prairie vegetation. Average slope at the location was 6.4%.

The experimental design for the study consisted of two randomized complete blocks, with the first block being located immediately upslope from the second.

Table 2.7 Regression equation for roughness coefficients for surface residue on croplands versus percent cover and Reynolds number.

Residue type	Regression coefficients* a	b	c	Coefficient of determination r^2
Corn	6.30×10^{-2}	1.53	2.34×10^{-1}	0.911
Cotton	8.88×10^{-2}	1.02	7.88×10^{-2}	0.731
Peanut	2.61×10^{-1}	1.56	5.06×10^{-1}	0.924
Pine needles	8.71×10^{-5}	3.63	6.52×10^{-1}	0.874
Sorghum	5.24	7.96×10^{-1}	4.55×10^{-1}	0.960
Soybeans	9.28×10^{-2}	2.84	1.02	0.919
Sunflower	1.66	8.87×10^{-1}	3.51×10^{-1}	0.916
Wheat	2.98×10^{-4}	3.27	6.28×10^{-1}	0.938
All residue types combined**	1.27×10^{-1}	1.55	3.88×10^{-1}	0.648

* Regression coefficients a, b and c are used in the equation

$$f_{cr} = a\ (\text{percentage cover})^b\ /\ (Re)^c$$

where f_{cr} is the roughness coefficient for surface residue on croplands and *Re* is Reynolds number.

** For the "All residue types combined" analysis, Darcy–Weisbach roughness coefficients ranged from 0.17 to 18.7, while Reynolds number varied from approximately 500 to 16,000.

Each experimental block consisted of six tillage operations performed at random locations within the block. The tillage operations included an anhydrous applicator, chisel plough, disk, field cultivator, moldboard plough and planter. These implements were chosen to provide a wide range of random roughness conditions.

Existing wheat residue was first removed from the study area by burning and hand raking. Selected tillage operations were then performed parallel to the contour at the study site. Plots, of an area 1 m^2, were established within each tillage treatment using galvanized sheet metal borders for the top and both sides of the plots. A trough, located at the bottom of the plots, was used to collect runoff. When not in use, the plots were covered with plywood which was placed several centimetres above the plots. The plywood covering prevented

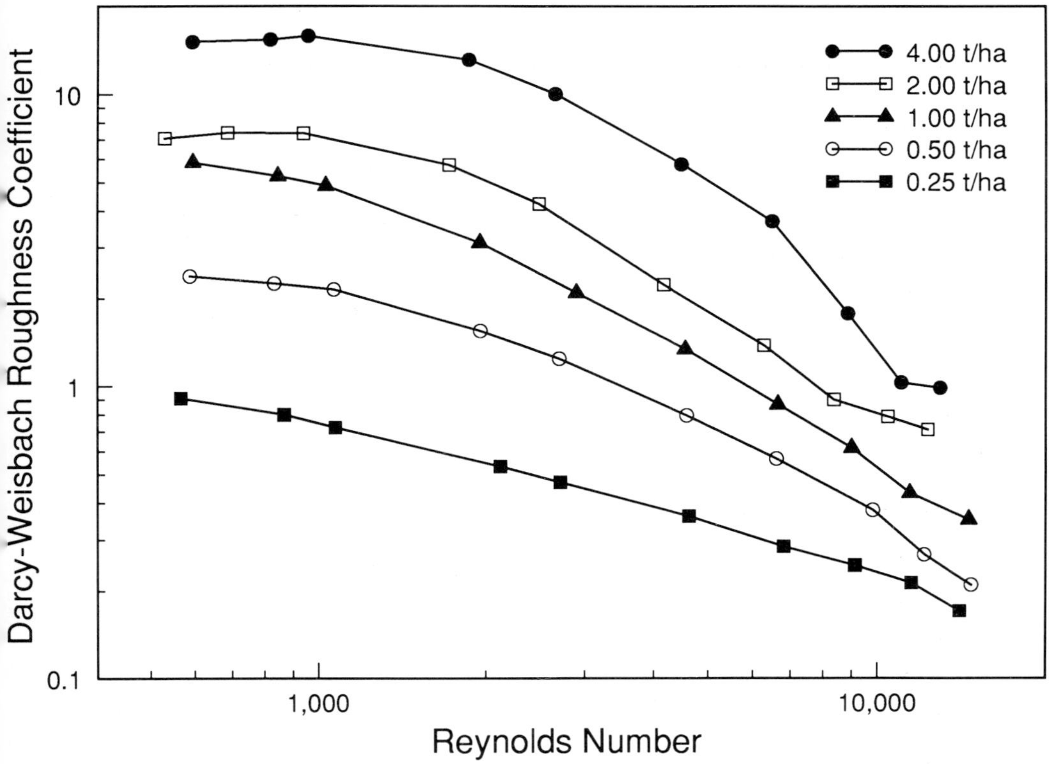

Figure 2.4 Darcy–Weisbach roughness coefficients vs. Reynolds number for selected rates of wheat residue.

disturbance of the soil surface by natural rainfall.

Differences in soil surface height were recorded using a mechanical profile meter. The surface profile meter, similar to the device described by Allmaras et al. (1967), could be easily rolled above the entire plot surface on a rectangular support frame. The support frame was of variable height and was levelled in the horizontal plane. The rectangular frame was supported by four 250 mm steel stakes which were securely anchored into the soil to provide a horizontal reference. The upper left corner of each plot border as viewed from the bottom of the plots was used as a vertical bench mark, creating a three-dimensional referencing system.

The profile meter consisted of a single row of equal length, 3.2 mm diameter steel pins positioned at a spacing of 6.4 mm. When lowered onto the soil surface, the top of the pins formed a nearly continuous line which was traced onto a strip of paper located behind the pins. The profile meter and frame were oriented so that surface elevations were measured parallel to the contour of the study area. Transects were spaced every 50 mm along the slope and transect traces were later digitized at 25 mm spacings. A total of 629 surface elevations were used for determination of random roughness for each plot.

Several tests were to be performed on each plot under identical soil conditions. Thus, soil-surface stabilization was required to prevent destruction of soil-form roughness during test procedures. After measurements for random roughness were obtained, the plot surfaces were stabilized using a biodegradable, latex-based soil stabilizer. The stabilizer was sprayed over the entire soil surface using a hand sprayer. The stabilizing material penetrated the soil approximately 5 mm, effectively binding the soil particles together in a water-permeable layer.

Following application and drying of the latex-base soil stabilizer, flow was added uniformly across the top of each plot at 12 selected rates. Flow inlet energy was dissipated at the top of the plots using an artificial turf carpet. Runoff was diverted into an HS flume with a stage recorder for measurement of flow rate.

Flow velocity was determined using dye tracing techniques. Approximately 0.2 l of fluorescent dye was uniformly injected across the width of the plot, 0.76 m upslope from the lower boundary. A peristaltic pump was used to continuously withdraw flow at four points spaced equally along the collection trough. Discharge was then circulated through a fluorometer which provided a visual display of dye concentration. Average time of travel was calculated as the length of time required for the dye concentration peak to reach the lower boundary. Five measurements of travel time were obtained at each of 12 inflow rates. The mean of the five readings was used to calculate flow velocity at a particular inflow rate.

Random roughness values

Random roughness was calculated using the procedure outlined by Allmaras et al. (1967). Table 2.8 presents random roughness measurements obtained in the present study, and values reported by Zobeck & Onstad (1987) in a review of available literature. Random roughness values in the present investigation ranged from 6 mm for the planter to 32 mm for the moldboard plough treatment.

The anhydrous applicator and planter caused little disturbance to the relatively smooth surface which existed at the study site. Random roughness values for these two operations were less than those reported previously. For the other tillage operations, random roughness measurements obtained in the present study were in close agreement with values reported by Zobeck & Onstad (1987).

The addition of rainfall may serve to reduce random roughness. To quantify this reduction, a relative random roughness term, *RRR*, was defined by Zobeck & Onstad (1987) as

Table 2.8 Random roughness values for selected tillage operations used to measure roughness coefficients on interrill areas.

Tillage operation	Random roughness* mm	Random roughness (present study) mm
Large offset disk	50	
Moldboard plough	32	32
Lister	25	
Chisel plough	23	21
Disk	18	16
Field cultivator	15	14
Row cultivator	15	
Rotary tillage	15	
Harrow	15	
Anhydrous applicator	13	8
Rod weeder	10	
Planter	10	6
No-till	7	
Smooth surface	6	

*Zobeck & Onstad (1987).

$$RRR = \frac{RR}{RR_o} \tag{2.16}$$

where RR is random roughness of a surface following rainfall, and RR_o is random roughness immediately after tillage. From published data on relative random roughness, Zobeck & Onstad (1987) developed the following equation

$$RRR = 0.89\,e^{-0.026\,cumulative\,rainfall} \tag{2.17}$$

where cumulative rainfall is given in cm. Equations 2.16 & 2.17 can be used to estimate random roughness of a surface following rainfall using information on cumulative rainfall since the last tillage operation.

Roughness coefficient equations

Darcy–Weisbach hydraulic roughness coefficients at varying Reynolds numbers for the moldboard plough and planter treatments are presented in Figure 2.5. The trends presented for the moldboard plough and planter operations are also characteristic of the other experimental treatments. In general, hydraulic roughness coefficients can be seen to decrease with greater Reynolds number.

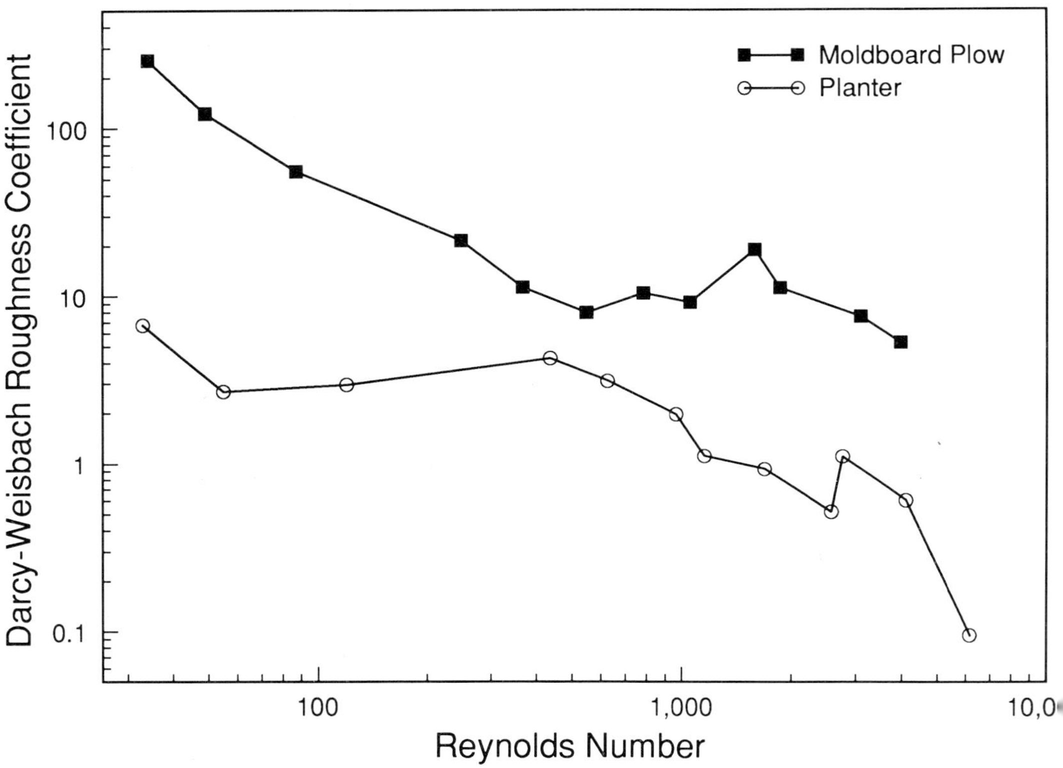

Figure 2.5 Darcy–Weisbach roughness coefficients vs. Reynolds number for selected tillage operations.

The moldboard plough and planter treatments produced the largest and smallest random roughness values, respectively. The largest hydraulic roughness coefficients usually occurred on those plots with the greatest random roughness. The planter treatments with relatively low random roughness values produced the smallest hydraulic roughness coefficients.

Within the same tillage operation, substantial variations in hydraulic roughness coefficients were found. These variations may have been caused by several factors. The range of selected flow rates produced conditions where the heights of the roughness elements were initially greater than and then less than flow depth. Also, as Reynolds number increased, differences in flow patterns sometimes occurred. Finally, transition from laminar to turbulent flow conditions may have resulted during a given test series.

Information from the six tillage treatments was used to derive the following regression equation for estimating roughness coefficients for interrill areas

$$f_{si} = \frac{6.30(RR_o)^{1.75}}{Re^{0.661}} \qquad (2.18)$$

where RR_o is given in mm. In deriving Equation 2.18, RR_o values varied from 6 to 32 mm while Reynolds number ranged from 20 to 6000. If rainfall has occurred since the last tillage operation, *RR* should be substituted for RR_o in Equation 2.18 to obtain the new roughness coefficient. Roughness coefficient values varying from 0.10 to 254 were used in the derivation of Equation 2.18. The relatively large roughness coefficients correspond with small Reynolds numbers. Reynolds number values used in this study were substantially less than those found in some of the other investigations.

Roughness coefficients for plants on cropland areas

Experimental procedures

Cox & Palmer (1948) conducted tests to measure roughness coefficients for alfalfa planted in test channels 0.61 m wide and 30.5 m long. Roughness coefficients for cotton, sorghum and wheat were determined by Ree & Crow (1977) using test channels with a bottom width of 6.1 m and a length of 183 m. In both studies, hydraulic measurements were collected under steady-state conditions. In addition, selected measurements were made to identify plant characteristics.

Roughness coefficient equations

Most of the hydraulic tests were performed at relatively large discharge rates which caused the vegetative materials to become submerged. Few of the tests were run at Reynolds number values which could be considered representative of overland flow conditions. From the available data, a maximum roughness coefficient value of 0.3 was assigned for cotton and sorghum. A maximum roughness coefficient of 3 was estimated for wheat, while alfalfa was assigned a maximum roughness coefficient value of 12. The following equation can be used to estimate the roughness coefficient for plants on cropland areas:

$$f_{st} = \frac{canopy\,height}{maximum\,canopy\,height} f_{stm} \qquad (2.19)$$

where f_{stm} is the maximum value of the roughness coefficient for selected plants on cropland areas. It should be noted that Reynolds number is not included as an independent variable in Equation 2.19.

Equation 2.19 was derived from limited experimental data. Thus, calculated roughness coefficient values should be considered as best estimates. If roughness coefficients are required for other plants on cropland areas, data for the crop reported above that is most like the material under consideration should be used.

Roughness coefficients for rangeland areas

Experimental procedures

A rotating boom rainfall simulator was used to supply rainfall to selected rangeland sites located throughout the western United States (Laflen et al. 1991). Rainfall was applied simultaneously to two plots having dimensions of 3.1 by 10.7 m. During the initial run, rainfall was applied for a 60 min duration at an intensity of approximately 65 mm h^{-1}. A wet run having a 30 min duration occurred approximately 24 h later. A very wet run with varying rainfall intensity and added inflow began about 30 min after completion of the wet run.

An optimization procedure similar to that used by Engman (1986) was employed by Weltz et al. (1992) to identify roughness coefficients for rangeland areas. The requirements for use of this procedure are that: (a) an equilibrium hydrograph must be achieved; (b) the infiltration rate is approximately uniform; and (c) the rainfall rate must be constant until the runoff hydrograph reaches an equilibrium condition.

Roughness coefficient equations

Weltz et al. (1992) used optimization procedures to develop an equation for estimating the roughness coefficient for litter and organic residue on rangelands

$$f_{lt} = 114 r_l^{3.0} \qquad (2.20)$$

where r_l is the fraction of the surface covered with litter and organic residue. Optimization procedures were also used by Weltz et al. (1992) to develop an equation for estimating the friction coefficient for plants on rangeland areas

$$f_{pb} = 39.0C_c^{0.8} + 126B_a^{0.8} \tag{2.21}$$

where C_c and B_a are the fraction of canopy cover and basal plant cover, respectively. Reynolds number is not included as an independent variable in either Equation 2.20 or 2.21.

Future needs for estimating roughness coefficients

Roughness coefficients for plants on croplands are not as well defined as some of the other factors contributing to hydraulic resistance. Additional experimental data for a wide variety of crops are needed. The data should include information on the effects of Reynolds number on roughness coefficients. The prediction equations should also include selected plant characteristics. Generalized equations should be developed which allow roughness coefficients to be estimated for plants not included in the experimental data sets. Many of these concepts are presently incorporated into procedures used in the design of grassed waterways (Temple et al. 1987).

Computer optimization techniques were employed to identify roughness coefficients for rangeland areas. Iteration procedures were used to achieve a best fit to the rising side of the hydrograph, resulting in a single roughness coefficient being identified for a particular site. Consequently, Reynolds number was not included in the regression equations obtained for estimating roughness coefficients for rangeland areas. Field experimental tests should be performed to determine the effects of Reynolds number on roughness coefficients. Again, generalized equations should be developed which relate roughness coefficients to selected characteristics of rangeland plants. Kao & Barfield (1978) related flow resistance parameters to Reynolds number and selected vegetation factors.

The additive property of roughness coefficients has been successfully demonstrated for the components of Equations 2.1 & 2.2. Equations 2.9–2.12 each contain four factors which may contribute to total hydraulic resistance. Procedures used to identify roughness coefficients for each of these components have been developed and tested. However, the ability to add these individual factors to obtain total hydraulic resistance for a particular site has not been verified. Field and laboratory tests should be conducted to determine whether individual roughness coefficients are additive. Measured and calculated roughness coefficients should be compared for a wide variety of surfaces.

Summary and conclusions

Darcy–Weisbach roughness coefficients are used to analyse overland flow. Total hydraulic resistance on a site may be caused by several factors. Equations were identified to estimate roughness coefficients for: (a) rills; (b) gravel and cobble materials; (c) surface residue on croplands; (d) interrill areas; (e) plants on cropland areas; and (f) rangeland areas.

A rainfall simulation study was conducted at 11 sites located throughout the eastern United States to measure rill hydraulic characteristics. Roughness coefficients were calculated from experimental measurements of flow rate, flow velocity and flow width. Regression equations were developed which related roughness coefficients for rills to Reynolds number.

Roughness coefficients for gravel and cobble materials were identified in a laboratory investigation. Selected rates of flow were introduced into a flume in which a given size class of gravel or cobble material had been securely attached. The laboratory data were used to develop regression equations which related roughness coefficients for gravel and cobble materials to surface cover and Reynolds number. The regression relations were tested using hydraulic data collected on surfaces containing a distribution of size classes. Close agreement between predicted and measured roughness coefficients was obtained by adding the roughness contributions of individual size classes.

A laboratory study was also performed to determine roughness coefficients for surface residue on croplands. Roughness coefficients were determined for corn, cotton, peanut, pine needles, sorghum, soybeans, sunflower and wheat residue. Regression equations were developed which related roughness coefficients for residue on croplands to surface cover and Reynolds number.

Roughness coefficients on interrill areas were identified in a field investigation. A random roughness parameter is frequently used to characterize surface microrelief. Six selected tillage operations were performed which produced a range of random roughness parameters. Hydraulic roughness coefficients corresponding with the random roughness parameters were then determined. The experimental data were used to derive regression relationships which related hydraulic roughness coefficients on interrill areas to a random roughness parameter and Reynolds number.

Field studies have been performed to determine roughness coefficients for plants on croplands (Cox & Palmer 1948, Ree & Crow 1977). Roughness coefficient measurements were made for alfalfa, cotton, sorghum and wheat. An equation was presented to relate roughness coefficients for plants on cropland to canopy height.

Roughness coefficients for rangeland areas were identified using data collected during rainfall simulation tests. Optimization techniques were used by Weltz et al. (1992) to determine roughness coefficients using the rising side of runoff hydrographs. Regression equations were then identified which related roughness coefficients on rangeland areas to surface cover, canopy cover and basal plant cover.

Roughness coefficients for plants on croplands and rangelands are not as well defined as some of the other factors contributing to hydraulic resistance. Additional experimental data for a wide variety of cropland and rangeland plants are needed. Generalized equations should be developed which relate roughness coefficients to selected plant characteristics and Reynolds number. Our ability to understand and accurately model upland flow hydraulics will improve as additional information on roughness coefficients becomes available.

Acknowledgement

This contribution is from the USDA-Agricultural Research Service, in co-operation with the Agricultural Research Division, University of Nebraska, Lincoln.

Notation

Symbol	Definition	Units
A	cross-sectional flow area	m^2
b	flow width	m
B_a	fraction of basal plant cover on rangeland areas	
C_c	fraction of canopy cover on rangeland areas	
f	Darcy–Weisbach roughness coefficient	
f_b	roughness coefficient associated with bed-form roughness	
f_{cr}	roughness coefficient for surface residue on croplands	
f_g	roughness coefficient associated with grain roughness	
f_i	total roughness coefficient for interrill cropland areas	
f_{ir}	total roughness coefficient for	

	interrill rangeland areas	
f_{lt}	roughness coefficient for litter and organic residue on rangelands	
f_{pb}	roughness coefficient for plants on rangeland areas	
f_r	total roughness coefficient for rills on croplands	
f_{ra}	roughness coefficient associated with rainfall	
f_{rk}	roughness coefficient for gravel and cobble materials	
f_{rr}	total roughness coefficient for rills on rangelands	
f_{si}	roughness coefficient for interrill areas	
f_{sr}	roughness coefficient for rills	
f_{st}	roughness coefficient for plants on cropland areas	
f_{stm}	maximum value of the roughness coefficient for selected plants on cropland areas	
f_{wo}	roughness coefficient without rainfall	
g	acceleration due to gravity	$m\ s^{-2}$
P	wetted perimeter	m
Q	flow rate	$m^3\ s^{-1}$
r_l	fraction of the rill surface covered with litter and organic residue on rangelands	
R	hydraulic radius	m
Re	Reynolds number	
RR	random roughness of a surface following rainfall	m
RR_o	random roughness immediately after tillage	m
RRR	relative random roughness	
s	average slope	
SF	shape factor	
V	mean flow velocity	$m\ s^{-1}$
y	flow depth	m
ν	kinematic viscosity	$m^2\ s^{-1}$

References

Abrahams, A. D., A. J. Parsons, S.-H. Luk 1986. Resistance to overland flow on desert hillslopes. *Journal of Hydrology* **50**, 343–63.

Alam, A. M. Z. & J. F. Kennedy 1969. Friction factors for flow in sand-bed channels. *Journal of the Hydraulics Division, American Society of Civil Engineers* **95**, 1973–92.

Allmaras, R. R., R. E. Burwell, R. F. Holt 1967. Plow-layer porosity and surface roughness from tillage as affected by initial porosity and soil moisture at tillage time. *Soil Science Society of America Proceedings* **31**, 550–6.

Chow, V. T. 1959. *Open-channel hydraulics*. New York: McGraw-Hill.

Cox, M. B. & V. J. Palmer 1948. *Results of tests on vegetated waterways, and methods of field application*. Oklahoma Agricultural Experiment Station Miscellaneous Publication Number MP-12.

Dunne, T. & W. E. Dietrich 1980. Experimental study of Horton overland flow on tropical hillslopes. 2. Hydraulic characteristics and hillslope hydrographs. *Zeitschrift für Geomorphologie Supplementband* **35**, 60–80.

Einstein, H. A. & N. L. Barbarossa 1952. River channel roughness. *Transactions of the American Society of Civil Engineers* **117**, 1121–32.

Emmett, W. W. 1970. *The hydraulics of overland flow on hillslopes*. United States Geological Survey Professional Paper 662-A. Washington, DC: US Government Printing Office.

Engelund, F. 1966. Hydraulic resistance of alluvial streams. *Journal of the Hydraulics Division, American Society of Civil Engineers* **92**, 315–26.

Engman, E. T. 1986. Roughness coefficients for routing surface runoff. *Journal of Irrigation and Drainage Engineering, American Society of Civil Engineers* **112**, 39–53.

Gilley, J. E. & S. C. Finkner 1991. Hydraulic roughness coefficients as affected by random roughness. *Transactions of the American Society of Agricultural Engineers* **34**, 897–903.

Gilley, J. E., E. R. Kottwitz, J. R. Simanton 1990. Hydraulic characteristics of rills. *Transactions of the American Society of Agricultural Engineers* **33**, 1900–6.

Gilley, J. E., E. R. Kottwitz, G. A. Wieman 1991. Roughness coefficients for selected residue materials. *Journal of Irrigation and Drainage Engineering, American Society of Civil Engineers* **117**, 503–14.

Gilley, J. E., E. R. Kottwitz, G. A. Wieman 1992. Darcy–Weisbach roughness coefficients for gravel and cobble surfaces. *Journal of Irrigation and Drainage Engineering, American Society of Civil Engineers* **118**, 104–12.

Guy, H. P. 1969. Laboratory theory and methods for sediment analysis. *United States Geological Survey Book 5*, Chapter C1, 23–30. Washington, DC: US Government Printing Office.

Hubbard, E. F., F. A. Kilpatrick, L. A. Martens, J. F. Wilson 1982. Measurement of time of travel and dispersion in streams by dye tracing. *Techniques of water-resources investigations of the United States Geological Survey, Book 3 (Applications of Hydraulics)*, Chapter A9. Washington, DC: US Government Printing Office.

Kao, D. T. Y. & B. J. Barfield 1978. Prediction of flow hydraulics for vegetated channels. *Transactions of the American Society of Agricultural Engineers* **21**, 489–94.

Laflen, J. M., J. L. Baker, R. O. Hartwig, W. F. Buchele, H. P. Johnson 1978. Soil and water loss from conservation tillage systems. *Transactions of the American Society of Agricultural Engineers* **21**, 881–5.

Laflen, J. M., W. J. Elliot, J. R. Simanton, C. S. Holzhey, K. D. Kohl 1991. WEPP: Soil erodibility experiments for rangeland and cropland soils. *Journal of Soil and Water Conservation* **46**, 39–44.

Lovera, F. & J. F. Kennedy 1969. Friction factors for flat-bed flows in sand channels. *Journal of the Hydraulics Division, American Society of Civil Engineers* **95**, 1227–34.

Meyer-Peter, E. & R. Müller 1948. Formulas for bed-load transport. *International Association for Hydraulic Research* **2**, 39–64.

Parsons, A. J., A. D. Abrahams, S.-H. Luk 1990. Hydraulics of interrill overload flow on a semi-arid hillslope, southern Arizona. *Journal of Hydrology* **117**, 255–73.

Phelps, H. O. 1975. Shallow laminar flows over rough granular surfaces. *Journal of the Hydraulics Division, American Society of Civil Engineers* **101**, 367–84.

Ree, W. O. & F. R. Crow 1977. Friction factors for vegetated waterways of small slope. *Agricultural Research Service* S-151. Washington, DC: US Government Printing Office.

Replogle, J. A., L. E. Meyers, K. J. Brust 1966. Flow measurements with fluorescent tracers. *Journal of the Hydraulics Division, American Society of Civil Engineers* **92**, 1–14.

Roels, J. M. 1984. Flow resistance in concentrated overland flow on rough slope surfaces. *Earth Surface Processes and Landforms* **9**, 541–51.

Savat, J. 1980. Resistance to flow in rough supercritical sheet flow. *Earth Surface Processes and Landforms* **5**, 103–22.

Shen, H. W. & R. M. Li 1973. Rainfall effect on sheet flow over smooth surface. *Journal of the Hydraulics Division, American Society of Civil Engineers* **99**, 771–92.

Swanson, N. P. 1965. Rotating-boom rainfall simulator. *Transactions of the American Society of Agricultural Engineers* **8**, 71–2.

Temple, D. M., K. M. Robinson, R. M. Ahring, A. G. Davis 1987. *Stability design of grass-lined open channels*. Agriculture Handbook Number 667. Washington, DC..: U.S. Government Printing Office.

Weltz, M. A., A. Arslan, L. J. Lane 1992. Hydraulic roughness coefficients for native rangelands. *Journal of Irrigation and Drainage Engineering, American Society of Civil Engineers* (in press).

Woo, D. C. & E. F. Brater 1961. Laminar flow in rough rectangular channels. *Journal of Geophysical Research* **66**, 4207–17.

Zobeck, T. M. & C. A. Onstad 1987. Tillage and rainfall effects on random roughness: a review. *Soil and Tillage Research* **9**, 1–20.

3

The control of headwater area on channel runoff in a small arid watershed

Aaron Yair

Abstract

Hydrological monitoring of hillslope and channel runoff at the Sede Boqer experimental watershed in the Negev Desert, Israel, revealed that the areas responding quickest to rainfall are represented by bedrock outcrops. Colluvial or alluvial sediments, due to their unconsolidated and porous nature, tended to absorb rainwaters and produced little or no runoff. Data obtained also indicate that on valley-side slopes, runoff is frequently generated within the upper rocky sections, whereas infiltration occurs downslope within the colluvial section with little or no runoff contributed to channel flow. Streamflows were mainly initiated near the headwater area where extensive rocky outcrops frequently generated high-magnitude flows.

Introduction

Hydrological monitoring carried out within humid regions has revealed that runoff generation is spatially non-uniform even within small watersheds carved out in lithology which is homogeneous. Runoff generation, responsible for storm channel flow, was found to be limited to the channel and riparian areas where saturated or nearly saturated conditions eliminate or inhibit infiltration losses (Hewlett & Hibert 1967, Whipkey 1967, Ragan 1968, Betson & Marius 1969, Dunne & Black 1970, Weyman 1970, Anderson & Burt 1978, Dunne 1978, Anderson & Kneale 1980). Similar conditions are unlikely to occur in dry areas. The paucity of rainfall and limited rainfall amounts, coupled with high infiltration losses in the dry channels, prevents the development of saturated or nearly saturated conditions in semi-arid or arid watersheds. Hillslope runoff is thus regarded as the main contributor to storm channel runoff, representing an essential stage in the initiation and development of any channel flow.

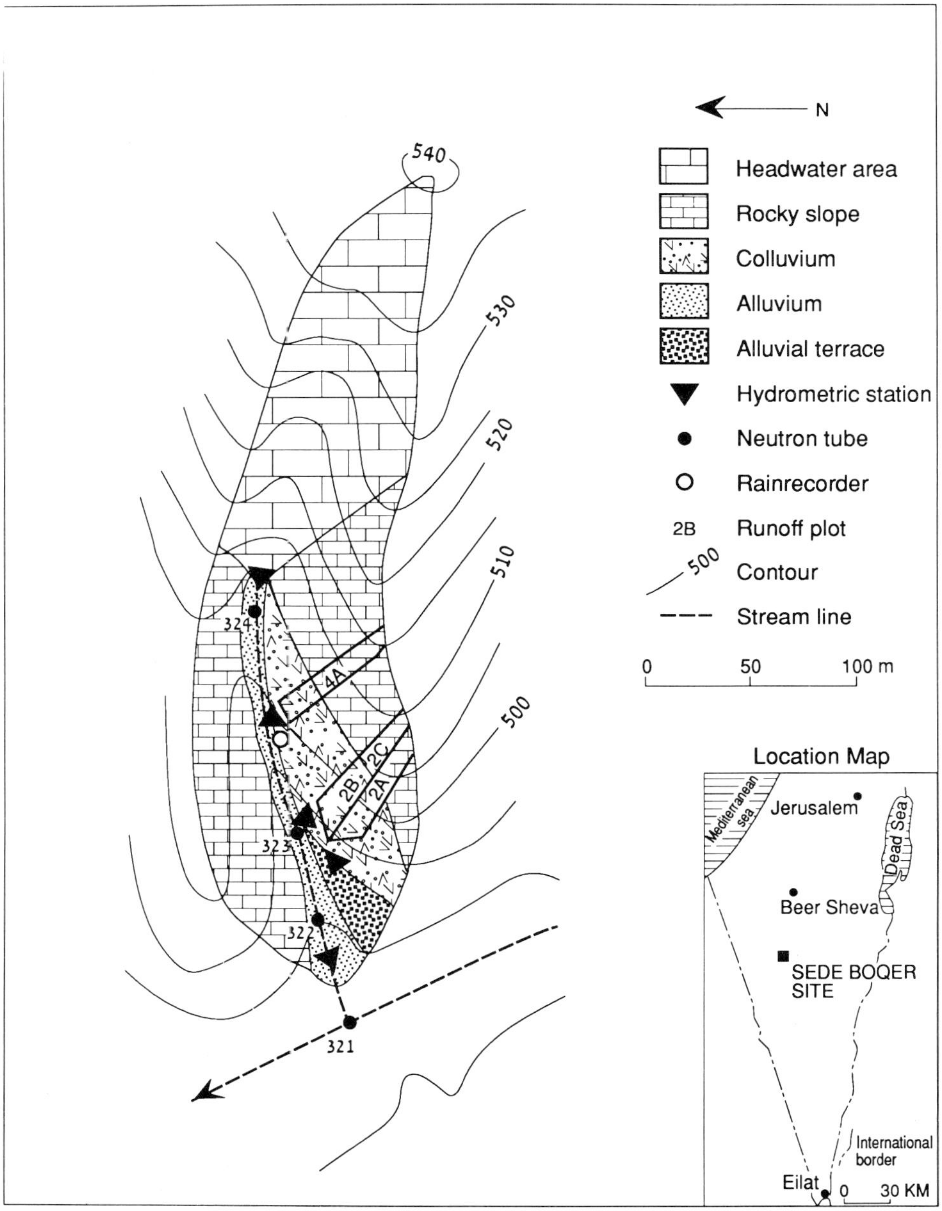

Figure 3.1 The Sede Boqer experimental watershed.

Long-term monitoring of hillslope runoff at the Sede Boqer experimental watershed, located in the Negev Desert of Israel (Fig. 3.1), revealed that hillslope areas responding quickest to rainfall are represented by bedrock outcrops with a low infiltration capacity. Runoff is generated for any rainstorm whose maximum intensity, for a period of 3 min, exceeds 5 mm h^{-1}. At the same time, unconsolidated colluvial or alluvial sediments absorb all rainwater for most rainstorms, thus producing little or no runoff (Yair et al. 1978, Yair 1983, Yair & Lavee 1985). Data obtained also indicate that where valley-side slopes have a colluvial mantle, runoff generated over their upper rocky section is normally absorbed on its path downslope within the colluvium, and contributes little or no runoff to channel flow (Fig. 3.2). The above observations raise a basic hydrologic question. If valley bottoms and colluvial valley-side slopes play a minor role in storm channel flow generation, then where does channel runoff originate? The hypothesis advanced here is that channel runoff is mainly derived from the channel headwater area. In this area extensive bedrock outcrops, almost completely devoid of any soil cover, will respond quickly to rainfall and frequently generate high runoff rates per unit area.

Aim of study

The present study had two main objectives:

(a) identification of the hydrometeorological conditions for the initiation of surface runoff over the various geomorphic units identified within the Sede Boqer experimental watershed; and

(b) estimation of the relative contribution of each of these units to storm channel flow.

Description of study area

The Sede Boqer experimental site is situated in the Negev Highlands (Fig. 3.1). Average annual rainfall is 93 mm with extreme values of 30–183 mm. The number of rain days varies from 15 to 42. Rain is limited to the winter season. Mean monthly temperatures range from 9°C in January to 27°C in August. Potential evaporation is 2200 mm (Zangvill & Druian 1983). The experimental watershed represents a first-order drainage basin which extends over an area of 2.15 ha. Geologically it is composed of limestones and flinty limestones of Turonian age that dip gently (4°) towards the northwest. The watershed can be subdivided into five distinct geomorphic units:

Unit 1: Headwater area. This extends over 0.81 ha and is composed of

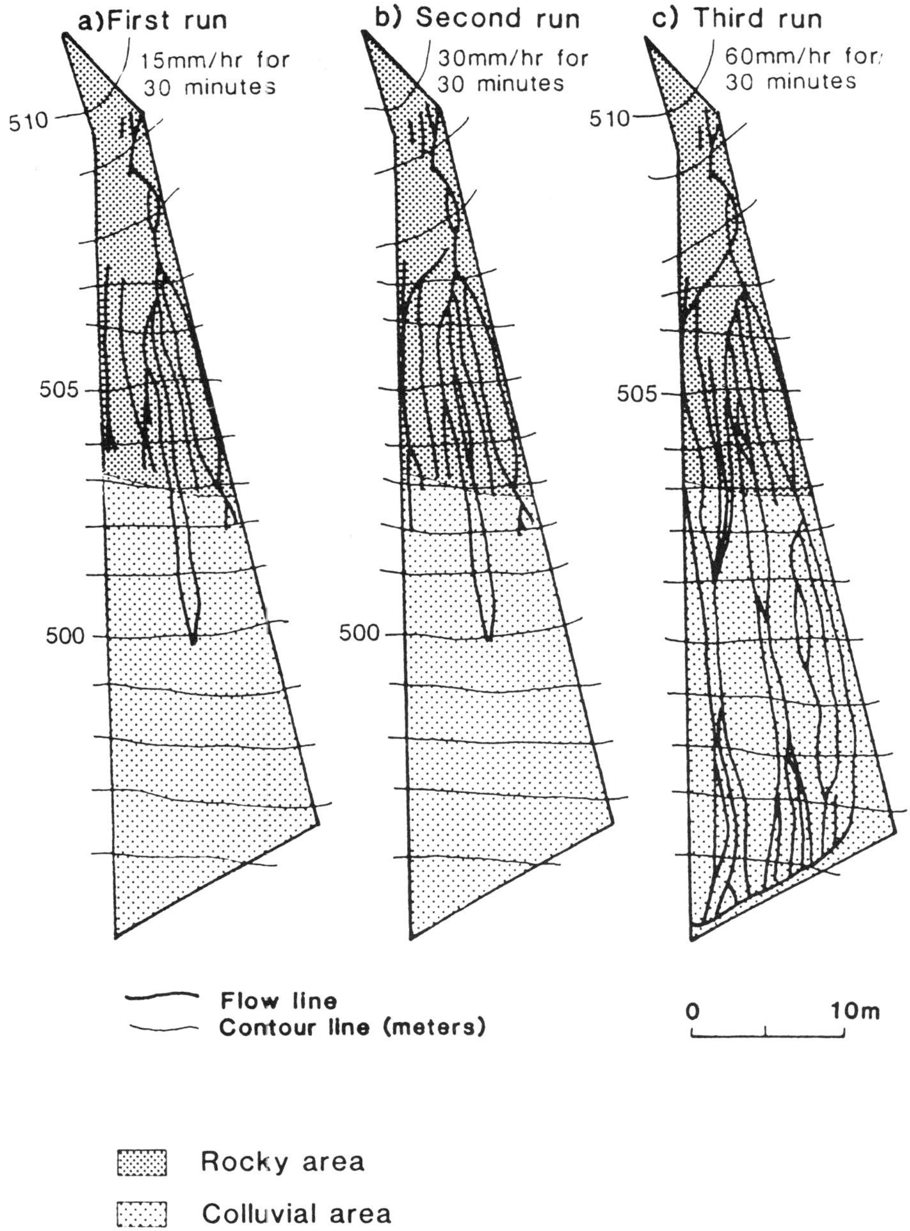

Figure 3.2 Flow lines mapped during sprinkling experiments

massive and flinty limestones. Very extensive rocky outcrops, almost devoid of any soil cover, form most of the surface.

Unit 2: Colluvial valley-side slopes. This unit extends over 0.64 ha with slopes ranging from 55 to 72 m in length. The upper part is made of massive limestone and displays a stepped topography with soil strips at the base of the steps. The massive limestone is underlain by thinly bedded and densely jointed limestone. Most of the latter formation is covered by a colluvial mantle which thickens quickly downslope.

Unit 3: Valley-side slopes with no colluvium. These slopes are much shorter than the opposite ones. They are carved in the thinly bedded and densely jointed limestone. The rock weathers into gravels and cobbles embedded in a thin soil cover. This unit covers 0.37 ha.

Unit 4: Alluvial terrace. This low terrace separates part of the colluvial area from the present-day active channel. It covers an area of 0.13 ha.

Unit 5: Alluvial reach. The alluvial fill extends from the upper rocky headwater area down to the mouth of the valley. It is mainly composed of fine-grained material of loessic origin with few gravels and cobbles. The alluvial fill covers 0.20 ha.

The experimental design

The Sede Boqer experimental watershed is equipped for automatic and simultaneous measurement of rainfall, and both hillslope and channel runoff (Fig. 3.1). Rainfall is measured with a rain recorder located at the lower part of the drainage basin. Hillslope runoff is measured at two plots. Plot 4A drains a whole slope whose upper part is rocky and lower part is colluvial. Plot 2 is subdivided into three subplots. Subplot 2C drains the upper rocky slope section, subplot 2B drains the colluvial slope section, and subplot 2A drains the adjoining whole slope from top to bottom. This design allows us to study the specific response of colluvial and rocky surfaces to rainfall as well as the response of a combined slope. The hydrometric stations for plot 4A and subplots 2A and 2B are located at the interface between the colluvium and the adjoining alluvial terrace.

Channel runoff is measured at two stations. The upper one is located at the transition from the headwater area into the alluvial area. The second station is located close to the mouth of the drainage basin. All hydrological stations are equipped with very sensitive stage recorders, the recording ratio being 2.5.

Table 3.1 Rainfall–runoff relationships (1988–89).

Date	Rain (mm)	Runoff (litres) Upper channel	Lower channel	Plot 2A	Plot 2B	Plot 2C	Plot 4A
13-12-88	1.9	0.0	0.0	0.0	0.0	0.0	0.0
24-12-88	16.1	269.0	0.0	0.0	0.0	0.0	0.0
25-12-88	10.3	2122.0	0.0	0.0	0.0	0.0	0.0
26-12-88	8.7	25945.0	3001.0	512.0	293.0	313.0	719.0
03-01-89	0.5	0.0	0.0	0.0	0.0	0.0	0.0
10-01-89	4.6	0.0	0.0	0.0	0.0	0.0	0.0
15-01-89	1.6	0.0	0.0	0.0	0.0	0.0	0.0
16-01-89	1.9	0.0	0.0	0.0	0.0	0.0	0.0
21-01-89	17.2	9605.0	0.0	219.0	232.0	226.0	476.0
27-01-89	3.9	3029.0	0.0	236.0	197.0	56.0	310.0
28-01-89	3.3	0.0	0.0	0.0	0.0	0.0	0.0
01-02-89	1.1	0.0	0.0	0.0	0.0	0.0	0.0
10-02-89	0.9	0.0	0.0	0.0	0.0	0.0	0.0
11-02-89	2.6	1077.0	0.0	0.0	0.0	0.0	0.0
12-02-89	6.5	0.0	0.0	0.0	0.0	0.0	0.0
21-02-89	9.0	18563.0	19825.0	1332.0	835.0	398.0	1177.0
22-02-89	0.6	0.0	0.0	0.0	0.0	0.0	0.0
09-03-89	0.9	0.0	0.0	0.0	0.0	0.0	0.0
14-03-89	3.8	0.0	0.0	0.0	0.0	0.0	0.0
Total	95.4	60610.0	23186.0	2299.0	1557.0	993.0	2622.0

Results

Results presented relate to the rainfall year 1988–89, for which hillslope and channel data are available.

Rainfall
The rainfall year 1988–89 began in October 1988 and ended in March 1989 (Fig. 3.3, Table 3.1). Total annual rainfall was very close to the long-term average and amounted to 95.4 mm. Nineteen storms of various magnitudes were recorded with eight amounting to less than 2 mm each. Three storms had rain amounts greater than 9 mm, with the largest storm amounting to 17.2 mm. Rainfall intensities were generally low to very low (Fig. 3.3). For most storms the maximum rain intensity for a period of 3 min did not exceed 10 mm h^{-1}. The highest intensity of 28.3 mm h^{-1} was recorded on 21 February 1989 during a rainstorm of 9 mm. All rainstorms are characterized by a high temporal variability in rain intensity and by an intermittent rain pattern.

Runoff
Runoff data for the two hydrological stations measuring channel runoff, and for the three hillslope plots, are contained in Tables 3.1 & 3.2. The area drained by each of the stations is given in Table 3.3. Data analysis shows that out of the 19 storms recorded, only seven generated runoff in one or more parts of the drained areas. Runoff occurred in three distinct periods; December 1988, and January and February 1989.

December 1988 The rainspell of December lasted three consecutive days (24–26 December). Total rainfall amounted to 35.1 mm with intermittent rainfall resulting in several separate flows (Fig. 3.4). Rain intensities were initially very low and increased towards the end of the storm, but never exceeded 10 mm h^{-1} (Table 3.2). The area most responsive to rainfall was the headwater area. Some runoff developed here on the first day, when 16.1 mm fell on the area. The limited runoff occurred during a short burst when maximum rain intensity reached the value of 4.5 mm h^{-1}. No runoff was recorded on the same day at any of the other drained areas (Table 3.1).

The following day, the area received 10.3 mm. Because of a slight increase in maximum rain intensity and an already wet surface, runoff developed again only over the rocky headwater area. Most of the runoff was generated on the third day when the highest rain intensity (9.5 mm h^{-1}) occurred on a very wet soil. At this stage the response of the area to rainfall was quite general. Very significant differences in specific runoff yields, however, can be observed. The highest runoff coefficients were obtained for the headwater area, followed by the rocky slopes with the lowest coefficients occurring for both the colluvial

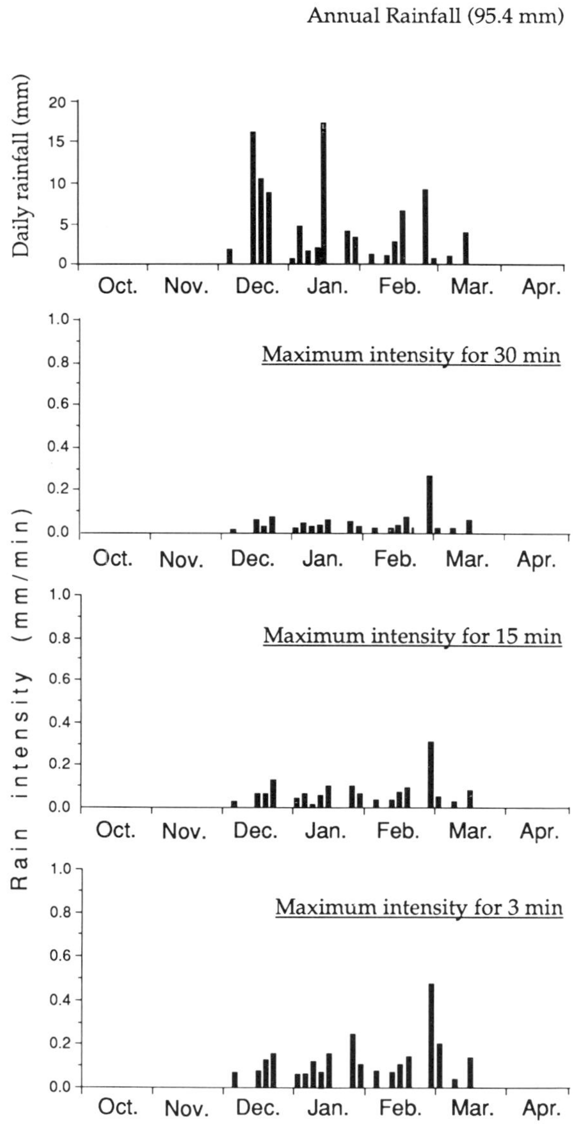

Figure 3.3 Rainfall year 1989–90: temporal distributions and intensities.

slope (2B) and the alluvial channel (Table 3.2).

January 1989 Two distinct flows occurred during this month. The first occurred on 21–22 January, when intermittent rainfall over a 24 h period totalled 17.2 mm. Intensities were low except for a few short showers of 7–9 mm h^{-1} which generated several distinct flows. Runoff occurred on the rocky

Table 3.2 Maximum rain intensities and runoff per unit area.

		Runoff per unit area (l m^{-2})					
Date	I max for 3 min mm h^{-1}	Upper channel	Lower channel	Plot 2A	Plot 2B	Plot 2C	Plot 4A
24-12-88	4.5	0.03	0.00	0.00	0.00	0.00	0.00
25-12-88	7.7	0.26	0.00	0.00	0.00	0.00	0.00
26-12-88	9.5	3.18	0.14	1.10	0.87	1.73	1.20
21-01-89	9.2	1.20	0.00	0.47	0.68	1.25	0.79
27-01-89	19.9	0.38	0.00	0.51	0.59	0.31	0.51
11-02-89	6.2	0.10	0.00	0.00	0.00	0.00	0.00
21-02-89	28.3	2.40	0.92	2.86	2.48	2.47	1.94
	Total	7.55	1.06	4.94	4.62	5.76	4.44

headwater area, over the rocky hillslopes and colluvial areas, but not at the lower channel. The second flow occurred on 27 January. Rain amount was low (3.9 mm) but with a high peak intensity of 20 mm h^{-1}. The runoff generation pattern was similar to that observed at the previous event (Table 3.1).

February 1989 Two flow events occurred during this month. During the first event (11 February), rainfall was limited to 2.6 mm with an intensity of 6.2 mm h^{-1}, which was sufficiently high enough to produce runoff at the headwater area. The second flow event took place 10 days later and was the major flow event of the year. Although the rain amount was not very high (9 mm), runoff was generated throughout the entire area. This is due to the fact that the storm had a convective character with an intensity of 28 mm h^{-1}

Table 3.3 Area of drained plots.

	Upper channel	Lower channel	Plot 2A	Plot 2B	Plot 2C	Plot 4A
Area (m^2)	8159	21450	465	337	181	607

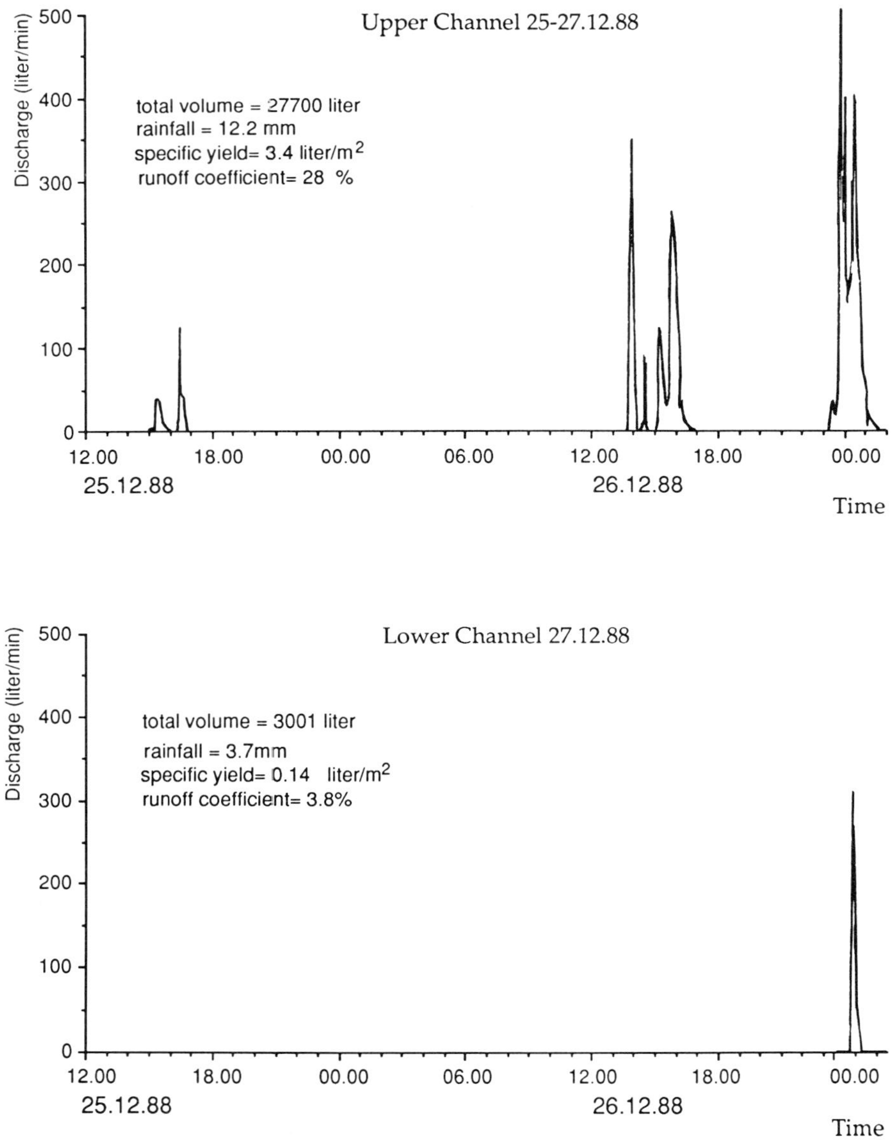

Figure 3.4 The storm of 25–27 December 1988 at the upper and lower channel.

occurring over several minutes.

Discussion

Runoff data presented here support previous data obtained in the area (Yair 1983, Yair & Lavee 1985) that the frequency and magnitude of runoff

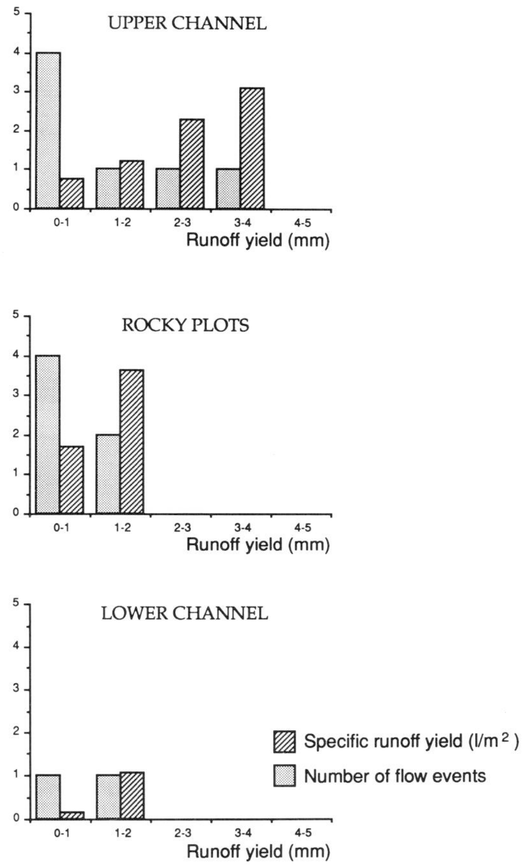

Figure 3.5 Frequency and magnitude of runoff events for various surface units.

generation are controlled by the ratio of bare bedrock outcrops and sediment-covered surfaces (Rbs). The area most responsive to rainfall was the headwater area, where this ratio is the highest within the studied watershed (Fig. 3.5). The threshold rain intensity for runoff generation over the headwater area was as low as 4.5 mm h^{-1}. A slightly higher threshold was required for runoff generation over the rocky section of the valley-side slopes where Rbs is lower owing to a more extensive soil cover. The lowest Rbs is found over the colluvial and alluvial parts of the watershed. The threshold rain intensity for runoff generation over these surfaces was estimated to be 10 mm h^{-1} for previously moistened surfaces and to be 15 mm h^{-1} for relatively dry surfaces. The difference in the response of rocky and sediment-covered surfaces was best expressed along the channel than over the hillslopes (Fig. 3.5). Seven daily flows were recorded at the upper channel but only two at the lower channel. The difference was more pronounced when runoff yields per unit area were considered (Table 3.2). For the hillslope area, little difference in runoff production per unit area was obtained for storms whose maximum rain intensity exceeded 20 mm h^{-1}. At lower intensities, the specific runoff yield of the rocky areas was twice as great as that of the sediment-covered surfaces.

The distribution of rocky and sediment-covered surfaces is not random in arid watersheds. Rocky outcrops always extend over the headwater area and upper parts of hillslopes, whereas colluvial mantles cover the lower part of the hillslopes and alluvial fills cover valley bottoms. Considering the difference in the hydrological behaviour of the two basic types of surfaces as well as their spatial distribution, the sediment-covered surfaces represent a very efficient buffer zone or filter which separates the source areas of runoff from the watershed outlet. A flow discontinuity is often observed on passing from the rocky into the colluvial slope section. This continuity is fully corroborated by biological (Yair & Danin 1980, Yair & Shachak 1982) and pedological (Wieder et al. 1985) findings which are indicative of the hydrological regime acting upon the hillslopes over a period of up to thousands of years.

The same trend is observed on passing from the rocky into the alluvial channel. The phenomenon of flow discontinuity is greatly enhanced by the intermittent character of the rain. The brief duration of the effective rainshowers causes runoff also to be short lasting. This duration is often shorter than the concentration time required for a continuous flow along an entire slope from top to bottom or along the entire channel length. Under these conditions the proximity of the rocky area to the channel becomes a crucial factor in storm channel runoff. In this respect, the extensive headwater area represents the area whose contribution to channel runoff can be expected to be the highest in both frequency and magnitude.

The efficient filtering effect of the colluvial and alluvial areas within the studied watershed was particularly well expressed on the 2 days when runoff did occur at the lower channel. On 26 December a runoff volume of 26 m^3 was recorded at the upper channel. Runoff also occurred over the rocky and colluvial valley-side slopes, being higher on the former than on the latter. Runoff yield from hillslopes on both sides of the channel was estimated to be 11 m^3 (1.1 l $m^{-2} \times 10^4$ m^2). However, flow at the lower channel was limited to 3 m^3. As most of the runoff for this event was generated over the headwater area, and as runoff from the colluvial slopes was probably absorbed within the alluvial terrace, one can assume that the main source area for the runoff recorded at the lower station had its origin in the headwater area. Nevertheless, only a small fraction of the flow recorded upstream did reach the lower channel, thus pointing to important losses between the two stations. These losses occurred even at peak flows which were higher for the upper than the lower channel (Fig. 3.4). A similar trend was obtained for the high magnitude event that occurred on 21 February. The runoff volume at the upper channel amounted to 18.5 m^3. Potential contribution by hillslopes was estimated to be 24 m^3 (2.4 $l\ m^{-2} \times 10^4\ m^2$). The total runoff volume produced during this event by the hillslopes and headwater areas amounted thus to some 42.5 m^3. At the same time, a runoff volume of only 19.8 m^3 was recorded at the lower channel, pointing again to the important infiltration losses into the alluvial terrace and channel fill.

It is very interesting to note that no flows were observed at the main channel, into which the instrumented watershed drains, during the entire rainfall season.

Conclusions

Data obtained for channel flow have several hydrological and geomorphological implications:

(a) The identification of rocky surfaces as areas whose response to rainfall is very rapid fits the concept of partial area contribution to runoff developed in humid areas. In arid regions rocky outcrops play the role reserved for saturated areas in the humid environment. When dealing with channel runoff, however, the spatial distribution of rocky and sediment-covered surfaces needs to be considered. As the latter surfaces act as efficient buffer zones, they tend to reduce runoff contribution from upper rocky slope sections to channel runoff. In other words, the proximity of rocky areas to the channel is crucial for channel runoff initiation and development. This can justify the important role that should be attributed to headwater areas in the development of channel runoff in arid areas.

(b) In more general terms, the ratio of bare bedrock outcrops and surficial sediments can be regarded as an important control of runoff generation in arid watersheds. Where this ratio is high, so are yields and rates of runoff. This promotes a well integrated drainage network. Where this ratio is low, runoff will be accordingly low, discontinuous in space and time, and result in a weakly integrated drainage system. Applying the Rbs model to landscape evolution, one can assume that geological changes in Rbs values will directly influence the degree of integration of fluvial systems in arid areas. A good example of this is provided by the recent evolution of the drainage network of the northern Negev (Yair & Enzel 1987). During the last glacial period aeolian loess penetrated into the area. The loess mantled rocky hillslopes, promoted infiltration losses, decreased runoff and resulted in a partial disorganization of the previously well-integrated drainage network.

(c) Base level is often regarded by geomorphologists as important in the control of channel processes and landscape evolution. The concept of base-level control rests on two assumptions. The first is that a continuous flow occurs quite often along the entire channel down to base level. The second assumption is that discharge, and accordingly flow energy, increases downstream during major flow events, permitting the base level to exert its influence in the upstream direction through the process of headwater erosion. Such a model is certainly valid in humid areas where the superimposition of storm flow over a perennial flow results in a downstream increase in discharge with minimal losses along the previously saturated channel.

Without denying the potential influence of base-level control on channel processes in arid areas, it appears that the pronounced filtering effect of the extensive sediment-covered surfaces might limit base-level control on landscape evolution in such areas. The frequent and rapid downstream decrease in channel discharge, observed in a small watershed of only 2.15 ha can be expected to be more pronounced in larger watersheds where the sediment storage is far more extensive. The discontinuous flow can be further enhanced in such watersheds by the spottiness of desert rainfall (Sharon 1981, Yair & Lavee 1985). Under such conditions, channel runoff generated in the headwater may die out before reaching the mouth of a remote basin. Such a process has been observed for a drainage basin extending over an area of 0.5 km^2 (Schick 1977) and for much larger watersheds in northern Africa and Australia (Dubief 1953, Williams 1971).

At the same time, sediment removed from the headwater active area is deposited downstream and thus contributes to raising the valley floor level. Sediment deposited in the channel promotes infiltration losses and leads to a further decrease in base-level control. The best expression of such a long-term evolution can be found in the closed basins and playas, which are characteristic of arid areas but absent in humid environments, where discharge increases in the downstream direction. Landscape evolution in the arid system described above can thus be regarded as controlled by processes active at the headwater area rather than by a remote base level.

Acknowledgements

The present study was conducted at the Sede Boqer experimental research station of the Arid Ecosystems Research Centre of the Hebrew University of

Jerusalem. The Centre provided the technical support for the fieldwork. The study is also part of a larger project on groundwater recharge supported by the German–Israel Foundation. I am grateful to Mr E. Sachs for his help with data collection and analysis, and Ms A. Bloch-Altman for drawing the illustrations.

References

Anderson, M. G. & T. P. Burt 1978. Experimental investigation concerning the topographical control of soil water movement on hillslopes. *Zeitschrift für Geomorphologie, Supplementband* **28**, 52–63.

Anderson, M. G. & P. E. Kneale 1980. Topography and hillslope soil water relationship in a catchment of low relief. *Journal of Hydrology* **47**, 115–28.

Betson, R. P. & J. B. Marius 1969. Source areas of storm runoff. *Water Resources Research* **5**, 574–82.

Dubief, J. 1953. Ruissellement superficiel au Sahara. *Colloques du Centre National de Recherche Scientifique* **35**, 303–14.

Dunne, T. 1978. Field studies of hillslope flow processes. In *Hillslope hydrology*, M. J. Kirkby (ed.) 227–90. Chichester, England: John Wiley.

Dunne, T. & R. D. Black 1970. Partial area contribution to storm runoff in a small New England watershed. *Water Resources Research* **6**, 1269–311.

Hewlett, J. H. & A. R. Hibert 1967. Factors affecting the response of small watersheds to precipitation in humid areas. In *Proceedings of the international symposium of forest hydrology*, W. E. Sopper & H. W. Lull (eds), 275–90. Oxford: Pergamon Press.

Ragan, R. M. 1968. An experimental investigation of partial area contributions. *International Association of Hydrological Sciences* **76**, 241–9.

Schick, A. P. 1977. A tentative sediment budget for an extremely arid watershed in the southern Negev. In *Arid geomorphology*, D. O. Doehring (ed), 139–63. Binghampton, New York: Publications in Geomorphology

Sharon, D. 1981. The distribution in space of local rainfall in the Namib Desert. *Journal of Climatology* **1**, 69–75.

Weyman, D. R. 1970. Throughflow on hillslopes and its relation to the stream hydrograph. *Bulletin of the International Association of Hydrological Sciences* **15**, 25–33.

Whipkey, R. Z. 1967. Subsurface stormflow on forested slopes. *Bulletin of the International Association of Hydrological Sciences* **10**, 74–85.

Wieder, M., A. Yair, A. Arzi 1985. Catenary soil relationships on arid hillslopes. *Catena Supplement* **6**, 41–57.

Williams, G. E. 1971. Flood deposits of the sand bed ephemeral streams of central Australia. *Sedimentology* **17**, 1–40.

Yair, A. 1983. Hillslope hydrology, water harvesting and areal distribution of some ancient agricultural systems, northern Negev, Israel. *Journal of Arid Environments* **6**, 283–301.

Yair, A. & A. Danin 1980. Spatial variations in vegetation as related to the soil moisture regime over an arid limestone hillside, northern Negev, Israel. *Oecologia* (Berlin) **47**, 83–8.

Yair, A. and Y. Enzel 1987. The relationship between annual rainfall and sediment yield in arid and semiarid areas: The case of the northern Negev. *Catena Supplement* **10**, 121–35.

Yair, A. & M. Shachak 1982. A case study of energy, water and soil flow chains in an arid ecosystem. *Oecologia* (Berlin) **54**, 389–97.

Yair, A. & H. Lavee 1985. Runoff generation in arid and semiarid zones. In *Hydrological forecasting*, M. G. Anderson & T. P. Burt (eds), 183–220. Chichester, England: John Wiley.

Yair, A., D. Sharon, H. Lavee 1978. An instrumented watershed for the study of partial area contribution to runoff in an arid area. *Zeitschrift für Geomorphologie Supplementband* **29**, 71–82.

Zangvill, A. & P. Druian 1983. *Meteorological data for Sede Boqer.* Desert Meteorological Paper, Series A, No. 8, Blaustein Institute for Desert Research, Ben-Gurion University.

4

Model KININF for overland flow on pervious surfaces

J. L. M. P. de Lima

Abstract

A soil water transport model based on the matric flux potential combined with the kinematic wave equations with interacting boundary conditions at the soil surface provided a numerical model for overland flow on an infiltrating surface (model KININF). The model has the advantages of having a relatively simple implementation on the computer and of requiring limited input data. It provides a simple way of visualizing overland flow on different infiltrating surfaces, and can be used for predicting soil moisture content profiles. Laboratory experiments were carried out to determine the performance of model KININF in predicting overland flow, infiltration and soil moisture movements. The flume was 1 m long, 0.5 m wide and 0.08 m deep. Two different soils were used: a loam from Limburg, The Netherlands, and a clay loam from Alentejo, Portugal. The rainfall simulator used oscillating nozzles sprinkling downwards and producing intermittent rainfall. Results show good agreement between the model and the experiments.

Introduction

General

Overland flow is a key process within the hydrologic cycle and a subject of great importance for people and their environment. Practical applications are found in the design of hydraulic structures, in irrigation, drainage, flood control, water erosion and sediment control, waste-water treatment, environment and wildlife protection, to mention only some of the most important fields of interest.

Not all rain reaching the surface is removed by overland flow. Part of it is involved in other processes, such as infiltration into the soil, interception by vegetation, accumulation of water in depression storage, evaporation and evapotranspiration. The interactions between these processes are not yet fully understood. In this chapter attention is given to some of these processes through the description of a model: model KININF (<u>KIN</u>ematic &

INFiltration). The model has a theoretical structure based on physical laws. However, it is simpler than those based on a rigorous physical approach. To complement the description of the model, a review is given on the factors and processes that affect overland flow. Laboratory experiments were carried out to test the model.

Overland flow and infiltration

Infiltration into the soil is usually the most important conditioning factor of rainfall loss, determining the balance between the gain of soil water and overland flow. A general approach to solve infiltration problems requires simultaneous solution of the equations describing the process of energy and mass transfer in a complex system embracing all the zones of water movement in the liquid and vapour phases.

The infiltration concepts applied in overland-flow models may be considered as a compromise, as they must be comprehensive and practical. Therefore factors of secondary importance, such as temperature and concentration of dissolved substances, must be neglected. Thus, simplified monophasic, unidirectional water-transport models on inert, homogeneous, isotropic and isothermic porous media are mostly employed.

The literature gives a distinct impression that overland flow and infiltration have been extensively studied as separate components (Woolhiser & Liggett 1967, Kibler & Woolhiser 1970, 1972, Singh 1975). The conventional approach has been through the rainfall-excess concept (Singh 1978).

Overland flow and infiltration occur simultaneously in nature during and after the occurrence of rainfall. What is required in overland-flow modelling is a combined study of these two components (Smith & Woolhiser 1971b). Infiltration is generally considered independent of overland-flow resulting in weak coupling of the two processes, i.e. infiltration influences overland flow, but not vice versa (Schmid 1989). Up to the time of ponding, computed infiltration rates, with and without accounting for overland-flow depth, are identical. Once ponding has occurred, there is a rise in water depth which may affect the infiltration rate (Schmid 1989). Several studies that combine physically based models of both overland flow and infiltration components can be found in the literature (including Foster et al. 1968, Smith & Woolhiser 1971a, b, Akan & Yen 1981, Akan 1985a, b, 1988, Lima 1989a).

Factors and problems affecting overland flow modelling

In modelling overland flow in a field situation there are many processes and conditions that should be taken into consideration. In the generation of overland flow two major mechanisms can be discerned. The first, often called Hortonian overland flow, occurs whenever the rainfall intensity exceeds the infiltration rate (Chorley 1978). The second, called saturation overland flow, is produced when the storage capacity of the soil is completely filled, so that all subsequent additions of water at the surface, irrespective of their rate of application, are forced to flow over the surface (Kirkby & Chorley 1967). The incorporation of these concepts into overland-flow modelling is still in progress (Kirkby 1988). For example, describing the unsaturated and saturated zones and the rising water table requires quite complex models and very detailed soil data.

Other problems are induced by the non-homogeneity of the soil and the presence of macropores. These are large continuous openings in the soil and may be very important in the movement of water (Germann & Beven 1985, Germann 1990). They may extend both vertically and horizontally, making the description of infiltration difficult.

The morphological factors that have been recognized to affect overland flow on slopes are the gradient, length and shape of the slope and its exposure to prevailing rain-bringing winds (Holy 1980, Lima 1988). The degree to which these morphological characteristics influence overland flow is closely related to other factors, such as soil characteristics, climate and vegetation (Dunne & Leopold 1978). The spatial variability of soil properties and soil moisture content in relation to the morphology of the catchment also causes modelling problems, as in the partial area concept (Smith & Hebbert 1983). Higher moisture contents are found in lower areas. This situation cannot be described accurately with only one spatial dimension.

Still related to the topography, the description of depression storage may be of considerable importance in the assessment of infiltration and on the overland flow retardation. The complexity of the surface depression storage makes it very difficult to find exact solutions of overland flow based on hydrodynamics, because of the complex boundary conditions. The infiltration behaviour of soils with disturbed surfaces (tilled soils, for example) is generally poorly understood (Moore et al. 1980). Interception losses to vegetation can also be significant.

Surface seal development strongly affects infiltration processes (Römkens et al. 1990). The dynamic interaction between solute transport in overland flow and physicochemical processes can cause a reduction of the infiltration rate on crusted saline and sodic soils and consequently increase overland flow during rainfall (Gerits & Lima 1991). The application of overland-flow models to these conditions may fail if that dynamic interaction is not considered. Seal development can also be caused by the dispersive and compactive action of raindrop impact and deposition of suspended sediment particles in pores and by filtration (Römkens et al. 1990).

Although rainfall is often accompanied by strong winds (Lyles et al. 1974, Lyles 1977), this fact is frequently neglected in overland-flow studies. Wind causes a change in the driving force of overland flow through tangential shear stress in the overland-flow water/air boundary, affecting water depths and velocities. Wind also affects overland flow indirectly, affecting the actual rainfall flux intercepted on the ground surface, which depends on the angle of incidence of the rain and the relative orientation of the sloping surface to the rain vector. Wind also affects the shape, size, angle of incidence and velocity of raindrops impact and respective splash shapes, which appear to have some effect on the flow resistance of overland flow (Lima 1989c, d, e).

The uptake of soil moisture by plants, evapotranspiration and evaporation from the soil surface and open water are important processes in the hydrologic cycle. They are highly complex processes and require a large amount of information for accurate modelling. Fortunately, when modelling overland flow for short periods of time, under heavy rainfall, these processes can usually be disregarded because of their low relative intensities.

In addition to all the processes and problems mentioned, there are many other problems associated with the conceptualization, development, computer implementation, calibration and validation of the mathematical models.

Description of the model

Introduction

Model KININF for overland flow on infiltrating surfaces is especially applicable to relatively short periods of time. Overland flow is assumed to constitute the runoff from a sloping natural surface during a heavy rainfall. For these short-term events the other runoff components, such as subsurface and groundwater runoff, and water uptake by plant roots, can be disregarded.

Figure 4.1 gives a schematic representation of the mathematical model KININF and of the notation used, where X is the space axis along the direction of flow; Y is a horizontal axis perpendicular to the X-axis; Z is the vertical axis; H is the axis perpendicular to both the X- and Y-axis; HOR is a horizontal axis; and θ is the volumetric water content axis. The slope is subdivided into compartments, between grid points, assuming uniform conditions of rainfall, infiltration and soil moisture profile within each compartment. For the grid point j (with a flow depth h_j and a flow velocity v_j) of the X-axis we have in the upper part of the figure an elementary area ($\Delta x.1$) with an area covered by mounds of A, and with an effective area for overland flow of $W = 1 - A/\Delta x$. In the lower part of the figure the semi-infinite homogeneous soil is subdivided into layers of equal thickness. In Figure 4.1 F is the water flux density, p is the rainfall rate, S is the slope gradient and n is a layer index.

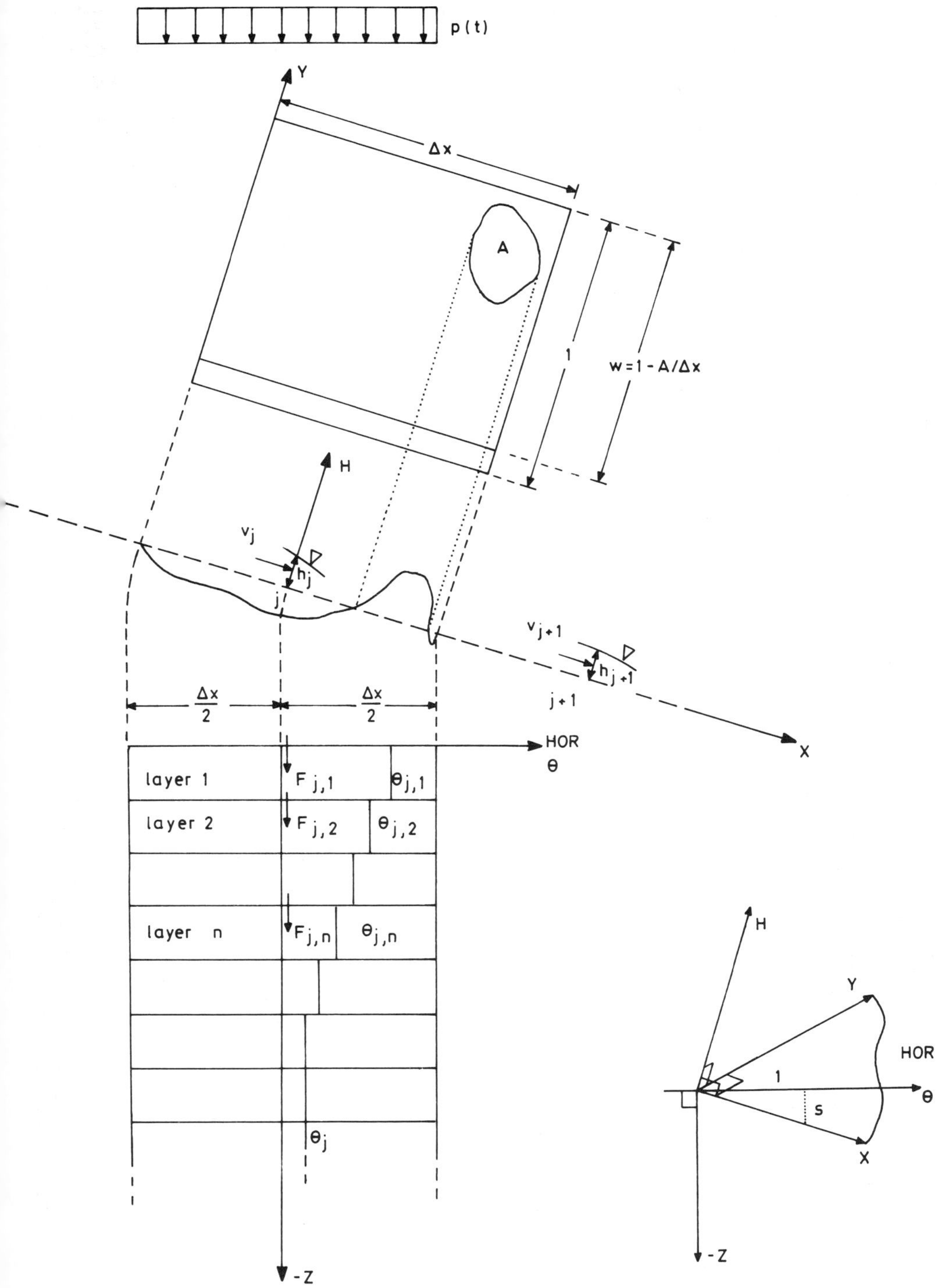

Figure 4.1 Schematic representation of model KININF with the notation used (after Lima 1989a).

The infiltration model

Vertical non-steady flow of water in unsaturated soil may be described by the one-dimensional continuity equation and Darcy's law in terms of the matric flux potential concept, respectively:

$$\frac{\delta\theta}{\delta t} = -\frac{\delta F}{\delta z} \tag{4.1}$$

$$F = -\frac{\delta\phi(\theta)}{\delta z} + K(\theta) \tag{4.2}$$

where F is the water flux density ($m^3\ m^{-2}\ s^{-1}$), θ is the volumetric water content ($m^3\ m^{-3}$), $K(\theta)$ is the hydraulic conductivity function ($m\ s^{-1}$), z is the vertical coordinate, positive upwards (m), t is the time (s), and $\phi(\theta)$ is the matric flux potential ($m^2\ s^{-1}$).

The matric flux potential ϕ (Kirchhoff transformation) is defined by:

$$\phi = -\int_{h}^{h_{sat}} K(h)\,dh = -\int_{\theta}^{\theta_{sat}} D(\theta)\,d\theta \quad \text{with } D(\theta) = K(\theta)\frac{\delta h}{\delta\theta} \tag{4.3}$$

where $D(\theta)$ is the soil water diffusivity ($m^2\ s^{-1}$), $h(\theta)$ is the pressure head (m) and the subscript "sat" refers to saturation.

This concept was adopted in the description of soil water transport by Klute (1952) and was called "matric flux potential" (Raats 1970) to indicate that the associated gradient vector field is the flux field as governed by matric forces only. Shaykewich & Stroosnijder (1977) indicated several advantages of the matric flux potential over the conventional formulation which typically evaluates the transport coefficient and the driving force for the matric flux separately. Among these advantages are the avoidance of numerical overshoot, especially in the simulation of situations involving strong gradients, and the reduction of computation time. Also, as a soil characteristic, the matric flux potential curve is more susceptible to interpretation in terms of fluxes than the soil water diffusivity or hydraulic conductivity functions (Berge et al. 1987).

For simplicity, the class of soils characterized by linear retention curves and exponential dependencies of the hydraulic conductivity upon the water content (Raats 1983) was used in the model:

$$h(\theta) = h_{sat} + \gamma(\theta - \theta_{sat}) \tag{4.4}$$

$$K(\theta) = K_{sat}\ e^{\xi(\theta - \theta_{sat})} \tag{4.5}$$

where γ and ξ are empirical constants.

The primary Equations 4.4 and 4.5 imply the following derived relationships:

$$D(\theta) = D_{sat} e^{\xi(\theta - \theta_{sat})} \tag{4.6}$$

where $D_{sat} = \gamma K_{sat}$, and

$$\phi(\theta) = D_{sat} \frac{[e^{\xi(\theta - \theta_{sat})} - 1]}{\xi}. \tag{4.7}$$

Equations 4.1 & 4.2 were solved numerically using an explicit finite-difference scheme (see appendix). $K(\theta)$ and $\phi(\theta)$ were calculated using Equations 4.5 & 4.7. The initial conditions are those for a semi-infinite homogeneous soil, with uniform volumetric water content (θ_0).

The overland flow model

The flow per unit width across a plane surface as a result of rainfall can be described by the one-dimensional continuity equation:

$$\frac{\delta h}{\delta t} + \frac{\delta q}{\delta x} = R(x,t) \tag{4.8}$$

where h is the flow depth at time t and position x *(m)*; x is the space coordinate along the direction of flow *(m)*; q is the volumetric water flux per unit plane width ($m^2\ s^{-1}$); and $R\ (x,\ t)$ is the rainfall excess ($m\ s^{-1}$).

By assuming that the bed slope equals the friction slope (kinematic wave assumption) and by using existing open-channel flow friction equations, the discharge at any point and time can be expressed as a function of water depth only, as follows:

$$q = \alpha h^m \tag{4.9}$$

where m is a parameter for the type of flow, and α is a hydraulic coefficient (units dependent on the m value).

To apply the kinematic wave model, the parameters α and m in the depth–discharge relationship (Eq. 4.9) must be assigned appropriate values (Stephenson & Meadows 1986, Gerits et al. 1990, Moore & Foster 1990). In this book, resistance to overland flow is treated in detail in Chapters 1, 2 and 5.

Equations 4.8 & 4.9 were solved numerically using the explicit finite-difference scheme known as the single-step Lax–Wendroff scheme (see Appendix). This finite difference scheme was recommended by Liggett & Woolhiser (1969) and has been used successfully by Kibler & Woolhiser (1970, 1972), Lane et al. (1975), Rovey et al. (1977), Freeze (1978), Wu et al. (1978) and Lima (1989a). Although the Lax–Wendroff method is very fast at each time step, stability conditions dictate a large number of time steps. An implicit scheme would probably involve less computation work.

In this work the upstream boundary is assumed to be at zero depth [$h\ (0,t) = v\ (0, t) = 0$ for $t \geq 0$, where v is the overland-flow velocity in m s^{-1}] and the downstream boundary is a continuing plane (along the direction of flow). In the first phase of modelling, zero depth initial conditions were assumed, i.e. an initially "dry" surface [$h\ (x,0) = 0$ for all x; $v\ (x,0) = 0$ for all x] with or without surface depression storage and an upstream boundary condition of no flow.

Depression storage, detention storage and obstacles

With respect to depression storage, the model is based on the Hortonian approach: all depression storage must be filled up to allow a thin sheet of overland flow to build up. Depression storage is thus defined as the amount of water held in surface depressions that may be evaporated or infiltrated.

Detention storage is the storage effect due to overland flow in transit. The water held by detention storage either runs off as overland flow or infiltrates. Obstacles like earth and vegetation mounds are taken into account in the overland flow process considering a reduction of the possible available flow path area. The infiltration into the obstacles is assumed to be equal to that of the remaining area.

The combined model of overland flow and infiltration

The mathematical model consists of simultaneous solution of Equations 4.1, 4.2, 4.8 & 4.9. Solution is simultaneous in the sense that solutions move concurrently in time, boundary conditions being interdependent.

The computational procedure for each time, t, and for each grid point, j, starts by solving the finite-difference scheme for vertical unsaturated flow and consequent determination of the infiltration rate (equivalent to the water flux density, F, at the soil surface) and the rainfall excess rate (see Appendix). These calculations are followed by solution of the explicit Lax–Wendroff scheme for overland flow (see Appendix).

The model KININF consists of a main computer program and three subroutines. It is possible to substitute the subroutine with the infiltration model with any other method, namely the Green & Ampt model or the empirical Horton infiltration equation. These models are currently being incorporated into overland-flow models, but are not considered here.

Figure 4.2 shows some possible outputs of the model, for a medium–fine sand, namely the advance of the wetting front at the middle point of a slope ($x = L/2$), overland flow hydrograph and infiltration at $x = L$, in time, and water depth and velocity in time and in space (input data is given in Table 4.1).

Table 4.1 Input data for the example of Figure 4.2.

Soil data	
$K_{sat} = 12.7 \times 10^{-6}$ m s^{-1}	$\theta_0 = 0.100$ m^3 m^{-3}
$D_{sat} = 18.3 \times 10^{-6}$ m^2 s^{-1}	$\theta_{sat} = 0.350$ m^3 m^{-3}
$\xi = 60.8$	
Surface data	
$S = 2\%$	$A = 0$ m^2
$L = 20$ m	$\alpha = 2.87$
$W = 1$ m	$m = 1.68$
depression storage = 0.001 m	
Rainfall data	
$p(1) = 0.15$ mm s^{-1}	$t(1) = 180$ s
$p(2) = 0.02$ mm s^{-1}	$t(2) = 240$ s
$p(3) = 0.10$ mm s^{-1}	$t(3) = 360$ s
Numerical grid data	
$\Delta t = 2$ s	$\Delta z = 0.02$ m
$\Delta x = 2$ m	total simulation time = 570 s

Laboratory experiments on soil flume under simulated rain

The laboratory set-up was mainly composed of two units: a soil flume and a rainfall simulator. The laboratory soil flume is 1 m long, 0.5 m wide and 0.8 m deep. The sides are made of Perspex up to a height of 0.12 m above the soil surface. The flume has no buffer zone for splash. The bottom of the flume consists of a metallic perforated plate to collect percolation water; a

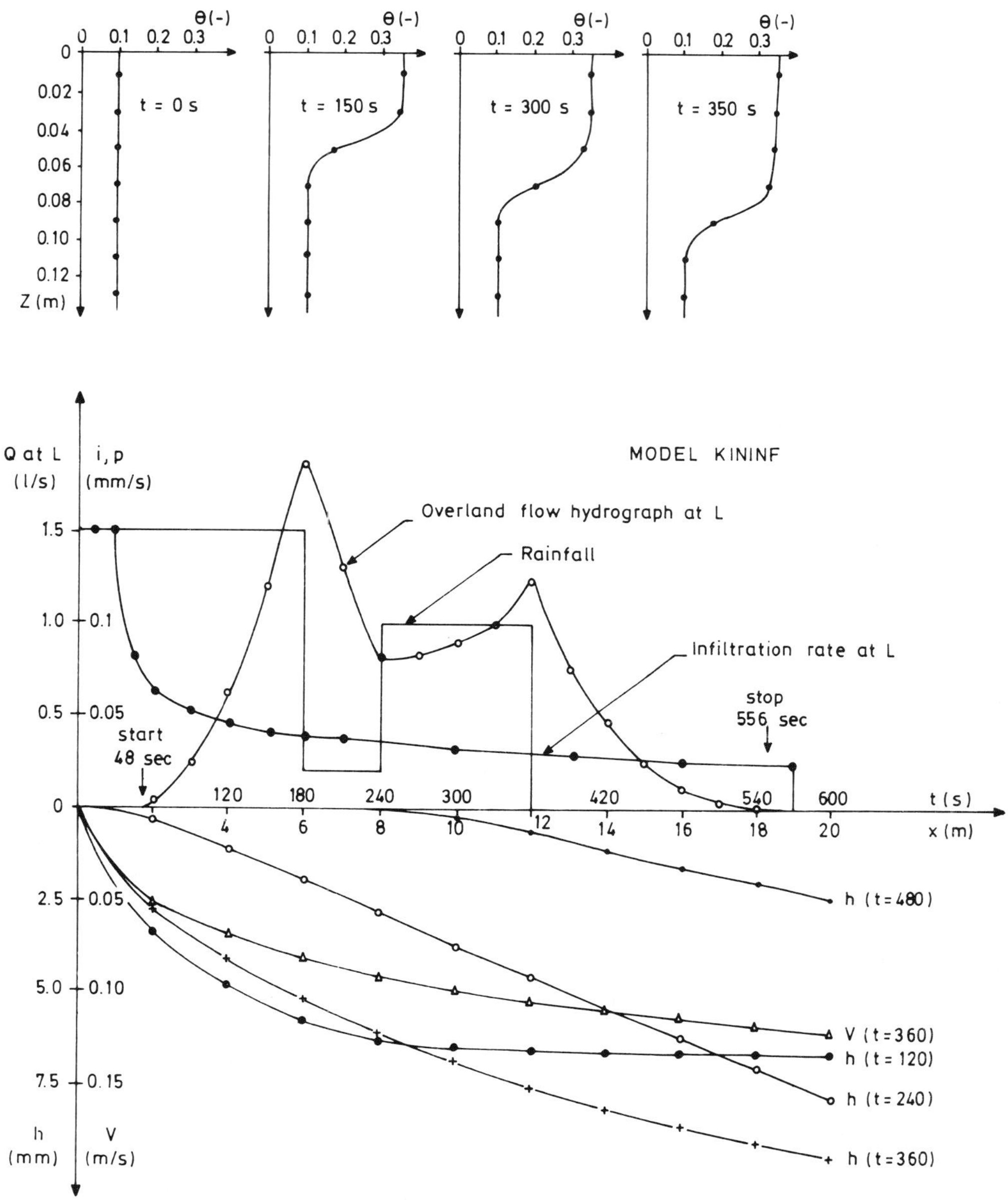

Figure 4.2 Outputs of model KININF for a medium–fine sand (Lima 1989a).

synthetic filter was placed over this perforated plate. Percolation discharge can be collected separately for the upper and downstream halves of the soil flume.

The programmable rainfall simulator used is similar to the one described by Neibling et al. (1981) and Foster et al. (1982). The simulator uses oscillating nozzles (continuous spraying with lateral oscillation across slope) sprinkling

Table 4.2 Measured and estimated physical data of the loam from Limburg and the clay loam from Alentejo.

Soil type:	Loam (Limburg)	Clay loam (Alentejo)
Measured (after Lima 1989a)		
θ_0 ($m^3\ m^{-3}$)	0.0107	0.006
θ_{sat} ($m^3\ m^{-3}$)	0.506	0.411
ρ_s ($kg\ m^{-3}$)	2640	2730
ρ ($kg\ m^{-3}$)	1150	1370
Estimated (after Rijtema 1969 and Stroosnijder 1976)		
K_{sat} ($\times 10^{-6}\ m\ s^{-1}$)	1.67	0.1134
D_{sat} ($\times 10^{-6}\ m^2\ s^{-1}$)	13.3	5.0
ξ (-)	25.9	66.8

downwards and producing intermittent rainfall. Four aluminium troughs, each with three nozzles, are installed across slope to cover an area of $6 \times 2.5\ m^2$. Nozzle height is set at about 2.4 m and nozzle spacing at 1.10 m apart in each trough. The spacing between troughs is 1.52 m. Different rainfall intensities are obtained by controlling the frequency with which the nozzle sweeps across an opening above the surface under study. The duration of off-periods between spray applications, and the frequency with which each nozzle sprays, significantly affect infiltration rates, uniformity of flow depth, and smoothness of the rising limb of the overland hydrograph (Sloneker & Moldenhauer 1974, Foster et al. 1982).

The rainfall simulator is claimed to generate a uniformly distributed rainfall pattern. Foster et al. (1982) found for the same type of rainfall simulator a coefficient of uniformity (Christiansen 1942) of approximately 86% for the rainfall intensity within an area bounded by four corner nozzles. This percentage was found to be independent of rainfall intensity over the range 25–125 mm h^{-1}. For the rainfall simulator used in the present study, Maanen & Vincentie (1986) obtained uniformity coefficients of around 85% in an area covered by 12 nozzles ($6 \times 2.5\ m^2$), over a rainfall intensity range of about 16–112 mm h^{-1}.

The impact velocity of the drops generated by this type of nozzle is nearly equal to impact velocities of natural raindrops when the nozzles are about 2.4 m above the surface (Neibling et al. 1981). However, it has been reported by

Foster et al. (1982) that the drops are slightly smaller than natural raindrops and, consequently, the impact kinetic energy of drops is about 75% of natural rainfall, the median drop size being 2.25 mm, with drop terminal velocity of 6.8 m s^{-1}. The experiments were carried out in stagnant air with a room temperature of approximately 18°C. Rainwater temperature was approximately 13°C.

Two soils were studied: a loam from Limburg (The Netherlands) and a clay loam from Alentejo (Portugal), as classified by the US Department of Agriculture texture classification system. Soil properties are presented in Table 4.2, where ρ_s is the average particle density (kg m^{-3}) and ρ is the average soil bulk density (kg m^{-3}).

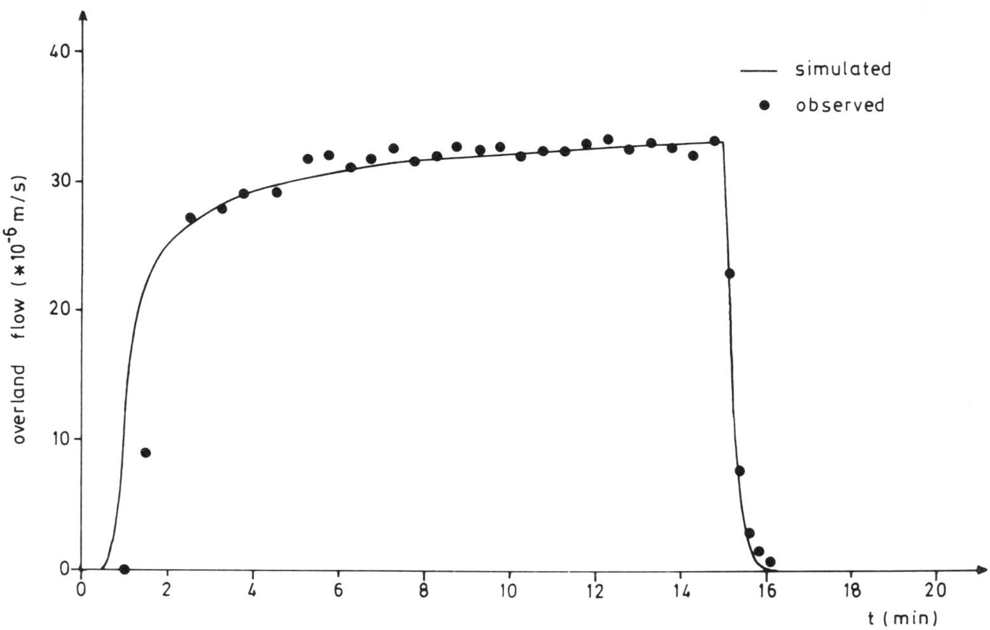

Figure 4.3 Measured and simulated overland-flow hydrographs for the clay loam soil from Alentejo for a constant rainfall of 0.03741 mm s^{-1} of 15 min duration (Lima 1989a).

The experiments were performed for initially dry soil conditions, i.e. using soil which was air-dried during several weeks. The soil was sieved through a 5.6 mm mesh screen, and placed loosely in the flume. Constant rainfall rates were used. The solid and liquid discharges from the experimental runs were collected in preweighed containers at variable time intervals (depending on discharge variations). Overland-flow velocity measurements were made using dye tracing (Abrahams et al. 1986). The advance of the soil moisture front in time was recorded by observing, at several locations, the advance of the wetting front through the Perspex wall, over the length of the flume. Boundary problems were negligible in comparison with spatial variation of the advance of the wetting front. Further details on the laboratory set-up and procedures are presented by Lima (1989b) and Giménez (1989).

Results

For both soils the physical properties θ_0, θ_{sat}, ρ_s and ρ were determined experimentally, and K_{sat}, D_{sat} and ξ were estimated (Table 4.2). The parameters α and m were estimated from overland-flow velocity measurements. No calibration procedure was applied to the parameters.

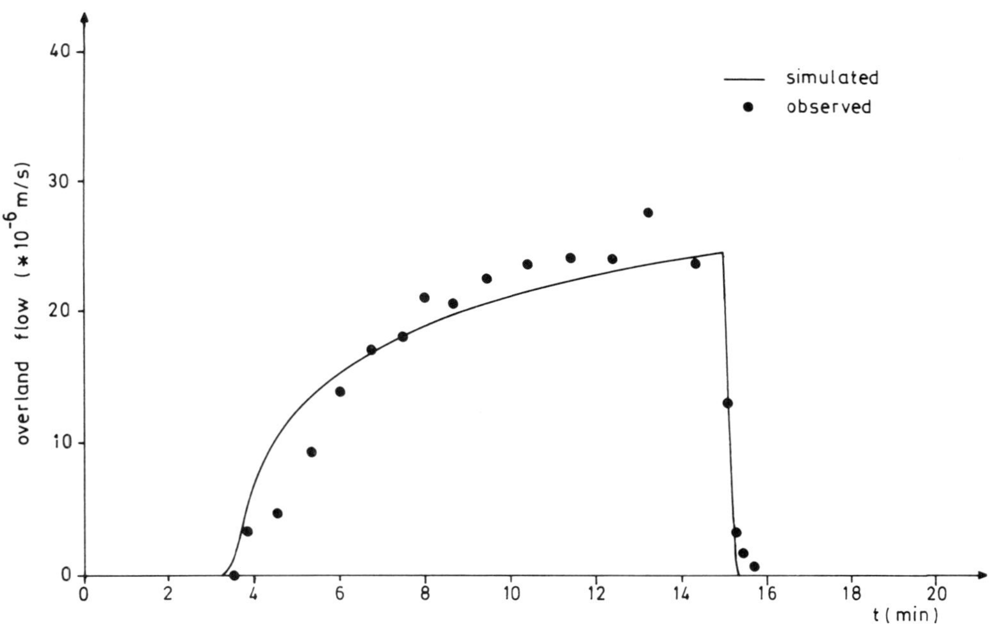

Figure 4.4 Measured and simulated overland-flow hydrograph for the loam from Limburg for a constant rainfall of 0.03741 mm s^{-1} of 15 min duration (Lima 1989a).

Experimental and simulated (predicted with model KININF) overland-flow hydrographs show good agreement (Figs 4.3 & 4.4) for a constant rainfall of 0.03741 mm s^{-1} of 15 min duration. The slope was 10%. With respect to the overland-flow hydrographs, the model efficiency is above 90% for both cases.

Figures 4.5 & 4.6 compare measured soil wetting fronts (average over the flume length) with KININF's estimate of the soil moisture profiles at different

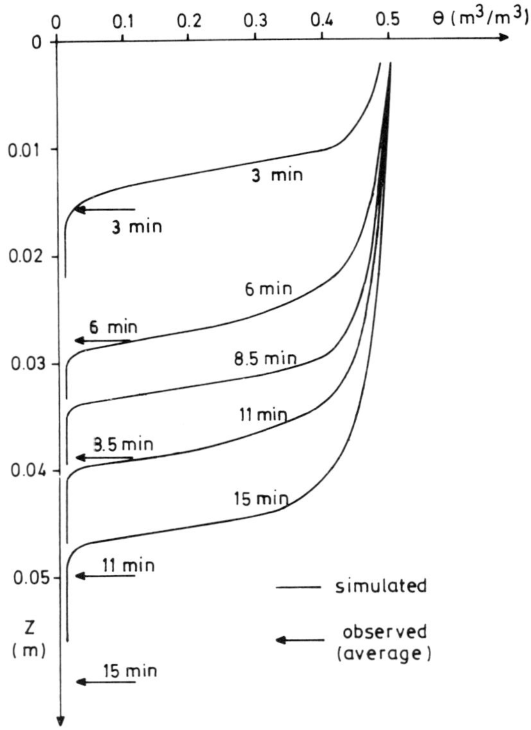

Figure 4.5 Comparison of average depths of soil wetting front over time and corresponding soil moisture profiles (model KININF) for the loam from Limburg (Lima 1989a).

times. For the soil from Limburg the model efficiency is, in this case, 72%; for the soil from Alentejo this analysis was not possible owing to the small number of observed data.

Differences between observed and simulated values may be caused by many factors. Although care was taken in placing the soil in the flume, uniform density was not achieved. Also, the measurements of the physical properties of the soil were not made exactly for the same conditions. Further, the estimated soil parameters may deviate from the real values.

Discussion and conclusion

Physically based mathematical models for overland flow that account for all the processes involved and their interactions can never be made. There are limitations due to the assumptions underlying the theoretical developments, and these may result in failure of the model to account for some specific features

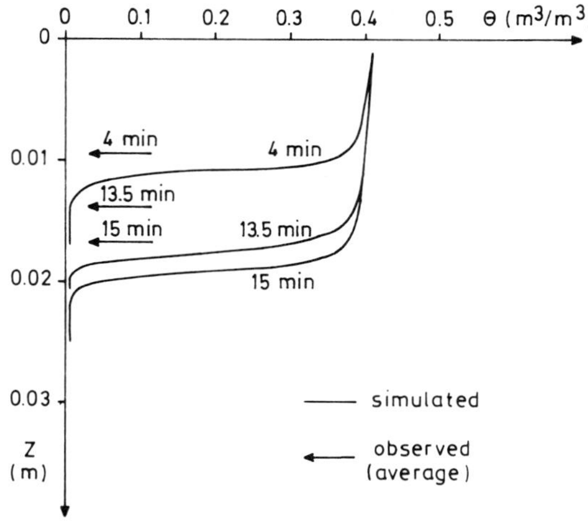

Figure 4.6 Comparison of average depths of soil wetting front over time and corresponding soil moisture profiles (model KININF) for the clay loam from Alentejo (Lima 1989a).

of the system and/or failure to represent the actual mechanisms at a fundamental level (Freeze 1978). However, simplifying assumptions should assure that the model still retains the basic characteristics of the physical system being modelled. Only then is the model of any use for practical purposes in engineering. Another limitation, closely related to the accuracy of the model in describing the overland flow and related processes, is the availability and reliability of the data. Physically based models require specification of both soil and surface properties in order to obtain average input data within defined grid points.

The combination of a soil-water transport model based on the matric flux potential concept with the kinematic wave equations with interacting boundary conditions at the soil surface provided a numerical model for overland flow on an infiltrating surface (model KININF) and was shown to be a good approximation. The model gave reliable prediction of overland flow from a laboratory soil flume. The model has the advantages of having a relatively simple implementation on the computer and of requiring limited input data. Since the model has no fitted parameters, it is especially appealing where field data are limited or difficult to obtain. It provides a simple way of visualizing overland flow on different infiltrating surfaces, and can be used for predicting soil moisture content profiles. The model can also be used to describe overland flow in small plots or microcatchments when appropriate physical parameters are known. This case has not yet been tested.

In interdisciplinary studies such as water erosion, environmental pollution or waste-water treatment, overland flow has to be related to other processes. For example, in water erosion studies, where attention should be paid to soil detachment and deposition, physically based overland-flow models should be coupled to sediment transport models. Definition of the purpose of the research and identification of the most important physical processes that need to be considered in the analysis should be the first essential steps. The refinement of the other models will eventually dictate the characteristics of the overland-flow model to be used in the combined study.

Acknowledgement

The description of the model KININF has been extracted from the Ph.D. thesis of the author, elaborated under the supervision of Prof. Dr W. H. van der Molen and Dr R. W. R. Koopmans, at the Wageningen Agricultural University (The Netherlands).

Appendix: Finite-difference equations used in model KININF

The finite-difference equations for the flow of soil water (from Eqs 4.1, 4.2, 4.5 & 4.7):

$$\theta_n^{t+\Delta t} = \theta_n - (F_{n+1} - F_n)\frac{\Delta t}{\Delta z} \tag{4.10}$$

$$F_n = -\frac{(\phi_n - \phi_{n-1})}{\Delta z} + \frac{K_n + K_{n-1}}{2} \tag{4.11}$$

$$K_n = K_{sat}\, e^{\xi(\theta_n - \theta_{sat})} \tag{4.12}$$

$$\phi_m = D_{sat}\frac{[e^{\xi(\theta_n - \theta_{sat})} - 1]}{\xi} \tag{4.13}$$

where the subscript n denotes a space level (layer index) and all non-superscripted variables are evaluated at the time t (see also Fig. 4.1).

The finite-difference Lax–Wendroff scheme for overland flow (from Eqs 4.8

& 4.9) is obtained by expanding h in Taylor's series:

$$h_j^{t+\Delta t} = h_j - \frac{\Delta t}{2\Delta x}[(\alpha h^m)_{j+1} - (\alpha h^m)_{j-1} - 2\Delta x R_j] + \frac{\Delta t^2}{4\Delta x^2}[(\alpha m h^{m-1})_{j+1} + (\alpha h^{m-1})_j][(\alpha h^m)_{j+1} - (\alpha h^m)_j - \Delta x R_j] - \frac{\Delta t^2}{4\Delta x^2}[(\alpha m h^{m-1})_j + (\alpha m h^{m-1})_{j-1}] \times [(\alpha h^m)_j - (\alpha h^m)_{j-1} - \Delta x R_j] + (R_j^{t+\Delta t} - R_j)\frac{\Delta t}{2} \quad (4.14)$$

where all non-superscripted variables on the right-hand side are evaluated at the time t, and j denotes a space level in the x direction (see also Fig. 4.1).

The following stability conditions were used respectively for the infiltration (Remson et al. 1971, Stroosnijder 1982) and the overland flow (Constantinides 1981, Stephenson & Meadows 1986) finite-difference schemes:

$$\Delta t < \frac{0.5(\Delta z)^2}{D(\theta)} \quad (4.15)$$

$$\frac{\Delta t}{\Delta x} < \frac{1}{m\alpha h^{m-1}}. \quad (4.16)$$

References

Abrahams, A. D., A. J. Parsons, S.-H. Luk 1986. Field measurement of the velocity of overland flow using dye tracing. *Earth Surface Processes and Landforms* **11**, 653–7.

Akan, A. O. 1985a. Predicting overland flow and infiltration. In *Hydraulics and hydrology in the small computer age*, W. R. Waldrop (ed.), Vol. 1, 79–83. Proceedings of the Specialty Conference, ASCE. Lake Buena Vista, Florida.

Akan, A. O. 1985b. Similarity solution of overland flow on pervious surface. *Journal of Hydraulics Division, Proceedings of the American Society of Civil Engineers* **111**, 1057–67.

Akan, A. O. 1988. Overland flow on pervious, convergent surface. *Nordic Hydrology Journal* **19**, 153–64.

Akan, A. O. & B. C. Yen 1981. Mathematical model of shallow water flow over porous media. *Journal of the Hydraulics Division, Proceedings of the American Society of Civil Engineers* **107**, 479–94.

Berge, H. F. M. ten, K. Metselaar, L. Stroosnijder 1987. Measuring of matric flux potential:

a simple procedure for the hydraulic characterization of soils. *Netherlands Journal of Agricultural Science* **35**, 371–84.

Chorley, R. J. 1978. The hillslope hydrologic cycle. In *Hillslope hydrology*, M. J. Kirkby (ed.), 1–42. Chichester, England: John Wiley.

Christiansen, J. E. 1942. *Irrigation by sprinkling*. University of California Agricultural Experiment Station Bulletin, No. 670.

Constantinides, C. A. 1981. Numerical techniques for a two-dimensional kinematic overland flow model. *Water SA* **7**, 234–48.

Dunne, T. & L. B. Leopold 1978. *Water in environment planning*. San Francisco: Freeman.

Foster, G. B., L. F. Huggins, L. D. Meyer 1968. Simulation of overland flow on short field plots. *Water Resources Research* **4**, 1179–88.

Foster, G. B., W. H. Neibling, R. A. Nattermann 1982. *A programmable rainfall simulator*. American Society of Agricultural Engineers, paper no. 82-2570.

Freeze, R. A. 1978. Mathematical models of hillslope hydrology. In *Hillslope hydrology*, M. J. Kirkby (ed.), 177–225. Chichester, England: John Wiley.

Gerits, J. J. P. & J. L. M. P. de Lima 1991. Solute transport and wind action in relation to overland flow and water erosion. *Catena* (in press).

Gerits, J. J. P., J. L. M. P. de Lima, T. M. W. van der Broek 1990. Overland flow and erosion. In *Process studies in hillslope hydrology*, M. G. Anderson & T. P. Burt (eds), 173–214. Chichester, England: John Wiley.

Germann, P. F. 1990. Macropores and hydrologic hillslope processes. In *Process studies in hillslope hydrology*, M. G. Anderson & T. P. Burt (eds), 327–63. Chichester, England: John Wiley.

Germann, P. F. & K. Beven 1985. Kinematic wave approximation to infiltration into soils with sorbing macropores. *Water Resources Research* **21**, 990–6.

Giménez, D. 1989. Effect of rainfall patterns and intensities on the surface of a silt loam soil: laboratory experiments. Unpublished M.Sc. thesis. Agricultural University Wageningen, The Netherlands.

Holy, M. 1980. *Erosion and environment*. Environment Sciences and Applications, Vol. 9. Oxford: Pergamon Press.

Kibler, D. F. & D. A. Woolhiser 1970. *The kinematic cascade as a hydrologic model*. Hydrology Paper No. 39, Colorado State University, Fort Collins, Colorado, USA.

Kibler, D. F. & D. A. Woolhiser 1972. Mathematical properties of the kinematic cascade. *Journal of Hydrology* **13**, 131–47.

Kirkby, M. J. 1988. Hillslope runoff processes and models. *Journal of Hydrology* **100**, 315–39.

Kirkby, M. J. & R. J. Chorley 1967. Throughflow, overland flow and erosion. *Bulletin of the International Association for Scientific Hydrology* **12**, 5–21.

Klute, A. 1952. A numerical method for solving the flow equation for water in unsaturated materials. *Soil Science* **73**, 105–17.

Lane, L. J., D. A. Woolhiser, V. P. Yevjevich 1975. *Influence of simplications in watershed geometry in simulation of surface runoff*. Hydrology Paper No. 81, Colorado State University, Fort Collins, Colorado.

Lima, J. L. M. P. de 1988. Morphological factors affecting overland flow on slopes. In *Land qualities in space and time*, Bouma J. & A. K. Bregt (eds), 321–4. Wageningen: PUDOC.

Lima, J. L. M. P. de 1989a. Overland flow under rainfall: some aspects related to modelling and conditioning factors. Unpublished Ph.D. thesis. Agricultural University Wageningen,

The Netherlands.
Lima, J. L. M. P. de 1989b. Water erosion in relation to stony soils, overland flow and sediment discharges: a laboratory experiment. Unpublished M.Sc. thesis. Agricultural University Wageningen, The Netherlands.
Lima, J. L. M. P. de 1989c. Overland flow under simulated wind-driven rain. In *Land and water use* (Proceedings of the Eleventh International Congress on Agricultural Engineering), 493–500. Rotterdam: Balkema.
Lima, J. L. M. P. de 1989d. The influence on the angle of incidence of the rainfall on the overland flow process. In *New directions for surface water modelling* (Proceedings of the IAHS Third Scientific Assembly), 73–82. Baltimore: IAHS Press.
Lima, J. L. M. P. de 1989e. Raindrop splash anisotropy: slope, wind and overland flow velocity effects. *Soil Technology* **2**, 71–8.
Liggett, J. A. & D. A. Woolhiser 1967. Difference solutions of the shallow-water equation. *Journal of the Engineering Mechanics Division, Proceedings of the American Society of Civil Engineers* **93**, 39–71.
Liggett, J. A. & D. A. Woolhiser 1969. Closure to Liggett & Woolhiser (1967). *Journal of the Engineering Mechanics Division, Proceedings of the American Society of Civil Engineers* **95**, 303–11.
Lyles, L. 1977. Soil detachment and aggregate disintegration by wind-driven rain. *Soil Conservation Society of America Special Publication* **21**, 152–9.
Lyles, L., J. D. Dickerson, N. F. Schmeidler 1974. Soil detachment from clods by rainfall: effects of wind, mulch cover, and initial soil moisture. *Transactions of the American Society of Agricultural Engineers* **17**, 697–700.
Maanen, J. van & R. Vincentie 1986. *De ijking van een programmeerbare regensimulator.* Department of Irrigation and Civil Engineering, Agricultural University Wageningen, Wageningen, The Netherlands.
Moore, I. D., C. L. Larson, D. C. Slack 1980. *Predicting infiltration and micro-relief surface storage for cultivated soils.* Water Resources Research Center, Bulletin 102. University of Minnesota, Minneapolis, Minnesota.
Moore, I. D. & G. R. Foster 1990. Hydraulics and overland flow. In *Process studies in hillslope hydrology*, M. G. Anderson & T. P. Burt (eds), 215–54. Chichester, England: John Wiley.
Neibling, W. H., G. R. Foster, R. A. Nattermann, J. D. Nowlin, P. V. Holbert 1981. Laboratory and field testing of a programmable plot-sized rainfall simulator. In *Erosion and sediment transport measurement*, Proceedings of the Florence Symposium, IAHS, Publication No. **133**, 405–14.
Raats, P. A. C. 1970. Steady infiltration from line sources and furrows. *Soil Science Society of America Proceedings* **34**, 709–14.
Raats, P. A. C. 1983. Implications of some analytical solutions for drainage of soil water. *Agricultural Water Management* **6**, 161–75.
Remson, I., G. M. Hornberger, F.J. Molz 1971. *Numerical methods in subsurface hydrology.* New York: Wiley-Interscience.
Rijtema, P. E. 1969. *Soil moisture forecasting.* ICW, Wageningen, The Netherlands.
Römkens, M. J. M., S. N. Prasad, F. D. Whisler 1990. Surface sealing and infiltration. In *Process studies in hillslope hydrology*, M. G. Anderson & T. P. Burt (eds.), 127–72. Chichester, England: John Wiley.
Rovey, E. W., D. A. Woolhiser, R. E. Smith 1977. *A distributed kinematic model of upland*

watersheds. Hydrology Paper No. 93, Colorado State University, Fort Collins, Colorado.

Schmid, B. H. 1989. On the overland flow modelling: can rainfall excess be treated as independent of flow depth? *Journal of Hydrology* **107**, 1–8.

Shaykewich, C. F. & L. Stroosnijder 1977. The concept of matric flux potential applied to simulation of evaporation from soil. *Netherlands Journal of Agricultural Science* **25**, 63–82.

Singh, V. P. 1975. A laboratory investigation of surface runoff. *Journal of Hydrology* **27**, 33–50.

Singh, V. P. 1978. Mathematical modelling of watershed runoff. In *Proceedings of International Conference on Water Resources Engineering*, Vol. II, 703–26. Bangkok, Thailand.

Sloneker, L. L. & W. C. Moldenhauer 1974. Effect of varying on-off time of rainfall simulator nozzles on surface sealing and intake rate. *Soil Science Society of American Proceeding* **38**, 157–9.

Smith, R. E. & R. H. B. Hebbert 1983. Mathematical simulation of interdependent surface and subsurface hydrologic processes. *Water Resources Research* **19**, 987–1001.

Smith, R. E. & D. A. Woolhiser 1971a. *Mathematical simulation of infiltrating watersheds*. Hydrology paper No. 47, Colorado State University, Fort Collins, Colorado.

Smith, R. E. & D. A. Woolhiser 1971b. Overland flow on an infiltrating surface. *Water Resources Research* **7**, 899–913.

Stephenson, D. & M. E. Meadows 1986. *Kinematic hydrology and modelling*. Amsterdam: Elsevier.

Stroosnijder, L. 1976. Infiltratie en herverdeling van water in grond. Unpublished Ph.D. thesis. Agricultural University Wageningen, The Netherlands.

Stroosnijder, L. 1982. Computer solutions of the mathematical models describing unsaturated soil water flow. Department of Soil Science and Plant Nutrition, Wageningen Agricultural University, Wageningen, The Nerherlands.

Woolhiser, D. A. & J. A. Liggett 1967. Unsteady, one dimensional flow over a plane – the rising hydrograph. *Water Resources Research* **3**, 753–71.

Wu, Y.-H., V. Yevjevich, D. A. Woolhiser 1978. *Effects of surface roughness and its spatial distribution on runoff hydrographs*. Hydrology paper no. 96, Colorado State University, Fort Collins, Colorado.

5

Modelling overland-flow hydrology for dynamic hydraulics

Helen Scoging

Abstract

A distributed mathematical model of hillslope overland flow, comprising three components, is presented. The first component uses the modified Green & Ampt equation to predict spatially variable times to runoff, and the decline in infiltration rate as a result of filling a fixed soil moisture store. The second component determines flowline characteristics (outflow direction from a cell and gradient) as a result of the differential force vectors operating across and down slope on a cell. This flowline routine provides macro flow routing controls for the accumulation of runoff down slope. The final component uses a finite difference solution to the kinematic wave equation for the spatial and temporal prediction of flow depth. An empirical expression, based on initial resistance and the rate of decline of resistance with flow depth, predicts the Darcy–Weisbach friction factor *f*, and is used with the flow equation to predict velocity. The model, distributed to the cell level, is able to integrate two important scales of process operation. The small scale is useful for distinguishing flowline concentrations and contributing interrill areas, and is thus the appropriate scale for analysis of the detailed mechanics of runoff erosion. The large scale reveals large-site dynamics of runoff and erosion potential, and is thus the appropriate scale for management strategies.

Introduction

The mathematical modelling of hillslope erosion by overland flow depends on three major sets of decisions, which have implications for model building, field experimentation and model testing:

(a) those that determine how runoff is generated at a point on the surface as a result of the interactions between rainfall, surface materials and infiltration;

(b) those that route flow over space, and thus are responsible for its hydraulics, temporal distribution and spatial concentration;

(c) those that are concerned in converting flow characteristics into erosion mechanics through consideration of the processes of erosion.

This chapter, and the following one, will address some of the issues associated with the first two sets of decisions, specifically within the context of scales of resolution, both temporal and spatial.

Until recently, two extreme scales of approach have dominated the literature on prediction of hillslope erosion: those of small-plot studies identifying between-site variations in plot properties and plot responses (e.g. Bork & Rohdenburg 1981), and those of larger hillslope studies based on gross indicators of runoff and sediment yield (e.g. Yair & Enzel 1987). Each of these approaches fails to account for within-site variability. The former studies, while accounting for variable responses to a single event as a result of different plot characteristics, do not incorporate the inflow that naturally occurs at any location due to context within the hillslope system, and hence fail to include flow concentration effects. In the latter studies, sources of variability cannot be disaggregated.

Combining both these scales, in an approach that models the hillslope as a set of distributed subareas, has identified the importance of variations in soil properties (Scoging 1982), infiltration (Sharma et al. 1980, Campbell et al. 1985), runoff concentration (Abrahams et al. 1990, Parsons et al. 1990) and runoff and soil-erosion patterns (Yair & Lavee 1985, Kirkby 1990) *within* a site as a result of differing local controls, the exceeding of specific thresholds in process mechanics and the spatial context of subareas. Such an approach allows processes to be modelled at the small scale, relying for operation on measured data, such as infiltration and runoff volumes, and calibration of runoff with hydraulic parameters, from point, or at best, small-area experiments, but also allows erosion predictions to be made at the hillslope scale, or larger.

Previous research on modelling overland flow in southern Spain (Scoging 1982, 1988, 1989, Scoging & Thornes 1980) identified the importance of spatially variable controls on infiltration, runoff initiation and flow routing in determining flow concentration downslope and erosion potential on four experimental hillslopes. The hillslope model, combining infiltration, runoff generation and dynamic flow routing developed in this research, is presented here to illustrate the integration of the two scales as a distributed mathematical model which behaves appropriately at both the micro-scale (flow hydraulics, erosion mechanics) and the macro-scale (runoff, sediment yield). The model is based upon a consideration of the influences of three main determinants of runoff. These are surface materials, slope gradient/direction and the friction factor (Darcy–Weisbach *f*). It takes into account the mechanics of their specific controls on infiltration (in determining the timing of runoff generation, and the volume of rainfall excess at a point), flow routing (in promoting dynamic flow concentration into "flowlines" distinguishable from contributing "interrill" areas) and the interaction between flow hydrology and environment through the friction factor in determining hydraulic behaviour.

Components of the dynamic overland-flow model

Infiltration controls

Most overland-flow models require an infiltration component to produce rainfall excess which can then be routed down slope. Rainfall excess is defined as that part of rainfall at a given time which cannot be infiltrated and which contributes to flow at the surface. Rainfall cannot infiltrate when rainfall intensity exceeds the capacity of a soil to infiltrate moisture supplied; when soil moisture storage is already full; and when surface changes, such as crusting, locally impede infiltration at the surface, despite an unfilled capacity to infiltrate below the surface.

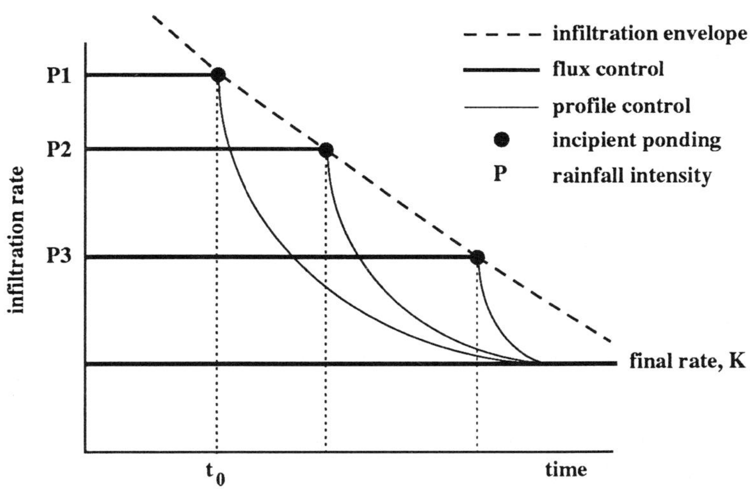

Figure 5.1 The infiltration envelope (from Smith 1972).

Three phases of infiltration can be recognized from the beginning of a storm (Fig. 5.1), described by the infiltration envelope (Smith 1972, Mein & Larson 1973).

(a) Pre-ponding or flux control ($P/K < 1$, where P = rainfall intensity and K = steady-state infiltration rate, related to soil conductivity) when infiltration occurs at the rainfall rate. The soil does not become saturated when the supply rate is less than the transmission rate, but approaches a limited moisture conductivity throughout the wetted zone. Where there is a fixed soil moisture store, ponding (and runoff) cannot occur until the soil moisture deficit is satisfied.

(b) Ponding, which occurs at $t = t_0$ when $P/K = 1$ and when soil water pressure at the surface equals atmospheric pressure. The amount of pre-ponding infiltration (or soil moisture storage capacity) is a function of both the rainfall intensity and initial soil water content, which together determine the moment when runoff is generated at the surface.

(c) Profile control, when $P/K > 1$, when infiltration is determined by hydraulic changes in the profile. A decreased hydraulic gradient ensues following extension of the transmission zone, thereby progressively restricting the surface entry of water and providing the physical basis for the monotonic decline of infiltration rate.

The infiltration envelope thus describes the conditions for runoff generation, in which for a given and fixed rainfall intensity there is a period after the onset of rainfall during which no excess is produced because the surface is unsaturated and infiltration is flux controlled (determined by the rainfall intensity), but once saturation has occurred the rate of infiltration falls systematically as a result of profile conditions.

Three commonly used expressions for modelling infiltration have been evaluated for semi-arid conditions in southern Spain (Scoging & Thornes 1980, Scoging 1982, 1988) under conditions of shallow, weakly developed soils and high rainfall intensities. The evaluation focused on the suitability of the expressions for meeting the requirements for a dynamic runoff generation model through the concept of a spatially variable time to runoff, t_0, controlled through the relationship between infiltration capacity and rainfall rate described by the infiltration envelope. These expressions are:

the modified Green & Ampt (1911) equations:

$$i = A + \frac{B}{t} \tag{5.1}$$

$$t_0 = \frac{B}{P - A} \tag{5.2}$$

the Philip (1957) equations:

$$i = A + \frac{B}{\sqrt{t}} \tag{5.3}$$

$$t_0 = \frac{B}{(P-A)^2} \tag{5.4}$$

the Kostiakov (1932) equations:

$$i = At^{-B} \tag{5.5}$$

$$t_0 = \left(\frac{A}{P}\right)^{\frac{1}{B}} \tag{5.6}$$

where i = infiltration rate;
t = time since beginning of rainfall;
P = rainfall intensity;
t_0 = time to ponding or time to runoff; and
A, B = parameters.

The Green & Ampt equations invariably provided a better fit to the observed infiltration series, and predicted times to runoff in closer agreement with observed runoff initiation times, than either the Kostiakov of Philip equations. It is argued that the Green & Ampt model, though based on empirical curve-fitting methods, has parameters which better reflect the process of infiltration under these physical conditions.

Assuming parameter *A* represents the final infiltration rate and parameter *B* is related to the fill rate of soil moisture storage (Eq. 5.1), the Green & Ampt equation, with increasing rainfall intensity, converges on a fixed volume or soil moisture storage needed to be filled before ponding, which reflects intrinsic soil properties and antecedent soil moisture. In contrast, the Philip and Kostiakov equations each predict storage volumes to ponding which decrease with increasing rainfall intensities, approaching the limit of zero-storage at very high rainfall intensities. In addition, the Philip equation, based on a truncated series in *t*, only approximates the early stages of infiltration, underestimating equilibrium infiltration rates and becoming asymptotic at ⅔ observed values. While the Kostiakov equation does provide an unrestricted series in *t*, the problem is that infiltration rate approaches zero as *t* becomes very large, rather than tending to a constant, positive final infiltration rate.

For a fully dynamic model which is able to predict runoff generation under different storm intensities and antecedent moisture conditions, a matrix of *A* and *B* parameters is required, representing the infiltration envelope for a given soil, but this is very rarely collected in the field (Bowyer-Bower 1992). More likely, small-plot experiments under simulated rainfall yield a pair of *A* and *B* parameters, which reflect both the soil condition and the experimental rainfall intensity. When applied to the larger scale, it is necessary to note the limitations placed on such predictions by this restricted infiltration envelope. In addition, it is important to be able to vary the *A* and *B* parameters across space. In the Spanish application of the model, the *A* and *B* values were related to properties of the soil, such as percentage gravel and median particle size, in order to generate distributed parameter values.

Topographic controls

One-dimensional runoff models commonly divide a slope profile (orthogonal to the contours) into subsections for flow routing. Flow convergence or divergence, however, involving concentration or dispersion of flow across and down slope, requires the second dimension, and therefore a mechanism of describing macro-routing directions. Numerous spatial units have been used in the literature to divide a large plot into basic units, or cells, for variable distribution and modelling (regular cells – Scoging 1988; equilateral triangles – Kirkby 1987; strips – Kirkby 1990, Marchington 1990).

In the model presented here, potential flowlines are adequately characterized for regular cells through vector analysis. It is assumed that a cell flowline is the resultant vector, *Fp*, of across, *Fa*, and downslope, *Fd*, vectors (Fig. 5.2a, b), each determined as the sum of opposite-boundary potential energy forces represented as height differences operating over boundary length.

With a constant reference framework allocated to each cell to define the highest corner (A, the flowline origin) and the appropriate across- (AB and CD) and downslope (AC and BD) boundary pairs, uniquely defining the zone of flowline destination (ACD), eight conditions fully specify cell topography and hence flowline exit categories *IRECT* 1–8 (Fig. 5.2c, d). Thus each cell is characterized by one outflow direction which is also an inflow direction for a neighbouring cell. In this way any cell can receive flow from its neighbours in a variety of combinations. Simplifying to consider only four vector directions (*IRECT* = 1–4), a cell may receive flow from any of 16 possible combinations (none; one-cell inflow 1, 2, 3, 4; two-cell inflow 13, 12, 14, 23, 24, 34; three-cell inflow 123, 124, 234, 134; and finally the special case of a hollow where inflow is received from all four neighbouring cells). In addition, the flowline routine determines *xyz* coordinates of flowline origin and

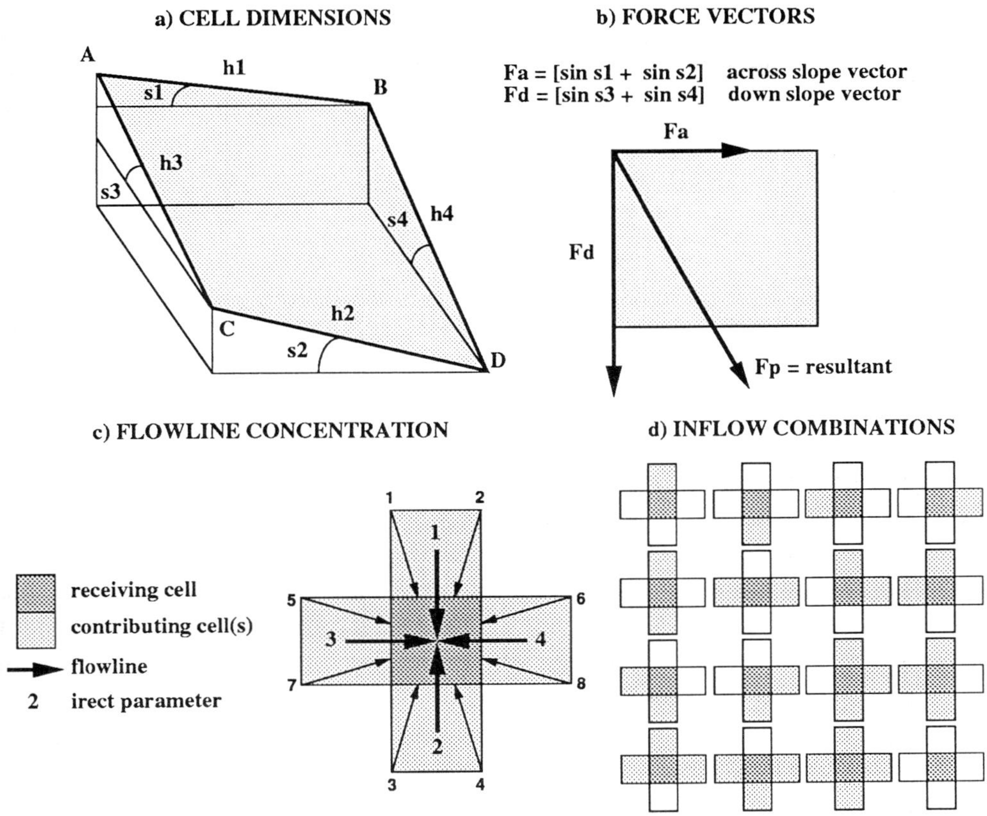

Figure 5.2 Cell flowline vector analysis.

destination, flowline slope gradient and length of flowline.

The flowline vector routine is a useful, simplified abstraction, providing two simple parameters for a dynamic flow model – macro flow direction (*IRECT*) and flowline slope – but it does have a number of drawbacks. First, flowlines are represented as originating from the highest corner of a cell and of having magnitude and direction which are determined from the relative heights of the four corners of a cell. Thus flowline direction is very sensitive to small differences in heights. Secondly, at the site scale, flowlines may converge, diverge or remain parallel, but at the cell scale routing can only occur as converging or parallel flow, since no splitting or divergence within a cell is accommodated. Thirdly, the assumption that the cells are rigid planes has two implications – one is the requirement to maintain a regular, orthogonal field survey design (suitable only for relatively gentle topography), the other is the problem of flexed cell planes (a situation that arises when diagonal corner heights are both greater than the remaining diagonal pair). The solution to this problem is to resolve the cell into two triangles and route the flow through each. Fourthly, where opposing vectors occur in adjacent cells (*IRECT* 1 and 2 in the downslope direction, and *IRECT* 3 and 4 in the across-slope direction), "hollows" can arise with no apparent outlet. Searching the heights of neighbouring cells will identify if these are real topographic hollows, or whether the flowline should be alternatively routed from the contiguous boundary. Finally, and implicit in some of the other drawbacks, it places high demands on field data collection.

The friction factor

Using the best-fit infiltration equation to predict time to runoff and infiltration through time (argued to be the Green & Ampt equation), and the cell flowline routine based on the relative topography of cell corner heights to route flow, the distributed dynamic flow model requires real-time hydrographs to be predicted for each cell. These are determined by a friction factor, in this case, the Darcy–Weisbach f.

Determination of f is of critical importance since this factor controls the shape of the hydrograph, and thus the timing of runoff contributions to neighbouring cells. Numerous studies have attempted to predict f from its relationship with discharge, Q, or Reynolds number, Re, ($= 4dV/\nu$, where d is flow depth, V is flow velocity, and ν is kinematic viscosity). There are two problems with this approach. The first concerns the use of small-plot studies, where inflow is absent (fixed-Q experiments), to assess the relationships between f and flow characteristics. Only experiments where flow can manifest the increasing concentration through time as a result of inflow from surrounding areas, due its position in the hillslope cascade, as well as the local onsite production of runoff due to the decline of infiltration, can the appropriate dynamics of the relationship between friction and flow be approached (e.g. Abrahams & Parsons 1991). The second problem is the nature of causality between f and Re or Q. A relationship which predicts f from Re assumes a knowledge of d and V and a causality from these to f. But if d and V are known, there is no need of the relationship, except in a descriptive sense, while the direction of causality should be from f (determined by surface roughness) and d (determined by inflow and rainfall excess) to V, as described by the Darcy–Weisbach flow equation:

$$V = \sqrt{\frac{8gdS}{f}} \tag{5.7}$$

where S is the energy slope and g is the gravitational constant.

However, if V is to be predicted through modelling, an expression which can predict f without recourse to Re is required. It is assumed here that there are two components in the consideration of a friction factor. The first is that of surface properties for a given cell in the absence of flow, that is initial resistance, denoted as fa, which is constant for a given surface condition. The second component, fb, is a dynamic one, reflecting the variability through time (or with stage) of flow depth (including rainfall excess generated on a cell and inflow from neighbouring cells). There are two approaches to combining these components into a dynamic friction factor. The first, derived from standard channel hydraulics, equates fa with a representative grain size or roughness height, and fb with flow depth, such that f = function (fa/fb), a relative roughness (e.g. Limerinos 1969):

$$\frac{1}{\sqrt{f}} = a + b\log\left(\frac{d}{d_{84}}\right) \qquad (5.8)$$

where d_{84} is the particle size for which 84% is finer.

In the absence of information concerning roughness heights, a second approach equates fa with initial roughness, and fb with the decline of roughness associated with increasing depth concentration:

$$f = fa - fb(d) \qquad (5.9)$$

Ideally both fa and fb should be spatially variable, reflecting the effect of surface properties in determining both initial resistance (fa) and the rate (fb) of friction loss.

Governing equations

Infiltration–runoff generation

Infiltration rate

$$i = A + \frac{B}{t} \qquad (5.1)$$

Time to runoff

$$t_0 = \frac{B}{A - P} \qquad (5.2)$$

Green & Ampt A

$$A = \textit{function (surface property)} \qquad (5.10)$$

Green & Ampt B $$B = \textit{function (surface property)} \tag{5.11}$$

Rainfall excess $$ex = (P - i)\,\Delta t \tag{5.12}$$

Friction factor

$$f = fa - fb(d) \tag{5.9}$$

Kinematic wave equations

Since Lighthill and Witham (1955), the kinematic approximation to the dynamic St Venant equations has been applied successfully to hillslopes (Woolhiser 1975, Cundy & Tento 1985). Its major assumption is that flow is uniform for a given spatial unit and hence the friction slope may be equated with the ground slope.

Continuity equation $$\frac{\partial q}{\partial x} + \frac{\partial d}{\partial t} = ex \tag{5.13}$$

Rating equation $$q = \alpha d^{m} \tag{5.14}$$

Flow equation $$V = \sqrt{\frac{8gdS}{f}} \tag{5.7}$$

Finite-difference solution (simple backward difference)

Equation 5.13 can be expressed in finite-difference form, where Δt and Δx are time and space increments:

$$\frac{q(x,t-\Delta t) - q(x-\Delta x,t-\Delta t)}{\Delta x} + \frac{d(x,t) - d(x,t-\Delta t)}{\Delta t} = ex(x,t-\Delta t) \tag{5.15}$$

Rearranging, and solving for $d(x,t)$, the unknown:

$$d(x,t) = (ex(x,t)\Delta t) + d(x,t-\Delta t) + \frac{\Delta t}{\Delta x}[q(x-\Delta x,t-\Delta t) - q(x,t-\Delta t)]\,. \tag{5.16}$$

Initially depth at $t = t_0 + \Delta t$ is determined entirely by $ex(x,t)$ since inflow discharge is zero. From depth, the friction factor can be determined from Equation 5.9, and velocity can be obtained from Equation 5.7, in finite-difference form:

$$V(x,t) = \sqrt{\frac{8gd(x,t)\,S(x)}{f(x,t)}} \tag{5.17}$$

and

$$q(x,t) = d(x,t)\,V(x,t) \tag{5.18}$$

$$Q(x,t) = d(x,t)\,V(x,t)\,W \tag{5.19}$$

$$Re(x,t) = \frac{4d(x,t)\,V(x,t)}{\nu} \tag{5.20}$$

where x = cell;
t = time; and
W = cell width.

Conclusion

A distributed flow model for dynamic hydraulics has been developed, comprising three main components (Fig. 5.3) with their field data parameterization requirements and model inputs summarized as follows:

(a) An infiltration–runoff generation component, based on a simplified Green & Ampt equation, describes the time at which runoff occurs, the decline of infiltration, and the rainfall excess for a given cell through time. Field data are required to test the validity of this infiltration model, and for the relationships between A and B parameters with respect to representative surface property conditions. Model inputs

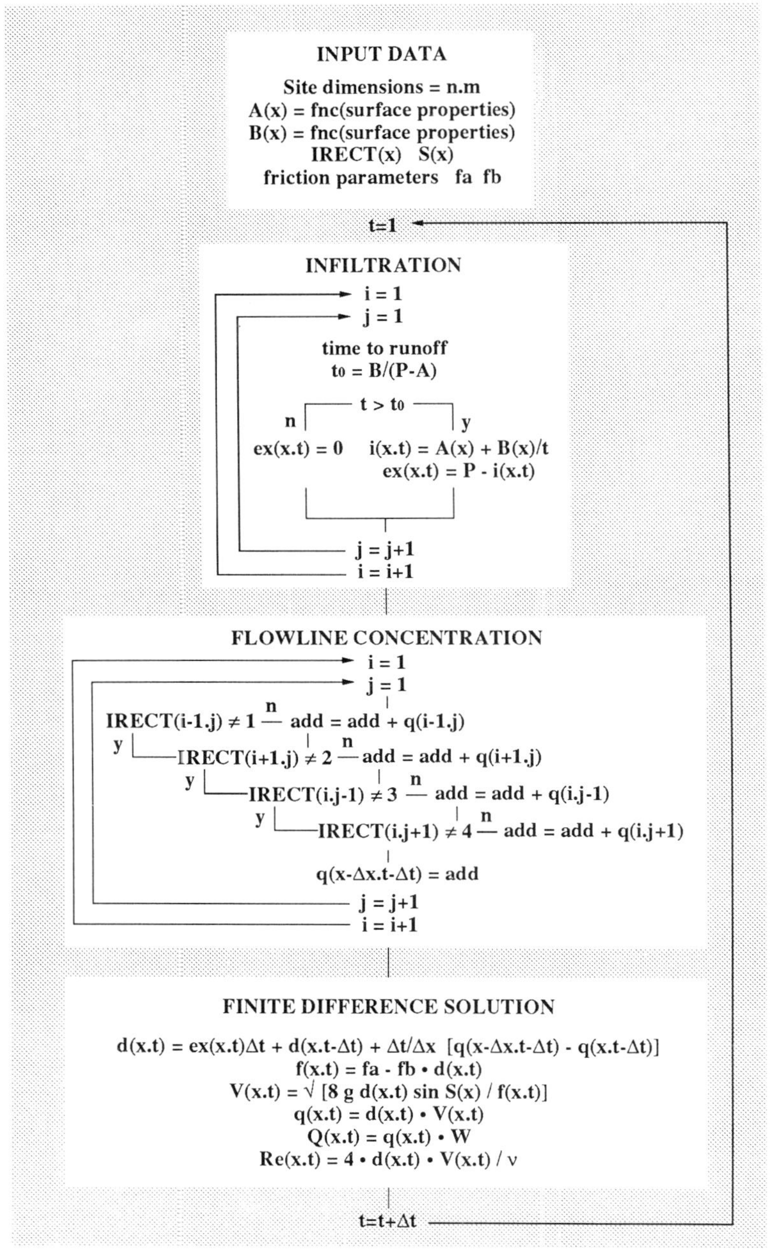

Figure 5.3 Flow diagram of dynamic overland flow model.

include the form of the infiltration model, the spatial distribution of surface property by cell, and the parameters and form of relationships between A and B and surface property.

(b) A flowline routine predicts macro flow routing in terms of cell outflow direction and gradient. Topographic height data are required for regular orthogonal cells. Model inputs are plot dimensions (*n*, *m* cells), *IRECT* flow direction parameter and flowline slope gradient for each cell.

(c) A finite-difference solution to the kinematic wave equation predicts flow depth as a result of rainfall excess on a given cell and inflow from neighbouring cells. A dynamic friction factor predicts f as a function of changing flow depth with respect to an initial roughness value. Velocity is determined from the Darcy–Weisbach equation. Field data are required for parameterization of the components of friction, *fa* and *fb*. Model inputs are *fa(x)*, *fb(x)* and their relationship to plot surface properties.

The model is based only on three key control variables: surface properties in determining infiltration and roughness, slope in controlling the macro-routing of flow, and a dynamic friction component which reflects flow depth in relation to initial roughness. In the next chapter the model will be parameterized for a semi-arid hillslope site at Walnut Gulch, Arizona, and the model's ability to predict spatial and temporal variations in flow hydrology and hydraulics at the cell, subsection, section and full site scales will be evaluated.

Symbols used in the text

A	infiltration parameter For Green & Ampt equation, A is the final infiltration rate	cm s^{-1}
B	infiltration parameter For Green & Ampt equation, B is the rate of soil moisture storage fill	cm
d	flow depth	cm
d_{84}	grain size for which 84% are finer	
ex	rainfall excess (P-i)	cm s^{-1}
f	Darcy–Weisbach friction factor	
fa	initial friction	
fb	rate of loss of friction with increasing depth	
Fa	across-slope cell vector	
Fd	downslope cell vector	
Fp	resultant cell flowline vector	
g	gravitational constant	981 cm s^{-2}

IRECT	cell flowline exit direction	
i	infiltration rate	cm s^{-1}
K	profile infiltration transmission rate	
m	coefficient reflecting flow type in kinematic equation	
P	rainfall rate	mm h^{-1}
Q	total discharge (dVW)	cm^3 s^{-1}
q	unit width discharge (dV)	cm^2 s^{-1}
Re	Reynolds Number ($4dV/\nu$)	
S	sine slope gradient	
t	time	s
t_0	time to ponding or time to runoff	s
V	flow velocity	cm s^{-1}
W	cell width	cm
(x,t)	space and/or time variable	
α	slope-friction coefficient in kinematic rating equation	
Δt	time increment	
Δx	space increment	
ν	kinematic viscosity	cm^2 s^{-1}

References

Abrahams, A. D. & A. J. Parsons 1991. Resistance to overland flow on desert pavement and its implications for sediment transport modelling. *Water Resources Research* **27**, 1827–36.

Abrahams, A. D., A. J. Parsons, S.-H. Luk 1990. The effect of spatial variability in overland flow on the downslope pattern of soil loss on a semiarid hillslope, southern Arizona. *Catena* **18**, 255–70.

Bork, H. R. & H. Rohdenburg 1981. Rainfall simulation in southeast Spain: analysis of overland flow and infiltration. In *Soil conservation problems and prospects.* R. P. C. Morgan (ed.), 293–301. Chichester, England: John Wiley.

Bowyer-Bower, T. 1992. Land surface response to rainfall in semi-arid systems: determinations using rainfall simulation in the Lowveld, Swaziland. Unpublished D.Phil. thesis. University of Oxford.

Campbell, S. Y., W. H. van der Molen, C. W. Rose, J. Y. Parlange 1985. A new method for obtaining a spatially averaged infiltration rate from rainfall and runoff rates. *Journal of Hydrology* **82**, 57–68.

Cundy, T. W. & S. W. Tento 1985. Solution to the kinematic wave approach to overland flow routing with rainfall excess given by the Philip equation. *Water Resources Research* **21**, 1132–40.

Green, W. H. & G. A. Ampt 1911. Studies on soil physics. 1: the flow of air and water through soils. *Journal of Agricultural Science* **4**, 1–24.

Kirkby, M. J. 1987. Modelling some influences of soil erosion, landslides and valley gradient on drainage density and hollow development. *Catena Supplement* **10**, 1–14.

Kirkby, M. J. 1990. A simulation model for desert runoff and erosion. Proceedings of the International Association of Hydrological Sciences workshop on erosion, transport and deposition processes, Jerusalem. *IAHS–AISH Publication* **189**, 87–104.

Kostiakov, A. N. 1932. On the dynamics of the coefficient of water percolation in soils and on the necessity for studying it from a dynamic point of view for purposes of amelioration. *Transactions of the 6th Commission of the International Society of Soil Science*, Part A, 17–21.

Lighthill, M. J. & G. B. Witham 1955. On kinematic waves. 1: Flood movement in long rivers. *Proceedings of the Royal Society* **A229**, 281–316.

Limerinos, J. T. 1969. *Relation of the Manning coefficient to measured bed roughness in stable natural channels.* United States Geological Survey, Professional Paper 650-D, 215–21.

Marchington, A. C. 1990. Towards a dynamic model of gully growth. Proceedings of the International Association of Hydrological Sciences workshop on erosion, transport and deposition processes, Jerusalem. *IAHS–AISH Publication* **189**, 121–134.

Mein, R. G & C. L. Larson 1973. Modelling infiltration during a steady rain. *Water Resources Research* **9**, 384–94.

Parsons, A. J., A. D. Abrahams, S.-H. Luk 1990. Hydraulics of interrill overland flow on a semiarid hillslope, southern Arizona. *Journal of Hydrology* **117**, 255–73.

Philip, J. R. 1957. Theory of infiltration: 1. The infiltration equation and its solution. *Soil Science* **83**, 345–57.

Scoging, H. M. 1982. Spatial variations in infiltration, runoff and erosion on hillslopes in semi-arid Spain. In *Badland geomorphology and piping.* Bryan, R. B. & A. Yair (eds), 89–112. Norwich: GeoBooks.

Scoging, H. M. 1988. A theoretical and empirical investigation of soil erosion in a semi-arid environment. Unpublished Ph.D. thesis. University of London.

Scoging, H. M. 1989. Runoff generation and sediment mobilisation by water. In *Arid zone geomorphology,* D. S. G Thomas (ed.), 87–116. London: Belhaven Press.

Scoging, H. M. & J. B. Thornes 1980. Infiltration characteristics in a semi arid environment. Proceedings of the International Association of Hydrological Sciences Symposium on the Hydrology of Areas of Low Precipitation, Canberra. *IAHS–AISH Publication* **128**, 159–168.

Sharma, M. L., G. A. Gander, C. G. Hunt 1980. Spatial variability of infiltration in a watershed. *Journal of Hydrology* **45**, 101–22.

Smith, R. E. 1972. The infiltration envelope: results from a theoretical infiltrometer. *Journal of Hydrology* **17**, 1–21.

Woolhiser, D. A. 1975. Simulation of unsteady overland flow. In *Unsteady flow in open channels, K.* Mahmood & V. Yevjevich (eds). 485–508. Fort Collins, USA: Water Resources Publications.

Yair, A. & Y. Enzel 1987. The relationship between annual rainfall and sediment yield in arid and semiarid areas: the case of the northern Negev. *Catena Supplement* **10**, 121–35.

Yair, A. & H. Lavee 1985. Runoff generation in arid and semi-arid zones. In *Hydrological forecasting, M. G.* Anderson & T. P. Burt (eds). 183–220. Chichester, England: John Wiley.

6

Application of a dynamic overland-flow hydraulic model to a semi-arid hillslope, Walnut Gulch, Arizona

Helen Scoging, Anthony J. Parsons, Athol D. Abrahams

Abstract

The model presented in Chapter 5 is applied to an 18 m wide by 35 m long runoff plot at Walnut Gulch Experimental Watershed, Arizona. Parameterization of the model includes: (a) a detailed topographic survey of the plot using a 2 ft by 2 ft cell grid and the measurement of the percentage of desert pavement in each cell; (b) small-plot experiments to determine the parameters of the infiltration equation, and their relationship with percentage of desert pavement; and (c) small-plot experiments to determine parameter estimates for the dynamic friction factor. The model is used to predict the flow hydraulics observed during a rainfall simulation on the runoff plot at several scales. The model successfully predicts site and cross-section hydrographs, cross-sectional distributions of discharge, depth and velocity and downslope changes in depth and velocity. However, the predicted relationships between friction factor and Reynolds number differ from those obtained from the field experiments. These differences may be due to the inadequacies of data obtained from small-plot experiments for parameterizing the model, or to different methods of calculation used in the field and in the model, or to shortcomings in the methods of field measurements. The application of the model highlights both the need for and benefits of closer links between field empiricists and modellers.

Introduction

The previous chapter has outlined a distributed, mathematical model of overland-flow hydraulics. This chapter parameterizes the model from both published and unpublished experimental data collected by Parsons, Abrahams & Luk, and applies the model to a semi-arid hillslope site at Walnut Gulch, southern Arizona (Parsons et al. 1990). In addition, a number of criteria are used to assess the ability of the model to predict the spatial and temporal variability in runoff hydraulic parameters at different spatial resolutions.

Site description
The site to which the model is applied consists of an 18 m wide by 35 m long runoff plot at Walnut Gulch Experimental Watershed, Tombstone, in southern Arizona (31°43′N, 110°41′W). The plot was located on a hillslope typical of the dissected piedmont in the watershed, with slope gradients ranging between 0° at the divide and 4.5° at the outlet. The site was underlain by Hathaway gravelly loam developed on well-cemented, coarse Quaternary alluvium. Vegetation comprised a sparse desert community, dominated by shrubby species which grow atop accumulations of fine sediments. The inter-mound surface was mantled by a desert pavement with a mean particle size of 10.4 mm. The size of the plot, which extended from the divide to beyond the rill head, enabled the examination of the full range of lateral and downslope variability of interrill overland flow on a shrub-covered, semi-arid piedmont hillslope. Further details of the field experiments are reported in Parsons et al. (1990).

Summary of model and data requirements

The model, distributed to the cell level, comprises three components with associated parameters for each cell, denoted by *(x,t)* for spatial and/or temporal variation. These components are cell infiltration and runoff generation, cell topography, flowline direction and gradient, and cell friction factor.

Infiltration and runoff generation
It has been argued (Scoging 1988) that the simplified Green and Ampt (1911) model (Eqs 6.1–6.3) is a better model to describe infiltration and runoff generation on Spanish semi-arid slopes than either the Philip (1957) or Kostiakov (1932) equations. In considering the soil moisture storage capacity as fixed for a given soil and antecedent mositure condition, runoff does not occur until that moisture deficit is satisfied, at a time, t_0, which depends on the rainfall intensity.

$$i(x,t) = A(x) + \frac{B(x)}{t} \qquad (6.1)$$

$$t_0(x) = \frac{B(x)}{P - A(x)} \qquad (6.2)$$

$$ex(x,t) = P - i(x,t) \qquad (6.3)$$

where $i(x,t)$ = infiltration rate;
t = time since beginning of rainfall;
$A(x)$ = parameter related to final infiltration rate;
$B(x)$ = parameter related to rate of fill of soil moisture deficit;
$t_0(x)$ = time to runoff;
P = rainfall intensity;
$ex(x,t)$ = rainfall excess.

This model requires further testing against the Philip (1957) (Eqs 6.4 & 6.5) and Kostiakov (1932) (Eqs 6.6 & 6.7) infiltration expressions to establish its validity for the Arizona plot.

$$i(x,t) = A(x) + \frac{B(x)}{\sqrt{t}} \tag{6.4}$$

$$t_0(x) = \frac{B(x)}{(P - A(x))^2} \tag{6.5}$$

$$i(x,t) = A(x)\,t^{-B(x)} \tag{6.6}$$

$$t_0(x) = \left(\frac{A(x)}{P}\right)^{\frac{1}{B(x)}}. \tag{6.7}$$

In addition, to distribute the *A* and *B* parameters, and hence time to runoff, infiltration and rainfall excess across the site, some relation between these parameters and surface properties of the cells is required. For hillslopes of which this site is typical, Abrahams & Parsons (1991a) have shown a relationship between surface stone cover and infiltration. Expressed simply, this relationship can be modified to one between percentage of desert pavement (*%P*) within a cell and the infiltration parameters, such that:

$$A(x) = \mathit{function}\ [\%\,P(x)] \tag{6.8}$$

$$B(x) = \mathit{function}\ [\%\,P(x)] \tag{6.9}$$

Cell topography, flowline direction and gradient
Topographic heights at the corners of a regular mesh of cells are required to determine flowline direction parameter *IRECT(x)* and flowline gradient *S(x)*.

Components of a friction factor, f(x,t)
Data on initial cell resistance, *fa(x)*, and the rate of loss of friction with increasing depth *fb(x)*, are required to parameterize the dynamic friction expression:

$$f(x,t) = fa(x) - fb(x) \cdot d(x,t) \tag{6.10}$$

where: *f(x,t)* = Darcy–Weisbach friction factor for cell *x* at time *t*;
fa(x) = initial cell friction;
fb(x) = rate of cell friction loss with increasing depth; and
d(x,t) = flow depth for cell *x* at time *t*.

Field data

It is essential that model parameters are derived from data that are different from those used to evaluate the model. Only static, contextual variables from the runoff plot are used to parameterize the model, the dynamic flow data obtained during rainfall simulator experiments on the plot being used to assess and evaluate independently the model's performance.

The model simulates runoff generation, infiltration and flow routing on the plot at the resolution of the cells. From these predictions, dynamic flowline depths, velocities and discharges, Darcy–Weisbach friction factors and Reynolds numbers can be obtained at resolutions ranging from the cell to the plot scale. The model is evaluated by comparing these predicted results with the data from the rainfall simulation experiments conducted on the plot.

Three sources of experimental data will be drawn on, some for model parameterization, while others will be used to evaluate the model. These are runoff plot data, data from infiltration experiments and small-plot experiments on overland flow hydraulics.

Runoff plot data
A detailed topographic survey of the plot was carried out using a 0.61 m by 0.61 m (2 ft by 2 ft) grid. For each cell in the grid, measurement was made of the percentage of its area covered by desert pavement. These data provided the model parameters *IRECT(x)*, *S(x)* and *%P(x)*.

Four rainfall simulation experiments (E1...E4) were conducted on the plot

(Parsons et al. 1990). Only one of these (E2) is used to evaluate the model since neither E1 nor E3 reached equilibrium conditions, and E4 was subject to possible crusting effects after the application of the three previous storms. For E2, which lasted for 20 min, average rainfall intensity was 97 mm h^{-1}. During the experiment four types of data were collected:

(a) Total outflow runoff was measured through time by an automatic stage recorder attached to a supercritical flume. Water reached this outlet either through concentration in the shallow rill flowing directly into the flume, or via boundary gutters in the tapered lower section of the plot into calibrated flumes and thence into the supercritical flume.

(b) On the plot, two sections, S1 and S2, respectively 12.5 m and 21 m from the upper boundary, were established. At these sections a number of miniature flumes (Parsons & Abrahams 1989a) were installed to measure flow depth and discharge, with little interference to flow.

(c) Across both sections, thin rods were emplaced at intervals of 0.5 m, adjacent to which flow depths were measured at minutes 4, 9, 14 and 19 during E2.

(d) At the end of E4, under equilibrium conditions, detailed analysis of flow through 10 m subsections of S1 and S2 (SS1 and SS2) were undertaken. Flow depth was measured at 5 cm intervals, the subsections were divided into zones of relatively uniform flow characteristics, termed partial sections, and surface velocity measured using dye injections (Abrahams et al. 1986b). For model evaluation it is assumed that had equivalent subsection data been obtained at the end of E2, it would have been comparable.

These data permit evaluation of the model's predictions of the spatial pattern of flowlines, hydrographs for the plot and for sections S1 and S2, depth distributions at S1 and S2, at-a-section and downslope hydraulic relationships, and equilibrium depth, velocity and discharge for subsections SS1 and SS2.

Infiltration experiments

Twenty-one infiltration experiments were undertaken on 1 m by 1 m plots located on the hillslope adjacent to the runoff plot. The design of these experiments was identical to that described in Abrahams & Parsons (1991a), except that rainfall was delivered at the design intensity of 80 mm h^{-1}, using a nozzle that was the same as those that supplied the rain for experiments on the large runoff plot. The infiltration plots had %P ranging between 20 and 100. These experiments are of the type where infiltration and runoff are generated onsite in response to a single rainfall intensity, without allowing for any modification in response due to inflow. Data from these experiments were used to assess the validity of the Green and Ampt model to control runoff generation, to provide A and B parameter values to simulate infiltration, and to characterize the relationships between %$P(x)$ and the infiltration parameters $A(x)$ and $B(x)$.

Small-plot experiments on overland-flow hydraulics
Abrahams & Parsons (1991b) undertook experiments on eight 0.61 m wide by 1.5 m long (2 ft by 5 ft) plots located on 100% desert pavement surfaces within the large runoff plot, to determine the relationships between the Darcy–Weisbach friction factor and plot characteristics. In these experiments runoff response to increasing input values was determined, imitating flow conditions on the plot through a consideration of both the onsite and inflow contribution to runoff hydraulics. These data were used to parameterize the model for the dynamic resistance terms $fa(x)$ and $fb(x)$.

Parameter estimation

Infiltration
Three commonly used infiltration equations were fitted to the data from the 21 1 m by 1 m rainfall–runoff plots.

Green & Ampt (1911) $$i = A + \frac{B}{t} \qquad (6.1)$$

Philip (1957) $$i = A + \frac{B}{\sqrt{t}} \qquad (6.4)$$

Kostiakov (1932) $$i = At^{-B} \qquad (6.6)$$

In all but one case (Table 6.1), Equation 6.1 performed better than either the Kostiakov or Philip equation, indicating the statistical superiority of the simplified Green and Ampt equation. For 20 plots significance levels of the regression had $p < 0.03$. The plot 14 significance level had $p = 0.068$.

Table 6.1 Estimates of small-plot infiltration parameters A and B from three equations:
Green and Ampt (1911) $i = A + B/t$
Kostiakov (1932) $i = At^{-B}$
Philip (1957) $i = A + B/\sqrt{t}$

Plot	Equation	df	r^2	$p(t)$	A mm min^{-1}	B mm	%P	Slope degrees	Rain mm h^{-1}	$t_0$obs min	$t_0$pred min	Final i mm h^{-1}
1	GA	20.1	0.330	0.005	0.393	1.797	88	1.417	74.3	2.867	2.126	28.97
	Ko	20.1	0.046	0.336	−0.074	−0.018					0.670	47.94
	Ph	20.1	0.261	0.015	0.196	2.478					2.281	28.38
2	GA	15.1	0.663	0.000	0.473	2.655	88	1.250	78.0	3.583	3.210	36.34
	Ko	15.1	0.533	0.001	0.263	−0.387					2.309	34.48
	Ph	15.1	0.646	0.000	0.145	3.872					2.902	34.67
3	GA	20.1	0.424	0.001	0.219	2.657	100	3.500	72.0	3.300	2.708	21.11
	Ko	20.1	0.088	0.181	−0.044	−0.291					0.574	22.67
	Ph	20.1	0.340	0.004	−0.059	3.547					2.238	20.25
5	GA	24.1	0.298	0.004	0.094	1.709	100	5.000	68.3	2.900	1.636	10.76
	Ko	24.1	0.031	0.378	−0.369	−0.286					0.151	10.89
	Ph	24.1	0.218	0.016	−0.082	2.289					1.537	10.34
6	GA	14.1	0.726	0.000	0.297	4.043	84	3.666	72.4	4.467	4.444	29.94
	Ko	14.1	0.586	0.001	0.487	−0.637					57.215	27.31
	Ph	14.1	0.724	0.000	−0.152	5.557					3.010	28.15
7	GA	17.1	0.769	0.000	0.447	3.626	88	2.250	82.5	4.576	3.907	37.69
	Ko	17.1	0.670	0.000	0.317	−0.403					2.831	37.22
	Ph	17.1	0.722	0.000	0.080	4.737					2.825	36.57
8	GA	19.1	0.528	0.000	0.494	2.129	88	2.000	81.8	3.617	2.449	36.02
	Ko	19.1	0.448	0.001	0.152	−0.297					1.083	34.97
	Ph	19.1	0.502	0.000	0.240	3.051					2.418	34.86
11	GA	17.1	0.631	0.000	0.796	1.925	68	4.500	93.4	3.033	2.531	53.53
	Ko	17.1	0.589	0.000	0.254	−0.250					1.288	50.92
	Ph	17.1	0.630	0.000	0.531	3.000					2.852	51.98
12	GA	7.1	0.890	0.000	0.661	2.068	72	6.250	69.4	4.000	4.172	45.86
	Ko	7.1	0.890	0.000	0.239	−0.290					2.202	43.63
	Ph	7.1	0.906	0.000	0.399	3.062					5.334	44.48
13	GA	5.1	0.682	0.022	1.300	0.719	20	7.833	93.8	3.400	2.730	80.15
	Ko	5.1	0.575	0.048	0.206	−0.065					1.032	79.35
	Ph	5.1	0.632	0.033	1.208	1.077					8.530	79.70
14	GA	5.1	0.519	0.068	0.852	1.531	72	7.000	73.5	5.650	4.105	55.71
	Ko	5.1	0.414	0.118	0.140	−0.134					1.177	55.43
	Ph	5.1	0.463	0.093	0.708	1.918					7.176	55.34
16	GA	3.1	0.953	0.004	0.780	2.654	64	8.580	69.0	7.550	7.173	54.76
	Ko	3.1	0.945	0.005	0.255	−0.232					2.145	53.86
	Ph	3.1	0.952	0.004	0.543	3.219					8.737	54.17
18	GA	4.1	0.857	0.008	0.835	1.785	56	8.250	70.9	5.500	5.149	55.45
	Ko	4.1	0.855	0.008	0.194	−0.182					1.530	54.37
	Ph	4.1	0.854	0.008	0.649	2.367					8.342	54.81
19	GA	4.1	0.783	0.019	0.660	4.062	60	4.000	71.3	7.667	7.688	51.78
	Ko	4.1	0.730	0.030	0.362	−0.332					4.136	51.07
	Ph	4.1	0.775	0.021	0.321	4.749					6.313	51.11
20	GA	9.1	0.741	0.001	1.037	2.025	44	4.333	93.8	3.833	3.847	68.29
	Ko	9.1	0.675	0.002	0.286	−0.184					1.382	66.79
	Ph	9.1	0.730	0.001	0.796	2.909					4.941	67.27
21	GA	5.1	0.832	0.004	1.075	1.528	48	2.167	84.0	5.083	4.702	69.08
	Ko	5.1	0.775	0.009	0.219	−0.124					1.250	68.52
	Ph	5.1	0.806	0.006	0.919	2.011					8.692	68.63
22	GA	12.1	0.533	0.003	1.002	1.831	44	6.666	98.2	2.983	2.877	65.61
	Ko	12.1	0.429	0.011	0.280	−0.199					1.270	62.98
	Ph	12.1	0.518	0.004	0.751	2.845					3.613	64.45
24	GA	4.1	0.691	0.040	0.817	1.157	52	7.167	66.0	5.000	4.088	52.49
	Ko	4.1	0.572	0.082	0.094	−0.120					1.173	52.00
	Ph	4.1	0.630	0.060	0.700	1.516					9.475	52.17
25	GA	11.1	0.825	0.000	0.825	1.656	64	8.500	82.1	3.083	3.048	54.46
	Ko	11.1	0.743	0.000	0.211	−0.205					1.318	52.77
	Ph	11.1	0.808	0.000	0.604	2.546					4.358	53.31
31	GA	14.1	0.895	0.000	0.446	3.194	96	4.333	76.5	3.600	3.853	36.34
	Ko	14.1	0.856	0.000	0.373	−0.476					6.315	34.03
	Ph	14.1	0.888	0.000	0.042	4.737					3.116	34.29
35	GA	2.1	0.947	0.027	0.842	1.851	60	7.833	66.8	7.083	6.822	56.07
	Ko	2.1	0.885	0.059	0.177	−0.162					1.546	55.51
	Ph	2.1	0.918	0.042	0.679	2.247					11.911	55.81
Means				Green and Ampt	0.683	2.219	69	5.071	78.0	4.417	3.965	47.64
				Kostiakov	0.192	−0.251					4.409	47.46
				Philip	0.439	3.035					5.267	46.69

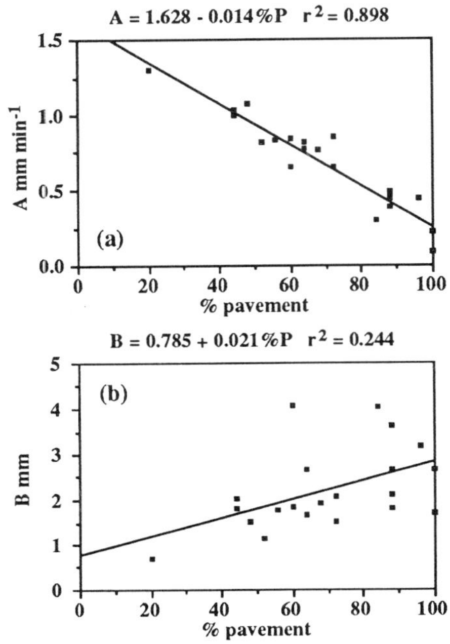

Figure 6.1 Relationship between (a) Green and Ampt *A* infiltration parameter and (b) Green and Ampt *B* infiltration parameter, and percentage pavement (1 m by 1 m plots).

Times to runoff, t_0, predicted by the Green and Ampt expression (Eq. 6.2) were in closer agreement with the observed values than times from either Philip or Kostiakov (Eqs 6.5 & 6.7, respectively). This result reinforces the argument for the suitability of this expression in semi-arid areas on the grounds that it better reflects the infiltration processes than either alternative. The derived Green and Ampt *A* and *B* plot parameters were then regressed against plot percentage pavement (*%P*). Figure 6.1a & b illustrates these relationships.

$$A(x) = 1.628 - 0.014\,\%P(x) \tag{6.11}$$

$$B(x) = 0.785 + 0.021\,\%P(x) \tag{6.12}$$

There are three potential problems for the application of these results to the large runoff plot. The first concerns the conditions under which these experiments were performed compared with the large-plot experiments. In the former, the soil surface was dry and dusty, typical of an Arizona summer and likely to promote high infiltration rates, whereas the large-plot experiment E2 occurred on surfaces which had already been wetted, and possibly crusted by earlier rainfall simulations. A second problem with this set of data is the fact that input to the 1 m by 1 m plots was not constant across all experiments, with input values ranging between 66 mm h^{-1} and 98 mm h^{-1} (averaging 78 mm h^{-1}, close to the design intensity). This means that the small-plot values of *A* and *B* are not strictly comparable with each other, nor with the estimated average simulated rainfall rate of E2, 97 mm h^{-1}. A third, and more general area of concern is the ability of such small-plot experiments to replicate the real processes of runoff generation and infiltration because they do not include the effects of inflow, and, furthermore, cannot reproduce the influence of spatial position in the hillslope hydrological cascade, for example the effect of contiguous areas of high *%P* on the large plot.

Flowline direction and gradient

The regular grid of 2 ft by 2 ft cells had a maximum of 29 cells across site and 57 down site. From row 46 down, the site narrowed to converge in a rill. Corner heights of cells were subjected to the flowline routine and cell flowline direction and gradient were determined, *IRECT(x)* and *S(x)*. Figure 6.2 illustrates the flowlines, derived entirely from height data, in association with the distribution of percentage pavement (*%P*) for the site. Three clear patterns can be identified: (a) areas of cells with low %P, associated with mounds of fines and vegetation clumps, form the source areas for flowlines; (b) cells with high %P values are associated with the dominant threads of flow; and (c) flowlines from the upper part of the site converge towards the lower, rill area.

A comparison of the predicted pattern of flowlines with photographs of dye tracing taken during the field experiments, suggests that the flowline routine is largely successful in replicating the pattern of flow threads, with the exception of two areas, namely around cell (08,35), where the small flowline should converge toward the rill, and around cell (23,25), where the flowline should join with the main thread entering the rill. In addition, field observations indicate that flow may "braid" within a local area, flowing out from a "cell" in more than one direction. The model is unable to replicate this within-cell divergence.

Friction factor

The data from the 5 ft by 2 ft plots were used to determine the relationship between *f* and depth. In these experiments, width and depth were measured over a 2 ft by 2 ft area over which input was increased systematically, allowing equilibrium conditions to be attained for each input level. These experiments and results have been reported in Abrahams & Parsons (1991b). Initially it was hoped that the *a* and *b* coefficients of the by-plot regression of *f* on *d* could be related to plot environmental parameters such as *%P*. Such relationships are

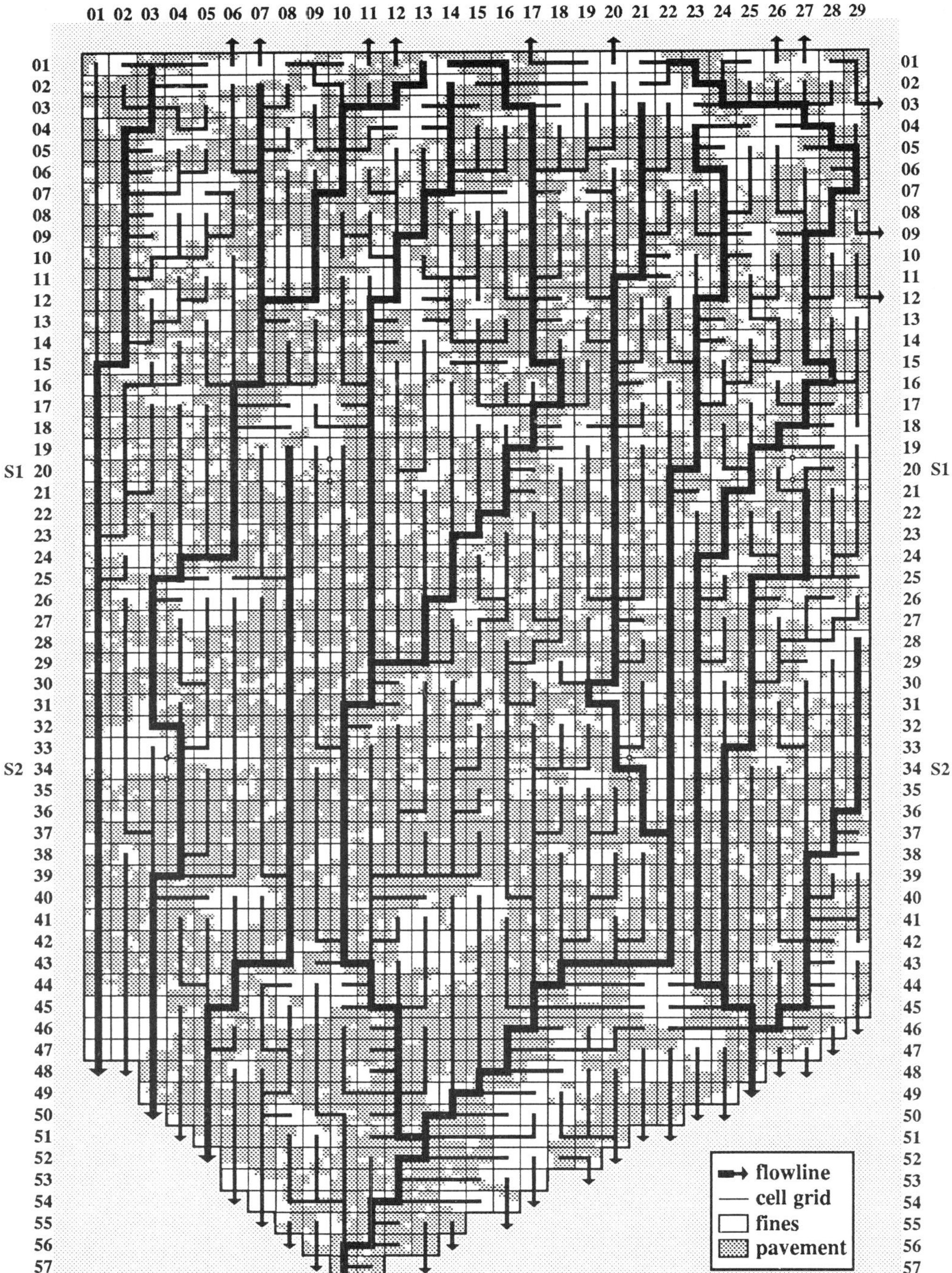

Figure 6.2 Flowlines predicted from vector routine and measured percentage pavement on Arizona runoff plot.

present in the data, and potentially very useful, but because there was no basis to extrapolate the results to small %*P* areas (experiments were conducted on plots with 100% desert pavement) their use is impossible in the present context. Table 6.2 illustrates the by-plot relationships between *f* and *d*, indicating that plot gradient and pavement gravel characteristics do have an influence on the nature of the relationship. Unfortunately, with too few experiments, these relationships cannot be disaggregated. Instead, the average of the *a* and *b* values for significant relationships was used, providing an overall expression for dynamic friction:

$$f(x,t) = 14.46 - 17.35d(x,t) \tag{6.13}$$

Table 6.2 Relationships between 5-ft by 2-ft plot characteristics and *f–d* parameters.

plot no.	a	b	sin S	D_G	%G
3	19.555	−18.890	0.0848	12.48	80.00
2	14.275	−17.628	0.0640	9.65	74.20
5	19.484	−28.585	0.0451	8.37	72.90
6	4.034	−3.968	0.0523	8.68	71.40
4	5.728	−6.930	0.0480	12.50	68.60
7	1.983	−0.701	0.0436	8.58	61.54
8	1.378	−0.381	0.0262	11.14	55.38
1	4.536	−4.312	0.0232	12.00	51.50

a, b parameters in $f = a - b(d)$ relationship
data from Abrahams & Parsons (1991b)
S plot slope
D_G mean gravel size, mm
%G % gravel

The lack of appropriate experimental data means that *fa* and *fb* are now constants, rather than parameter values related to site characteristics and distributed by cell according to its surface property. It was expected that low-%*P* and high-%*P* cells would behave differently in terms of the effect of friction on flow hydraulics. The former would be expected to have relatively low initial resistance, *fa,* compared with the latter; whereas the latter, which are flow-concentrating cells, would have rapid rates of friction decline with depth, compared with the former. One other potential problem with this expression is the limit of Equation 6.13, where at minimum friction, $f = 0$, depth is at a maximum of 0.83 cm. As it is likely that flow depths on the large site will exceed this value, the rate of decline of friction, *fb*, is probably overestimated.

Model results

Site and section S1, S2 hydrographs

The model was run under the conditions of E2, for 19 minutes, with $\Delta t = 1$ s, $\Delta x = 2$ ft and rainfall input of 97.0 mm h^{-1}. Because most of the experimental hydraulic data were collected for sections S1 and S2 (rows 20 and 34 in Fig. 6.2), having noted already the depth limitations of Equation 6.13, and because rill flow below row 45 is likely to require different model assumptions, it was decided to restrict model prediction to the rectangular part of the plot (rows 1–45). In order to compare model predictions of the site with observed total site discharge (which includes the rill area) it was assumed that the former could be multiplied by an area proportionality factor, bearing in mind that this is likely to be a slight underprediction of the rill discharge. Thus, outflow from the model was multiplied by 1.14 (the ratio of the whole site to model area).

Run 1 Infiltration parameters

$$A(x) = 1.628 - 0.014\,\%P(x) \tag{6.11}$$

$$B(x) = 0.785 + 0.021\,\%P(x) \tag{6.12}$$

Friction parameters

$$f(x,t) = 14.46 - 17.35\,d(x,t) \tag{6.13}$$

Using measured total site hydrograph and S1 and S2 section hydrographs as the criteria for assessment (Fig. 6.3, run 1), the model performed very badly initially, overpredicting infiltration and underpredicting discharge by up to 60%, particularly with respect to the early parts of the section hydrographs. This was due in part to overestimation of the *A* and *B* parameters, and hence t_0 and infiltration rates, from the 1 m by 1 m small-plot experiments as a result

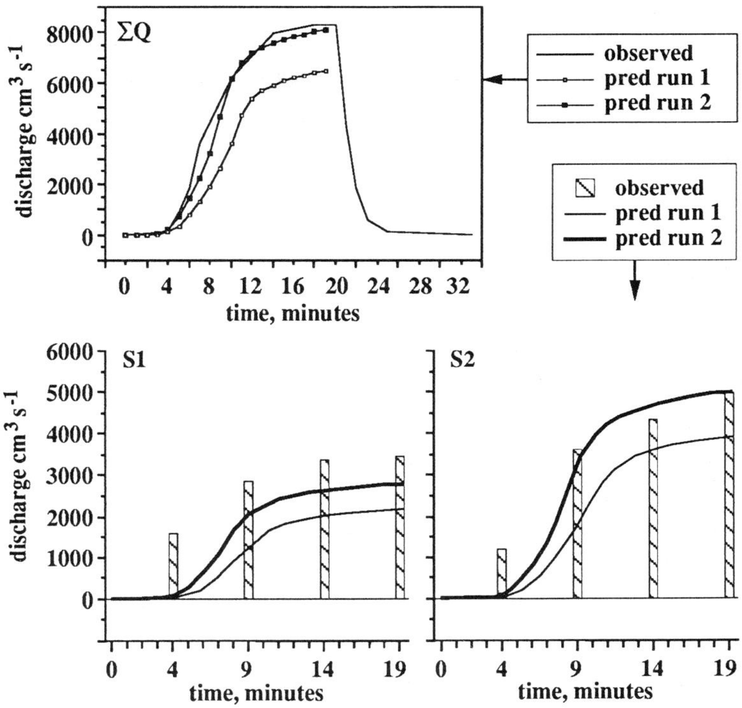

Figure 6.3 Observed and predicted (run 1 and run 2) hydrographs for whole runoff plot (ΣQ) and sections S1 and S2.

of differences in plot condition, antecedent moisture content and possible crusting effects.

However, it also seems likely that different processes are operating at the two scales. In contrast to the small-plot experiments where soil moisture deficit is filled only by onsite rainfall, this deficit will be filled from two sources on the runoff plot – onsite rainfall and runoff flowing onto a local area from surrounding parts. It is therefore expected that times to runoff on the large plot will be shorter, and hence the decline in infiltration will start earlier and result in more rainfall excess at the beginning of a storm.

In addition to these infiltration problems, the model failed below row 25 as a result of the depth limitations of Equation 6.13, predicting negative f values as flow depth increased beyond 0.83 cm.

In an attempt to solve these problems, due in each case to the inability of small-plot experiments to replicate the conditions that promote appropriate runoff volumes at the larger scale, two parameter modifications were introduced. The first involves decreasing the intercept value of 1.628 to 1.45 in the relationship between $A(x)$ and $\%P(x)$ in Equation 6.11 (Parsons & Abrahams 1989b, Appendix 1). This has two effects: final infiltration rates are reduced thereby increasing runoff, and times to runoff become smaller, thus initiating more runoff earlier. The second modification applies to the *fb* parameter in Equation 6.13. The limit of this equation in predicting $f = 0$ at $d = 0.83$ cm on the small plots is an unrealistic restriction for the large runoff plot, where maximum depths were observed between 1.7 cm and 1.9 cm. The *fb* parameter was therefore reduced to 8.0, allowing for a maximum depth of 1.8 cm to occur before the model failed. The model was run again using modified parameters.

Run 2 Infiltration parameters

$$A(x) = 1.45 - 0.014\,\%P(x) \tag{6.14}$$

$$B(x) = 0.785 + 0.021\,\%P(x) \tag{6.12}$$

Friction parameters

$$f(x,t) = 14.46 - 8.0d(x,t) \tag{6.15}$$

Figure 6.3 (run 2) illustrates the much improved performance of the model for total site discharge, and for S1 and S2 hydrographs. However, it was found impossible to correct for the runoff timing component in the first few minutes, with the result that comparisons between observed and predicted runoff are ignored for minute 4.

Section S1 and S2 distributions of hydraulic variables

During the analysis of model simulations of hydraulic variables for comparison with observed, published data, it became apparent that the results at the section scale were very sensitive to the method of calculation chosen. In addition, the determination of section discharge and hydraulic values is approached in different ways in the field experiments and in the model. Appendix 2 explains these differences.

It should be recognized that, as Appendix 2 indicates, the source data for derived hydraulic variables are almost entirely dependent on the measurement of flow depth, with no field data at the site or section level to indicate real variability in flow velocity or resistance. It is therefore essential that the simulated distributions of flow depths are consistent with field expectations, if subsequent extension of the model to consider variability in velocity and friction is to be valid.

Relative frequency distributions of simulated values of all depths (including

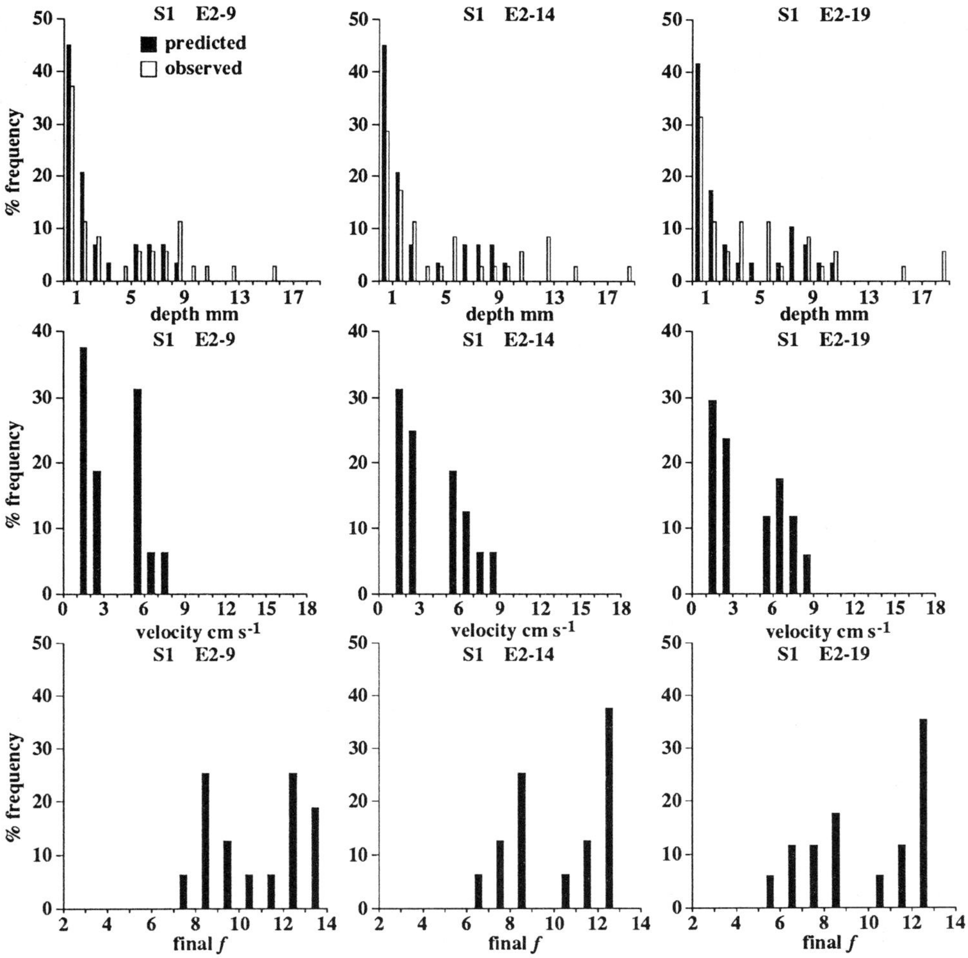

Figure 6.4 Distributions of observed and predicted (run 2) flow depth, velocity and final friction factor for section S1.

$d < 1$ mm) were compared with observed depth distributions for sections S1 and S2 (top graphs, Figs 6.4 & 5, respectively) for minutes 9, 14 and 19, showing reasonably similar overall characteristics, described by a negative exponential distribution. The model is less successful in predicting the very largest depths, especially on S1. Maximum predicted depth was 14 mm compared with 19 mm observed. However, the overall distributions are considered satisfactory, giving some confidence in the predicted distributions of velocity and final friction factor *f* (middle and lower graphs in Figs 6.4 & 5, respectively). Percentage frequency in these cases is relative to the total number of cells with $d \geq 1$ mm.

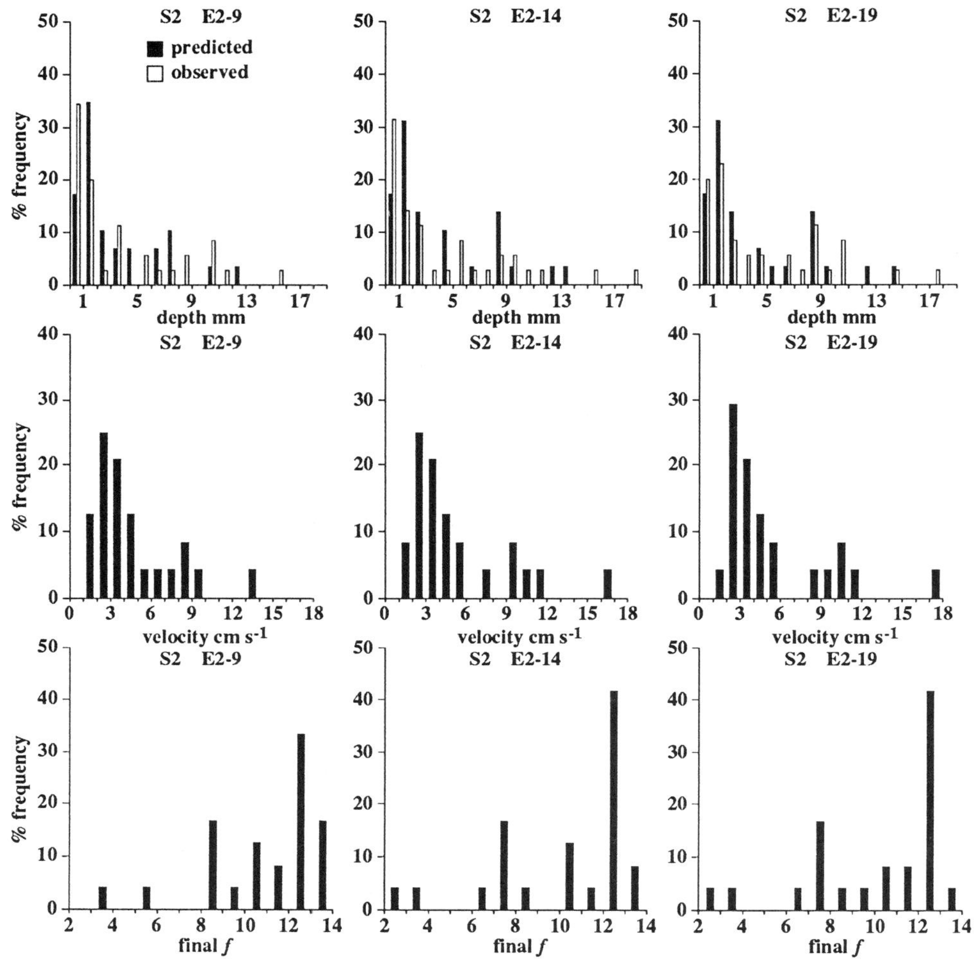

Figure 6.5 Distributions of observed and predicted (run 2) flow depth, velocity and final friction factor for section S2.

For both S1 and S2 there is a strong bimodality in velocity, which increases through time, and particularly so on S2, indicating the dual nature of flow on this site. Swifter, deeper threads of flow are interspersed with areas of shallower, slower runoff, corresponding to the main flowlines and their contributing areas. At S1 the final friction factor is also bimodal, with peaks in f around 7 (flowlines) and 12 (contributing areas). S2 also peaks at $f = 12$, but a significantly higher proportion of cells are in the contributing-area group than on S1, and there is a broader distribution of low f values. This is consistent with the field observations (Parsons et al. 1990, Abrahams et al. 1991), suggesting that in moving down slope from S1 to S2 the number of flowlines diminishes, as they begin to converge, increasing the area of higher resistance and shallow, slow flow.

At-a-section and downslope hydraulic relationships

Table 6.3 presents hydraulic variables for S1 and S2 through time (minute 9, 14, 19; see Appendix 2 for methods of calculation). The different methods used to determine section hydraulic values are presented for two main reasons: first, direct comparisons between observed and simulated data must be made using comparable methods of calculation (method 1 is the source data derived from the source method, and method 3 shows the directly comparable predictions using the source method). Secondly, the differences between source and model approaches to section hydraulics highlight important issues for the development of closer links between field empiricists and modellers in terms of the provision of data from the field or from model simulations (how estimates of section means are derived) and their interpretation (what does the concept of section hydraulics represent in describing flow dynamics?). A comparison of source method 1 and simulated, cell-based section means from method 2 illustrates both these issues through their different interpretations of hydraulic relationships at-a-section.

At-a-section hydraulic relationships are defined as section S1 or S2 means through time, while downslope hydraulic relationships identify changes in hydraulic variables between S1 and S2. Parsons et al. (1990) discuss the observed changes in hydraulic relationships both at-a-section and down slope, from data for sections S1 and S2, while Abrahams & Parsons (1991b) develop the arguments for non-linearity in friction effects. Although there are absolute differences between observed and simulated mean section values (Fig. 6.6 illustrates comparable methods 1 and 3), the general trends in depth, velocity and friction factor with increasing Reynolds number are adequately predicted by the model for both the at-a-section and downslope situations. On the other hand, the cell-based approach to calculating section values (Table 6.3, method 2) suggests an alternative interpretation. While at-a-section and downslope hydraulic trends are similar for observed and predicted *d–Re* and *V–Re*, and for downslope relationships between *f* and *Re*, the at-a-section *f–Re* relationship predicted by method 2 is negative and the reverse of the source method.

There are three possible explanations for this discrepancy, other than a poor model, associated with the source method employed by Parsons et al. (1990), while a fourth reflects a potential limitation of the model.

(a) By deriving mean section velocity from mean section depth and total discharge, and noting the strong non-normality in the depth distribution, any bias in the estimate of depth is thrice reflected in the estimate of the

Table 6.3 S1 and S2 hydraulic variables.

minute	method†	inundated width m	mean depth cm	mean velocity cm s^{-1}	total discharge cm^3 s^{-1}	mean *f*	mean *Re*
S1							
4	1	9.5	0.4168	4.024	1593.5	10.58	588
9	1	11.0	0.5984	4.362	2871.3	12.93	915
9	2	10.4	0.4343	3.804	2106.1	12.31	763
9	3	10.4	0.4419	4.474	2048.3	9.07	693
14	1	12.5	0.6090	4.464	3397.9	12.56	953
14	2	10.4	0.4897	4.199	2662.8	11.43	971
14	3	10.4	0.4982	4.560	2353.6	9.85	796
19	1	13.0	0.6053	4.386	3451.0	12.44	931
19	2	10.4	0.5060	4.319	2833.0	11.17	1032
19	3	10.4	0.5146	4.582	2444.0	10.08	827
S2							
4	1	8.0	0.2828	5.309	1201.0	4.82	526
9	1	11.5	0.5187	6.086	3630.5	6.72	1107
9	2	14.6	0.4003	4.565	3305.5	9.23	888
9	3	14.6	0.4070	6.097	3631.0	5.28	870
14	1	12.0	0.5829	6.224	4353.8	7.22	1272
14	2	14.6	0.4652	5.253	4729.4	8.11	1256
14	3	14.6	0.4732	6.231	4314.0	5.86	1034
19	1	13.0	0.6053	6.328	4979.1	7.26	1343
19	2	14.6	0.4797	5.412	5045.7	7.87	1342
19	3	14.6	0.4880	6.259	4468.1	5.98	1071

† See Appendix 2
1 Source data using source method (Parsons et al. 1990 and unpublished data)
2 Model method, using cell data to determine section means
3 Source method applied to model data, using modified mean depth rating curve discharge and average section V

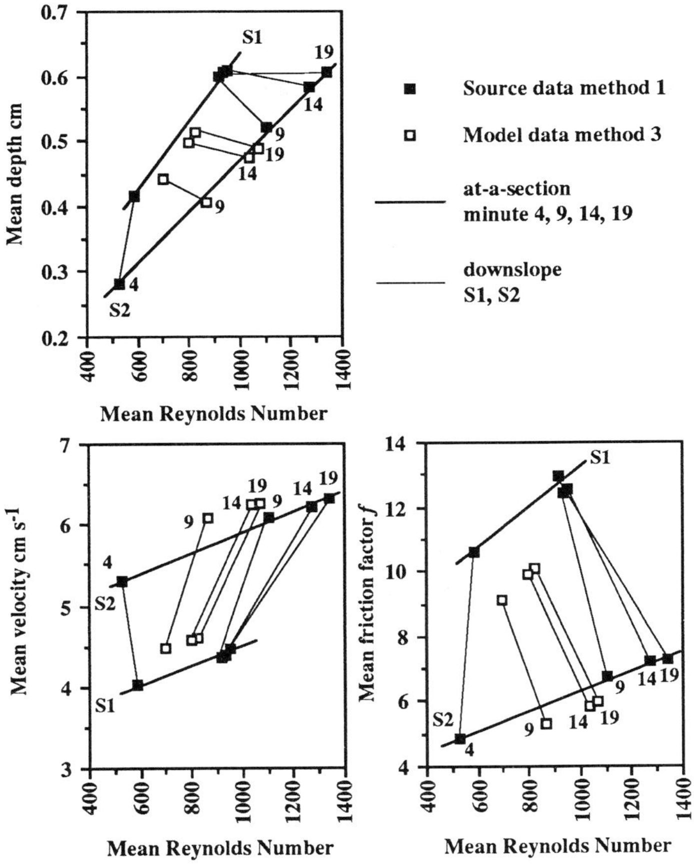

Figure 6.6 Observed and predicted (run 2) at-a-section (minutes 4, 9, 14, 19) and downslope (S1, S2) hydraulic relationships.

Darcy–Weisbach friction factor, through the dependence of velocity on depth, and of the friction factor on velocity and depth.

(b) Mean section velocity is a derivative of mean section depth, so the degrees of freedom associated with velocity are reduced to zero. In other words, mean velocity is overconstrained, and the mean friction factor is additionally so.

(c) The determination of mean section depth involves the inundated width through the ratio of inundated width to site width. The same inundated width is used again to estimate mean section velocity, thus entering the determination of the friction factor three times.

(d) The inundated width predicted by the model, for both S1 and S2, under the

> constraints of the infiltration parameters derived from the small-plot experiments, reaches a constant equilibrium width before 9 min under 97 mm h^{-1}, whereas the field measurements on the large plot show inundated width increasing up to minute 19 of E2. Consequently, the effect of increasing inundated width cannot be assessed in the model predictions of changes of friction factor on the rising limb of the hydrograph.

Since the model cannot fully reproduce the conditions for evaluating the observed at-a-section *f–Re* relations, nor can the observed relations be disaggregated from the method used to determine *f*, it can only be concluded that further field experiments are required to substantiate the positively trending *f–Re* relationship at the section scale. However, it is worth noting that there are physical reasons to expect positively trending *f–Re* relations (see Ch. 1).

Site variability of hydraulic variables

Despite the inconclusiveness of the model results at the section scale for at-a-section relationships through time, the model does appear to reflect the observed downslope hydraulics between S1 and S2. Extending these simulated results to consider all possible sections at equilibrium (rows 1–45 of Fig. 6.2), Figure 6.7 (top graph) illustrates the spatial pattern downslope of the inundated width, shedding some interesting light on the full nature of simulated downslope changes in hydraulics at the plot scale. Inundated width, as expected, is small at the upper part of the plot, increases to a maximum around row 27, marking a zone of more-or-less parallel flowlines, and then decreases down slope as flowlines converge with each other. This analysis, therefore, suggests that S1 and S2 belong to two different hydraulic zones. These predicted changes in inundated width are in accord with observed changes during the experiments on the large plot (see Ch. 13).

This two-zone pattern, suggested by the downslope differences in inundated width (Fig. 6.7), is reflected in all predicted section hydraulic variables (and is more evident for extreme values associated with flowline extension down slope): the rate of change of flow depth is greater up slope of row 27, becoming more gentle in the zone of flowline convergence, while velocity reverses this pattern, having a slightly gentler rate of change in the upper part. Friction declines more rapidly in the upper zone, while Reynolds numbers indicate the effect of flowline convergence by marked discontinuities at row 27 and again at row 43. In all cases a second-order polynomial fits the predicted data better than any linear relationship with distance down slope.

The relationships among all predicted row hydraulic variables are presented in Figure 6.8, and indicate that, at the plot scale, downslope flow hydraulic changes are accommodated more by increases in velocity (log–log regression

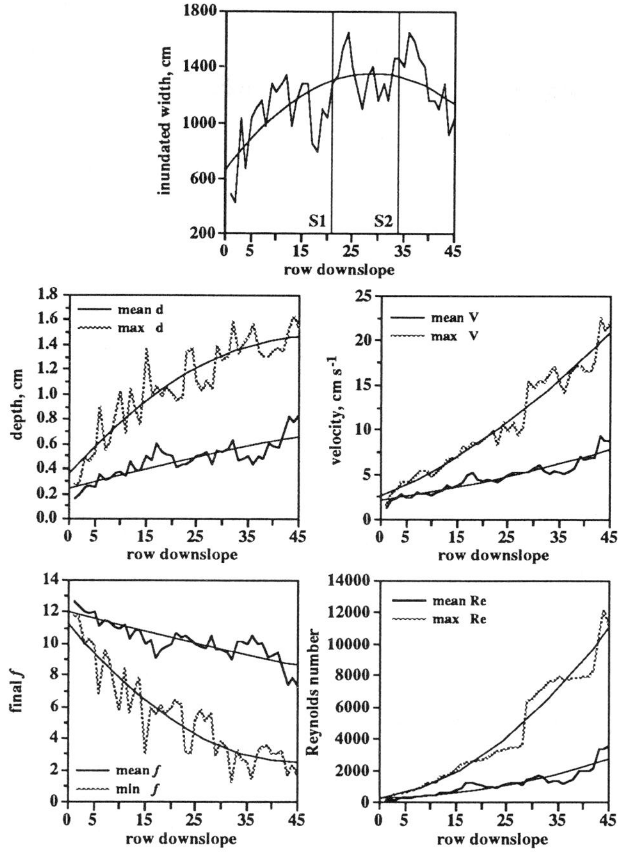

Figure 6.7 Relationships between predicted (run 2 at t = 19 min) mean hydraulic variables (width, depth, velocity, final friction factor and Reynolds number) and distance down slope.

slope = 0.488) than by increases in depth (regression slope = 0.399), as Reynolds number rises. In other words, a ten-fold increase in *Re* down slope is associated with a three-fold increase in velocity and a 2.5-fold increase in depth. Although these predictions underestimate the downslope increase in velocity obtained in the empirical data, which show that all the increase in *Re* is accommodated by increase in velocity (Parsons et al. 1990), they do support the notion that downslope concentration of flow results in increased downslope hydraulic efficiency.

With some confidence in the model's ability to predict spatial hydraulic variability which reflects the observed field data at equilibrium, the full site simulation results are presented. Figure 6.9 illustrates the predicted spatial distribution of the final friction factor $f(x,t)$ at equilibrium (t = 19 min), and Figures 6.10 & 6.11 show the spatial patterns of flow depth and velocity,

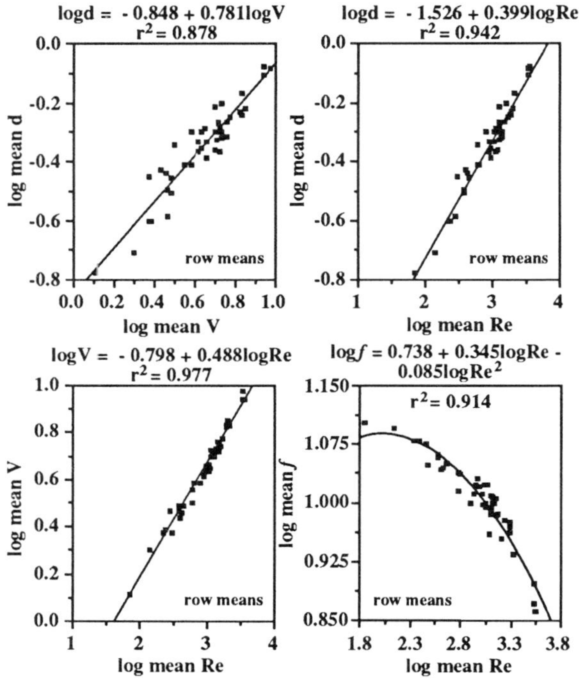

Figure 6.8 Log–log relationships between predicted (run 2 at t = 19 min) mean hydraulic variables (*d–V*, *d–Re*, *V–Re* and *f–Re*).

respectively. The distinctive pattern of an upper zone (up slope of row 27) of nine separate threads of subparallel flow, slower and shallower than the five converging threads of deeper, swifter flow in the lower zone, becomes manifest.

Flow regimes

The standard log*f*–log*Re* relationship for river flow in channels (e.g. Chow 1959) describing flow regime and bed-roughness influences on the *f–Re*

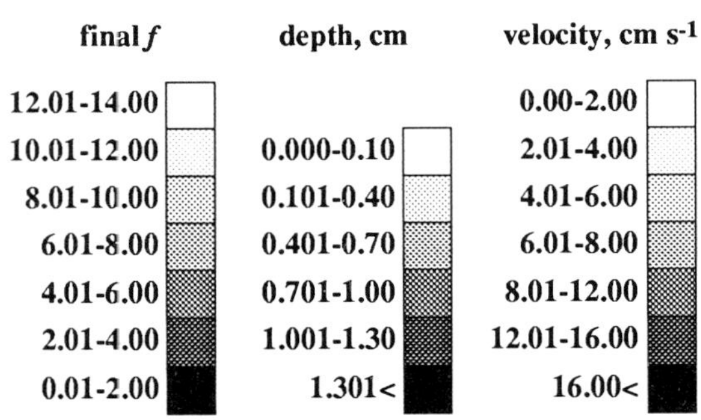

Key for Figures 6.9–6.11

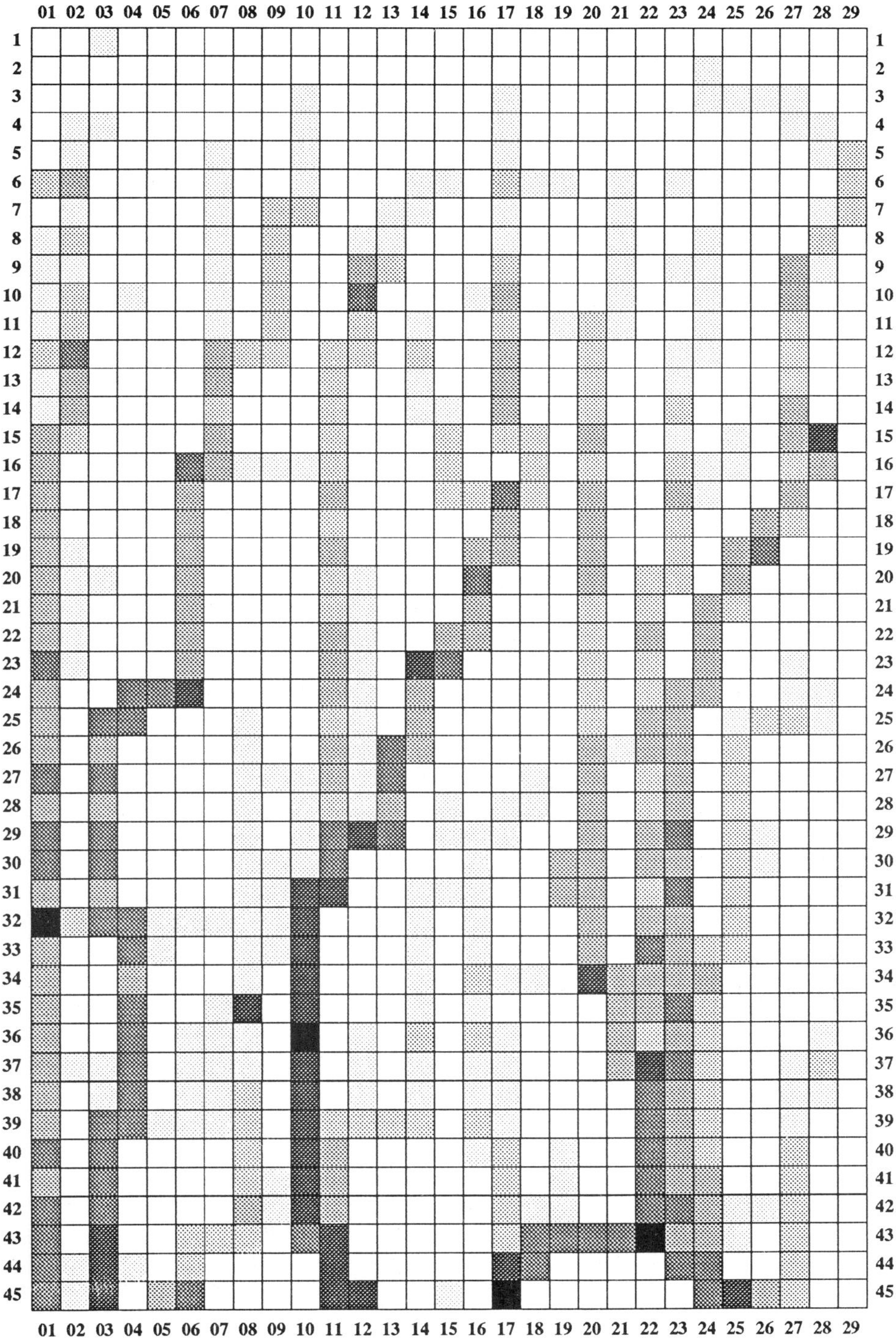

Figure 6.9 Spatial pattern of predicted (run 2 at t = 19 min) final friction factor f on Arizona runoff plot.

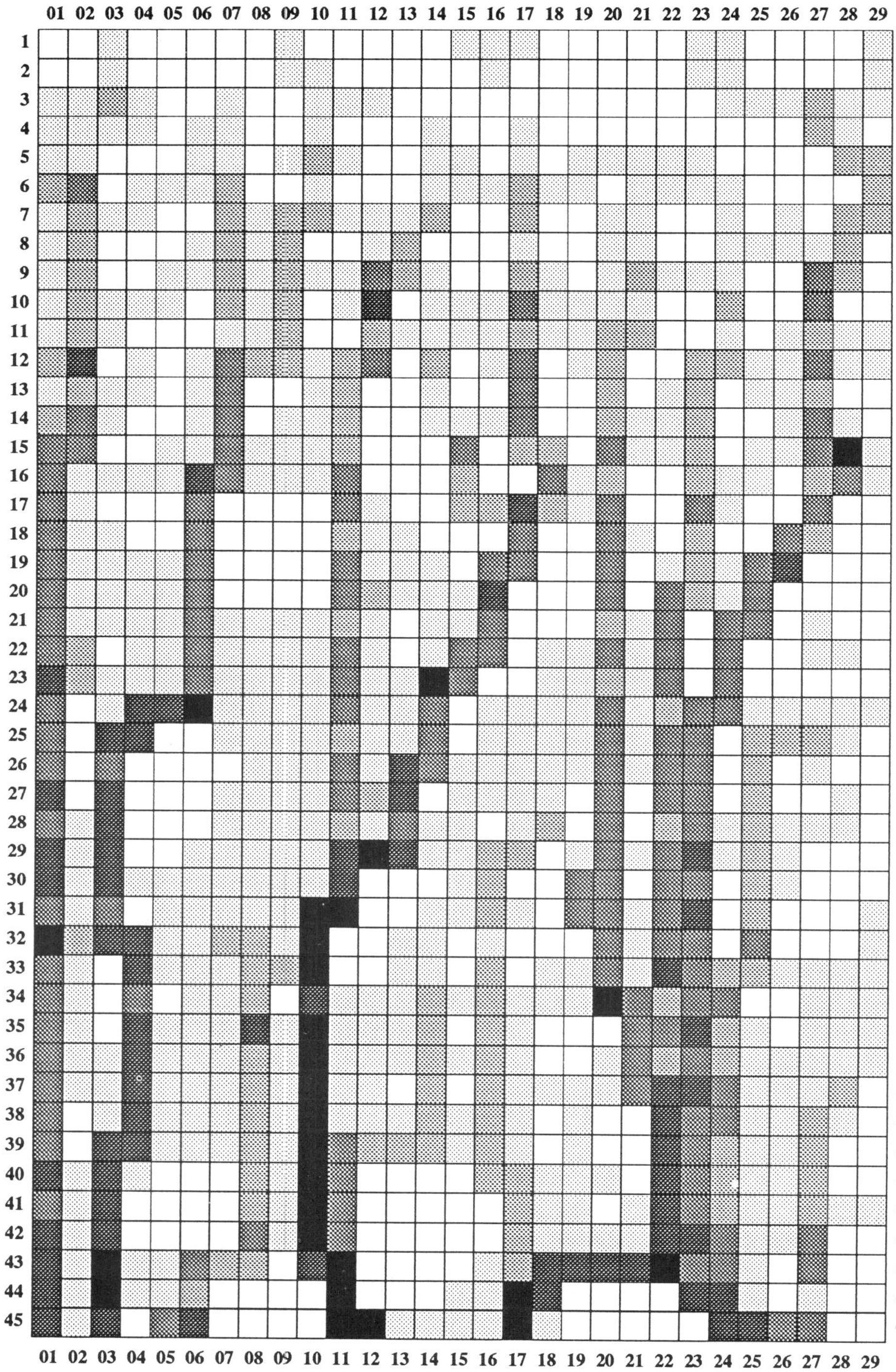

Figure 6.10 Spatial pattern of predicted (run 2 at t = 19 min) flow depth on Arizona runoff plot.

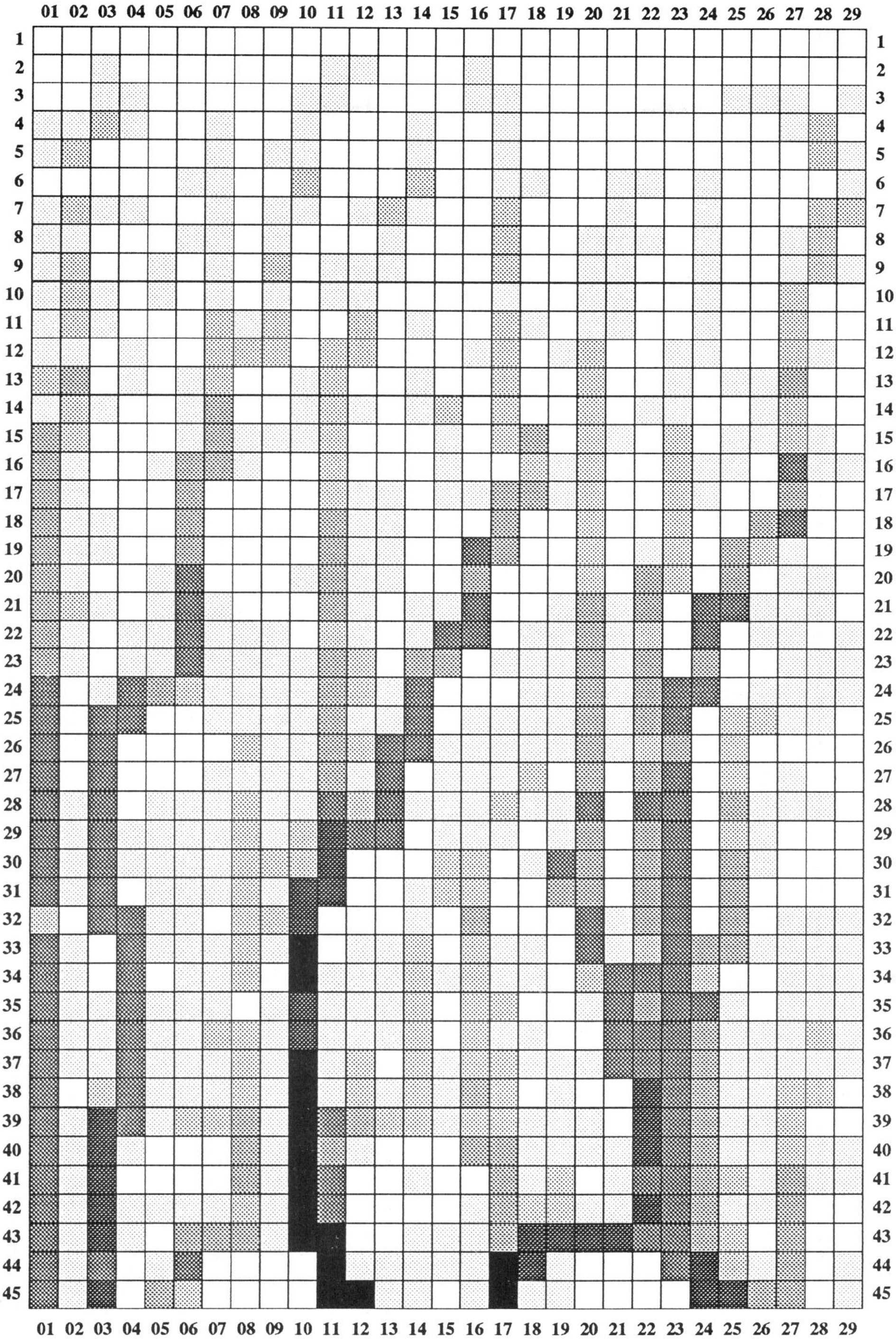

Figure 6.11 Spatial pattern of predicted (run 2 at t = 19 min) flow velocity on Arizona runoff plot.

relationship, is not supported by field observations nor by model simulation data. Figure 6.12 illustrates the relationship between predicted cell equilibrium friction factor and Reynolds number in log–log form (top graph), indicating a clear departure from channel flow relations. This shows a non-linear relationship which describes a continuum between a regime of high *f* values, which are relatively insensitive to increases in Reynolds numbers (associated with non-flowline or contributing cells), and a regime in which *f* falls rapidly with increasing *Re* (those cells that concentrate runoff spatially into flowlines). This pattern is reflected in the second graph of Figure 6.12, showing the log–log relationship between flow depth and flow velocity. At low flows, which occur in all contributing cells,depth increases more rapidly than velocity, but at high flows, which occur only in the flowline cells, velocity increases more rapidly than depth.

Rather than interpreting these *f*–*Re* regimes for overland flow according to the

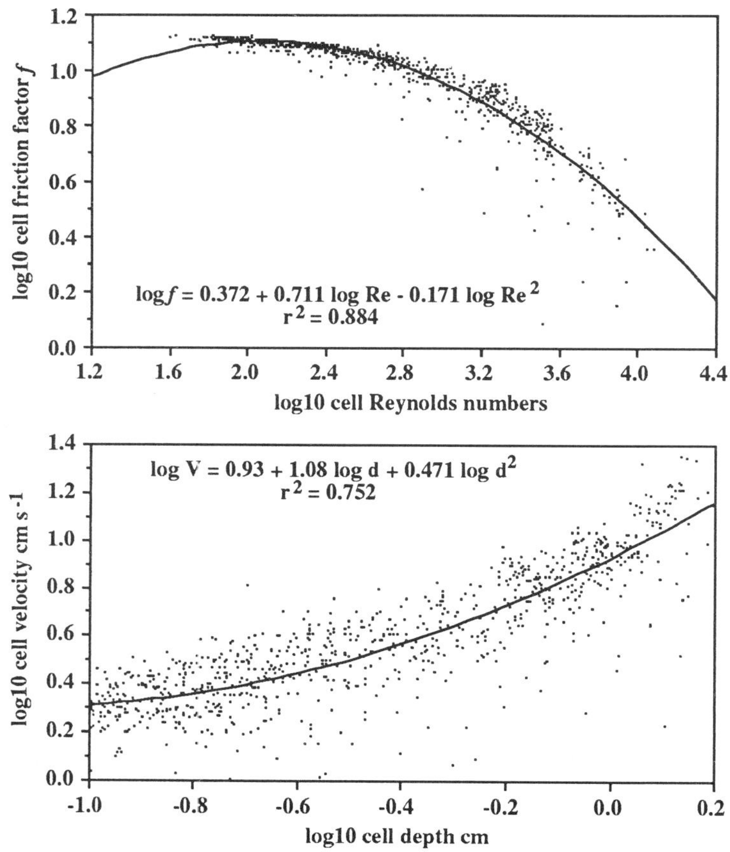

Figure 6.12 Log–log relationship between predicted (run 2 at t = 19 min) cell hydraulic variables (*f*–*Re*, *V*–*d*).

traditional channel model of laminar-turbulant hydraulics a different interpretation identifies low flow and high flow regimes as representing spatial distinctions between areas which contribute to a flowline and the deeper, faster threads or flowlines. One reason for this is that laminar conditions on rough gravel hillslopes under rainfall impact are highly unlikely. Instead the low-flow condition is represented by two situations: in the upper zone of many parallel flowlines, numerous, narrow intervening areas of shallow flow occur which are unable to overcome initial resistance and which, therefore, remain through a storm as essentially barely flowing, surface-depth concentrations or ponding; in the downslope zone, more pronounced spatial discontinuities are produced by larger, but fewer, converging flowlines, separated by fewer, but wider areas of shallow flow. In addition, in this low-regime zone the predicted relationship between *f* and *Re* is limited by the field data, and hence Equation 6.15, in so far as initial friction, *fa*, is a constant for all pavement and slope conditions, whereas in reality much more scatter is likely in this zone, as variability in initial resistance reflects variations in surface properties.

It is interesting to note that the significant, second-order polynomial fitted to the log*f*–log*Re* relationship indicates precisely this variability where the experimental data fall short. Below Reynolds numbers of about 100, an increase in *f* is predicted by the equation with increasing *Re*. This is equivalent to suggesting that for cells which do not yield much runoff (the low percentage pavement areas, specifically the mounds of fine material) initial friction is low ($f = 2.35$ when $Re = 1$), and increases as *Re* rises, as a result of depth concentration, to a maximum $f = 13$–14 at Reynolds numbers around 100–250. Of these cells, only those with high percentage pavement serve to concentrate runoff in flowlines, and in these cells *f* declines rapidly as depth, velocity and Reynolds numbers all increase in response to lower infiltration rates, flowline development and flowline convergence down slope.

The interpretation of this non-linear *f*–*Re* relationship explicitly identifies the significance of spatial position of a cell, not as a simple function of distance down slope but as a functional location within the hydraulic cascade, a function that depends both on its local characteristics (onsite) and on its spatial context with its neighbours. Thus non-flowline or contributing cells may occur at any downslope location, while flowline cells require both lower onsite infiltration rates and inflow from upslope areas in order that they may develop. Hence it is evident that small-plot experiments, which do not account for this functional position, cannot be used successfully to aid our understanding of overland-flow hydrology and hydraulics; and that erosion predictions that are based on average estimates of flow character will fail entirely to distinguish spatial discontinuities resulting from the differences between unconcentrated and concentrated flow hydraulics.

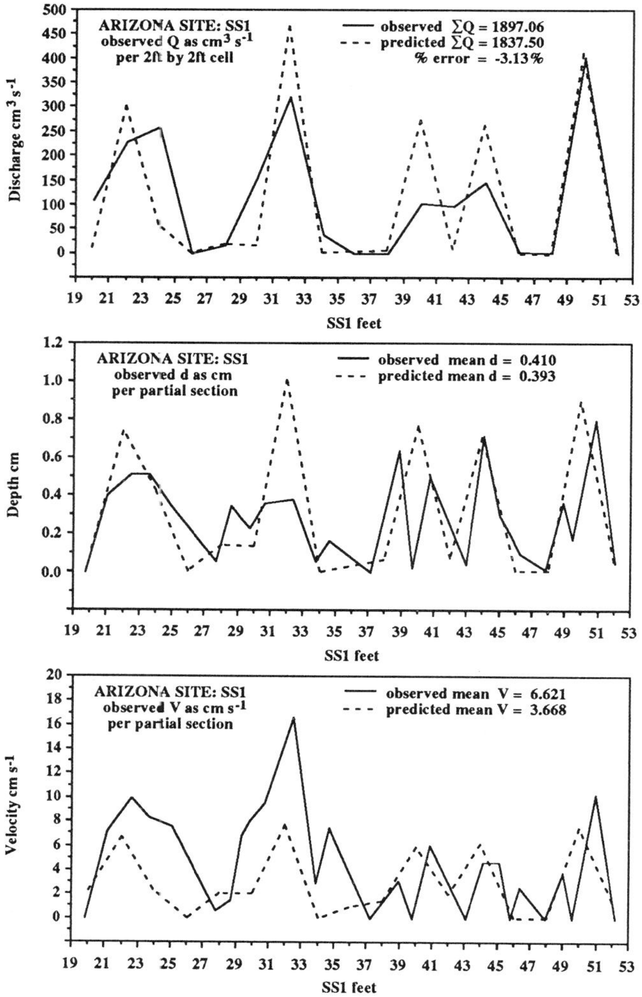

Figure 6.13 Observed and predicted (run 2 at t = 19 min) discharge, depth and velocity across subsection SS1.

Subsection SS1 and SS2 spatial distribution of hydraulic variables

Detailed equilibrium depth, velocity and discharge data for two 10 m subsections (SS1 and SS2), forming part of S1 and S2, were collected in the field at the end of E4. Despite the reported simulated rainfall input of 68 mm h^{-1} for this experiment, compared with model rainfall of 97 mm h^{-1}, these data are compared for three reasons. The first suggests either that rainfall input to E4 may have been higher than measured, or, more likely, that the effects of crusting on the site, despite a lower rainfall input, resulted in similar conditions, since there is good agreement between model and observed results. The second reason is to show the considerable strength of the model in

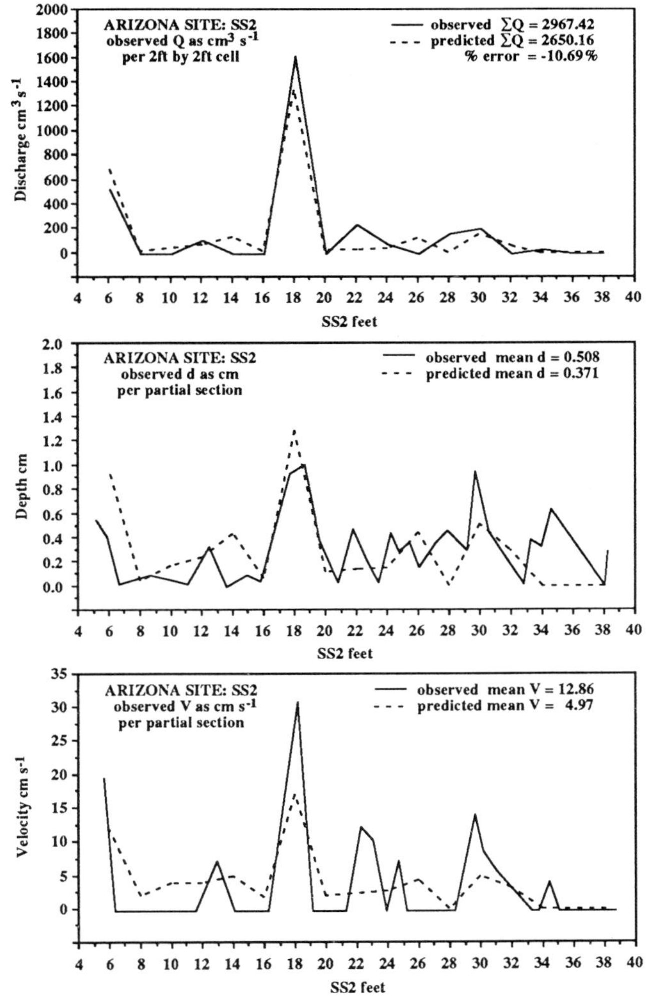

Figure 6.14 Observed and predicted (run 2 at t = 19 min) discharge, depth and velocity across subsection SS2.

predicting variable runoff across a subsection, drawing attention to the pattern of dominant flow threads interspersed with areas of low or no flow, and the change in the ratio of flowline area to non-flowline area between S1 and S2. Figures 6.13 & 6.14 illustrate observed and predicted discharge, depth and velocity patterns for SS1 and SS2, respectively. Thirdly, these experimental data are unique in providing some idea of measured velocities, against which the model predictions may be judged.

Errors in total discharge predictions across the subsections are −3% and −10% for SS1 and SS2. However, absolute subsection mean depths and velocities are significantly underestimated as a function of the observed data being collected for very small partial areas (see Table 2 in Parsons et al. 1990), between 6 cm and 100 cm, the majority being less than cell width, 61 cm. In order to compare these data, observed discharge was aggregated to the cell level, but measured velocity and depth data remain at the disaggregate level. The spatial resolution of the model is unable to predict the extreme values at subcell widths (e.g. SS1 at 32 ft; SS2 at 18 ft), but where partial section widths approach cell dimensions, the predictions are fairly convincing. More important, perhaps, than absolute values is the ability to predict the spatial distribution of flowlines across each subsection.

Conclusions

Conventionally, overland flow on interrill areas has been modelled as a sheet of flowing water, of unit width, focusing on downslope increases in depth, velocity and erosivity. The field experiments on the large Arizona site challenged this view, demonstrating that the sheet-flow approach is quite inappropriate for natural surfaces, where water forms rivulets which join in a dendritic pattern. Abrahams et al. (1986a) described flow as "a shallow sheet of water with threads of deeper, faster flow diverging and converging around surface protuberances, rocks and vegetation". Parsons et al. (1990) argued that "such lateral and downslope variations in overland flow need to be incorporated into models of interrill overland flow so that these models may become more realistic and, hence, better predictors of runoff hydrographs and sediment yield", while Abrahams et al. (1989, 1991) assessed the significance of such variability for soil erosion, and hence pointed the way for flow model development to predict erosion.

The model presented here - based only on three key control variables: percentage pavement in determining infiltration rates and the timing of runoff initiation; slope gradient in controlling the macro-routing of flow; and a dynamic friction factor which reflects flow depth in relation to initial roughness - appears to have some success in simulating the reality of overland flow on desert pavements described by these authors. However, the results suggest a number of limitations, in part associated with model conceptualization, others related to parameterization from field experimental data. The model is demanding in its field data requirements; in particular it has attempted to predict beyond the validity of data collected by Parsons, Abrahams and Luk. Rather than being a criticism, this allows identification of the required range of field experimentation. Specifically, more information is needed on infiltration processes and the influence of surface properties on friction, within an appropriate spatial context. Small-plot studies provide only limited approximation to the large-scale dynamics of strongly converging flowlines. The model's major problem lies in its inability to marry small-plot data with the large-plot dynamics.

Although the results give some confidence in the model's ability to predict the spatial and temporal patterns of converging flow, this lack of coherence between the two scales means that it is impossible to disentangle errors in the model from process differences between small plots and the large site. Nowhere is this more apparent than in the infiltration and runoff generation model component. It is suggested that the Green and Ampt infiltration parameters derived from the 1 m by 1 m plots actually fail to represent infiltration processes on the main site. In the latter, runon from upslope flowlines combined with spatial contiguity of high percentage pavement areas will substantially reduce runoff generation times and steepen the rate of decline of infiltration, leading to a much more rapid hydrograph rise in the early stages.

The model is capable of predicting spatially and temporally dynamic flowline hydrologic and hydraulic behaviour. In addition, it suggests an *f*–*Re* relationship which is very different from the traditional channel model. At the at-a-section scale, increased resistance to increasing Reynolds numbers is indicated, resulting from lateral spreading of flow, while downslope resistance falls with increasing Reynolds numbers due to increasing flowline concentration. At the site scale at equilibrium, resistance falls rapidly only for cells that are able to concentrate and route flow efficiently as deep, high-velocity flowlines, characterized by their high percentage pavement cover and connectivity with other such cells in an effective hydrologic and hydraulic cascade. Such dynamics reflect the differences between unconcentrated and concentrated overland flow, or between a flowline and its contributing area. Furthermore, the model predicts a non-linear, equilibrium log*f*–log*Re* relationship, where friction increases initially for non-flowline cells (where flow depth accumulation is the major process), reaches a plateau of maximum resistance for the majority of cells where flowline development is inhibited (through high infiltration or lack of pavement contiguity), and which only falls steeply for those cells which become the deep, swiftly flowing threads when flowline development is maximized and velocity increases become more important.

Further development of the dynamic flowline model, however, is limited in three main areas. First, by the constraints of the existing field data, since there is no adequate (i.e. independent) assessment of the friction factor for a true rising hydrograph. Data derived from both sets of small-plot experiments are particularly prone to problems because neither has included low values of $\%P$

and, in the case of the infiltration plots, has not replicated the processes of infiltration that occur on the large site, resulting in *ad hoc* adjustments of regression parameters. The second limitation is the dependence of the model results on an arbitrary basic spatial unit. Data from the field, which are collected at a different spatial scale, must be "smoothed" in the case of subsection variables, and model predictions are limited to that resolution in the assumption of uniform flow over the full cell width. Finally, the conceptual constraints are recognized in a pseudo-dynamic model, where infiltration is only loosely coupled with runoff, subsurface processes are ignored, and area-specific depths and velocities have no dynamic interaction with gravel–rough surfaces through micro-topography.

Acknowledgements

The first author is indebted to Athol Abrahams and Tony Parsons for allowing their data to be scrutinized in order to apply and evaluate the model. Earlier development of the model was supported by a Natural Environment Research Council postgraduate studentship.

Appendix 1

Parsons & Abrahams (1989b) reported that their original small-plot infiltration experiments overpredicted final infiltration rates, fc in mm h^{-1}, when applied to equilibrium discharges on the large runoff plot. Their solution was to reduce the intercept value in the relationship between final infiltration rate, fc, and percentage stone cover on a plot, $\%SC$.

$$fc = 132.6 - 1.434\%SC \quad \text{original} \quad (6.16)$$
$$fc' = 96.54 - 1.434\%SC \quad \text{revised.} \quad (6.17)$$

With a close relationship ($r^2 = 0.888$) between percentage stone cover and percentage pavement $\%P$, described by

$$\%SC = 4.77 + 0.6881\%P \quad (6.18)$$

it was possible to introduce a modification to the intercept value in the relationship between the Green and Ampt A (final infiltration rate parameter) and $\%P$ which, in combination with the $B - \%P$ relationship, predicted modified final infiltration rates, i', in close agreement with these authors' estimates.

$$A = 1.628 - 0.014\%P \quad \text{original} \qquad (6.11)$$
$$A = 1.450 - 0.014\%P \quad \text{revised} \qquad (6.14)$$
$$B = 0.785 + 0.021\%P \qquad (6.12)$$
$$i' = A + B/t. \qquad (6.1)$$

It should be noted that the modified intercept in the fc'–$\%SC$ relationship results in negative infiltration when $\%SC > 67\%$ or when $\%P > 95\%$. Ignoring these negative values, Figure 6.15 shows the effect of both modifications in predicting final infiltration rates across a range of pavement conditions, and their comparability.

$\%P$	fc'	i'
0	89.7	87.3
50	40.8	44.7
100	−7.9	3.0

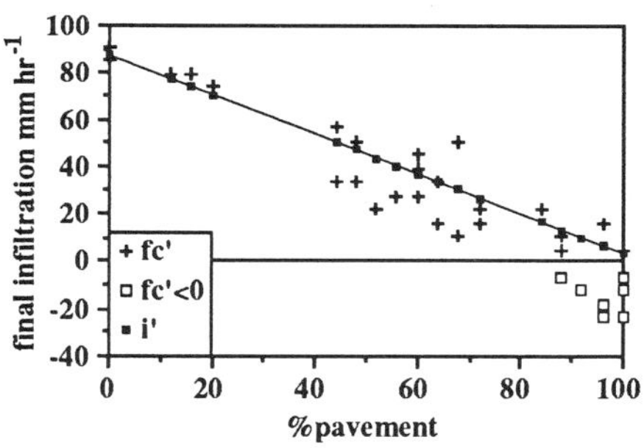

Figure 6.15 Comparison of final infiltration rate and $\%P$ using modified fc'–$\%P$ intercept (Parsons & Abrahams 1989b) and modified Green and Ampt A–$\%P$ intercept (Eq. 6.14) in $i' = A+B/t$.

At the design intensity of 80 mm h^{-1} for the rainfall simulator, the fc' modification predicts a threshold pavement of 10%, less than which all rainfall infiltrates, while the A modification for i' predicts that threshold at 9$\%P$.

Appendix 2

In the field, sample depths were measured at 0.5 m intervals across a section, and are used to determine mean section depth, which is then used as the governing parameter in a negative exponential depth distribution. These generalized depths, in combination with a depth–discharge rating curve, developed for each section using data from miniature flumes, and inundated width measurements, yielded total section discharge. From a knowledge of total section discharge, section mean depth and inundated width, section mean velocity was determined through continuity principles as

$$\bar{V}' = \frac{\Sigma Q''}{\bar{d}'\Sigma W'}. \tag{6.19}$$

Further derivatives of these flow variables were determined at the section scale as

$$\bar{f}' = \frac{8g\bar{d}'\bar{S}'s\cdot}{\bar{V}'^2} \tag{6.20}$$

and

$$\bar{Re}' = \frac{4\bar{d}'\bar{V}'}{\nu}. \tag{6.21}$$

In contrast to this, the model simulates all flow and hydraulic variables at the cell scale, and section values are derived from the full distribution of cell values, as, for example

$$\bar{d} = \frac{\Sigma d}{n}. \tag{6.22}$$

In addition, there are two minor complications. First, the difficulty of measuring flow under field conditions imposed a measurement resolution of 1 mm flow depth. The model, however, has no such restriction. Since all cells with flow will contribute to predicted section discharge, all such cells should contribute to section means. The effect of this is to reduce substantially these estimates of section means. An alternative approach is to apply a 1 mm depth filter to the simulated data and calculate section means from only those cells where depth would have been measured in the field. This approach violates the continuity principle since $\bar{d}\bar{V}\Sigma W$, based on cells where flow depth ≥ 1 mm, no longer equals total section discharge. The second difficulty revolves around differences between the field measurement strategy, in which depth readings were taken at 0.5 m intervals along a section, taking no account of flow direction, and the model routing strategy, in which cells with *IRECT* values of 3 or 4 (horizontal routing within a section) must be discounted if they are not to be doubly included in the calculation of section discharge. In order to maintain comparability with the field data collection strategy, it was decided that these cells would be included in the determination of section means but not in the calculation of section discharge.

Table 6.3, concerned with mean section flow and hydraulic data, compares the source data based on field measurements of depth and inundated width during E2 with two sets of simulated section means. In method 1 the source data are determined from the source method. In method 2 simulated section discharges and section hydraulic means are obtained using the full distribution of cell values for depth, velocity and discharge. Section discharge is the sum of all cell discharges where *IRECT* = 1 or 2 (thereby avoiding double counting where flow is across a cell). Section hydraulic means are calculated as

$$\bar{X} = \frac{\Sigma X}{n} \tag{6.23}$$

where X denotes a hydraulic variable; n is the number of cells with depth $\geq$ 1 mm. Method 3 determines simulated section discharges and section hydraulic means using the source method applied to cell depth and inundated width data, applying a 1 mm depth filter.

Method 1 Source data and source method (Parsons et al. 1990), sections S1 and S2

Flow depth, d', was measured across each section at 0.5 m intervals and each value of d' was taken to be representative of the 0.5 m section in which it was centred. Hence total inundated width was determined as:

$$\Sigma W' = n_1 W_1 \tag{6.24}$$

and mean section depth $\bar{d}'$ was calculated as:

$$\bar{d}' = \frac{\Sigma d'}{n'} \frac{W_2}{\Sigma W'} \tag{6.25}$$

where d' = sample depth $\geq$ 1 mm;
n' = number of samples (= 35);
$\Sigma W'$ = inundated width (= $n_1 W_1$);
n_1 = number of samples with $d' \geq$ 1 mm;

W_1 = sample width (= 0.5 m);
W_2 = total plot width (= 18 m).

A negative exponential distribution with mean depth $\overline{d}'$ as parameter was fitted to the depth data for S1 and S2 at minutes 4, 9, 14 and 19 (Abrahams et al. 1989, Parsons et al. 1990) as a general model of the distribution of depths d'' across a section.

Using the negative exponential distribution to characterize the depths d'', and a rating equation for $Q^{\dagger}-d^{\dagger}$ derived from the miniature flume data for each section, total discharge $\Sigma Q''$ was determined by substitution of $d^{\dagger}$ by d''. For S1 the rating curve is given by:

$$\log Q^{\dagger} = 1.1587 \log d^{\dagger} - 0.4825 \tag{6.26}$$

and for S2 it is given by:

$$\log Q^{\dagger} = 1.141 \log d^{\dagger} - 0.3179. \tag{6.27}$$

From a knowledge of total section discharge $\Sigma Q''$, inundated width $\Sigma W'$ and mean section depth $\overline{d}'$, mean section velocity $\overline{V}'$, Reynolds number $\overline{Re}'$ and friction factor $\overline{f}'$ were derived (the last using mean section gradient):

$$\overline{V}' = \frac{\Sigma Q''}{d' \Sigma W'} \tag{6.28}$$

$$\overline{Re}' = \frac{4\overline{d}'\,\overline{V}'}{\nu} \tag{6.29}$$

$$\overline{f}' = \frac{8g\overline{d}'\,\overline{S}'s\,\cdot}{\overline{V}'^{2}} \tag{6.30}$$

where ν = kinematic viscosity;
g = gravitational constant;
$\overline{S}'S1$ = S1 mean sine slope;
$\overline{S}'S2$ = S2 mean sine slope.

Method 2 Model simulations (cell-based)
Full site discharge:

$$\Sigma Q(t) = [\Sigma Qo(x,t)]\frac{As}{Ap} \tag{6.31}$$

where $\Sigma Q(t)$ = total site discharge;
$Qo(x,t)$ = cell discharge from 25 outlet cells;
Ap = number of cells for predicted area;
As = number of cells for whole site.

Section discharge:

$$\Sigma Qs \cdot (t) = \Sigma Qc(x,t) - \Sigma Qa(x,t) \tag{6.32}$$

where $\Sigma Qs \cdot (t)$ = section S1 or S2 discharge;
$Qc(x,t)$ = cell discharge from 29 cells;
$Qa(x,t)$ = discharge from cells with *IRECT* values of 3 or 4.

Section depth, velocity, friction factor and Reynolds number In the field, measurement resolution was 1 mm. Hence this value is used as a filter for comparison between observed and predicted section values. These variables are predicted for each cell by the model as $d1(x,t)$, $V1(x,t)$ and $f1(x,t)$, respectively. Reynolds number for each cell is derived as:

$$Re1(x,t) = \frac{4d1(x,t)\,V1(x,t)}{\nu} \tag{6.33}$$

Section hydraulic variables are derived in the following way:

$$\bar{d}(t) = \frac{\Sigma d1(x,t)}{n(t)} \tag{6.34}$$

$$\bar{V}(t) = \frac{\Sigma V1(x,t)}{n(t)} \tag{6.35}$$

$$\bar{f}' = \frac{8g\bar{d}(t)\bar{S}'s \cdot}{(\bar{V}(t))^2} \tag{6.36}$$

$$\bar{Re}' = \frac{4\bar{d}(t)\bar{V}(t)}{\nu} \tag{6.37}$$

where $n(t)$ = number of cells where $d \geq 1$ mm. For subsections SS1 and SS2, where there was more time for field data collection, the field resolution, and hence prediction filter, for depth was 0.5 mm.

Method 3 Source methods applied to the model data at section scale
This method uses cell depths $d1(x,t)$ modified by source method for mean depth $\bar{d}'$, and rating curve estimate of cell discharge $\Sigma Q''$.

Inundated width
$$\Sigma W(t) = n(t)\,W. \tag{6.38}$$

Mean depth
$$\bar{d}' = \frac{\Sigma d1(x,t)}{m}\frac{W_2}{\Sigma W(t)}. \tag{6.39}$$

Total section discharge $\Sigma Q''$ is derived from the $d^{\dagger}$– $Q^{\dagger}$ rating curve, substituting $d1(x,t)$ for $d^{\dagger}$.

Mean velocity
$$\bar{V}' = \frac{\Sigma Q''}{\bar{d}'\Sigma W(t)} \tag{6.40}$$

Mean f
$$\bar{f}' = \frac{8g\bar{d}'\bar{S}'s\cdot}{\bar{V}'^2} \tag{6.41}$$

Mean *Re*
$$\bar{Re}' = \frac{4\bar{d}'\bar{V}'}{\nu} \tag{6.42}$$

where m = total number of cells in section (= 29);
W_2 = total section width (= 18 m);
W = cell width (= 60.96 cm).

Symbols used in the text

General symbols

A	infiltration parameter	cm s^{-1}
B	infiltration parameter	cm
d	flow depth	cm
ex	rainfall excess $(P\text{-}i)$	cm s^{-1}
f	Darcy–Weisbach friction factor	
fa	initial friction	
fb	rate of decline of friction with depth	
fc'	final infiltration rate	mm h^{-1}
g	gravitational constant	981 cm s^{-2}
$IRECT$	cell flowline exit direction	
i	infiltration rate	cm s^{-1}
i'	final infiltration rate	mm h^{-1}
P	rainfall rate	mm h^{-1}
Q	total discharge (dVW)	cm^3 s^{-1}
q	unit width discharge (dV)	cm^2 s^{-1}
Re	Reynolds number $(4dV/\nu)$	
S	sine slope gradient	
t	time	s
t_0	time to ponding or time to runoff	s
V	flow velocity	cm s^{-1}
W	width	cm
$\%P$	percentage pavement	
Δt	time increment	
Δx	space increment	
ν	kinematic viscosity	cm^2 s^{-1}

Source data symbols

d'	0.5 m sample depths $\geq$ 1 mm	mm
d''	depths derived from negative exponential	mm
$d^{\dagger}$	miniature-flume depths	mm
$\overline{d}'$	section mean depth	cm
$\overline{f}'$	section mean friction factor	
n'	number of samples, 35	
n_1	number of samples with $d' \geq$ 1 mm	
$Q^{\dagger}$	miniature flume discharge	
$\Sigma Q''$	total section discharge	
$\overline{Re}'$	mean section Reynolds number	
$\overline{Ss1}$	mean section S1 sine slope 0.0524	

$\overline{S}s2$	mean section S2 sine slope 0.0612	
$\overline{V}'$	mean section velocity	cm s^{-1}
W_1	sample width, 0.5 m	m
W_2	total site width, 18 m	m
$\Sigma W'$	inundated section width	m

Model data symbols

$A(x)$	Green and Ampt cell infiltration parameter	cm s^{-1}
Ap	number of cells in predicted area	
As	number of cells in whole site area	
$B(x)$	Green and Ampt cell infiltration parameter	cm
$d(x,t)$	cell depth	cm
$d1(x,t)$	cell depth where $d(x,t) \geq 1$ mm	cm
$\bar{d}(t)$	mean section depth	cm
$ex(x,t)$	cell rainfall excess	cm s^{-1}
fa	initial resistance	
fb	rate of loss of friction with flow depth	
$f1(x,t)$	cell friction factor where $d(x,t) \geq 1$ mm	
$i(x,t)$	cell infiltration rate	cm s^{-1}
$IRECT(x)$	cell flowline direction parameter	
m	number of cells in section	
$n(t)$	number of cells in section where $d(x,t) \geq 1$ mm	
P	rainfall intensity, 97 mm h^{-1}	mm h^{-1}
$\%P(x)$	cell percentage pavement	
$Qa(x,t)$	discharge from outlet cells	cm^3 s^{-1}
$Qc(x,t)$	discharge from section cells	cm^3 s^{-1}
$Qo(x,t)$	discharge from section cells with $IRECT = 3$ or 4	cm^3 s^{-1}
$\Sigma Q(t)$	total site discharge	cm^3 s^{-1}
$\Sigma Qs \cdot (t)$	total section discharge	cm^3 s^{-1}
$Re1(x,t)$	cell Reynolds number where $d(x,t) \geq 1$ mm	
$\overline{Re}'$	mean section Reynolds number	
$t_0(x)$	cell time to runoff	s
$V1(x,t)$	cell velocity where $d(x,t) \geq 1$ mm	cm s^{-1}
$\overline{V}(t)$	mean section velocity	cm s^{-1}
W	cell width, 60.96 cm	cm
$\Sigma W(t)$	inundated section width	m

References

Abrahams, A. D. & A. J. Parsons 1991a. Relation between infiltration and stone cover on a semiarid hillslope, southern Arizona. *Journal of Hydrology* **122**, 49–59.

Abrahams, A. D. & A. J. Parsons 1991b. Resistance to overland flow on desert pavement and its implications for sediment transport modelling. *Water Resources Research* **27**, 1827–36.

Abrahams, A. D., A. J. Parsons, S.-H. Luk 1986a. Resistance to overland flow on desert hillslopes. *Journal of Hydrology* **88**, 343–63.

Abrahams, A. D., A. J. Parsons, S.-H. Luk 1986b. Field measurement of the velocity of overland flow using dye tracing. *Earth Surface Processes and Landforms* **11**, 653–7.

Abrahams, A. D., A. J. Parsons, S.-H. Luk 1989. Distribution of depth of overland flow on desert hillslopes and its implications for modeling soil erosion. *Journal of Hydrology* **106**, 177–84.

Abrahams, A. D., A. J. Parsons, S.-H. Luk 1991. The effect of spatial variability in overland flow on the downslope pattern of soil loss on a semiarid hillslope, southern Arizona. *Catena* **18**, 255–70.

Chow, V.-T. 1959. *Open-channel hydraulics*. Tokyo: McGraw-Hill.

Green, W. A. & G. A. Ampt 1911. Studies on soil physics. 1: the flow of air and water through soils. *Journal of Agricultural Science* **4**, 1–24.

Kostiakov, A. N. 1932. On the dynamics of the coefficient of water percolation in soils. *Transactions of the Commission of the International Society of Soil Science* **A**, 17–21.

Parsons, A. J. & A. D. Abrahams 1989a. A miniature flume for sampling interrill overland flow. *Physical Geographer* **10**, 96–105.

Parsons, A. J. & A. D. Abrahams 1989b. Predicting infiltration and runoff on semi-arid hillslopes, southern Arizona. Abstract in *Landschaftgenese und Landschaftőkologie* **16**, 37.

Parsons, A. J., A. D. Abrahams, S.-H. Luk 1990. Hydraulics of interrill overland flow on a semi-arid hillslope, southern Arizona. *Journal of Hydrology* **117**, 255–73.

Philip, J. R. 1957. Theory of infiltration. 1. The infiltration equation and its solution. *Soil Science* **83**, 345–57.

Scoging, H. M. 1988. A theoretical and empirical investigation of soil erosion in a semi-arid environment. Unpublished Ph.D. thesis. University of London.

7

Effect of land-surface configuration on catchment hydrology

Rodger B. Grayson & Ian D. Moore

Abstract

The representation of overland flow in distributed-parameter, physically based hydrologic models is critically important from the point of view of both the structure and parameterization of the models. Flow may occur in rills, ephemeral gullies and as sheet flow, with quite different hydraulic responses. The effect of these three flow representations on distributed flow characteristics (i.e. flow depth and velocity) and catchment outflow hydrographs is analysed by application of a terrain-based hydrologic model to a 7 ha catchment in New South Wales, Australia. The choice of surface roughness parameters and the way in which overland flow is represented has a profound effect on both predicted outflow hydrographs and distributed flow characteristics. The latter are particularly important if the output from a hydrologic model is to be used to predict sediment or nutrient transport, since these processes are sensitive to flow velocity and depth. Results show that calibrating a model using only an outflow hydrograph can lead to very different estimates of distributed flow characteristics, depending on the representation of the underlying hydrologic processes.

Introduction

There are many underlying or implicit assumptions involved in the development of deterministic, physically based hydrologic models and their application to real catchments. These range from the model structure to the constituent algorithms and their parameterization. Physically based models usually consist of linked process equations derived from physics principles (conservation of mass, momentum and energy) at a point, laboratory or plot scale. These models are developed with the belief (or hope) that the model parameters are measurable. In contrast, empirical or functional models can be thought of as simple input–output models whose parameters can rarely be interpreted in terms of physical processes. Another term used with physically based hydrologic models is "distributed-parameter" which refers to the ability to represent the spatial variability of parameters and/or processes within catchments. The application of distributed-parameter, physically based models to real catchments presupposes: (a) that physical processes can be represented in a deterministic way and that the overall catchment response can be represented by the combined action of the constituent process algorithms; (b) that the spatial variability of a catchment can be represented by distributed values of the model parameters; and (c) that model parameters can be meaningfully measured or estimated at the element scale of the model.

These assumptions do not stand up to close scrutiny. Deterministic is defined in the *Oxford Dictionary* as "the doctrine that everything that happens is determined by a necessary chain of causation". This definition correctly describes the processes by which rainfall becomes runoff in the real world. However, in relation to the modelling of those processes, there is a presumption that the algorithms represent the sequence of causes well enough to determine the outcome uniquely. At small scales, and in ideal soils, this may be possible, but in the field it is not. The equations describing the underlying physics of ideal soil cores cannot be extrapolated reliably to the field or catchment scale, as the conditions under which many process-based equations are derived, such as a laboratory column of homogeneous soil, are different from those in the field.

Model scale is determined by the model structure and the size of the elements, cells, grids or hydrologic units within which the model parameters are assumed to be uniform. The scale at which homogeneity is assumed in models is usually larger than the scale at which directly measurable parameters (e.g. hydraulic conductivity) are determined in the field, but smaller than that represented by outflow hydrographs. Therefore, measured parameter values do not integrate the response of the "elemental" area and there is an inconsistency in scale between that used in the measurement of field variables and the way in which they are applied in models. This conflict led Klemeš (1986) to remark: "It also seems obvious that search for new measurement methods that would yield areal distributions, or at least reliable areal totals or averages, of hydrologic variables such as precipitation, evapotranspiration, and soil moisture would be a much better investment for hydrology than the continuous pursuit of a perfect massage that would squeeze the nonexistent information out of the few poor anaemic point measurements ...". This sentiment could be extended to include infiltration parameters and Manning's roughness coefficient. The result of these approximations is that the parameters lose their physical significance and the model becomes limited in terms of predictive capability because parameter values cannot be determined a priori (James & Burges 1982, Beven 1983). The model parameters also become a function of the model structure.

These assumptions and limitations led Beven (1989) to refer to so-called distributed-parameter, physically based models as lumped, conceptual models.

The issues are discussed in greater depth by Beven (1989), Moore & Hutchinson (1991) and Grayson et al. (1992).

Distributed parameter models that give estimates of flow over a catchment are the basis of many sediment and pollutant transport models. The flow velocity and depth predictions are used in equations to estimate sediment transport capacity and sediment detachment that are then used, for example, to estimate the amount of soil removed from an area (e.g. ANSWERS developed by Beasley et al. 1980). It is apparent that the accuracy of the final estimates of soil loss or nutrient movement are directly related to the flow estimates. The flow estimates from models are subject to errors as a result of uncertain antecedent conditions and poor resolution in data, as well as limitations imposed by the model structure and algorithms.

The manner in which overland flow is represented in these distributed-parameter hydrologic models also gives rise to major uncertainties in predicted flow characteristics. Do the model algorithms explicitly differentiate between sheet, rill, ephemeral gully and channel flow and between laminar and turbulent flow, and if so, how is the surface represented and how does it relate to reality? This chapter examines the way surface flow is represented in distributed-parameter hydrologic models and the likely effect of different assumed surface configurations on catchment discharge and distributed flow characteristics such as flow depth and velocity.

Modelling framework

For this study we wanted to use a model that could simulate three-dimensional runoff processes including saturation and Hortonian overland flow and also one that could simulate sheet flow and flow in microchannels (i.e. rills), ephemeral gullies and stream channels. The model had to be flexible enough to allow different combinations of flow configurations and runoff-producing mechanisms while still maintaining fidelity with the terrain.

Several terrain-based overland flow models have recently been reported in the literature, including the triangulated irregular network (TIN)-based models of Jones et al. (1990), Goodrich et al. (1991) and Vieux (1991); grid-based models such as SHE (Abbott et al. 1986) and those of Freeze (1980); and the contour-based model of Moore & Grayson (1991) (see also Grayson 1990, Moore & Foster 1990, Moore et al. 1990). These different approaches are described by Moore et al. (1991). Only the Hortonian overland-flow mechanism has been represented in the three TIN-based models and most grid-based models are computationally intensive. Moore & Grayson (1991) describe separate contour-based hydrologic models for simulating saturation overland flow and Hortonian overland flow. Grayson (1990) integrated these two runoff-producing mechanisms into the dynamic, terrain-based hydrologic model called THALES. The current study is based on the application of the THALES hydrologic model in conjunction with Moore & Grayson's (1991) TAPES-C (Topographic Analysis Programs for the Environmental Sciences-Contour) terrain analysis programs. THALES is used here simply as a vehicle for studying the impact of surface configuration on overland flow.

TAPES-C: Topographic Analysis Programs for the Environmental Sciences-Contour

TAPES-C (Moore & Grayson 1991) is a set of computer programs that generate a network of interconnected elements on a surface defined by elevation isolines. There is a companion set of programs called TAPES-G that are based on gridded elevation data. TAPES-G provides a more efficient method for obtaining topographic attributes but is not well suited to dynamic hydrologic modelling. TAPES-C computes element networks using the "stream-tube" approach first proposed by Onstad & Brakensiek (1968) and builds upon the earlier work of O'Loughlin (1986) and Moore et al. (1988a). Each element is connected to upslope and downslope elements, facilitating the application of a finite element-type model to the stream-tube network. Elements can have more than one upslope connecting element but only one downslope connecting element. In addition, saddle points are better represented by relaxing the need for streamlines to terminate at high points and allowing ridge lines to be defined. This form of catchment partitioning allows the complex two-dimensional flow equations to be reduced to a system of coupled one-dimensional equations. The TAPES-C partitioning of a 7 ha catchment at Wagga Wagga in New South Wales, Australia is illustrated in Figure 7.1.

TAPES-C calculates the element network and the following attributes for each element: (a) element area; (b) total upslope contributing area; (c) element number of the upslope and downslope connecting elements; (d) x, y, z coordinates of the element centroid; (e) x, y, z coordinates of the midpoint on the downslope boundary of the element; (f) the average slope of an element; (g) the width of the upslope and downslope boundaries of the element ; (h) the length of the element in the flow direction; (i) the aspect or azimuth of the element; and (j) plan curvature. Channel networks can be defined using the criteria of Band (1989) in which an element with an upslope contributing area above a threshold value is assumed to be a channel element. The stream channel is assumed to cross the upper and lower boundaries of the element at the points of maximum curvature.

TAPES-C requires either a file of digitized contour lines in the form of a digital line graph (DLG) or access to a digitizing table to provide the input

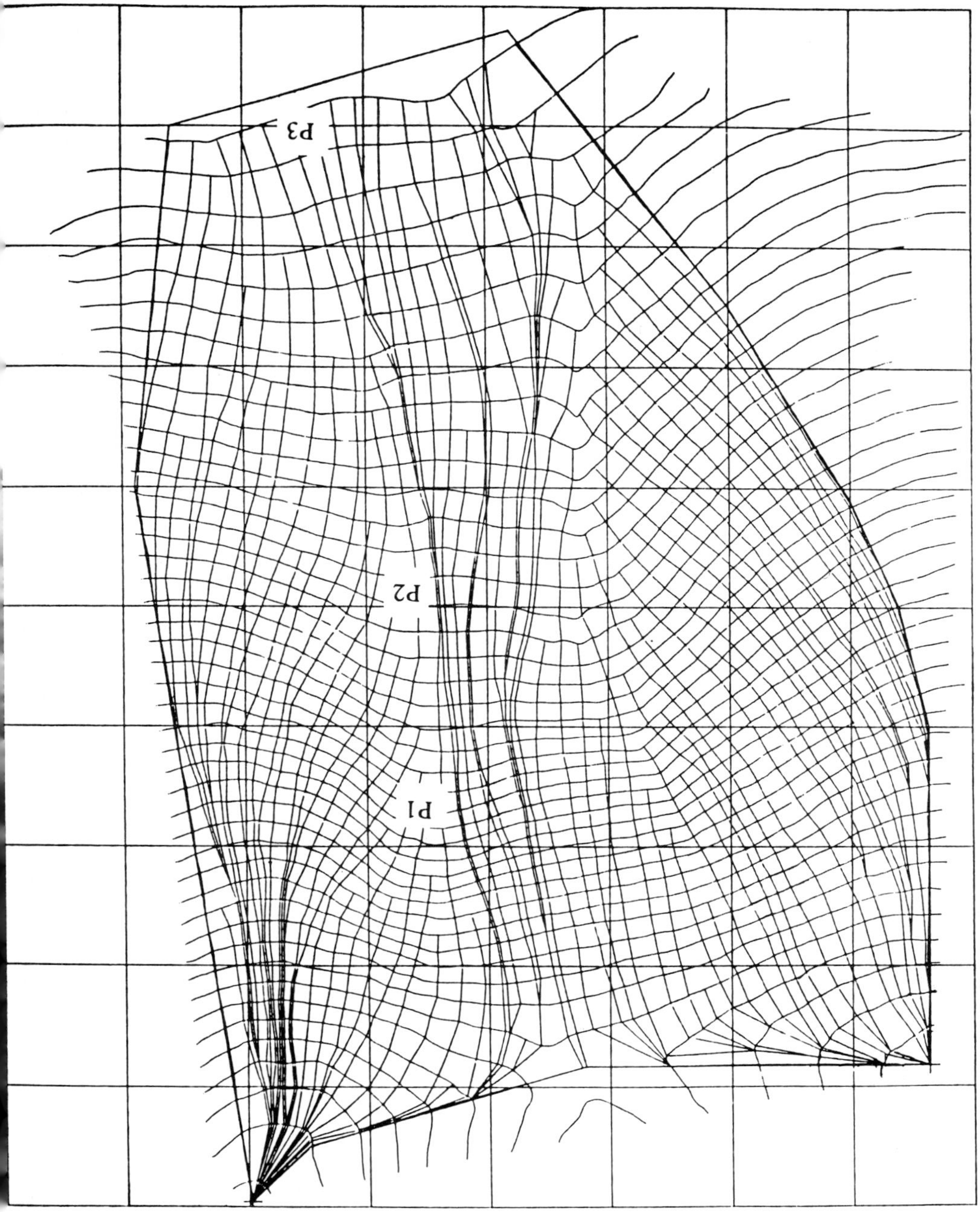

Figure 7.1 TAPES-C partitioning of the Wagga Wagga catchment: 1 m contour interval.

data. Boundaries of the catchment can be digitized by the user, or the program can calculate the boundary, based on flow paths from points at the catchment outlet.

Hydrologic model — THALES

THALES is named after Thales of Miletos, a Greek philosopher who recognized the influence of topography on runoff generation. The TAPES-C catchment partitioning and terrain analysis provides the basic model structure for THALES. A wide range of hydrologic processes can be represented within the model, allowing the physical characteristics of a catchment and storm behaviour to determine the actual response. Figure 7.2 is a conceptual diagram of THALES. By modelling the subsurface movement of water, as well as the infiltration and surface runoff processes, a continuum of hydrologic responses, excluding baseflow from deep aquifers, can be simulated. An early version of THALES (Grayson et al. 1988) had the same basic structure but different process algorithms.

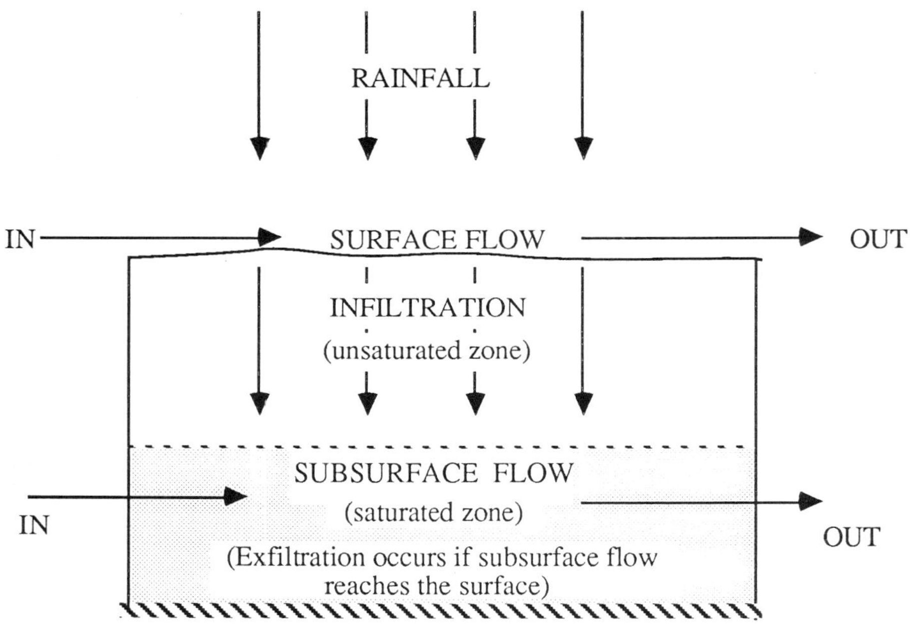

Figure 7.2 Conceptual diagram of THALES.

The philosophy behind the development of THALES was to minimize the number of parameters and processes represented while maintaining the basic components of a "physically based" model. The rationale was that any additional algorithms simply provided "more knobs to twiddle" and would offer no additional information about the true catchment hydrologic response. Our intention was not to develop yet another model but, rather, to make use of the structure that TAPES-C provides to investigate some issues related to the modelling of hydrologic response, and particularly those related to topography.

By keeping the model as simple as possible, the "subtle interdependence between the 'sensitive' and 'insensitive' parameters" (Hornberger et al. 1985) is minimised and the effect of the underlying assumptions is easier to identify. Inclusion within a model of processes in addition to the key processes often exacerbates the problem of over parameterization and gives a false impression of a model's capability (Klemes 1983, Anderson & Rogers 1987, Beven 1989). It is common in applications of complex models for many of the subprocess descriptions to be insensitive, and similar results can be obtained by simpler models (e.g. Mein & Brown 1978). For these reasons, processes such as depression storage and lateral unsaturated subsurface flow are not represented in THALES.

Model parameters are assumed to be constant in each element, so that only the meso-scale variations in soil properties and their effects on catchment hydrology are represented. The limitations of this assumption were briefly discussed earlier. The connectivity of individual elements, determined by the TAPES-C catchment partitioning, provides the element network for overland flow and subsurface routing of water. The model is based on the assumption that infiltrated water flows down slope in a saturated layer overlying an impermeable base. If the subsurface flow rate exceeds the capacity of the soil profile to transmit the water, surface saturation occurs and rain falling on the saturated areas becomes direct runoff. Runoff can also be produced by exfiltration of subsurface water and infiltration excess (i.e. Hortonian) overland flow.

Infiltration Accurate prediction of the Hortonian response of a catchment relies on an effective representation of the infiltration process. In THALES the user has the option of computing the infiltration rate and volume and the rainfall excess for each element, using either a relationship based on an equivalence between the Green and Ampt and the Horton infiltration capacity equations (Morel-Seytoux 1988a, b), or the relationship of Smith & Parlange (1978). Variable rainfall-intensity infiltration is simulated, but the rainfall intensity in any time increment is assumed constant. During drainage, the deepening and change in soil-water content of the transmission zone is calculated using the procedures described by Morel-Seytoux (1988a), assuming a rectangular-shaped wetted zone.

The rainfall excess for each element is $i_e = i - f + R_o$, where i is the net rainfall intensity, f is the infiltration capacity and R_o is the runon per unit area to the element from upslope, provided $i + R_o > f$. Infiltration continues at the infiltration capacity rate, even after $i + R_o$ drops below the saturated hydraulic conductivity, provided there is a non-zero flow depth at the outlet of the element.

Water that enters the soil profile is added to the unsaturated zone which then discharges into the underlying saturated zone; i.e. only vertical flow is modelled in the unsaturated zone. The rate at which water enters the saturated zone is approximated by the unsaturated hydraulic conductivity, $K(\theta)$, calculated as a function of the average soil-water content of the unsaturated zone, using the Brooks & Corey (1964) relationship. This assumes that the hydraulic gradient is unity. The soil-water content is determined from a volume balance of the unsaturated zone. The total unsaturated soil volume for an element changes over time as the saturated depth varies, and a water balance is used to redistribute the unsaturated soil-water content as a result of a change in total unsaturated soil volume. Lateral flow enters and exits the element via the saturated zone. If the saturated zone reaches the surface, saturated source area runoff occurs and exfiltration of subsurface flow is possible. More complete details of the methods for simulating infiltration and the soil-water balance are given in Grayson (1990), Moore & Foster (1990) and Moore & Grayson (1991).

Subsurface and overland flow Saturated subsurface flow and overland flow are modelled using the kinematic wave equations (Moore & Foster 1990). These equations have been shown to be a good approximation for many subsurface flow and overland flow conditions (Woolhiser & Liggett 1967, Morris & Woolhiser 1980, Beven 1982). The resistance equations, or kinematic forms of the momentum equations, can be expressed as:

$$Q = K\frac{A}{\gamma}\sin\beta \quad \textit{for} \quad 0 \leq A < \omega\gamma D \tag{7.1}$$

and

$$Q = \alpha A^{m} \tag{7.2}$$

for subsurface and overland flow, respectively, where Q is the discharge, A is the flow cross-sectional area, K is the effective hydraulic conductivity of the soil, ω is the width of the element orthogonal to the streamlines, γ is the effective porosity (total porosity − field capacity soil-water content), β is the local slope (in degrees), D is the thickness of the hydrologically active soil profile (above a restricting layer), and α and m are parameters related to the overland flow characteristics (turbulent or laminar flow), slope and surface roughness. For turbulent broad, sheet flow; $\alpha = n^{-1}\omega^{-2/3}\tan^{1/2}\beta$ and $m = 5/3$; while for laminar broad, sheet flow: $\alpha = g\omega^{-2/3}\tan\beta/k\nu$ and $m = 3$; where n is Manning's roughness coefficient, ν is the kinematic viscocity, k is a dimensionless roughness parameter, and g is the acceleration of gravity. Laminar flow occurs when the Reynolds number ($Re = vR/\nu$, where v is the flow velocity and R is the hydraulic radius) is less than some critical value. For channelized flow we assumed that only turbulent flow occurs, in which case $\alpha = n^{-1}C^{2/3}\tan^{1/2}\beta$ and $m = 4/3$, where C is the coefficient in the assumed relationship between hydraulic radius, R, and cross-sectional area of a channel, $R = CA^{1/2}$ (Moore & Burch 1986, Moore & Foster 1990).

High hydraulic conductivities are indicated by the small response times of many catchments to subsurface flow, particularly forested catchments. It is thought that this is the result of water flowing through interconnected macropores created by roots, other organic matter and cracks between aggregates and rocks, rather than through the soil matrix (e.g. Mosley 1979, Sloan & Moore 1984). For these reasons the hydraulic conductivity used to characterize the subsurface flow might be different from that measured from small soil cores. The parameter K should be considered as an "effective" hydraulic conductivity of the soil profile rather than the measured value of hydraulic conductivity for the soil matrix.

The subsurface and overland-flow kinematic wave equations are solved numerically using Brakensiek's (1967) four-point implicit finite-difference scheme. The finite-difference form of the continuity equation can be expressed as:

$$A_i^j - A_i^{j+1} + A_{i+1}^j - A_{i+1}^{j+1} + 2\frac{\Delta t}{\Delta x_e}[Q_i^{j+1} - Q_{i+1}^{j+1}] + 2qA_e\frac{\Delta t}{\Delta x_e} = 0 \quad (7.3)$$

where the subscripts refer to time and the superscripts to position, Δt is the time interval, A_e is the plan area of the element, Δx_e is the flow distance along a streamline through the element, and q is the rainfall excess or inflow to the saturated zone for the element. This assumes a weighting factor of unity on the finite difference approximation of the $\partial Q/\partial x$ term in the continuity equation, and is used because it is unconditionally stable and dispersion of the numerical scheme overcomes the problems of kinematic shocks that can occur on cascading planes. Numerical dispersion causes an underprediction and delay in the magnitude and timing of peak runoff with this method (Kibler & Woolhiser 1970). However, the magnitude of this problem varies with applications. On the study site, problems with numerical dispersion were not significant. Equation 7.3 is an implicit equation and is solved using the Newton–Raphson method. The values of A_e, Δx_e, ω and β for each node point in the solution domain (i.e. each element) are estimated by TAPES-C.

Surface runoff may continue after rainfall ceases and when the rainfall intensity is less than the prevailing infiltration capacity. This surface runoff can infiltrate in a downslope element if the soil profile of the downslope element is not saturated, i.e. where Hortonian overland flow occurs. This runoff–runon problem is likely to be important where broad, sheet flow occurs, but relatively unimportant when the flow concentrates in defined rills, channels and streams.

Study site - Wagga catchment

The Wagga catchment is maintained by the New South Wales Department of Conservation and Land Management for research purposes and is located 5 km south of the Wagga Wagga township in New South Wales, Australia (35°06′S, 147°22′E). The catchment is 7.03 ha in area and has an easterly aspect. Slopes vary from 5 to 20% and average 2.5%. The vertical relief is about 45 m. The soil is a Yellow solodic soil (Dy 2.42; Northcote et al. 1976) with a hard-setting, fine sandy loam A-horizon overlying a coarse, angular, blocky clay B-horizon (Adamson 1976). Soils grade from 1 m deep, rock-free profiles in lower slope depositional areas to shallow soils with high rock content on eroded upslope and ridge areas.

Originally the catchment was severely eroded and gullied. In the late 1940s the gullies were filled and terraces were built on a 1 m vertical interval. An improved pasture was established and maintained until April 1986, when the contour furrows were filled and the catchment was ploughed and cropped to wheat. In 1987 the catchment was ploughed and deliberately left bare to make it vulnerable to erosion (Moore et al. 1988a).

Average annual rainfall between 1952 and 1974 was 675 mm and ranged from 353 mm to 949 mm (Adamson 1976). In the same period, the annual runoff and soil loss averaged 11.2 mm, ranging from 0 to 12.2 mm, and 0.03 Mg ha^{-1}, ranging from 0 to 0.35 Mg ha^{-1}, respectively. The hydrologic response of the catchment is dominated by subsurface flow and saturation overland flow. Rainfall erosivity peaks in February, April and October, with the highest erosivity occurring in October (Adamson 1978, Aveyard et al. 1983). During winter, from May to August, rainfall erosivity is normally low.

The hydrologic simulations on which the subsequent results are based are for a series of events that occurred in June and July, 1987. The total rainfalls for the months of June and July were 111 and 64 mm, respectively, compared to the long-term average values (1948–1974) of 50 and 53 mm, respectively. These rainfalls followed a very dry period from March to May 1987 (Moore et al. 1988a). Details of the hydrologic response of the catchment are discussed in Grayson (1990).

Model simulations and the representation of surface flow

Methods

Simulations were carried out assuming two different overland-flow producing mechanisms (Table 7.1). The first set of simulations assume that runoff occurs via saturation overland flow and consist of three surface-flow representations, each with two different surface-roughness conditions (Manning's *n* values). They included: (a) sheet flow over the entire saturated area (A and B); (b) rill flow over the entire saturated area (C and D); and (c) ephemeral gully networks as measured in the field at the end of the storm, and sheet flow in the saturated, non-channelled areas (E and F).

Table 7.1 Simulation identification

Identification	Flow representation	Manning's *n*
Saturation Overland Flow		
A	overland sheet flow	0.07
B		0.025
C	rill flow, 2 rills per metre	0.07
D		0.025
E	combined sheet and channel flow	0.07 - surface and channel
F		0.025 - surface and channel
Hortonian Overland Flow		
G	spatially invariant infiltration parameters	spatially variable

The second set of simulations assume that only Hortonian overland flow occurs, although field observations over a 5 year period indicate that this is a rare phenomenon on the Wagga catchment. Storms are generally of low intensity and long duration. These simulations were performed to see whether, paraphrasing the words of Klemes (1986), the "right" answers could be obtained for the "wrong" reasons, i.e. use incorrect process descriptions to obtain similar results to those using correct process descriptions. Only one simulation is reported here and this is based on a constant infiltration rate and variable Manning's *n* (G). There was no attempt to optimize the solution by including depression storage or multiple combinations of *n* and infiltration parameters. Subsurface flow was not modelled.

All model parameters used in the simulations, except those associated with the hydraulics of overland flow, were based on direct field measurements. The methods of measurement and values of the adopted parameters are discussed in Grayson (1990). Simulations C and D, using rills, were intended to simulate the concentration of flow over rough surfaces into small rivulets. The rill flow equations require a value of width-to-depth ratio, number of rills per unit width and a Manning's *n* value for the rill. The influence of the width-to-depth ratio is relatively small, so the most hydraulically efficient value was chosen (Moore & Burch 1986).

Selection of the number of rills per unit width is rather subjective. Theoretically there should be a rill density at which the sediment delivery rate is maximized, and Hirschi & Barfield (1988) estimated this to be approximately 1.3 rills per metre for their experimental conditions. However, Parsons (1987) found rill density to be positively correlated with slope and uncorrelated with sediment properties on rilled hillslopes in Spain and England. The application of rill-flow equations in THALES is not concerned with sediment transport, since, in general, the flow on the Wagga catchment is not of sufficient power to erode the surface. That is, there is little soil movement on the catchment except in ephemeral gullies. On natural hillslopes the density of flow concentrations is approximately 1 to 4 per metre (Emmett 1970). The relevance of these values to the cultivated soil at Wagga is questionable, but they do indicate that there is likely to be a wide range of values. The frequency of preferred flow paths observed in stereo photographs of the surface could be used to obtain actual values, but these depend on the discharge relative to the surface roughness. No such photo pairs were available at the time these simulations were undertaken. A value of 2 rills per metre was chosen, somewhat arbitrarily. The Manning's *n* value for the rills is also difficult to determine. Given that the primary purpose of this study is to investigate the influence on simulated flow characteristics of the approach to surface-flow modelling, the same value as for the overland-flow simulation was used (0.07). Loch et al. (1989) indicate that the actual value is likely to be lower than 0.07. A value of 0.025 was also used and this is comparable to simulation B for the sheet-flow assumption. This is slightly lower than the values suggested by Loch et al. (1989) but they observed a decrease in *n* as discharge increased and their maximum discharge of 2.0 $l\ s^{-1}$ is much less than the maximum discharges recorded at the Wagga catchment.

The third approach to modelling surface flow was to assume sheet flow on all but the elements within which the ephemeral gullies were known to exist. The dimensions of the channels were measured after the event and these *measured* values were used in the simulation by overlaying the positions of the ephemeral gullies on the elements and assigning the appropriate channel dimensions to these "channel" elements (see Barling et al. 1988 for details of the survey of ephemeral gullies). Two simulations were performed, the first using a Manning's *n* of 0.07 for both the surface and channels and the second using a value of 0.025 for both the surface and channels.

Results and discussion

Saturation overland-flow assumption Figure 7.3 shows the observed hydrograph versus simulation A, which assumes sheet flow and uses measured soil parameters. Manning's *n* was taken as 0.07, based on the recommendations of Engman (1986) for a chisel-ploughed surface without mulch. Figures 7.4–7.8 show the predicted hydrographs for the three surface-flow assumptions and the two Manning's *n* values compared to simulation A (overland sheet flow and $n = 0.07$). Table 7.2 is a summary of the estimated peak flows for the three main peaks and the six simulations. There are large

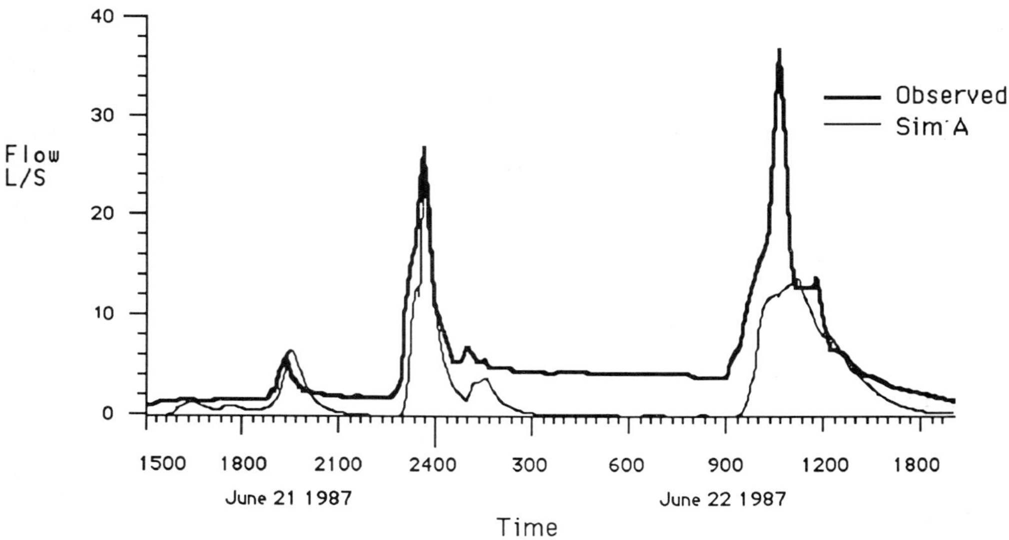

Figure 7.3 Comparison of observed catchment discharge and that predicted by simulation A (sheet flow, $n = 0.07$).

differences in the simulated responses, both in the peak values and, to a lesser extent, in their timing. Figures 7.4–7.6 and Table 7.2 indicate that when overland sheet flow or rill flow is assumed, a reduction in Manning's *n* from 0.07 to 0.025 almost doubles the maximum peak flow. For ephemeral gully flow the difference is less pronounced because the distance travelled by overland flow before it reaches a channel is short and the routing down the

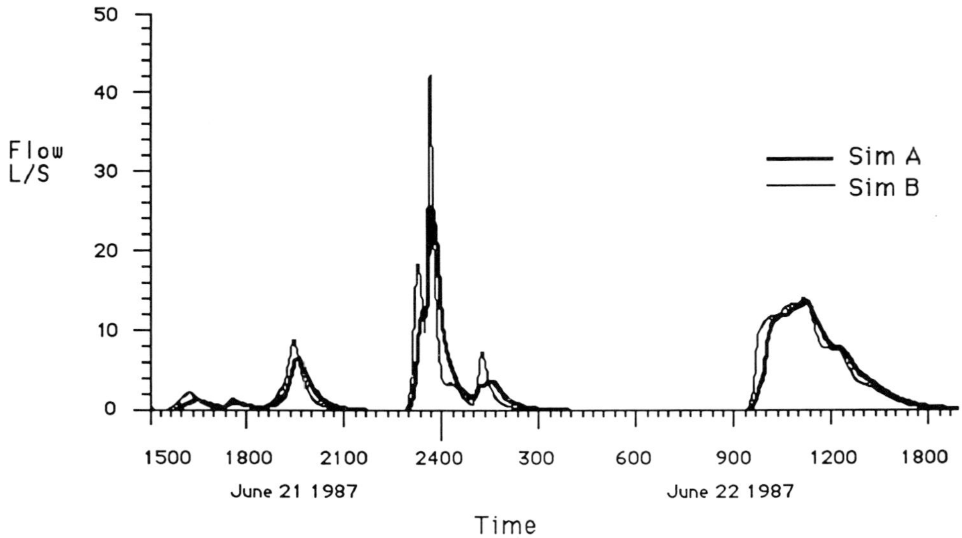

Figure 7.4 Comparison of predicted catchment outflow hydrographs for simulation B (sheet flow, n = 0.025) and A (sheet flow, n = 0.07).

channel dominates the hydrograph. Figures 7.5 & 7.6 show the effect of representing the flow as sheet or rill flow. The rill flow assumption increases the flow velocities and so increases the rate of rise of the hydrograph and the peak flow estimates. The ephemeral gully flow representations shown in Figures 7.7 and 7.8 have similar effects.

Table 7.2 Peak flow rates for the three major runoff peaks.

Identification	Flow Peak 1 ($l\ s^{-1}$)	Flow Peak 2 ($l\ s^{-1}$)	Flow Peak 3 ($l\ s^{-1}$)
A	6.7	25.7	13.9
B	9.0	42.1	14.1
C	9.0	36.3	14.2
D	10.0	63.7	14.2
E	9.1	46.3	14.2
F	9.9	57.1	14.2
Measured	6.2	27.0	37.2

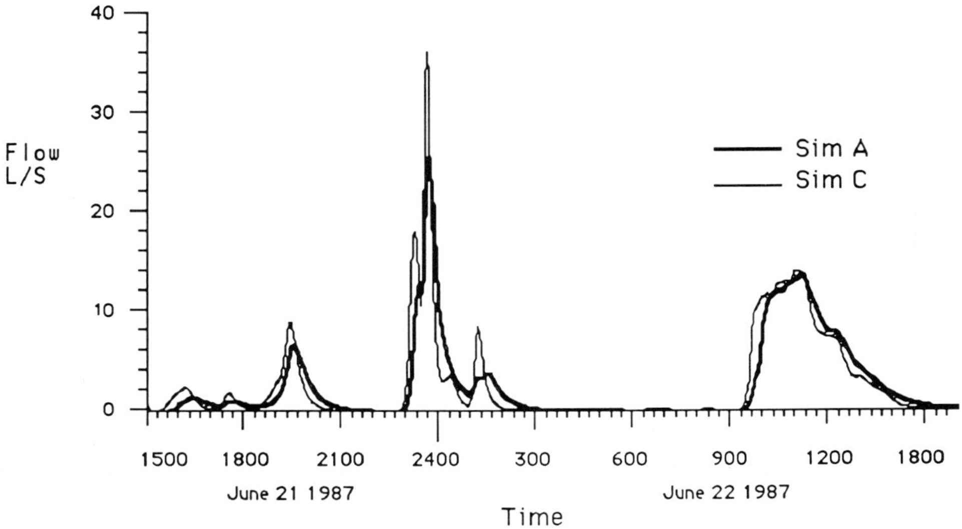

Figure 7.5 Comparison of predicted catchment outflow hydrographs for simulation C (rill flow, $n = 0.07$) and A (sheet flow, $n = 0.07$).

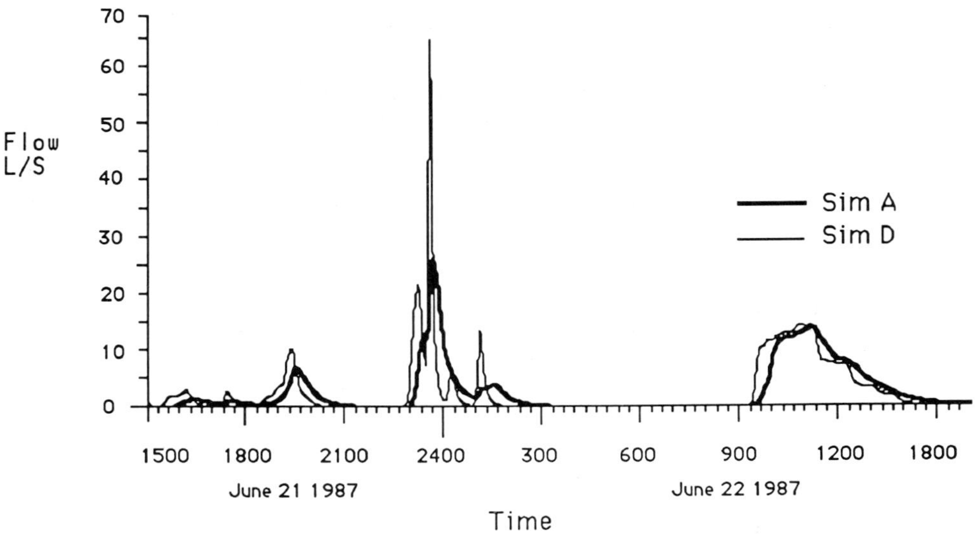

Figure 7.6 Comparison of predicted catchment outflow hydrographs for simulation D (rill flow, $n = 0.025$) and A (sheet flow, $n = 0.07$).

If THALES were used as the hydrologic component of a sediment and nutrient transport model, the simulated flow characteristics on the catchment would be used to calculate the distributed soil movement or nutrient transport. The predicted spatially distributed overland-flow depths and velocities at the time of peak surface runoff for the six simulations are presented in Figures 7.9–7.14, for simulations A to F, respectively. As would be expected from the

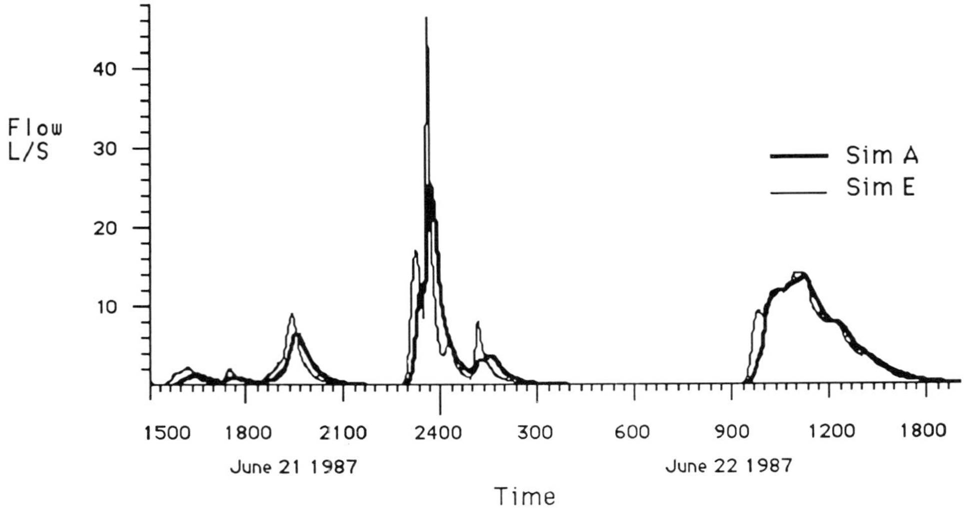

Figure 7.7 Comparison of predicted catchment outflow hydrographs for simulation E (ephemeral gully flow, $n = 0.07$) and A (sheet flow, $n = 0.07$).

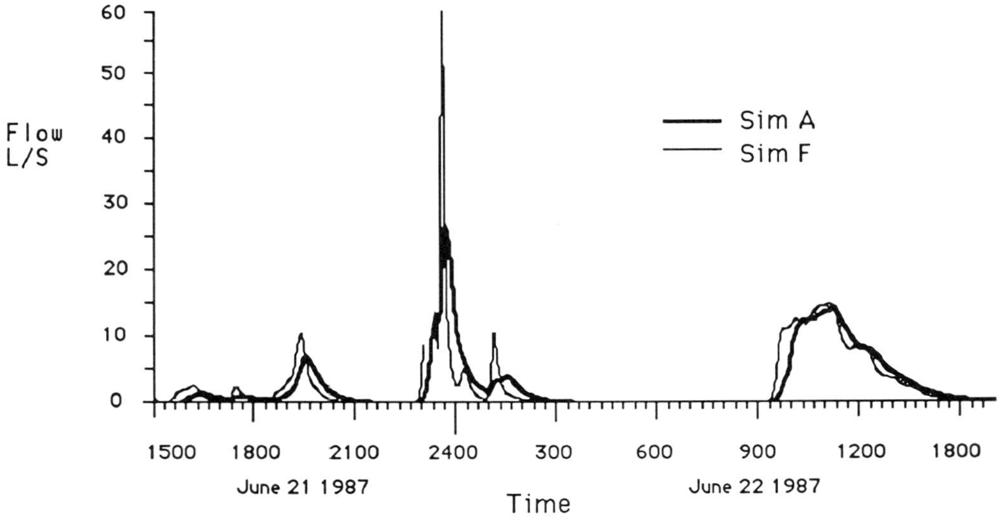

Figure 7.8 Comparison of predicted catchment outflow hydrographs for simulation F (ephemeral gully flow, $n = 0.025$) and A (sheet flow, $n = 0.07$).

hydrographs, the differences are great. Were such diagrams constructed for all simulations, decreases in Manning's n for all flow types would show increased flow velocities, and for the sheet and channel flow, decreased flow depth. In the case of rill flow, the reduction in attenuation of runoff from the upper areas due to the lower Manning's n value is such that both the flow depth and velocity increased, so instantaneous discharge greatly increased. This trend is

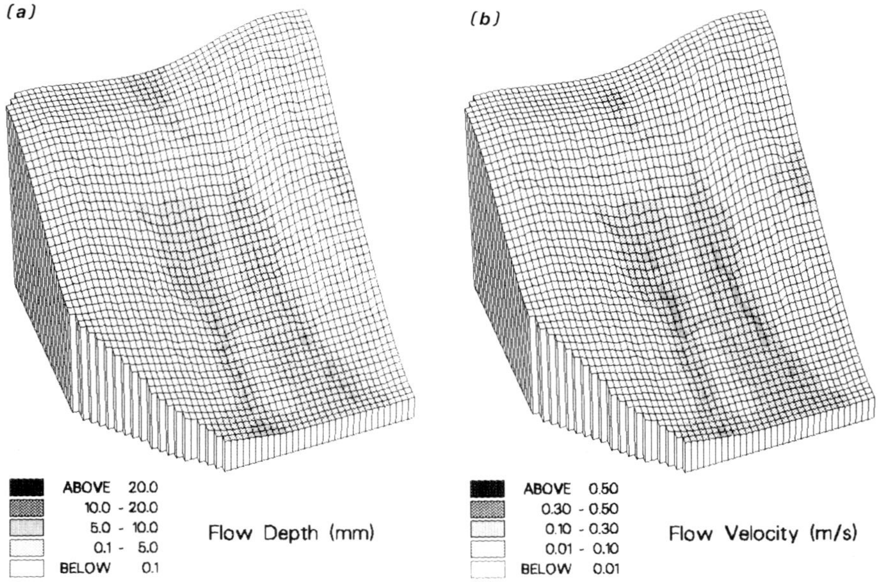

Figure 7.9 Predicted distributed flow depths (a) and velocities (b) for simulation A.

reflected in the discharge at the catchment outlet where the peak flow of C is 36 l s^{-1} compared to 63.7 l s^{-1} for D. Flow velocities in the channels were greater than for the rilled surface, while flow depths were similar in the lower areas. In the upper areas, the rilled surface had shallower flow depths than the channel representation because, outside the channels, the flow was modelled as sheet flow.

The flow characteristics of three elements on the catchment were selected for detailed examination. These are marked P1, P2 and P3 on Figure 7.1 and the simulated flow depth, flow velocity, Froude number (F), Reynolds number (Re) and kinematic flow number (k_e) for the six simulations at each location are presented in Table 7.3. The three dimensionless numbers can be expressed as:

$$F = \frac{v}{\sqrt{gh}} \tag{7.4}$$

$$Re = \frac{vR}{\nu} \tag{7.5}$$

$$k_e = \frac{\tan\beta L}{F^2 h} \tag{7.6}$$

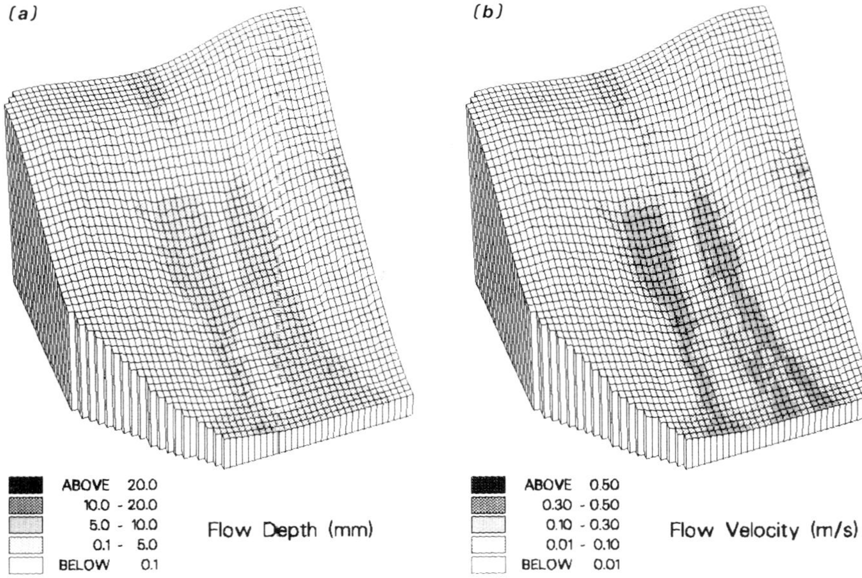

Figure 7.10 Predicted distributed flow depths (a) and velocities (b) for simulation B.

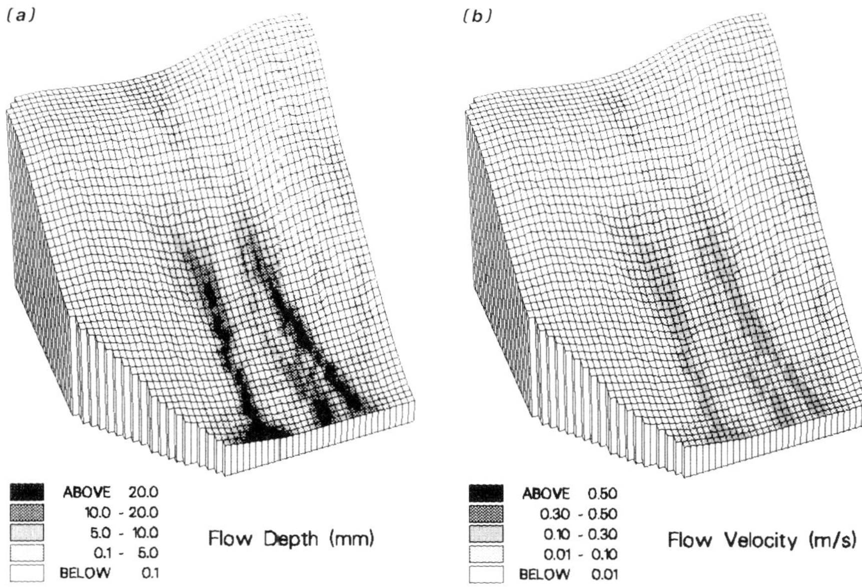

Figure 7.11 Predicted distributed flow depths (a) and velocities (b) for simulation C.

where v and h are the flow velocity and depth, g is the acceleration of gravity, ν is the kinematic viscosity, L is the length of the flow plane, R is the hydraulic radius and β is the slope of the surface. There are major differences in the

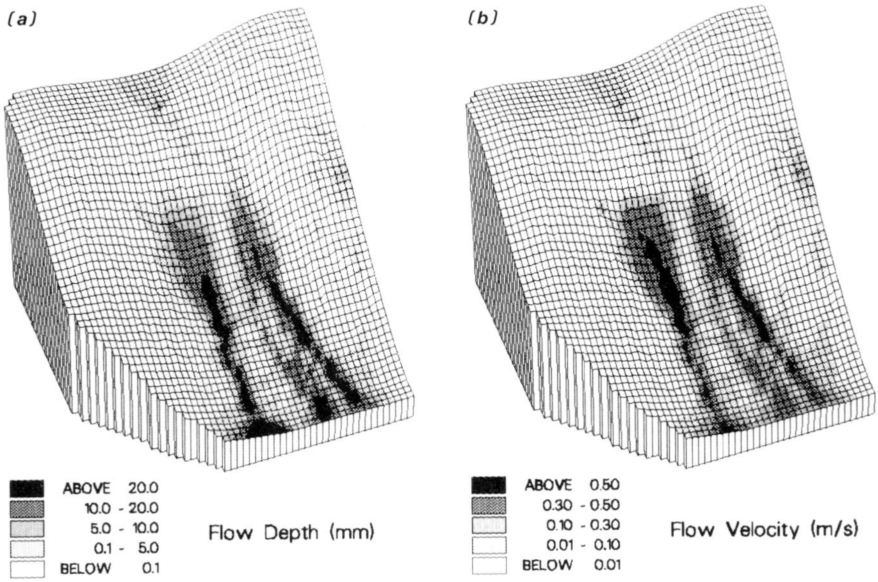

Figure 7.12 Predicted distributed flow depths (a) and velocities (b) for simulation D.

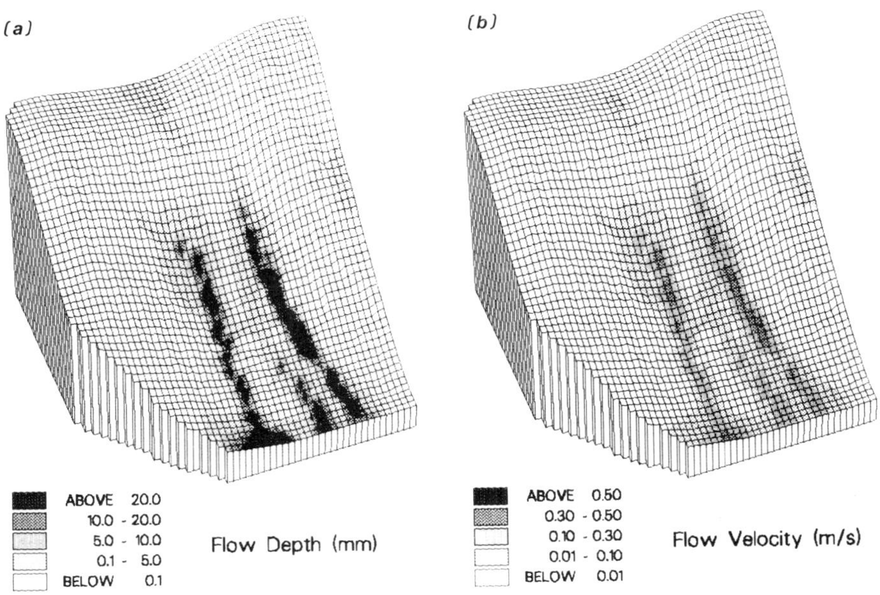

Figure 7.13 Predicted distributed flow depths (a) and velocities (b) for simulation E.

simulated flow characteristics, with flow depths varying by a factor of up to 20 and flow velocities varying by a factor of up to 9. The Froude numbers show that flow is generally simulated as being tranquil, except for the low Manning's

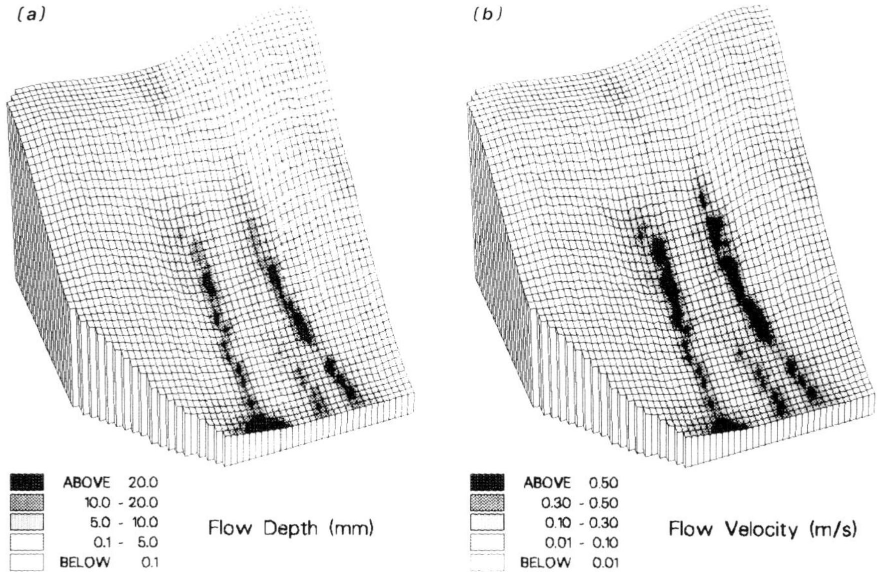

Figure 7.14 Predicted distributed flow depths (a) and velocities (b) for simulation F.

n values in B and D. Reynolds number estimates indicate turbulent flow in all but the uppermost element for all the flow representations. Manning's equation was derived for turbulent flow so, theoretically, the exponent and coefficient in the depth discharge equation should be altered for the calculations on these elements. However, the other assumptions related to surface flow are so gross that this would provide no extra "physical basis".

The similarity of estimated runoff rates for Peak 3 (Table 7.3) between all the flow representations indicates that only for short intense storms does the difference in flow representation become important. The measured rainfall is of long-enough duration for the catchment to be virtually at steady state, so the dominant feature of the model becomes the estimated saturated area rather than the method of representing the surface flow. The comparison of response for Peaks 2 and 3 and the different flow representations highlight the interaction between the model structure and output.

This also raises difficulties for sensitivity analysis, as the sensitivity of the parameters is strongly related to the event used for the analysis. Anderson & Rogers (1987) discussed many problems with the application and interpretation of sensitivity analysis, and noted that the sensitivity of a particular parameter is dependent on the values of the other model parameters, as well as on the input data.

Table 7.3 Flow characteristics for selected points on the catchment.

Flow Characteristic	A	B	C	D	E	F
Position P1						
Depth (mm)	0.6	0.4	2.2	6.4	11.4	6.0
Velocity (m s^{-1})	0.04	0.08	0.06	0.32	0.31	0.68
Froude No.	0.49	1.28	0.38	1.27	0.92	2.79
Reynolds No.	914	146	240	4100	10850	14230
Kinematic No.	5050	1010	2330	70	76	16
Position P2						
Depth (mm)	7.4	6.3	33.6	31.8	48.3	25.2
Velocity (m s^{-1})	0.15	0.37	0.26	0.69	0.46	0.97
Froude No.	0.55	1.50	0.45	1.24	0.66	1.95
Reynolds No.	4440	9390	17250	44170	66640	83450
Kinematic No.	320	50	110	15	34	8
Position P3						
Depth (mm)	7.6	5.6	41.2	34.8	56.9	44.1
Velocity (m s^{-1})	0.12	0.28	0.24	0.60	0.40	0.99
Froude No.	0.45	1.20	0.38	1.03	0.54	1.51
Reynolds No.	3710	6400	19800	41800	76900	152600
Kinematic No.	470	90	120	20	40	7

It must be borne in mind that the validity of the flow depth and velocity predictions is unknown and cannot be compared to real values because none were measured. It is not even possible to say which of the representations is the most realistic. The simulations do, however, highlight the major effect on flow characteristics of the way in which surface flow is represented. The ephemeral gully network used for simulation was measured at the end of the storm sequence so would actually have been developing during the storms. The most physically realistic representation should take account of the changing channel network. Surface flow in non-channel areas could be represented by

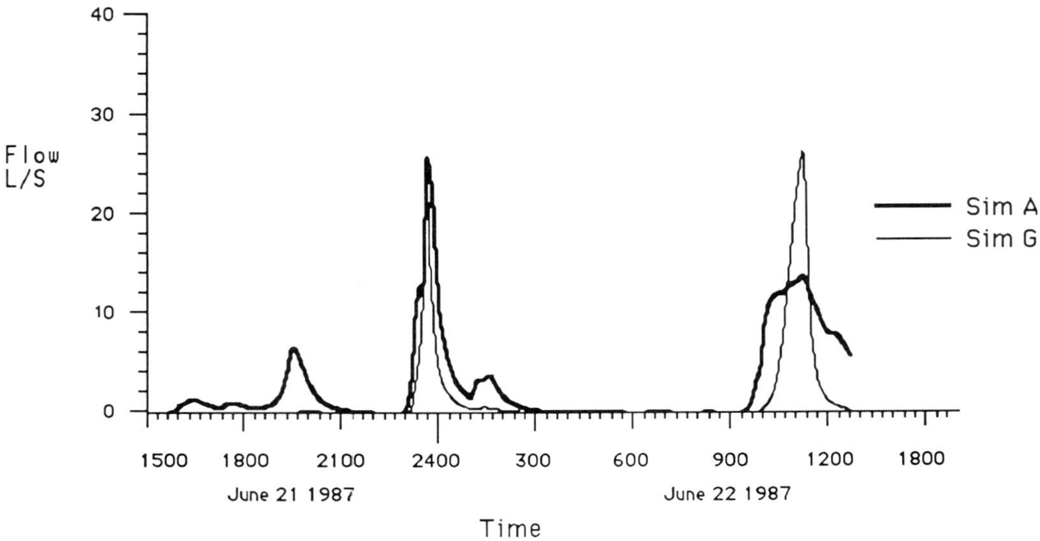

Figure 7.15 Comparison of predicted catchment outflow hydrographs for simulation G (Hortonian overland flow) and A (saturation overland flow, sheet flow, n = 0.07).

either sheet or rill flow, but neither would be physically correct. The surface is not smooth, so sheet flow does not occur exclusively, and neither is there a uniform number of rills per metre. Published Manning's n values reflect not only surface roughness but also the effects of flow channellization, form drag of large obstacles, variations in local flow regime, the impact of raindrops and the transport and erosion of soil (Engman 1986), further complicating the issue of flow representation.

Hortonian overland flow assumption The hydrograph fitted by varying Manning's n for a constant infiltration rate (G) is compared to simulation A in Figure 7.15. Figure 7.16 shows the resulting flow depth and flow velocity estimations at the time of peak runoff. There is no subsurface flow algorithm operating, so baseflow is not simulated. The predicted runoff rate for Peak 2 is 28 l s^{-1}, compared to the measured value of 27 l s^{-1} and simulation A of 26 l s^{-1}.

The results presented in Figure 7.15 show that the outflow hydrograph based on the Hortonian mechanism is similar to those based on saturation overland flow (remembering that there is no baseflow algorithm in the Hortonian simulation). The comparison highlights the danger associated with using the outflow hydrograph alone as a measure of a model's performance. The distributions of simulated flow depths and velocities are vastly different (Figs 7.9–7.14 & 7.16). In addition, they illustrate the ease with which an

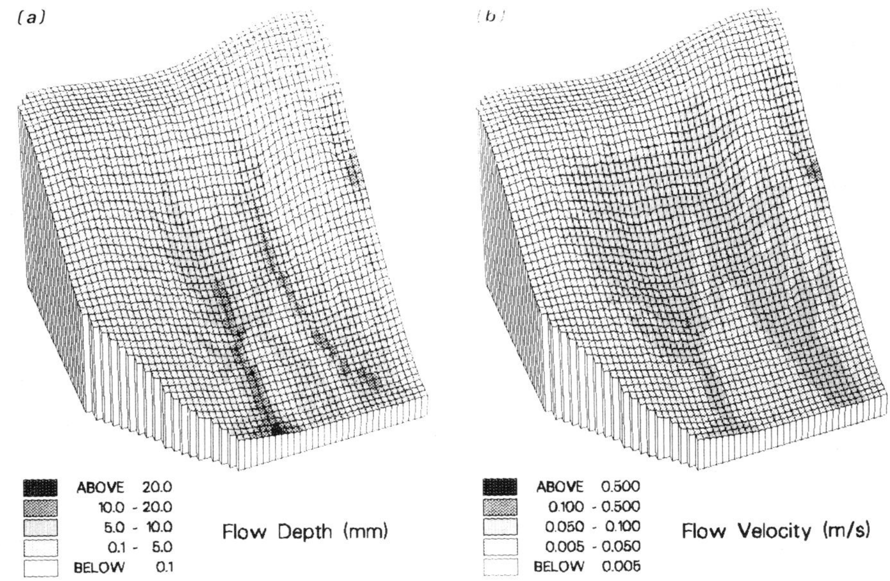

Figure 7.16 Predicted distributed flow depths (a) and flow velocities (b) for simulation G.

inappropriate model (in the physical sense) can give results similar to those from a model that incorporates the processes known to be occurring. In most cases where models are fitted to outflow hydrographs, more parameters are optimized than were used here and the extra degrees of freedom almost ensure "good fits". Given the uncertainty of parameter values such as Manning's n, this generally means that the optimized values will fall within the "acceptable range".

General discussion

The appropriate representation of surface flow is complicated, catchment specific and dependent on the model structure. While much research has been carried out on the hydraulics of overland flow, few useful improvements in the practical application of equations have occurred, and this is proving to be a barrier to the improvement of current models.

An additional question in the use of estimated flow characteristics in sediment and nutrient transport models is whether the mean flow depth of surface flow is an appropriate estimate for flow detachment algorithms. Field measurements of flow depth have shown that a negative exponential distribution, requiring an estimate of mean flow depth and a single parameter (equal to the reciprocal of the mean), characterizes the variation in flow depth on natural surfaces

(Abrahams et al. 1989). When this distribution was used in a sediment detachment equation (Du-Boys), much higher detachment rates were calculated than when using the mean flow depth. If the stage is reached where models are capable of predicting mean flow depths on various parts of a catchment at the element scale, the results of studies such as that of Abrahams et al. (1989) will enable the within-element variability of flow depth to be represented. A similar study of flow velocities may also provide methods for interpreting the "sheet flow" velocity as calculated by the Manning equation, into a real velocity distribution.

The most important conclusion that can be made about the preceding simulations is that the underlying assumptions relating to representation of the surface flow have as large an effect on the flow characteristics as do the parameter values. It should be noted that the Manning's *n* values used in all simulations were within the range suggested by Engman (1986) for a surface of this type (0.006–0.17). Some models compute the overland flow by solving the diffusion wave approximation to the St. Venant equations rather than using simplifications such as the kinematic approximation, but still model overland flow as sheet flow (e.g. SHE, Abbott et al. 1986). The preceding analysis suggests that the variation in simulated flow characteristics due to parameter choice and representation of the flow itself is so great that argument over the complexity of the flow equations is futile. Indeed, there is no "physical basis" for the representation of surface flow on natural surfaces by broad sheet flow (Dunne 1982). There is, therefore, no "physical basis" to the estimated flow characteristics, nor to any subsequent estimates of sediment or nutrient transport derived from those characteristics. The flow characteristics are simply an artefact of the model structure and assumptions that must be made, especially when the models are being applied for predictive purposes.

If the equations used to calculate sediment and nutrient transport were derived from measurements in the field at the scale of a model element, and were the flow depth and velocity (or any other required input) also measured at the same scale, validation of detailed models would be possible and the algorithms would be "physically correct" in that they would integrate the catchment response at the model scale. Instead, the equations generally used in sediment transport models are based on laboratory or river studies where the flow depth and velocity are relatively well known. These equations need relatively accurate estimates of the flow depth and velocity if the parameters in these relationships are to remain "physically realistic". Such estimates are not available because we do not know the appropriate relationships or parameters to represent the flow processes (Dunne 1983). Were these processes to be well represented, it would still be necessary to include the "within element" variation by methods such as that of Abrahams et al. (1989).

It could be argued that the results described above relate only to the model used herein. However, the implications for other models cannot be denied as the underlying assumptions used in the development of other models are similar, if not even more crude, particularly in the representation of topography and flow on the surface.

Conclusions

The preceding discussion demonstrates that our ability to model hydrologic response in a deterministic and precise way is poor. The models used, while appearing conceptually sophisticated, are based on assumptions that are often invalid or questionable, on algorithms that are crude representations of reality and generally derived at a scale different from that to which they are applied and on field data that is insufficient to estimate spatial variability of parameters or even to fully validate a model (Dunne 1982, James & Burges 1982, Beven 1983, Klemes 1983, Burges 1986, Anderson & Rogers 1987). Generally, the models have enough variables to enable close fits between simulated and observed data to be obtained (e.g. Stephenson & Freeze 1974, Sorooshian & Gupta 1983, Beven 1989). If these models are based on fitted parameters, how does one interpret the distributed catchment behaviour (because the parameter values influence those distributed estimates)? Philip (1975) stated the problem more generally: "The process of inferring the spatial (or sometimes temporal) variation of the parameters of the system from the output runs into the problem that any one of a large (even an infinite) number of assumed modes of variation may yield approximately the same output ...the investigator who believes in the physical reality of the parametric values he infers does so at his own peril."

It is possible to obtain "good fits" when representing the wrong processes, as shown previously where Hortonian flow assumptions produced a similar fit to a hydrograph simulated using a saturated overland-flow model – "right results for the wrong reasons" in the terminology of Klemes (1986). There appear to be some major problems in the modelling of surface hydrology by so-called distributed-parameter, physically based models. They provide us with an enormous amount of information and have the theoretical potential to provide a universal tool for the representation of hydrologic response, but they are somewhat removed from reality. This is particularly true when the models are used for predictive purposes where parameter values must be chosen a priori. Even when parameters are fitted to available information, the values are unlikely to be unique nor to solely represent the influence of the process they are intended to describe. In addition, the basic model structure and assumptions regarding surface flow drastically alter the estimates of catchment outflow and distributed flow characteristics.

Acknowledgements

This study was funded in part by grants SES-8912042 and SES-8912938 from the National Science Foundation (USA) and by the University of Minnesota Agricultural Experiment Station, the Centre for Environmental Applied Hydrology, the Centre for Resource and Environmental Studies and the Water Research Foundation of Australia. This paper is dedicated to the memory of Mr A. Keith Turner, Reader in Agricultural Engineering at the University of Melbourne, who made important contributions to the understanding of overland flow.

References

Abbott, M. B., J. C. Bathurst, J. A. Cunge, P. E. O'Connell, J. Rasmussen 1986. An introduction to the European Hydrological System – Système Hydrologique Européen. SHE 2: Structure of a physically-based, distributed modelling system. *Journal of Hydrology* **87**, 61–77.

Abrahams, A. D., A. J. Parsons, S. H. Luk 1989. Distribution of depth of overland flow on desert hillslopes and its implications for modelling soil erosion. *Journal of Hydrology* **106**, 177–84.

Adamson, C. M. 1976. Some effects of soil conservation treatment on the hydrology of a small rural catchment at Wagga Wagga. *Journal of the Soil Conservation Service of New South Wales* **32**, 230–49.

Adamson, C. M. 1978. Conventional tillage systems as they affect soil erosion in southern New South Wales. *Journal of the Soil Conservation Service of New South Wales* **34**, 199–202.

Anderson, M. G. & C. C. M. Rogers 1987. Catchment scale distributed hydrologic models – A discussion of research directions. *Progress in Physical Geography* **11**, 28–51.

Aveyard, J., G. J. Hamilton, I. J. Packer, P. J. Barker 1983. Soil conservation in cropping systems in southern New South Wales. *Journal of the Soil Conservation Service of New South Wales* **39**, 113–20.

Band, L. E. 1989. A terrain-based watershed information system. *Hydrological Processes* **4**, 151–62.

Barling, R. D., R. B. Grayson, I. D. Moore, A .K. Turner 1988. Factors affecting the position and size of rills on cultivated land. *Conference on Agricultural Engineering, Hawkesbury, September 25–27*. Institution of Engineers Australia, National Conference Publication 88/12, 238–42.

Beasley, D. B., L. F. Huggins, E. J. Monke 1980. ANSWERS: A model for watershed planning. *Transactions of the American Society of Agricultural Engineers* **23**, 938–46.

Beven, K. 1982. On subsurface storm flow: predictions with simple kinematic wave theory for saturated and unsaturated flows. *Water Resources Research* **18**, 1627–33.

Beven, K. 1983. Surface water hydrology – runoff generation and basin structure. *Reviews of Geophysics and Space Physics* **21**, 723–9.

Beven, K. 1989. Changing ideas in hydrology – the case of physically-based models. *Journal of Hydrology* **105**, 157–72.

Brakensiek, D. L. 1967. Kinematic flood routing. *Transactions of the American Society of Agricultural Engineers* **10**, 340–3.

Brooks, R. H. & A. T. Corey 1964. *Hydraulic properties of porous media.* Hydrology Paper No. 3, Colorado State University. Fort Collins: Colorado.

Burges, S. J. 1986. Trends and directions in hydrology. *Water Resources Research* **22**, 1S–5S.

Dunne, T. 1982. Models of runoff processes and their significance. In: *Scientific Basis of Water Resource Management,* 17–30. National Academy Press.

Dunne, T. 1983. Relation of field studies and modeling in the prediction of storm runoff. *Journal of Hydrology*, **65**, 25–48.

Emmett, W. W. 1970. *The hydraulics of overland flow on hillslopes.* Geological Survey Professional Paper 662-A. Washington, DC: US Government Printing Office.

Engman, E. T. 1986. Roughness coefficients for routing surface runoff. *Journal of Irrigation and Drainage Division, Proceedings of the American Society of Civil Engineers* **112**, 39–53.

Freeze, R. A. 1980. A stochastic-conceptual analysis of rainfall runoff processes on a hillslope. *Water Resources Research* **16**, 391–408.

Goodrich, D. C., D. A. Woolhiser, T. O. Keefer 1991. Kinematic routing using finite elements on a triangulated irregular network (TIN). *Water Resources Research* **27**, 995–1003.

Grayson, R. B. 1990. Terrain-based hydrologic modelling for erosion studies. Unpublished Ph.D. thesis, Department of Civil and Agricultural Engineering, University of Melbourne, Australia.

Grayson, R. B., I. D. Moore, E. M. O'Loughlin, A. K. Turner 1988. The development of a hydrologic model for erosion prediction. *Conference on Agricultural Engineering, Hawkesbury, September 25-27.* Institution of Engineers Australia, National Conference Publication 88/12, 234–7.

Grayson, R. B., I. D. Moore, T. A. McMahon 1992. Physically based hydrologic modelling II. Is the concept realistic ? *Water Resources Research* (submitted).

Hirschi, M. C. & B. J. Barfield 1988. KYERMO-A physically based research erosion model. Part II Model sensitivity analysis and testing. *Transactions of the American Society of Agricultural Engineers* **31**, 814–20.

Hornberger, G. M., K. J. Beven, B. J. Cosby, D. E. Sappington 1985. Shenandoah watershed study: Calibration of a topography-based, variable contributing area hydrological model to a small forested catchment. *Water Resources Research* **21**, 1841–50.

James, L. D. & S. J. Burges 1982. Selection, calibration and testing of hydrologic models. In *Hydrological modelling of small watersheds*, C. T. Haan, H. P. Johnson, D. L. Brakensiek (eds), 435–72. St. Joseph, Michigan: American Society of Agricultural Engineers.

Jones, N. L., S. G. Wright, D. R. Maidment 1990. Watershed delineation with triangular-based terrain models. *Journal of the Hydraulics Division, Proceedings of the*

American Society of Civil Engineers **116**, 1232–51.

Kibler, D. F. & D. A. Woolhiser 1970. *The kinematic cascade as a hydrologic model.* Hydrology Paper No. 39, Colorado State University. Fort Collins, Colorado.

Klemes, V. 1983. Conceptualization and scale in hydrology. *Journal of Hydrology* **65**, 1–23.

Klemes, V. 1986. Dilettantism in hydrology: Transition or destiny? *Water Resources Research* **22**, 177S–188S.

Loch, R. J., J. C. Maroulis, D. M. Silburn, 1989. Rill erosion on a self mulching black earth II. Sediment transport equations. *Australian Journal of Soil Research* **27**, 536–45.

Mein, R. G. & B. M. Brown 1978. Sensitivity of parameters in watershed models. *Water Resources Research* **14**, 299–303.

Moore, I. D. & G. J. Burch 1986. Sediment transport capacity of sheet and rill flow: application of unit stream power theory. *Water Resources Research* **22**, 1350–60.

Moore, I. D. & G. R. Foster 1990. Hydraulics and overland flow. In *Process Studies in Hillslope Hydrology,* M. G. Anderson and T. P. Burt (eds), 215–54. Chichester, England: John Wiley.

Moore, I. D. & R. B. Grayson 1991. Terrain-based catchment partitioning and runoff prediction using vector elevation data. *Water Resources Research* **27**, 1177–91.

Moore, I. D. & M. F. Hutchinson 1991. Spatial extension of hydrologic process modelling. *Proceedings of the International Hydrology and Water Resources Symposium, Perth, 2-4 October, 1991*. Institution of Engineers Australia, *National Conference Publication* 91/22, 803–8.

Moore, I. D., G. J. Burch, D. H. Mackenzie 1988a. Topographic effects on the distribution of surface soil water and the location of ephemeral gullies. *Transactions of the American Society of Agricultural Engineers* **31**, 1098–107.

Moore, I. D., E. M. O'Loughlin, G. J. Burch 1988b. A contour-based topographic model for hydrological and ecological applications. *Earth Surface Processes and Landforms* **13**, 305–20.

Moore, I. D., R. B. Grayson, J. P. Wilson 1990. *Runoff modelling in complex three-dimensional terrain. Hydrology in Mountainous Regions I – Hydrological Measurements; the Water Cycle.* International Association of Hydrological Sciences, Publication No. 193, 591–8.

Moore, I. D., R. B. Grayson, A.R. Ladson 1991. Digital terrain modelling: a review of hydrological, geomorphological and biological applications. *Hydrological Processes* **5**, 3–30.

Morel-Seytoux, H. J. 1988a. Recipe for simple but physically based modelling of the infiltration and the local runoff process. In *Proceedings of the eighth annual American Geophysical Union Front Range Branch Hydrology Days*, H. J. Morel-Seytoux & D. G. De Coursey (eds), 226–47. Fort Collins, Colorado: Colorado State University.

Morel-Seytoux, H. J. 1988b. Equivalence between infiltration parameters in the Horton and Morel-Seytoux formulae. In: *Proceedings of the Eigth Annual American Geophysical Union Front Range Branch Hydrology Days*, H. J. Morel-Seytoux & D.G. De Coursey (eds.), 248–59. Fort Collins, Colorado: Colorado State University.

Morris, E. M. & D. A. Woolhiser 1980. Unsteady one-dimensional flow over a plane. partial equilibrium and recession hydrographs. *Water Resources Research* **16**, 355–66.

Mosley, M. P. 1979. Streamflow generation in a forested watershed, New Zealand. *Water Resources Research* **15**, 795–806.

Northcote, J. H., G. D. Hubble, R. F. Isbell, C. H. Thompson, E. Bettenay 1976. *A Description of Australian Soils.* Clayton, Victoria, Australia: CSIRO, Wilke.

O'Loughlin, E. M. 1986. Prediction of surface saturation zones in natural catchments by topographic analysis. *Water Resources Research* **22**, 794–804.

Onstad, C. A. & D. L. Brakensiek 1968. Watershed simulation by stream path analogy. *Water Resources Research* **4**, 965–71.

Parsons, A. J. 1987. The role of slope and sediment characteristics in the initiation and development of rills. In *Processes and Measurement of Erosion*, A. Godard & A. Rapp (eds), 211–20. Paris: Editions du Centre National de la Recherche Scientifique.

Philip, J. R. 1975. Some remarks on science and catchment prediction. In *Prediction in catchment hydrology,* T. G. Chapman & F. X. Dunin (eds), 23–30. Canberra, Australia: Australian Academy of Science.

Sloan, P. G. & I. D. Moore 1984. Modeling subsurface stormflow on steeply sloping forested watersheds. *Water Resources Research* **20**, 1815–22.

Smith, R. E. & J. Y. Parlange 1978. A parameter-efficient hydrologic infiltration model. *Water Resources Research* **14**, 533–8.

Sorooshian, S. & V. K. Gupta 1983. Automatic calibration of conceptual rainfall-runoff models: The question of parameter observability and uniqueness. *Water Resources Research* **19**, 260–8.

Stephenson, G. R. & R. A. Freeze 1974. Mathematical simulation of subsurface flow contributions to snowmelt and runoff, Reynolds Ckreek Watershed, Idaho. *Water Resources Research* **10**, 284–94.

Vieux, B. E. 1991. Geographic information systems and non-point source water quality and quantity modelling. *Hydrological Processes* **5**, 101–13.

Woolhiser, D. A. & J. A. Liggett 1967. Unsteady one-dimensional flow over a plane – the rising hydrograph. *Water Resources Research* **3**, 753–71.

8

Deterministic chaos in surface runoff

Jonathan D. Phillips

Abstract

Runoff and overland flow often exhibit high degrees of spatial and temporal variability. This complexity may be due in part to deterministic chaos. Chaos in surface runoff would imply that a portion of the observed complexity is due to inherent, deterministic dynamics of runoff rather than environmental heterogeneity. This is investigated for two aspects of surface runoff by establishing a link between chaos and asymptotic stability, and testing for the latter. Runoff generation is analysed via an interaction matrix describing the inter-relationships between soil moisture, infiltration capacity and runoff. When soil moisture is viewed as having a positive influence on runoff by promoting saturation overland flow, the system is stable and non-chaotic. When soil moisture and runoff are viewed as having a competitive relationship with respect to precipitation disposal (as would be the case where infiltration-excess runoff prevails), the system is likely to be unstable and potentially chaotic. The hydraulic geometry of overland flow is assessed by examining the stability of an equation system, derived from the Darcy–Weisbach equation, which describes the mutual adjustments of velocity, depth, slope and resistance. This system is asymptotically unstable and potentially chaotic. Complex mutual adjustments can thus arise from simple, deterministic kinematic flow dynamics. Neither surface runoff generation nor flow hydraulics are generally chaotic, implying that chaos will not be observed routinely. It appears that chaos is possible but not ubiquitous in surface runoff, and may occur with regard to overland flow hydraulics and the generation of Hortonian runoff.

Introduction

The generation of surface runoff and the hydraulics of overland flow are, like many hydrologic and geomorphic phenomena, characterized by high levels of spatial and temporal variability. Marked variation in hillslope runoff over short distances has been observed by many workers, including Emmett (1970), Roels (1984), Abrahams et al. (1986) and Julien & Moglen (1990). Beven (1987) and Pearce (1990) have noted that, even after considerable research has been devoted to the topic, the sources of runoff and their spatial distribution is still a subject of debate. The influence of spatial and temporal complexity on hydrologic responses of drainage basins, especially in the form of spatial variability of hydrologic parameters, is one of the most active and critical areas of hydrologic research today (see Beven 1987, Wood et al. 1990).

The observed complexity of surface runoff is typically ascribed to the cumulative outcomes of numerous individual events, and to the inability or unfeasibility of modelling or describing in detail every factor that may influence runoff response or flow hydraulics. However, there is reason to believe that there may be deterministic complexity - chaos - in surface runoff. If this is the case, then at least some of the complexity is inherent in system dynamics and would occur independently of environmental variability.

Chaos is complex, apparently random behaviour arising from the non-linear dynamics of (sometimes very simple) deterministic systems. The deterministic complexity of chaos differs from the traditional view of complexity in environmental systems (termed stochastic complexity) in that the former is inherent in the system and cannot be eliminated by any level of reductionist analysis.

It is reasonable to suspect chaos in surface runoff for several reasons. First, chaos has been shown to exist in two closely related phenomena, the mechanics of turbulent flows (for example, Eckmann 1981) and atmospheric dynamics including precipitation (for example, Rodriguez-Iturbe et al. 1989). Secondly, it has been suggested that the existence of chaos in fundamental physical phenomena such as turbulence is likely to impart chaos into landscape processes in general (Culling 1987, 1988). Finally, chaos has been proposed as a possible explanation for the complex spatial patterns observed in a variety of earth surface phenomena (Slingerland 1989, Malanson et al. 1990, Turcotte 1990).

The purpose of this chapter is to determine whether surface runoff may be chaotic (or under what conditions chaos is likely) in two contexts: the generation of surface runoff, and the hydraulic geometry of overland flow. Chaotic behaviour is characterized by sensitive dependence on initial conditions and increasing divergence over time. Thus the existence of chaos would have important implications for long-term runoff predictions, in effect making such predictions in a deterministic mode impossible. Chaos does not preclude short-term deterministic predictions or longer-term prediction in a stochastic or probabilistic framework. Chaos would imply that a reductionist approach to understanding hydrologic variability has limited potential and that system-level, holistic approaches - essentially hydrologic pattern recognition - are necessary for understanding system behaviours and evolution. In the context of modelling and prediction, chaos would mean that increasingly detailed and sophisticated deterministic models are likely to do little in terms of reducing uncertainty and improving predictability.

Chaos and instability

Chaos is generally agreed to refer to complex, pseudorandom behaviour arising from non-linear deterministic systems. However, there is no formal physical or mathematical definition of chaos and thus no general or global analytical test for chaos. While no formal definition of chaos exists, there is consensus that the distinguishing characteristic of chaos is sensitive dependence on initial conditions and increasing divergence over time. This is true both in the mathematical literature (for example, Thompson & Stewart 1986, Baker & Gollub 1990, Wiggins 1990) and in the literature on chaos in earth surface processes (Culling 1987, Slingerland 1989, Malanson et al. 1990).

There are three basic approaches for determining whether a dynamical system exhibits chaotic behaviour and for analysing the nature of that behaviour. First, numerical simulations are used to map system behaviour in phase space over time. Secondly, time series data of a single realization of a system (for example, time series of daily runoff as a realization of a snowmelt runoff system; Wilcox et al. 1991) can be analysed using methods such as those of Grassberger and Procaccia (1983) or Wolf et al. (1985) (see Ruelle 1987 for a review) to detect chaotic behaviour. Finally, the growth rate of disturbances or perturbations of a system can be analysed to determine whether increasing divergence over time exists (Lorenz 1965, Puccia & Levins 1985, Wiggins 1990). Readers are referred to numerous recent textbooks for a full treatment of chaos theory and other aspects of non-linear dynamic systems (Thompson & Stewart 1986, Schuster 1988, Baker & Gollub 1990, Rasband 1990, Wiggins 1990).

The latter approach – perturbation analysis – is adopted here because it can be linked directly to asymptotic stability. Unlike chaos, asymptotic stability has a formal mathematical definition (for example, Wiggins 1990, 6–10). A system is asymptotically stable if, after a small perturbation away from an equilibrium, the system returns arbitrarily close to its pre-disturbance state, approaching it asymptotically. An asymptotically unstable system will deviate from the pre-disturbance equilibrium at an exponential rate. The concept of asymptotic stability is valid for mathematically small perturbations in local phase space. In hydrologic terms this translates to perturbations, which are not so severe as to add or obliterate system feedback mechanisms, and spatial or temporal scales over which external environmental controls, such as climate and lithology, are reasonably constant. A system that is unstable in the asymptotic sense will be unstable in response to perturbations of any magnitude. Asymptotic stability is closely linked to chaos, and can be tested for analytically. The approach described below was chosen because of its great generality. The analysis and interpretations depend only on the qualitative nature of fundamental hydrologic and hydraulic relationships; not on the characteristics of a particular data set or equation system.

The diagnostic features of deterministic chaos are described by a system where small differences in initial states (Δ_0) diverge exponentially. These evolutionary trends are called trajectories. The difference between two trajectories at time t (Δ_t) is given by

$$\Delta_t \sim \Delta_o e^{Lt} \tag{8.1}$$

where L is the Lyapunov exponent, which clearly indicates the presence of chaotic behaviour. If $L = 0$ then $\Delta_0 = \Delta_t$, and if $L < 0$ then the difference is exponentially damped over time and $\Delta_t \rightarrow 0$ (i.e. asymptotic stability). But if $L > 0$ then there exists the increasing divergence which indicates chaos.

A description of the underlying hydrologic processes is given by a (perhaps unknown) set of non-linear partial differential equations. These may be transformed to a set of n ordinary differential equations

$$\frac{dx_i}{dt} = f_i(x_1, x_2, \ldots x_n), \; i = 1, 2, \ldots n \tag{8.2}$$

with n variables x_i.

The rate of growth of perturbations (δx) of such a system is governed by a set of linear differential equations

$$\frac{d\delta x_i}{dt} = (\sum_{j=1}^{n} A_{ij}) x_j, \quad i = 1, 2, \ldots n. \tag{8.3}$$

Here A_{ij} are elements of the Jacobian matrix of $f = (f_1, \ldots f_n)$ defined by

$$A_{ij} = \frac{\partial f_i(x_1, x_2, \ldots x_n)}{\partial x_j} \Big|_{x = x_0} \tag{8.4}$$

where x_o is the initial equilibrium state.

The Lyapunov exponents of the system are the eigenvalues of the Jacobian with elements A_{ij} (Fraedrich 1987, Wiggins 1990). The Jacobian is exactly equivalent to an interaction matrix A whose elements a_{ij} signify the positive, negative or zero influences of the ith component on the jth component of the system as reflected in the dynamic equations of the system (Puccia & Levins 1985, 246–8).

The interaction matrix **A** is of special interest because it is possible to determine stability for quite generalized hydrologic systems. The question of whether there are any positive L can be determined from the characteristic polynomial of an interaction matrix where all that is known is whether each entry a_{ij} is positive, negative or zero. Because the sum of the diagonal elements of the matrix equals the sum of the eigenvalues (and thus of the Lyapunov exponents) the interaction matrix can also reveal whether the average Lyanpunov exponent is positive (the significance of this is discussed below). The average L will be positive if at least one self-effect term (a_{ii}) is positive and if the sum of $a_{ii} > 0$ is greater than the sum of negative a_{ii}.

The eigenvalues of the system are the roots of the characteristic polynomial of A. The signs of the real parts of the eigenvalues can be determined from the coefficients of the characteristic polynomial using the Routh–Hurwitz criteria (Cesari 1971). The coefficients α_k of the characteristic equation are equivalent to the feedback F_k at level k of the system (Puccia & Levins 1985),

$$F_k = \Sigma(-1)^{m+1} Z(m,k) \qquad (8.5)$$

where $Z(m,k)$ is the product of m disjunct loops with k system components. A disjunct loop is a sequence of one or more a_{ij} which have no component i or j in common. At level zero, feedback is set to $F_o = \alpha_o = -1$. The characteristic equation is then

$$F_0\lambda^n + F_1\lambda^{n-1} + F_2\lambda^{n-2} + \ldots + F_{n-1}\lambda + F_n = 0. \qquad (8.6)$$

Necessary and sufficient conditions for all real parts of the eigenvalues to be negative, according to the Routh–Hurwitz criteria, are that $F_i < 0$ for all i and that successive Hurwitz determinants are positive, though only alternate determinants have to be tested (Puccia & Levins 1985, 167–70). The second condition can be expressed algebraically, and for $n = 3$ or $n = 4$ is

$$F_1F_2 + F_3 > 0. \qquad (8.7)$$

The Routh–Hurwitz criteria apply to the signs of the roots of any polynomial. If both Routh–Hurwitz criteria are met, all eigenvalues have negative real parts, all Lyapunov exponents are negative and the system is asymptotically stable. If the criteria are not met, the system is unstable.

Instability cannot be strictly related to chaos. Asymptotic instability can exist in linear systems, whereas chaos derives only from non-linear systems. However, instability in the asymptotic sense of a non-linear system (assessed by linearizing via a Taylor expansion; see Puccia & Levins 1985) is both a necessary and sufficient condition to show that chaotic behaviour is at least *possible*. The chaos literature is unclear as to what conditions are necessary for a system to be called chaotic. Some workers, such as Wolf et al. (1985) hold that a system is chaotic if there are any positive Lyapunov exponents. There is little disagreement that if one $L > 0$ a system could behave chaotically in certain circumstances, but Berryman & Millstein (1989) maintain that for an ecological system to be termed chaotic the average L must be positive. An unstable system can clearly exhibit behaviours other than deterministic chaos (Puccia & Levins 1985, Wiggins 1990). May (1976) has demonstrated quite clearly that some systems may behave chaotically under certain circumstances but not under all circumstances.

A threefold classification will be employed here. An asymptotically stable system is non-chaotic. An unstable system with average positive Lyanpunov exponents is termed generally chaotic, implying that chaotic behaviour is a routine feature. An unstable system where the average L is less than or equal to zero is termed potentially chaotic, meaning that deterministic chaos is possible but not routine or ubiquitous. In a hydrologic context, if a system is non-chaotic then observed complex, apparently random behaviour can safely be assumed to derive from stochastic forcings or environmental heterogeneity. In a generally chaotic system, apparently random hydrologic phenomena are likely to derive from deterministic chaos. In a potentially chaotic system, the hydrologist should consider chaos as one of the possible explanations of observed random-like behaviour.

Because it deals with a linear system, the use of the Routh–Hurwitz method to determine asymptotic (in)stability may initially appear inappropriate in the context of chaos theory. Note, however, that the analysis is applied to a non-linear system which has been linearized by examining time-behaviour of perturbations or by a Taylor analysis (Cesari 1971, Puccia & Levins 1985). Thompson & Stewart (1986, 200–4), among others, explicitly establish the relationship between the stability analysis of the linearized system and chaotic behaviour of the non-linear system. The stability properties of a linearized system are exactly the same as those of the non-linear "parent" system.

Qualitative asymptotic stability analysis has been applied to several problems in hydrology and fluvial geomorphology. Slingerland (1981) determined that river at-a-station hydraulic geometry is asymptotically unstable, based on a Routh–Hurwitz analysis of Hey's (1979) hydraulic geometry model. Phillips (1990) performed a similar analysis of the problem, this time deriving an interaction matrix for the system from a standard kinematic flow resistance equation, and also found hydraulic geometry to be unstable. Phillips & Steila (1984) used the approach to assess the stability of wetland hydrologic systems disturbed by artificial drainage, and found that asymptotic stability was contingent upon the frequency of artificial channel maintenance. Geomorphic applications of the approach are presented by Scheidegger (1983) and Phillips (1987).

Generation of surface runoff

The surface runoff response to a given precipitation input is determined by soil moisture (both antecedent moisture and soil moisture storage capacity) and infiltration. Runoff, soil moisture and infiltration are mutually interdependent.

Infiltration has a negative influence on runoff, because higher infiltration will reduce runoff, and vice versa. Infiltration has a positive influence on soil moisture, as soil moisture cannot be recharged unless precipitation infiltrates, and any increases or decreases in infiltration rates or capacities would produce corresponding increases or decreases in soil moisture.

Soil moisture has a negative influence on infiltration, as reflected in the well-known relationship whereby infiltration rates are at a maximum in a dry soil and decrease curvilinearly with increasing moisture content, reaching a constant value when the soil is saturated. Changes in soil moisture storage will thus result in changes of infiltration capacity in the opposite direction. Soil moisture also has a negative, self-limiting effect. Water depletion is limited by matric potentials and does not exceed the wilting point in most circumstances. Moisture storage is limited not only by a finite capacity, but by the tendency for saturation to facilitate both percolation and throughflow.

Runoff has a negative influence on soil moisture, as the greater the proportion of precipitation that runs off, the less water there is available for soil moisture recharge. Likewise, slower runoff rates promote soil recharge. The effect of runoff on storage could also be zero or negligible during steady-state saturation-excess runoff. Runoff has a positive influence on infiltration. Higher runoff rates have been correlated with greater infiltration rates in field studies, with increased pressure heads being the apparent mechanism involved (Lane et al. 1987, Rawls et al. 1990).

The effect of soil moisture on runoff could be positive or negative in various situations. On one hand, increased soil moisture storage would promote saturation-excess runoff (and decreased soil water would prevent such runoff), implying a positive relationship. On the other hand, where the soil is not near saturation or is rarely or never saturated, soil moisture has a negative influence on runoff rates and volumes, because soil water becomes a "competitor" with surface runoff in the partitioning of precipitation inputs. Note that the influence of soil moisture on runoff via infiltration has already been accounted for.

The relationships outlined above are expressed in the form of Equation 8.2 by writing $dq/dt = f_1(S,F)$; $dS/dt = f_2(q,F)$; $dF/dt = f_3(q,S)$; where q is total

Table 8.1 Interaction matrix for surface runoff generation for given precipitation inputs. The influence of soil moisture on runoff may be positive or negative (see text).

	S	F	q
Soil moisture	$-a_{11}$	$-a_{12}$	$\pm a_{13}$
Infiltration (F)	a_{21}	0	$-a_{23}$
Runoff (q)	$-a_{31}$	a_{32}	0

runoff during a precipitation event, S is soil moisture storage and F is cumulative infiltration. The exact form of f_1, f_2, f_3 need not be known, and in fact a number of specific equations or models could be used. The positive, negative or negligible influences of q, S and F on each other are all that is necessary to apply the Routh–Hurwitz criteria. These are shown in Figure 8.1 and in an interaction matrix in Table 8.1. A number of specific non-linear dynamic equations could be used to model the runoff–soil moisture–infiltration relations, but by linearizing around an equilibrium and considering the time behaviour of a small perturbation, the problem is reduced in the qualitative case to Table 8.1.

For the case where the influence of soil moisture on runoff is positive ($a_{13} > 0$),

$$F_1 = -a_{11} < 0 \tag{8.8a}$$

$$F_2 = -a_{12}a_{21} + a_{13}(-a_{31}) + (-a_{32}a_{23}) < 0 \tag{8.8b}$$

$$F_3 = (-a_{23})(-a_{31})(-a_{12}) + a_{13}a_{32}a_{21}. \tag{8.8c}$$

Feedback at level three will be negative if the loop linking infiltration to runoff to soil moisture to infiltration is stronger than the loop connecting soil moisture to runoff to infiltration to soil moisture (i.e. if $[-a_{23}][-a_{31}][-a_{12}] > [a_{13}a_{32}a_{21}]$). This will hold true if the effects of infiltration on runoff are stronger than those of runoff on infiltration, which will virtually always be the case. Because F_1 and F_2 are more strongly negative than F_3, the second stability criterion (Eq. 8.7) is easily met. The system is thus stable and non-chaotic.

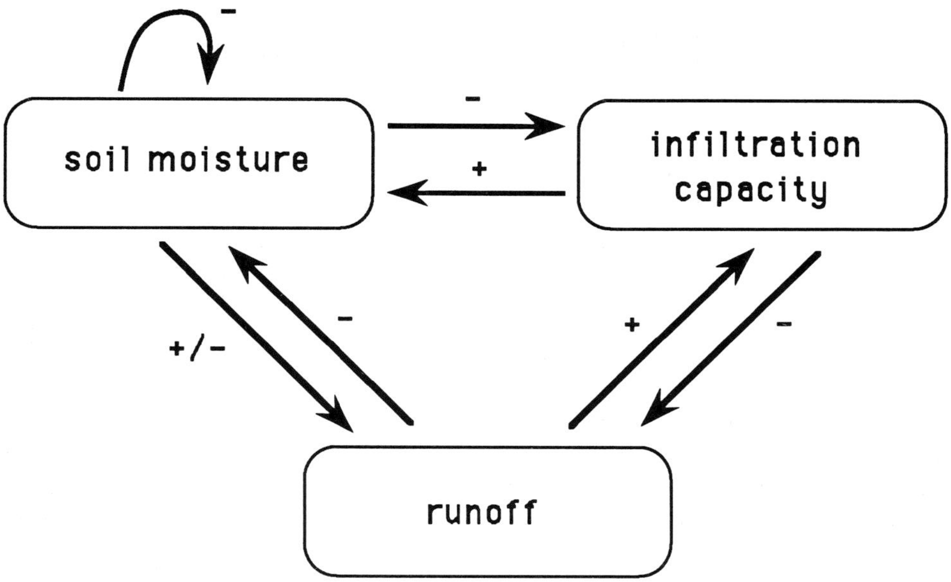

Figure 8.1 Loop model of inter-relationships among soil moisture, infiltration capacity and runoff for a given precipitation input.

When $a_{13} < 0$ and soil moisture has a negative effect on runoff, F_1 is unchanged and

$$F_2 = -a_{12}a_{21} + (-a_{13})(-a_{31}) + (-a_{32}a_{23}) \quad (8.9a)$$

$$F_3 = (-a_{23})(-a_{31})(-a_{12}) + (-a_{13})a_{32}a_{21} < 0. \quad (8.9b)$$

Feedback at level two will be negative if the combined effects of the infiltration–soil moisture and runoff–infiltration relationships ($-a_{12}a_{21}$ and $-a_{23}a_{32}$) are stronger than the two-way relationship between soil moisture and runoff ($[-a_{13}][-a_{31}]$). Because in this case ($a_{13} < 0$) soil moisture and runoff are competitive, their relationship is fundamentally one of precipitation partitioning and thus operates almost instantaneously. The soil-moisture–infiltration-capacity and runoff–infiltration capacity relationships may involve some time lags with respect to the influence of soil moisture and runoff on infiltration. Therefore, the stability condition that $F_3 < 0$ may not be met in many cases. Further, because F_2 is likely to be only weakly negative, if it is negative at all, the stability condition given in Equation 8.7 is unlikely to be met. The system is thus not likely to be asymptotically stable and will be potentially chaotic (but not generally chaotic) in most instances.

The implications are that where soil moisture has a direct relationship with surface runoff, the system is likely to be stable. Where soil moisture has a negative influence on runoff, the system is more likely to be unstable and potentially chaotic. The former situation is probable where or when saturation-excess runoff is the dominant runoff-producing mechanism. The latter (unstable and chaotic) situation is more prevalent when infiltration-excess (Hortonian) overland flow is dominant.

Stability analysis shows that surface runoff generation will not, as a general rule, be characterized by deterministic chaos. Saturation-excess flow is non-chaotic. The higher the precipitation intensity and the lower the infiltration capacity, the greater the likelihood of infiltration-excess flow, instability and potentially chaotic behaviour. In the case of Hortonian flow, deterministic chaos should not be routinely observed, but cannot be ruled out as a possible source for complex behaviour.

Examples from field data

Wilcox et al. (1991) tested for chaos in an 8800-day time series of snowmelt runoff in southwestern Idaho using the Grassberger and Procaccia (1983) algorithm. They found no evidence of chaotic behaviour. Snowmelt runoff would be characteristic of a situation where soil moisture has a direct relationship with surface runoff, producing saturation-excess overland flow. Therefore the results are consistent with the prediction of the generalized model above. The apparently random time series of snowmelt runoff was attributed to the complex interaction of numerous factors influencing runoff, rather than to deterministic chaos (Wilcox et al. 1991).

Chaos in infiltration-excess runoff has not been examined in a field setting. Smith & Bretherton (1972) did address the issue of whether sediment transport as a function of Hortonian sheet flow is unstable, via a perturbation analysis. It was found that Hortonian sheet flow is inherently unstable and will separate into streams which will incise channels. This expectation was confirmed by field experiments on hillslopes in Kenya by Dunne & Aubry (1986). Loewenherz (1991) conducted a more complete stability analysis of Hortonian overland flow, confirming the general instability while eliminating the problematic implication of Smith & Bretherton's analysis of an infinite number of infinitely small rills. While these results are consistent in a very general way with the finding in this paper that Hortonian runoff generation may be asymptotically unstable, the runoff-generation and flow-separation phenomena are not necessarily linked.

Examples from runoff plots

If surface runoff generation via infiltration-excess flow is unstable and potentially chaotic, as implied by this analysis, then relatively small variations in initial conditions (precipitation intensities, soil characteristics, etc.) should produce much greater variations in the system state in the future when the system is behaving chaotically. This behaviour should be occasionally, though not regularly, observed in data. A review of published runoff plot data found two instances where behaviour consistent with deterministic chaos can be observed, though data are insufficient for a rigorous test of chaos.

Mutchler & Greer (1980) established four sets of paired runoff plots of varying length but identical in other respects. These were subjected to simulated rainfall in a study of the effect of slope length on erosion from low slopes. All plots were on the same soil type (Leeper silty clay loam; Vertic Haplaquept) at an experimental station in Mississippi. Plots were levelled to a grade of 0.2% and the surfaces uniformly disked. Plots were 23 m to 183 m long and 3 m wide. Simulated rainfall was applied at a design intensity of 76 mm h^{-1} in two 1 h increments about 5 h apart. Runoff was measured with an HS flume and water stage recorder. A more detailed description is given in the original source (Mutchler & Greer 1980).

The test plot soils were apparently not saturated, and given the relatively high simulated rainfall intensity, runoff generation can be assumed to be dominantly Hortonian. Despite efforts to control simulated rainfall intensities, actual measured application rates at each plot varied from 59.9 mm h^{-1} to 85.3 mm h^{-1}. Antecedent soil moisture also varied slightly. These differences in application rates in most cases produced much greater differences in total measured runoff.

Figure 8.2 shows the percentage differences in runoff and simulated rainfall at each pair of plots (computed as the difference between each W and E plot, divided by the smaller value of rainfall or runoff). The histogram shows that in five of eight cases, proportional differences in runoff were greater than the proportional differences in application rates. In general, results from the Mississippi plots exhibit the sensitive dependence on initial conditions (minor variations in rainfall intensity and antecedent moisture) and divergence over time characteristic of chaos.

Plot data from Ewing & Mitchell's (1986) runoff and sediment transport simulation studies also show the time-divergence and sensitive dependence on initial conditions characteristic of chaos. In this case, groups of 3–6 plots (48 plots total) of 3 m by 11 m were established at newly reclaimed surface mine sites in Illinois and Indiana. Each group of sites was established on identical soils or materials, with only minor (1% or less) variation in slope gradient. Final infiltration capacities and other infiltration-equation parameters were

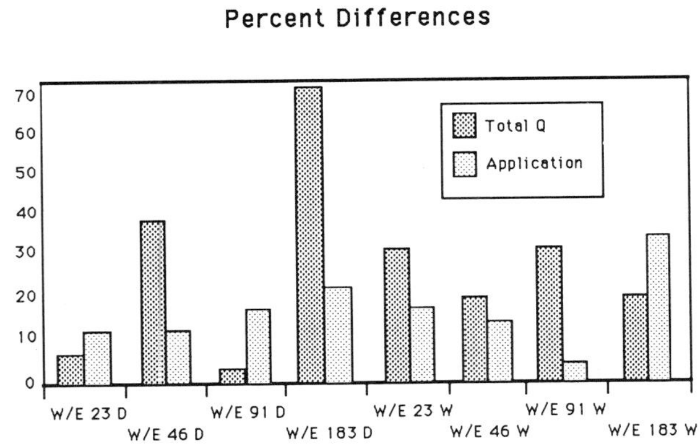

Figure 8.2 Proportional differences in total runoff (*Q*) associated with proportional differences in simulated runoff application rates at identical paired runoff plots (data from Mutchler & Greer 1980).

determined from three rainfall-simulator storms at each plot (for more details see Ewing & Mitchell 1986).

Results showed significant – and often dramatic – variations in final infiltration capacity and in the ratio of maximum to final infiltration within groups of similar plots. Because only minor, local variability of soil and geomorphic properties exists within each group of plots, the clear divergence of infiltration rates over time and within each group is interpreted as evidence of sensitive dependence on initial conditions. An example plot of final infiltration rates at a group of five similar sites in western Illinois from the Ewing & Mitchell (1986) data is shown in Figure 8.3.

Hydraulic geometry

Hydraulic geometry studies are concerned with the mutual adjustments of stream channels and the flows of water and sediment they convey. The hydraulic geometry approach may also be applied to overland flow, as hillslope surfaces both influence and are influenced by the water flowing across them.

At-a-station hydraulic geometry in this case is concerned with how changes in imposed flow (discharge from upslope and locally generated, surface runoff) are accommodated at a given location on a hillslope. While a number of individual parameters influencing hydraulic geometry may vary at a given site (see Hey 1979 in the context of channels) there are basically four ways an increase or decrease in imposed flow can be accommodated, i.e. by increasing or decreasing velocity, depth (hydraulic radius), frictional resistance or energy

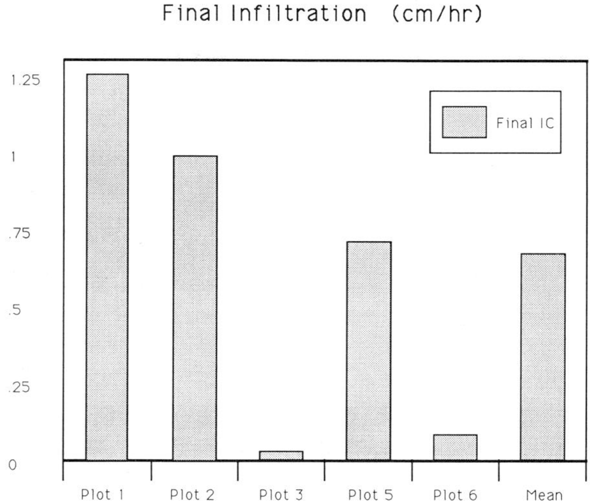

Figure 8.3 Differences in final infiltration capacity (cm h^{-1}) at runoff plots under simulated rainfall, with only minor differences in soil properties (data from Ewing & Mitchell 1986, plots 1–6).

grade slope. Each of these fundamental hydraulic variables is accounted for in the Darcy–Weisbach equation for uniform, turbulent, kinematic sheet or rill flow. Written to solve for discharge, this equation is

$$Q = wd\left(\frac{8gRs}{f}\right)^{0.5} \tag{8.10}$$

where Q is discharge, w is width of the slope element, d is mean depth, g is the gravity constant, R is hydraulic radius, s is the energy grade slope, and f is the Darcy–Weisbach friction factor. Where flows are wide relative to their depth, $R = d$. This is the case for sheet flow, and for composite or aggregate rill flow.

The mutual interdependencies of the fundamental hydraulic variables can be illustrated by rewriting the Darcy–Weisbach equation to solve for V, s, R and f in the following form so that positive and negative influences are quickly seen from the exponents:

$$V = R^{0.5} s^{0.5} f^{-0.5} (8g)^{0.5} \tag{8.11a}$$

$$s = V^2 R^{-1} f^1 (8g)^{-1} \tag{8.11b}$$

$$R = V^2 s^{-1} f^1 (8g)^{-1} \tag{8.11c}$$

$$f = V^{-2} R^1 s^1 (8g)^1 . \tag{8.11d}$$

Note that the qualitative relationships between the variables would be identical for equations describing a laminar flow regime, though in some cases laminar flow resistance is accounted for by a viscosity term. The equation system above has been shown to be asymptotically unstable (Phillips 1990) and is thus potentially chaotic.

Instability and deterministic chaos in this context require the existence of multiple modes of adjustment, i.e. qualitatively different ways in which the system can respond to a given change in imposed flow (Phillips 1990, 1991). A mode of adjustment in hydraulic geometry is defined as a particular combination of increases, decreases or relative constancy of hydraulic variables in response to changes in Q. Theoretically, one would expect V, s and R to change in the same direction as Q, and f to change in the opposite direction. Existence of qualitatively different modes of adjustment thus depends on the possibility of opposite-from-expected behaviour: V, s and/or R changing in the opposite direction from discharge, and f in the same direction. It has already been shown that multiple modes of adjustment exist in river channel hydraulic geometry and that their presence is consistent with traditional hydraulic geometry theory (Phillips 1991). Opposite-from-expected behaviour has not otherwise been addressed in the context of system stability and chaos, but is well known in many specific cases. For example, the friction factor might well increase as discharge increases as the cross-sectional shape changes.

The general asymptotic instability of the hydraulic geometry system (Eqs 8.11) suggests that the hydraulic geometry of overland flow is also unstable and possibly chaotic. One important difference is that in channels energy gradient is readily adjusted, while in surface runoff the energy gradient is essentially fixed by, and approximated by, the land-surface slope. Certainly all studies of overland flow make this assumption (Abrahams et al. 1986). Does the exclusion of slope from the hydraulic geometry model by assuming its short-term constancy influence its stability properties?

An interaction matrix linking V, R and f, derived from the signs of the exponents in Equations 8.10, is given in Table 8.2. For this system

$$F_1 = 0 \tag{8.12a}$$

$$F_2 = a_{12}a_{21} + (-a_{13})(-a_{31}) + a_{23}a_{32} > 0 \tag{8.12b}$$

$$F_3 = a_{12}a_{23}(-a_{31}) + (-a_{13})a_{32}a_{21} < 0. \tag{8.12c}$$

Because the absence of any self-stabilizing loops (which results in $F_1 = 0$) predetermines that the stability criteria cannot be met, a negative self-effect was assumed for energy grade slope in the channel geometry model (Phillips 1990). But since $F_2 > 0$ in any case, the stability criteria cannot be met even if a self-stabilizing loop were added to the model. The hydraulic geometry of overland flow is asymptotically unstable.

Table 8.2 Interaction matrix for overland flow hydraulic geometry.

	V	*R*	*f*
Velocity	0	a_{12}	$-a_{13}$
Hydraulic radius	a_{21}	0	a_{23}
Friction factor	$-a_{31}$	a_{32}	0

Evidence from field data

As is the case with channel flow, asymptotically unstable and chaotic overland flow hydraulic geometry means that there must be multiple modes of adjustment and opposite-from-expected responses of *R*, *V* and *f* to changes in discharge. Such behaviour can, in fact, be observed in field data.

Abrahams et al. (1986) published hydraulic data from six runoff plots at Walnut Gulch Experimental Watershed, Arizona. Runoff was generated using simulated rainfall, and hydraulic variables were measured at three cross-sections on each plot for six runs. There are thus 18 hydraulic geometry stations, each with hydraulic data for a sequence of six increasing discharges. At six of 18 stations mean depth (hydraulic radius) behaved in an opposite-from-expected manner at least once, decreasing in response to increased flow. In two cases at two stations *R* stayed constant as flow

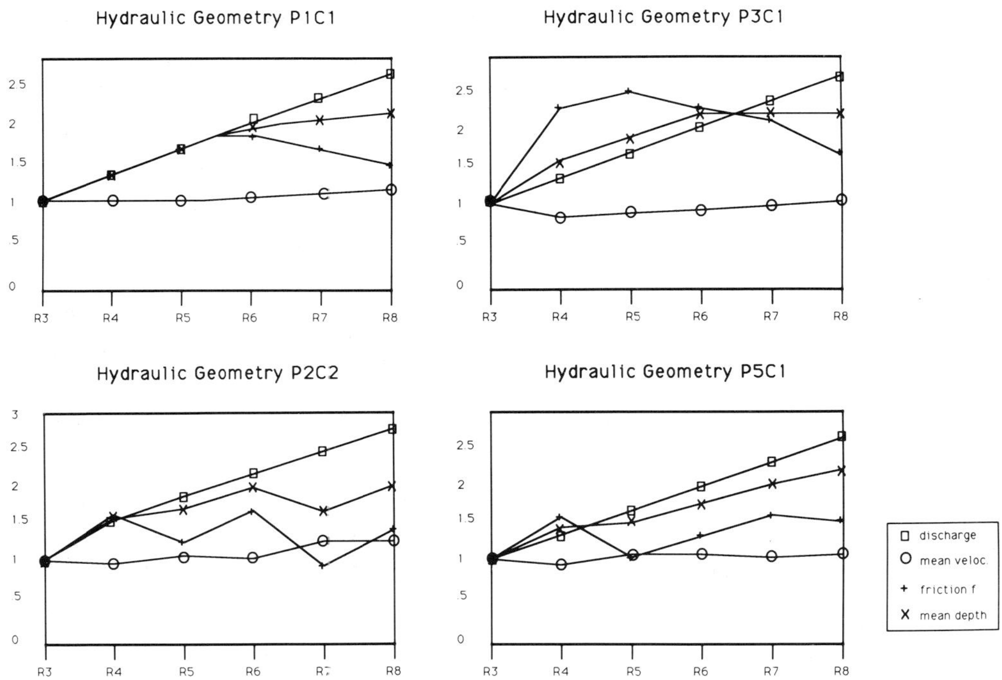

Figure 8.4 Trends in discharge, mean depth, mean velocity and friction factor at four representative cross sections from the data of Abrahams et al. (1986).

increased. Mean velocity exhibited opposite-from-expected behaviour at least once at 13 stations, declining in response to increased discharge. The friction factor frequently showed such behaviour, increasing in conjunction with discharge at all but one station. Plots of relative increases or decreases of Q, d, V and f for four typical cross-sections are shown in Figure 8.4.

Discussion

The complicated spatial and temporal patterns often associated with surface runoff generation and related hydrologic parameters may not always be attributable to environmental heterogeneity. Some of the apparent complexity may be deterministic chaos. The analysis above suggests that chaos is possible with regard to runoff generation in general, and more likely where and when Hortonian runoff dominates. But, most importantly, the analysis shows that deterministic chaos is unlikely to be regularly observed in runoff generation.

To the extent that chaos does characterize runoff generation, its existence implies that efforts to reduce uncertainty in rainfall–runoff relationships by increasingly detailed monitoring or modelling may have limited promise. This suggests that probabilistic or statistical approaches to characterizing spatially variable hydrologic responses (often within physical deterministic models) are more fruitful strategies for reducing uncertainty than are reductionist approaches. Because spatial and temporal complexity is sometimes an inherent property of runoff–soil-moisture–infiltration relationships, "holistic" efforts to identify representative or characteristic scales of runoff generation (see Wood et al. 1990) or overland flow lengths (Julien & Moglen 1990) may provide more insight into rainfall–runoff dynamics than reductionist studies.

Deterministic chaos in overland flow hydraulic geometry implies that the response to changes in imposed flow is predictable in a deterministic sense only when the relative rates and intensities of the responses of velocity, depth and frictional resistance are known. The multiple modes of adjustment in runoff hydraulics may be especially important with regard to erosion and sediment transport. If slope is considered fixed, variation in unit stream power (for example) at a station is controlled by velocity. Because velocity may not always increase or decrease in conjunction with discharge in the asymptotically unstable system, the relationship of erosion and sediment transport to runoff volume may be quite variable, independently of the effects of sediment supply or detachment rates.

As is the case for any chaotic dynamic system, chaos in overland flow bodes ill for long-term deterministic prediction, though it does not preclude short-term prediction or probabilistic prediction. The effectiveness of a stochastic modelling technique is unaffected by whether random elements are real or apparent. Chaos bodes well for explaining complex phenomena, however. Its presence indicates that complicated, irregular phenomena need not be written off as being controlled by environmental details whose measurement would require impossible or unreasonable levels of detail. These phenomena may have relatively simple deterministic explanations.

A major problem with respect to chaos analysis in hydrology (and in the Earth and environmental sciences in general) is that some level of stochastic complexity is likely to be present in addition to any deterministic chaos which may be present. While random (stochastic), regular or periodic, and chaotic dynamics all have clear signals in time series data, it is difficult to confirm the presence and extent of chaos when both chaotic and stochastic signals are present. Culling (1987) believed that this fact would inhibit applications of chaos theory in physical geography.

Here, for example, while the runoff plot data show behaviour consistent with chaos, there is no way to determine whether there may be elements of stochastic complexity as well. With regard to the hydraulic geometry data of Abrahams et al. (1986) presented here, it is possible that opposite-from-expected behaviour may be, at least partially, an artefact of measurement error inherent in the data-collection techniques (A. J. Parsons and A. D. Abrahams, personal communication). It is also worth noting that four cross sections were explicitly excluded from analysis by Abrahams et al. (1986, 351–2) because of their unusual behaviour.

In addition to problems associated with identifying chaos in field data, there is the issue of the extent to which chaos is a property of equations as opposed to a property of hydrologic systems. The findings of analyses based on stability or perturbation theory such as those here (or findings based on numerical simulations) are applicable to real hydrologic systems only to the extent that the equation system, box-and-arrow model or interaction matrix is a true representation of the real system.

Because chaos is so widely suggested to occur in geophysical phenomena, the finding that a hydrologic system is non-chaotic has obvious utility. Such a result justifies continued efforts to improve predictability via reductionist approaches, and reaffirms stochastic forcings and environmental heterogeneity as the primary sources of complexity. If a system is found to be generally or potentially chaotic, many hydrologists may well ask: So what?

In a generally chaotic system the implication is that observed complexity can be explained deterministically, but that predictability cannot be substantially improved by a reductionist approach or by increasingly sophisticated deterministic modelling. Prediction for practical purposes must rely on probabilistic or stochastic approaches. An understanding of system dynamics should be based on attempts to identify the number and nature of strange attractors and the system's behaviour relative to them. Generally chaotic systems are unlikely in hydrology because they require the presence of self-ehancing positive feedback, which is rare in hydrologic systems.

In a potentially chaotic system, the hydrologic implication is that deterministic chaos might be an explanation for observed complexity. In other words, chaos in such a system must be considered (along with stochastic forcings and environmental heterogeneity) as one of the potential sources or causes of apparent randomness. In this case the task is to determine under what conditions chaotic behaviour may occur and the extent to which these conditions will obtain in the field.

The models presented here show that chaos can occur in simple abstract representations of surface runoff systems. Supporting data show results that are consistent with the model predictions. Results provide less insight into the extent of chaos. Future studies should be directed at determining the general extent of chaos in surface runoff and other hydrologic phenomena and at distinguishing between deterministic and stochastic complexity.

Conclusions

A simple model reflecting the mutual influences of soil moisture, infiltration and runoff suggests that runoff generation in response to a given precipitation input is likely to be stable where saturation-excess runoff is predominant, and may be unstable when infiltration-excess runoff dominates. Data from runoff plots show that behaviour consistent with deterministic chaos does occur in runoff generation, as evidenced by large differences in total runoff or final infiltration capacity associated with small differences in initial conditions of site properties or simulated rainfall rates.

The hydraulic geometry of overland flow is also shown to be asymptotically unstable and potentially chaotic, based on the relationships between velocity, depth and friction factor given in the Darcy–Weisbach equation. The chaotic behaviour is manifested as multiple modes of adjustment. These are qualitatively different combinations of increases, decreases or constancy of hydraulic factors in response to changing imposed flows. Field data illustrate this type of behaviour.

Chaos is unlikely to be a common feature of surface runoff and it is therefore unlikely that chaos theory will occupy a central role in the study of overland flow. However, chaos in runoff cannot be dismissed, and application of chaos theory may be necessary for better understanding of the observed complexity in surface runoff.

References

Abrahams, A. D., A. J. Parsons, S.-H. Luk 1986. Resistance to overland flow on desert hillslopes. *Journal of Hydrology* **88**, 343–63.

Baker, G. L. & J. P Gollub 1990. *Chaotic dynamics: an introduction*. Cambridge: Cambridge University Press.

Berryman, A. A. & J. A. Millstein 1989. Are ecological system chaotic – And if not, why not? *Trends in Ecology and Evolution* **4**, 26–8.

Beven, K. 1987. Towards a new paradigm in hydrology. In *Water for the future: hydrology in perspective*, J. Rodda & N. C. Matalas (eds), 393–403. International Association of Hydrological Sciences Publ. 164, Wallingford, England.

Cesari, L. 1971. *Asymptotic behavior and stability problems in ordinary differential equations*. Berlin: Springer.

Culling, W. E. H. 1987. Equifinality: Modern approaches to dynamical systems and their potential for geographical thought. *Institute of British Geographers Transactions New Series* **12**, 57–72.

Culling, W. E. H. 1988. A new view of the landscape. *Institute of British Geographers Transactions New Series* **13**, 345–60.

Dunne, T. & B. F. Aubry 1986. Evaluation of Horton's theory of sheetwash and rill erosion on the basis of field experiments. In *Hillslope processes*, A. D. Abrahams (ed.), 31–53. London: Allen and Unwin.

Eckmann, J. P. 1981. Roads to turbulence in dissipative dynamical systems. *Reviews of Modern Physics* **53**, 655–71.

Emmett, W. W. 1970. *The hydraulics of overland flow on hillslopes*. US Geological Survey Professional Paper 662A. Washington: US Government Printing Office.

Ewing, L. K. & J. K. Mitchell 1986. Overland flow and sediment transport simulation on small plots. *Transactions of the American Society of Agricultural Engineers* **29**, 1572–81.

Fraedrich, K. 1987. Estimating weather and climate predictability on attractors. *Journal of Atmospheric Sciences* **44**, 722–8.

Grassberger, P. & I. Procaccia 1983. Measuring the strangeness of strange attractors. *Physica* **D9**, 189–208.

Hey, R. D. 1979. Dynamic process–response model of river channel development. *Earth Surface Processes and Landforms* **4**, 59–72.

Julien, P. Y. & G. E. Moglen 1990. Similarity and length scale for spatially varied overland flow. *Water Resources Research* **26**, 1819–32.

Lane, L. J., J. R. Simanton, T. E. Hakonson, E. M. Romney 1987. Large-plot infiltration studies in desert and semi-arid rangeland areas of the Southwestern USA. In *International conference on infiltration development and application*, 365–76. Honolulu: University of Hawaii-Manoa.

Loewenherz, D. S. 1991. Stability and the initiation of channelized surface drainage: A reassessment of the short wavelength limit. *Journal of Geophysical Research* **96B**, 8453–64.

Lorenz, E. N. 1965. A study of the predictability of a 28-variable atmospheric model. *Tellus* **27**, 321–33.

Malanson, G. P., D. B. Butler, S. Walsh 1990. Chaos in physical geography. *Physical Geography* **11**, 293–304.

May, R. M. 1976. Simple mathematical models with very complicated dynamics. *Nature* **261**, 459–67.

Mutchler, C. K. & J. D. Greer 1980. Effect of slope length on erosion from low slopes. *Transactions of the American Society of Agricultural Engineers* **23**, 866–9.

Pearce, A. J. 1990. Streamflow generation: An Austral view. *Water Resources Research* **26**, 3037–47.

Phillips, J. D. 1987. Sediment budget stability of the Tar River, North Carolina. *American Journal of Science* **287**, 780–94.

Phillips, J. D. 1990. The instability of hydraulic geometry. *Water Resources Research* **26**, 333–42.

Phillips, J. D. 1991. Multiple modes of adjustment at unstable river channel cross-sections. *Journal of Hydrology* **123**, 39–49.

Phillips, J. D. & D. Steila 1984. Hydrologic equilibrium status of a disturbed eastern North Carolina watershed. *GeoJournal* **9**, 350–6.

Puccia, C. J. & R. Levins 1985. *Qualitative modeling of complex systems*. Cambridge, Mass.: Harvard University Press.
Rasband, S. N. 1990. *Chaotic dynamics of nonlinear systems*. New York: John Wiley.
Rawls, W. J., D. L. Brakensiek, J. R. Simanton, K. D. Kohl 1990. Development of a crust factor for a Green Ampt model. *Transactions of the American Society of Agricultural Engineers* **33**, 1224–8.
Rodriguez-Iturbe, I., B. F. dePower, M. B. Sharifi, K. P. Georgakakos 1989. Chaos in rainfall. *Water Resources Research* **25**, 1667–75.
Roels, J. M. 1984. Flow resistance in concentrated overland flow on rough slope surfaces. *Earth Surface Processes and Landforms* **9**, 541–51.
Ruelle, D. 1987. *Chaotic evolution and strange attractors: the statistical analysis of time series for deterministic non-linear systems*. New York: Cambridge University Press.
Scheidegger, A. E. 1983. Instability principle of geomorphic equilibrium. *Zeitschrift für Geomorphologie* **27**, 1–19.
Schuster, H. G. 1988. *Deterministic chaos: an introduction*. Weinheim: Physik Verlag.
Slingerland, R. L. 1981. Qualitative stability analysis of geologic systems, with an example from river hydraulic geometry. *Geology* **9**, 491–3.
Slingerland, R. L. 1989. Predictability and chaos in quantitative dynamic stratigraphy. In *Quantitative dynamic stratigraphy*, T. A. Cross (ed.), 45–53. Englewood Cliffs, New Jersey: Prentice-Hall.
Smith, T. R. & F. P. Bretherton 1972. Stability and the conservation of mass in drainage basin evolution. *Water Resources Research* **8**, 1506–29.
Thompson, J. M. T. & H. B. Stewart 1986. *Nonlinear dynamics and chaos*. New York: John Wiley.
Turcotte, D. L. 1990. Implications of chaos, scale-invariance and fractal statistics in geology. *Global and Planetary Change* **89**, 301–8.
Wiggins, S. 1990. *Introduction to applied nonlinear dynamical systems and chaos*. Berlin: Springer.
Wilcox, B. P., M. S. Seyried, T. H. Matison 1991. Searching for chaotic dynamics of snowmelt runoff. *Water Resources Research* **27**, 1005–10.
Wolf, A., J. B. Swift, H. L. Swinney, J. A. Vastano 1985. Determining Lyapunov exponents from a time series. *Physica* **D16**, 285–317.
Wood, E. F., M. Sivapalan, K. J. Beven 1990. Similarity and scale in catchment storm response. *Reviews of Geophysics* **28**, 1–18.

9

Extending overland-flow models to problems of slope evolution and the representation of complex slope-surface topographies

Andrew J. Baird, John B. Thornes, Glenn P. Watts

Abstract

The representation of overland flow as a kinematic wave is well established and has been used with success by many hydrologists. However, overland flow models that use the kinematic wave have generally been developed for very simple slope-surface topographies and no attempt has been made to use these models to simulate three-dimensional slope evolution by erosion. The usual formulation of existing overland-flow models has important weaknesses when used to simulate erosion. The most critical is the adoption of a uniform water depth and an average roughness over the reach under consideration. To some extent these weaknesses can be overcome by partitioning the flow into rills and interrills, but this approach fails to recognize the very dynamic nature of this partitioning and underestimates the role of concentrated flow in the so-called interrills.

We explore two ways of extending existing approaches. First, we present a model that has been designed to apply to three-dimensional slope evolution by erosion. Secondly, we look at an alternative way of modelling topographically complex surfaces in overland-flow models. In the first model, we show how even small rates of erosion can change flow patterns on a slope surface dramatically. In the second model, flow in rills is partitioned dynamically over time and space. This dynamic partitioning affects the outflow hydrograph both in its time to peak and duration: this is demonstrated in a number of example hydrographs generated using the model.

Introduction

In this chapter we address the well-known problem of modelling erosion in three dimensions at the hillslope scale and the appropriate hydrological routing method for this purpose. We also look at an alternative approach to conventional kinematic wave overland-flow models that accounts for rilled or gullied topography, and that avoids the usual preordained division into rills and interrills and the need for an arbitrary threshold of rill initiation. We start by considering the kinematic wave approach to overland flow routing on hillslopes and its limitations when used for topographically complex slopes.

Unsteady, gradually varied, open channel flow is described by the St. Venant equations (the dynamic wave model), which consist of the unsteady continuity equation and the equation of motion with appropriate boundary conditions. To apply kinematic wave theory, the shallow-water wave is assumed to be long and flat so that the friction slope is nearly equal to the bed slope. In other words, the gravitational forces are assumed to be balanced by the frictional resistance due to inertial and viscous forces, so that the net force is zero and is not included in the formulation. It is the absence of force considerations that distinguishes kinematic waves from dynamic waves. The equation of motion is therefore replaced by an equation describing uniform flow. Thus the kinematic wave model is given by the unsteady continuity equation and a uniform flow formula:

$$\frac{\partial A}{\partial t} + \frac{\partial Q}{\partial x} = i \tag{9.1}$$

$$V = U_c C \sqrt{RS} \tag{9.2}$$

where A is area (cm^2), Q is discharge (cm^3 s^{-1}), t is time (s), i is the lateral inflow at distance x (cm^2 s^{-1}), V is the channel mean velocity (cm s^{-1}), R is the hydraulic radius (cm), U_c is the unit adjustment factor in the Chezy equation (Eq. 9.2) (cm$^{1/2}$ s^{-1}) ($U_c = 5.52$ in the CGS system of units), C is the Chezy roughness or resistance coefficient (dimensionless), and S is the channel slope (cm cm^{-1}).

The assumptions of the kinematic wave model have been shown to apply to the problem of overland-flow routing by many authors (see, for example, Lighthill & Whitham 1955, Henderson 1966, Kibler & Woolhiser 1970, Ponce et al. 1978). The model was first applied to simple sloping planes by Henderson & Wooding (1964) while Kibler & Woolhiser (1970) modelled overland flow as a kinematic cascade. Today the model is widely used to route overland flow in hillslope hydrology and erosion models (for recent examples see Rose et al. 1983, Yair & Lavee 1985, Kirkby 1990). Miller (1984, 20) notes that "in overland flow routing where the routing distances are short, the slopes steep, and the lateral inflow large, the kinematic wave model will probably produce adequate results".

In an erosional context, the kinematic wave equations can be applied to relatively smooth slopes and are used to provide values of the water depth and flow velocity for use in sediment concentration balance equations (see, for example, Croley 1982). However, despite the evident success of the kinematic wave model in hydrology, its usual formulation has some important weaknesses when used in erosion studies and when applied to topographically complex surfaces.

The most critical limitation of existing models, for our purposes, is the assumption of a uniform water depth and an average roughness over the slope reach under consideration. A further limitation of existing models is that they have been applied to geometrically very simple surfaces. Lastly, none of the models have been applied to the problem of three-dimensional dynamic slope evolution by erosion. These limitations of existing models are addressed in this chapter. We look first at a research model used to simulate slope evolution and secondly at an alternative means of accommodating a topographically complex surface in an overland-flow model that uses kinematic assumptions.

Three-dimensional dynamic slope evolution by erosion

The principal reason for existing models ignoring three-dimensional dynamic slope evolution by erosion may be the mathematical and computational difficulties involved in solving the flow equations in two dimensions. One way around this problem is to assume that overland flow on a hillslope is always orthogonal to the contours (see Ch. 7). This implies that flow follows the steepest path available to it, and therefore that the path of a drop of water arriving at the slope surface can be determined by consideration of the topography of the slope alone. This assumption is consistent with kinematic wave theory where the friction slope is assumed to be equal to the bed slope. On relatively steep slopes with negligible microtopography, flow paths do tend to be fixed for a given geometry. However, on shallower slopes the micro-divides can be overtopped and flow may be routed according to the water-surface slope rather than the bed slope. It is important to realize that the calculation of orthogonals is critically dependent on the accuracy with which the contour map has been prepared and on the level of detail present in the map. However, in general, the drawing of orthogonals to the contours provides a good approximation of the route taken by overland flow. The flow strips thus formed can be considered to be independent of each other and therefore can be treated as one-dimensional kinematic cascades with the cell boundaries described by the orthogonals and the contours. Clearly, there may be convergence or divergence within each strip and the governing flow equations must be written to accommodate this.

Thornes (1984) realized that by taking this approach to modelling hillslope overland flow, a difficult problem in semi-arid landform development could be tackled. The origin and subsequent propagation of gully headcuts is a subject that has engaged the curiosity of geomorphologists for many years (for example, Ireland et al. 1939, Kirkby 1980, Dietrich et al. 1986, Howard 1988, Willgoose et al. 1991). One important feature of many gully systems is that they tend to form well-ordered dendritic networks, suggesting that while stochastic processes may play a part in gully growth and bifurcation, there are deterministic processes involved as well. Most studies of semi-arid hillslope hydrology assume either a fixed geometry or make an a priori distinction between rill and interrill areas (see later this chapter): these assumptions do not allow gullies to form and develop. Modelling hillslope overland flow by flow strips defined by orthogonals and contours allows this problem of gully initiation to be tackled without the need to specify an arbitrary threshold at which a rill or gully is initiated.

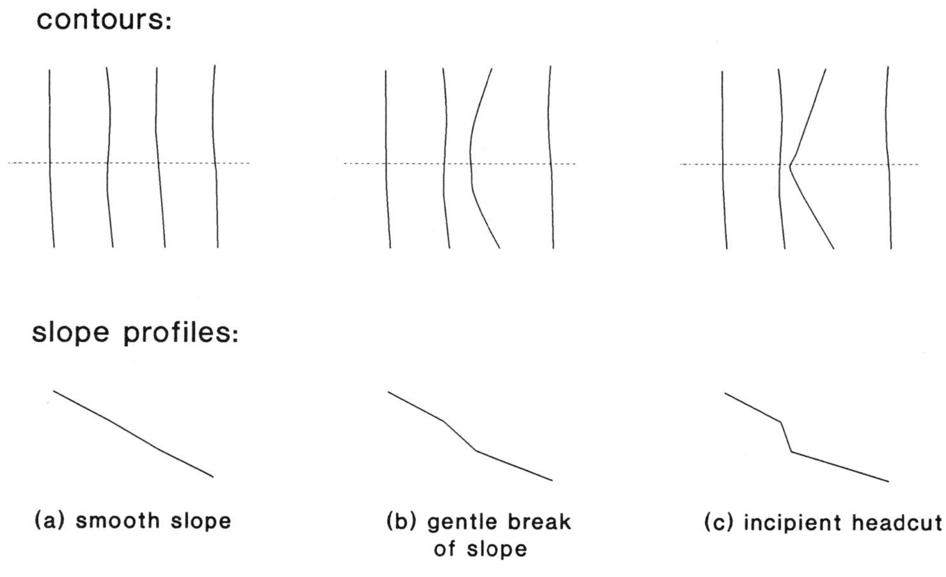

Figure 9.1 Contour migration by erosion leads to the development of a headcut.

The key to the problem is that downcutting by erosion can also be considered as contour migration: changes to the ground surface can be represented by relocating the relevant contour. In this way the locations of the cells used in the model are not fixed but may change in shape and average slope angle through time. In simple analytical models of hillslopes represented in this way, a velocity of migration can be calculated for each point on the original contours. Depending on slope angle, roughness and convergence, different contours may have different upslope velocities. If a contour is moving faster than the next contour upslope, it will eventually catch up with it. At this point a knickpoint will have formed, because there is a vertical step in the long profile of the hillslope (Fig. 9.1). When applied across complex hillslopes, this approach provides a versatile way of examining three-dimensional evolution. It also allows headcuts to bifurcate if the flow conditions are suitable. For bifurcation to occur, flow strips around the edges of an existing headcut must erode more rapidly than those in the centre. Indeed, gully headcuts may also disappear by a similar process: the disappearance of unfavourably located headcuts may be part of the explanation for the well-conditioned ordering of many gully systems.

Marchington (1987) and Kemp (née Marchington) (1990) presents a numerical model of overland flow on hillslopes which makes use of the properties of contours and their orthogonals to investigate the hydrological behaviour of hillslopes whose complex topography and time-dependent inputs makes the analytical solution of the governing flow equations intractable. This model, known as SAFMO (Semi-Arid Flow MOdel), takes a contour map of the slope in question and calculates orthogonals in order to define the flow strips. Cell characteristics such as slope, and convergence and divergence factors, are then determined. The kinematic wave equations are then solved by the method of characteristics (see, for example, Cundy & Tento 1985). Water depth in the kinematic wave equations is a function of both space and time. However, it is possible for an observer to move down the system with the water so that the water depth appears to be a function of time alone. The method of characteristics makes use of this by defining the characteristic path along which the observer would have to travel and the time-dependent depths along this path. Thus the solution is given throughout the flow domain. In SAFMO the water depths at the ends of each of the cells are calculated by interpolation from the characteristic curves. In a version of SAFMO developed by Watts and Thornes, rainfall of varying intensity may be applied to the soil surface. Infiltration is calculated by the equation of Kirkby (1985):

$$i = a + \frac{b}{S} \tag{9.3}$$

where i is instantaneous infiltration rate (cm s^{-1}), S is soil storage already occupied (cm), and a (cm s^{-1}) and b (cm^2 s^{-1}) are empirically derived parameters.

With an initially dry soil profile and an excess of water, Equation 9.3 takes a similar form to that of Philip (1969). However, the advantage of Equation 9.3 over that given by Philip (1969) is that it takes into account the wetting history of the cell in question so that, at a given time, cells where there has been more water availability have lower infiltration rates. Thus, when included in a distributed catchment model, Equation 9.3 allows different parts of the soil surface to have different infiltration rates.

Recent development of SAFMO by Watts has allowed the application of the contour migration ideas outlined above. Many factors control erosion by overland flow. These include soil properties, the detaching power of the flow and the transporting capacity of the flow. In the examples illustrated here, erosion is calculated in SAFMO according to the instantaneous velocity of the water alone. This approach is very simplified but is suitable for illustrative purposes. For slopes such as those in the Soil Erosion Facility of the Department of Geography, University of Bristol (SERFUB), very little redeposition of sediment occurs and, therefore, ignoring the transporting capacity of the flow may not be too serious. SAFMO calculates an erosion depth for each node for each timestep. This is interpolated across the contour to provide the depth of downcutting at a large number of points (currently 1000) on each contour. In this way the detailed information provided initially for each contour is not smoothed by the numerical procedure. At each of these points, downcutting is transformed into upslope contour migration. To avoid abrupt changes in contour locations, which can cause numerical instability, new contour locations are calculated regularly (either every timestep or when the maximum accumulated erosion in any single location exceeds a prespecified value). From the new contour locations a new set of orthogonals is generated, and the new cell characteristics are determined. Thus SAFMO has become a totally dynamic model of hillslope evolution by overland flow.

SAFMO's utility as a slope evolution model can be demonstrated by a simple example. Figure 9.2a shows the original topography of the slope used in this example, which is a real slope from the SERFUB. The slope is 7 m long and 2.5 m wide and 13 contours have been used to describe it. The contours have a 4 cm interval so that the drop from the top of the slope (at the top of Fig. 9.2a) to the base is 0.48 m, an average angle of about 4°. The original topography is relatively simple, notable features being a small hollow near the centre of the base of the slope and a slight ridge to the left of this. Eleven orthogonals have been drawn to define 10 flow strips. These orthogonals start at the base of the slope, and some converge very quickly. For example, flow strip 1 and flow strip 10 only reach about 1.5 m and 1 m up the slope, respectively. Only four strips reach the top of the slope: these are strips 2, 3, 5 and 6. It would be expected that these will exhibit most change because their contributing area is the greatest. Strips 2, 5 and 6 all show convergence so it might be expected that most change will occur on them.

Figure 9.2b shows the effect of erosion on slope form from a simulation of

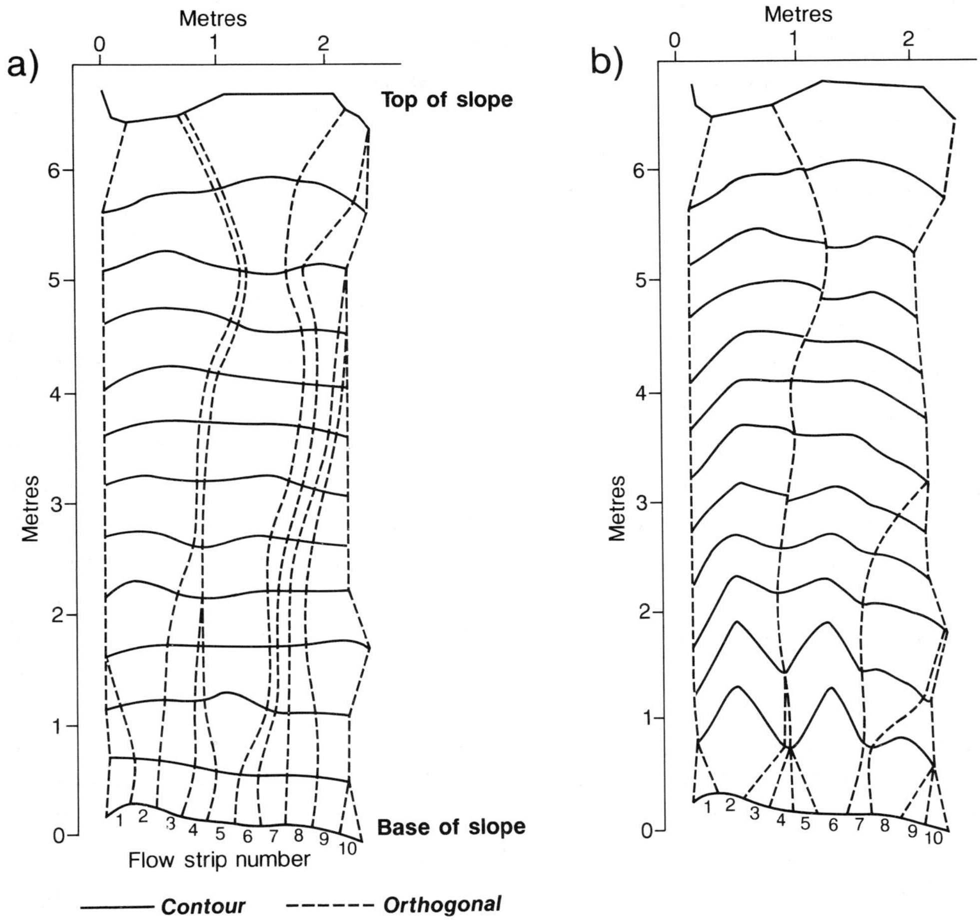

Figure 9.2 Simulated erosion of the SERFUB experimental slope: (a) original surveyed topography with 10 flow strips of equal basal width; (b) surface after simulated erosion, contour migration and flow-strip redefinition.

constant rainfall over a 2 h period. It must be emphasized that the erosion "law" used here is a simple power function that relates velocity to downcutting. It has not yet been parameterized so the example given here is suitable for illustration only and no inferences can be made about rates of slope evolution. However, it is clear that in this simulation the slope has changed dramatically. In some ways it has become much more simple: only two flow strips drain much of the slope. These are strips 2 and 6. Strip 2 has not grown very much, but strip 6 has captured all the area which used to be drained by strip 5. At the top of the slope the contours are almost as they were before: there will be little erosion here because there is relatively little overland flow. However, accumulations of flow downslope have caused quite large changes near the base of the slope. In particular, two hollows have formed and, especially in strip 2, the lower contours seem to be catching up with the upper ones. This is a potential location for a headcut.

This simulation demonstrates how important dynamic changes in hillslope form can be to the overall hydrology of the slope and flow patterns on the slope surface. The modelling of dynamic slopes in this way has also demonstrated how even very slow processes can lead to hydrological changes which are important at this scale. Consider the situation shown in Figure 9.3. An orthogonal drawn from point A on contour 1 will reach point B on contour 2. However, if over time the contour around B migrates, there will come a time when B is no longer the closest point on contour 2 to point A and the orthogonal will flip to point C. Even if erosion is extremely slow, the flip of the orthogonal and the resulting dramatic change in the size of the flow strip will be instantaneous. Thus a very small rate of topographic change can lead to extremely rapid hydrological change: in this case a pulse of water would move down the flow strip. Whether this type of catastrophic response to very small amounts of erosion occurs in real hillslope systems requires field investigation, but it is well documented in channel avulsion in rivers. Such behaviour may explain the seemingly random nature of some overland flow events, and most definitely points to the need for overland-flow models to take account of slope evolution.

SAFMO provides a very flexible approach for modelling dynamic evolution of the slope form using overland flow patterns to control surface erosion. There is considerable scope for further development of SAFMO, especially by the inclusion of more realistic sediment erosion and transportation algorithms, and there is potential to apply the model to larger and more complex slopes. SAFMO is a flexible tool for simulating the change of real hillslopes and for providing a distributed picture of the hydrological behaviour of such a slope through time. It may also be used for hypothesis and scenario testing. For example, modification of the geometry of a gully headcut might provide a means of inducing the dispersion of the headcut rather than its propagation. Different headcut geometries can be modelled by SAFMO to examine how they behave. Alternatively, different climatic scenarios can be assessed for their effect on erosion, helping to point to slopes which will be sensitive to climatic change.

However, SAFMO has a number of weaknesses. It can deal only with flow where the depth can be assumed to be equal to the hydraulic radius, which confines it to relatively smooth-surfaced hillslopes (see below). For example, it cannot be used to simulate the direct effect of small micro-topographic features on flow, such as rills and clumps of vegetation. Further, while the dynamic redefinition of the hillslope makes SAFMO a very flexible tool for

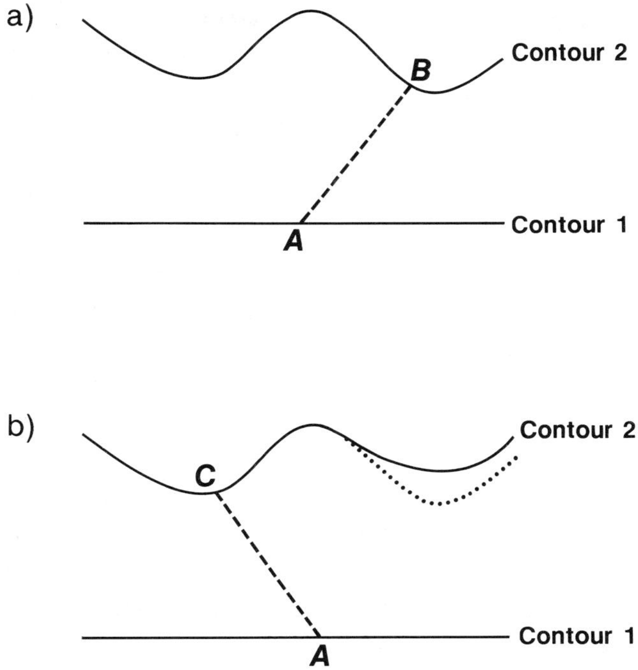

Figure 9.3 A small quantity of downcutting can lead to a dramatic change in orthogonal location: (a) the orthogonal from A on contour 1 reaches contour 2 at B; (b) migration of contour 2 in the area around B results in the orthogonal now meeting contour 2 at C.

evaluating hillslope change, it is very expensive in terms of computing time. Because of this, the application of the model to very long time periods or very large slopes may be prohibitive. Finally, SAFMO is very demanding in terms of data input requirements. The contour-based approach to erosion modelling needs a very detailed survey of the hillslope, since the accuracy of the slope representation directly affects the orthogonal location and therefore the operation of hydrological and erosional processes represented in the model. Thus, while SAFMO is a very useful research tool, there is a need for a model which can deal with slopes that have a range of micro- and macro-topographic elements and which has parsimonious data requirements.

The need for an alternative approach to routing overland flow on slope surfaces with complex micro- and macro-topography

Most slopes in areas of significant erosion are topographically complex across the slope. This micro- and macro-relief arises from vegetation (such as clump grasses), partially buried and freely standing stones, and from rills and gullies. These protuberances cause concentration of the flow and produce greater flow depths and velocities, which cause local increases in erosion rates (see Thornes et al. 1990). From a number of carefully controlled flow experiments on desert surfaces, Abrahams et al. (1989) recorded these variations in flow depths and velocities and showed that the use of mean water depth rather than the distribution of flow depths can result in a severe underestimation of flow detachment of sediments. Approximations to flow routing which average out these variations assume that sheet flow conditions occur, which, as Emmett (1970) has shown, is rarely the case on natural hillslopes.

To some extent the problem of variable depths across the slope can be avoided by partitioning the flow into rill and interrill areas and routing each separately. Rills typically appear as numerous, small flow concentrations that tend to be parallel across the slope except where the land form converges or diverges. While runoff and sediment tend to move laterally on interrills towards the rills by both splash and overland flow, runoff and sediment move directly downslope in rills unless the rills are forced across the slope by relatively large surface features, such as rocks and plant hummocks. Rills thus tend to operate as absorbing barriers in a Markovian sense to the random walks which the fine-scale roughness imposes on overland flow.

A rill–interrill dichotomy was developed for modelling by Foster & Meyer in 1975 and has been popular in flow and sediment routing models since then. However, although it is possible to describe separate flow and erosion processes in rills and interrills, very little is known about how we should realistically and accurately partition water and sediment between the two. Early in the process of rill initiation and formation the rill–interrill dichotomy should be a dynamic rather than a static phenomenon that depends on rainfall rate, overland flow intensity and slope-forming materials. However, this dichotomy is rarely treated as dynamic so that while it can be of use for generating relative volumes of splash, wash and concentrated flow, it cannot allow for changes in rill location and geometry under varying flow conditions, and fails to recognize that these features are created and destroyed during actual flows and that they are dynamically partitioned down slope as well as through time during a storm.

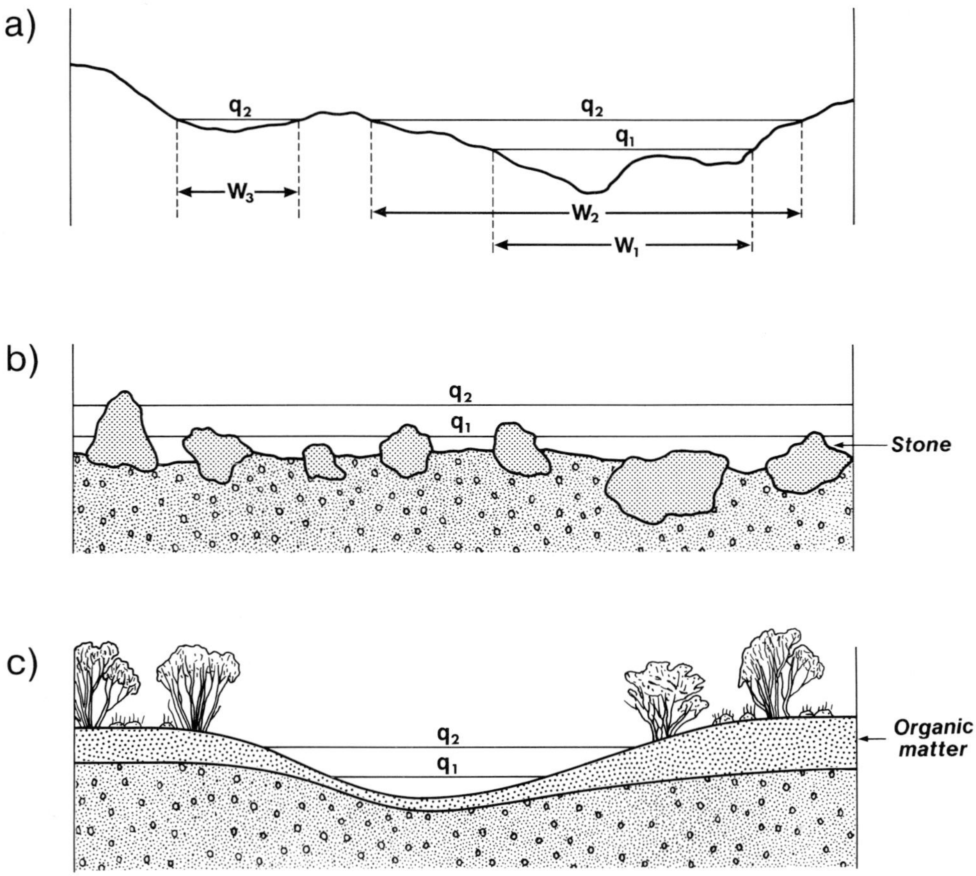

Figure 9.4 Causes of reticular flow on surface slopes. Flow concentrates into a channel network in (a) because of an uneven topography, in (b) because of stones, and in (c) because of plant mounds. With reticular flow, surface properties change with different discharges. In (a), increasing discharge (q_1 to q_2) leads to occupancy of another channel, a reduction in hydraulic efficiency and an increase in infiltration because of the greater wetted perimeter (defined as the wetted surface area across the width w). In (b), overtopping of partially buried stones leads to greater hydraulic efficiency. In (c), an increase in discharge in a single channel also leads to greater hydraulic efficiency and increased infiltration losses.

Croley (1982) presents a model in which rills are assumed to be prismatic, evenly spaced, and of a constant width so that sediment concentration and flow with respect to time and distance will be the same in each rill for a spatially uniform rainfall. A similar approach is used by Kirkby (1990) in which rill geometry is idealized as a set of equally spaced linear troughs, and rill cross section, depth and spacing are estimated from the proportion of the slope area covered by rills. In the model of Kirkby (1990) the geometry of interrill surfaces is at first assumed and subsequently depends on erosion of, or deposition in, the rill. While these approaches are suitable for surfaces in which the rill spacing is even and clearly defined (for example, ploughed fields), they fail to recognize the very dynamic nature of the partitioning of flow and sediment between rills and interrills mentioned above, and tend to underestimate the role of concentrated flow in the interrills.

The problems associated with the assumptions of uniform sheet flow and a rill–interrill dichotomy suggest that an alternative approach is needed which can adequately represent the dynamic relationships between micro- and macro-topographic elements on slope surfaces. In this chapter we avoid the formal differentiation between processes in rills and interrills on the grounds that for some soils there already exists a continuum between processes in micro-rills and larger rills which cannot be modelled dynamically under these assumptions. The idea of the continuum between processes in micro-rills and larger rills is based on reticular flow (see below) rather than sheet flow or a simple rill–interill dichotomy. Below we present a new model (RETICular flow model) based on this reticular-flow concept.

Reticular flow refers to the well-known tendency of flows to be generated and proceed under the control of microtopographic features such as stones and vegetation, first described in detail by Emmett (1970), modelled by Thornes et al. (1990) and investigated experimentally by Abrahams, Parsons and Luk in several papers (for example, 1989, 1990, Ch. 1 of this volume). In particular reticular flow implies that the flow is concentrated in relatively narrow parts of the surface, as a consequence of which flow depths, velocities and erosion may vary across the surface by orders of magnitude. A further consequence is that topographic relief has to be dealt with either as a series of detailed maps or in a statistical fashion if erosion rates are to be meaningfully predicted from models. Thirdly, reticular flow involves the idea that surface properties, especially infiltration and roughness characteristics, may change with different discharges, as shown in Figure 9.4.

Basic approach used in the RETIC model

The starting point for this exploration is the notion of a topographic free or active volume discussed by Thornes et al. (1990). It is assumed that water arriving from an upslope reach over a short time period is distributed on a topographically complex surface from the deepest point upwards. The depth to which this water can reach (shown as "fill depth" in Fig. 9.5) is determined

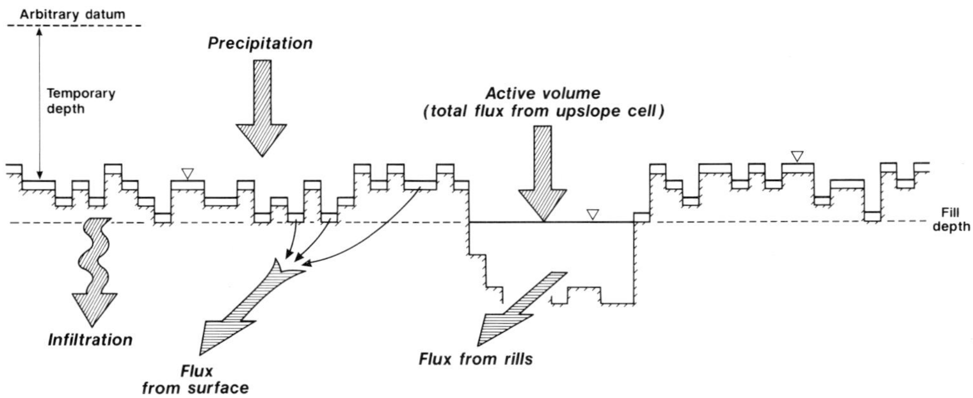

Figure 9.5 Basic structure of the RETIC model. The slope surface consists of a number of depths of unit width below an arbitrary datum. The total flux from the upslope cell for the previous timestep fills the surface from the deepest point upwards. The depth to which the active volume fills the surface is called the fill depth.

by calculating the functional relationship between depth and active volume. This calculation is invoked for every timestep to account for changes in the ground and water-surface topography. An example of the functional relationship between depth and active volume is illustrated in Figure 9.6. The form of the function will depend on the topography and the amount of water stored on the slope surface (see below). The second basic idea is that the surface can be represented by a series of profile depths or "flow strips" of unit width which operate independently of their neighbours. This is similar to the conceptual infiltration model of Moore & Clarke (1981). There is assumed to be no dynamic interaction, such as turbulence and sediment exchange, between adjacent strips. In the applications of the model so far, the strips have been assumed to be 1 cm in width. The strip depth may be expressed relative to the mean topographic surface or to any arbitrary datum. To avoid using a map to give the depth of each strip, the depths can be obtained from known depth distributions.

These two basic ideas are incorporated in an explicit finite difference solution to the kinematic wave equations. The topographic data can be provided for specific slope cross sections, drawn from empirically derived frequency distributions, or provided from theoretical distributions. In *theoretical* simulations, each cell is assigned a number of depths drawn from a negative exponential density function. Evidence for a negative exponential density of depths on surfaces without rills has been given by Abrahams et al. (1989). If rills or gullies are present, their depths are taken from a normal density function. In simulating flow on *real slopes* the actual profiles measured at

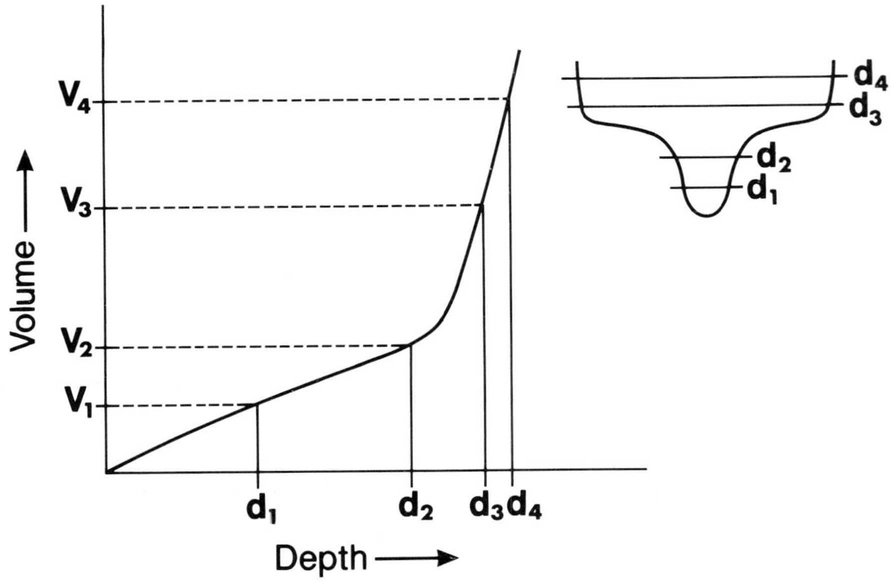

Figure 9.6 Example relationship between depth and active volume.

fixed intervals are used. When the model is run, the depths in both cases are then determined by the erosion and deposition on each strip.

For any timestep the dynamic partitioning of water is as follows. Residual water left on each profile in each cell provides a dynamic water-surface topography (shown as "temporary depth" in Fig. 9.5). The relationship between depth and active volume for this temporary surface is then calculated. Since we know the total flux arriving from upslope for the previous timestep, the depth to which this active volume will fill the surface (the fill depth) can be found from the depth–active volume function. This water is then allocated to every strip below the fill depth. Rainfall additions and infiltration losses are then allocated to every strip, whether it is below or above the fill depth. The procedure for water partitioning is summarized in Figure 9.7.

Standard kinematic routing models assume that hydraulic radius is equal to the water depth. In a model that allows for variation of water depths across the slope and the development of dynamic flow concentrations, this assumption is unrealistic and we need to consider the hydraulic radius of the strips in which flow concentrates. In the RETIC model this problem is dealt with in two ways:

(a) Water above the fill depth is assigned a hydraulic radius equivalent to the flow depth on the assumption that flow on the surface is very shallow, so that the hydraulic radius approaches the flow depth.

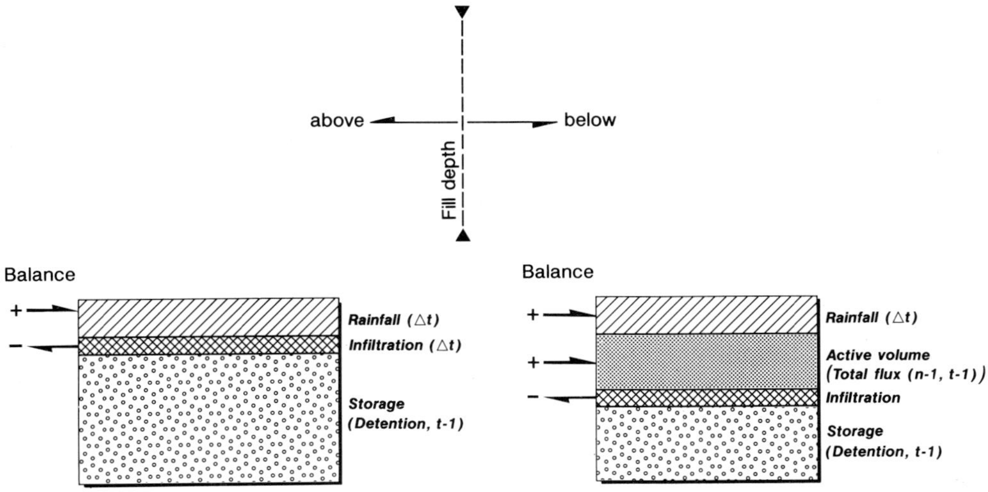

Figure 9.7 Partitioning of water between strips on a cell during each timestep.

(b) Where flow concentrates, such as the channels denoted by the letters A, B, C, D and E in Figure 9.8a when the inundation of the surface is at level 8, the hydraulic radius for the *group* of strips forming the flow concentration is calculated. The flow for each flow concentration is then calculated using the uniform flow equation. Thus, when flow concentrates as flow concentrations the strips are not treated as strictly independent of each other.

The calculated fluxes for each strip are collected as a reservoir. This reservoir is used to provide the active volume for the next downslope cell in the next iteration.

The calculated routing velocities will be sensitive to the positioning of the depths in a flow concentration. For the same set of depths, for example 1, 2, 3, 4, 5 and 6 cm, the hydraulic radius varies according to how they are arranged. When the depths are arranged in monotonically increasing order, as in this list, the hydraulic radius is minimized. However, if the depths are drawn in a random manner, there is a set of possible hydraulic radii (and hence velocities) according to how they are configured. We have evaluated the effects of different configurations of the same set of depths on hydraulic radius. If we assume six depths (as above) there are 720 permutations of depth location. As average depth increases, the coefficient of variation of hydraulic radius decreases (from about 14% at an average depth of 3.5 cm to 2.6% at an average depth of 93.5 cm). In other words, the predicted velocities will be more consistent with deeper rills or greater discharges. As the variation in depth increases, the coefficient of variation of the hydraulic radius is more consistent, but generally higher at about 20%, the increase in variation in depth being offset by the increase in mean depth. The results indicate that in the range of values common on slopes, the coefficient of variation of hydraulic radius due to the position of depths is likely to be around 15% if the depth values are selected at random. As erosion progresses this variability will reduce markedly.

In the RETIC model, the value of Chezy's *C* is held constant for each strip in each cell, on the assumption that it represents the skin friction. Variations in the efficiency of overland flow are then controlled by the hydraulic radius and slope factors in the uniform flow equation.

Figure 9.8 exemplifies the variation in efficiency due to changes in discharge, the wetted profile and hence hydraulic radius, for a hypothetical slope section. Efficiency is defined as the "section velocity", which is the ratio of the sum of the discharges from each flow concentration on the slope to the sum of the wetted area of the flow concentrations. For the surface illustrated in Figure 9.8a, mean velocity for each flow concentration at each level of inundation is calculated using Equation 9.2 assuming a gradient (*S*) and skin friction (*C*) of 0.0699 and 10, respectively. From Figure 9.8 it can be seen that the efficiency of flow from the section increases but at a decreasing rate as the flow concentration denoted by the letter A is filled. If the steady-state water level is raised to level 7, three other channels or flow concentrations (B, C and D) are present. When these channels are occupied by flowing water the rate of increase in hydraulic efficiency of the surface falls markedly (see Fig. 9.8). The hydraulic efficiency of the surface falls at inundation level 8 and again at level 9, and then rises as all of the strips become submerged. The central part of this plot, when hydraulic efficiency falls and then rises again with increasing discharge, is similar to plots of discharge verses the Darcy–Weisbach friction factor for semi-arid surfaces given by Abrahams et al. (1990).

Abrahams et al. (1990) show that an increase in discharge on a surface with a complex micro-topography will result in a rise and then a fall in friction. They attribute the initial rise in roughness to the successive inundation of roughness elements, which causes the wetted upstream projected area of the elements to increase rapidly and so increase the Darcy–Weisbach friction factor. However, a point is soon reached when most of the elements are largely or wholly submerged, and any further increase in discharge is accompanied by a progressively slower rate of increase in wetted upstream projected area of the elements (see Fig. 9.9). They suggest that existing models need an explicit form of this relationship. However, this is a misidentification of the problem. Every slope has a different configuration and

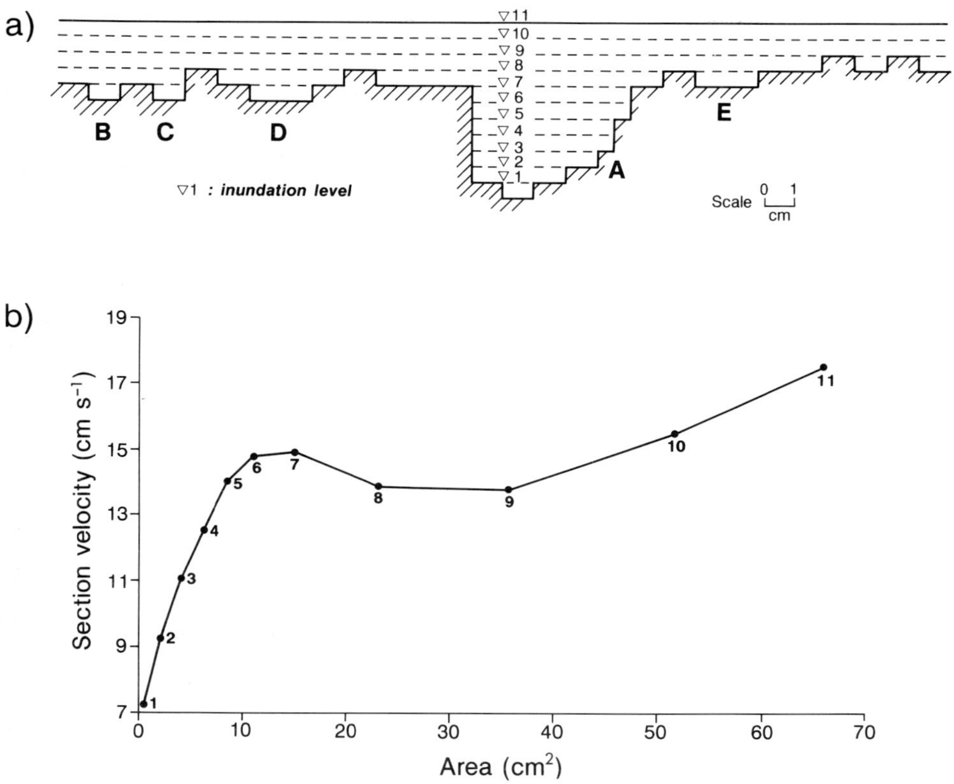

Figure 9.8 (a) Hypothetical slope cross section consisting of a series of depths of unit width below an arbitrary datum. (b) As the steady-state water level is raised, the section velocity first increases and then decreases before increasing again in a monotonic fashion.

as long as the hydraulic radius is properly estimated, rather than being replaced with average depth, the problem of a variable cross section is accommodated as it is in conventional river models.

Two clear advantages of the new model over traditional approaches can be identified. The first is that the model allows flow concentrations to change in shape, size and in number both over time and down slope, as indicated in Figure 9.8. The second advantage is that it gives the depth and velocity of the flow for each strip. By simulating varying depths of water across the slope, the model can more easily approximate real-world spatial variations of erosion on rilled and gullied slopes during a storm event. Moreover, these variations affect the runoff hydrograph itself and so the model is better able to simulate the impact of erosion on hillslope hydrology. We have not yet included an

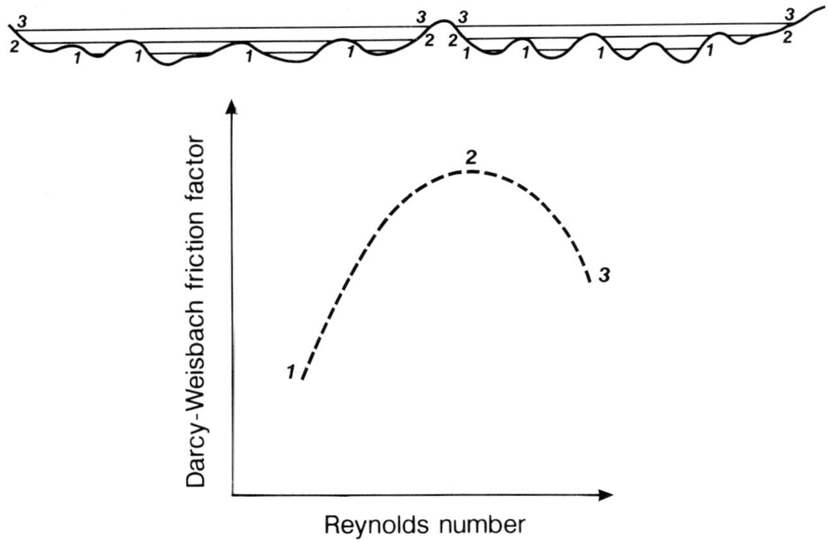

Figure 9.9 Hypothetical ground surface of intermediate-sized roughness elements of varying height, showing how the progresive inundation of such a surface gives rise to a convex-upward relationship between the Darcy–Weisbach friction factor and the Reynolds number of the flow (after Abrahams et al. 1990). Reproduced with permission of IAHS Press.

erosion routine in the model. However, the model has been developed in a form in which an erosion component can be readily incorporated so that, in its complete form, the model will simulate entrainment, deposition and routing of sediment.

Example model output for non-rilled and rilled surfaces

To illustrate the operation of the model, we consider three scenarios. In the first, we compare the model's output with output from a conventional model in which the water depth across the surface is assumed to be uniform. In the second we vary the number of rills across the slope surface, while in the third we look at the effect on the storm hydrograph of having two rills of different dimensions on the slope. In each case we assume a uniform input of rain to the slope surface. In the first scenario rain falls at a constant intensity of 0.001 cm s^{-1}. In the second and third scenarios rain falls at an intensity of 0.002 cm s^{-1} for 300 s at the start of each simulation. For these last two scenarios mean rill depth increases linearly down slope. In each case infiltration was set to zero.

Before the results are discussed, it should be noted that finite difference solutions can introduce appreciable amounts of numerical diffusion and dispersion. Numerical diffusion manifests itself as the attenuation of the simulated hydrograph, while numerical dispersion appears as the steepening or flattening of the rising limb of the hydrograph (see Kibler & Woolhiser 1970, Ponce et al. 1979, Ponce 1991). Numerical diffusion and dispersion are functions of the cell size and timestep used in the solution. If the cell size and timestep are reduced, the diffusion is also reduced, so that as cell size and timestep approach zero so does the diffusion.

Diffusion and dispersion effects in the model were analysed heuristically and it was found that for fixed cell sizes and timesteps the effects of diffusion and dispersion in the example simulations described below were small and of a similar magnitude. Differences in model output can therefore be attributed principally to changes in the model input and therefore changes in the functional relationships of the model.

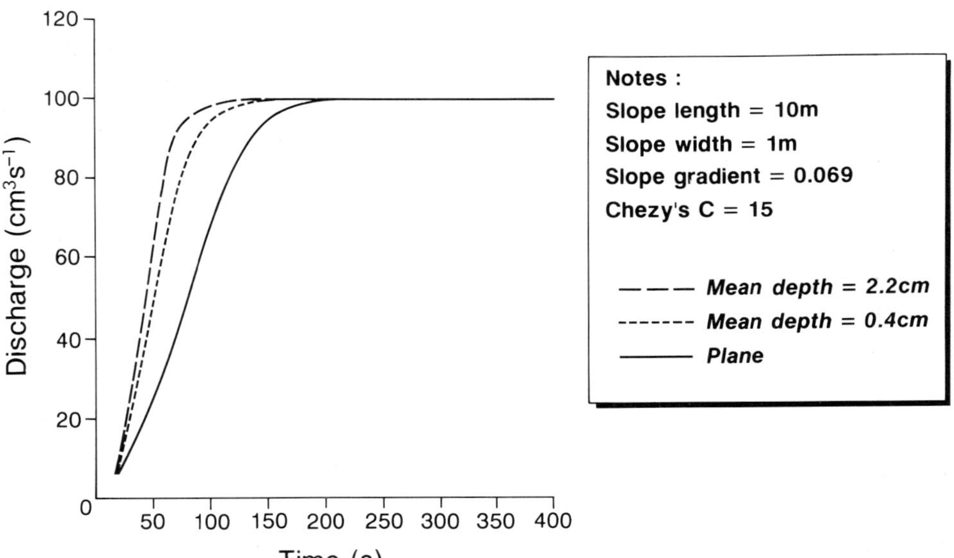

Figure 9.10 Example model output for non-rilled surfaces and steady rainfall input.

In the first situation we consider three surfaces; one is smooth, while the other two have a variety of depths derived from a negative exponential density function. The mean depth of the second surface is 0.4 cm while the mean depth of the third surface is 2.2 cm. The distribution of depths remains constant down slope in both cases. The slope in the simulations is 10 m in length, and the gradient is 0.069 (about 4°). The skin friction (Chezy's *C*) is the same for all three surfaces and is given a value of 15. The hydrographs for each surface under steady rainfall are shown in Figure 9.10. Discharge in each case is that from the base of the slope. From Figure 9.10 we can see that the slowest response to rainfall comes from the smooth surface. This might at first seem counter-intuitive until we remember that flow will concentrate in the deeper flow strips in the second and third surfaces and therefore produce a more rapid response to rainfall. This type of behaviour would have to be catered for in conventional models by lowering the value of roughness (i.e. increasing the value of C) in Equation 9.2. However, by keeping the skin friction constant, the new model allows different surface configurations to produce changes in the discharge hydrograph. In addition to producing a hydrograph that is only partly dependent on the value of roughness, the new model also gives the flow depths and velocities across the surface (see above), although these are not reproduced here.

In the second set of simulations we consider four cross-slope profiles. In the first there are no rills, in the second there is a single rill, while in the third and fourth there are two and three rills, respectively. Rill dimensions in each case are identical so all that is varied is the number of rills. The interrill surface in each case is produced from a negative exponential distribution of depths with a mean depth of 0.4 cm. Mean rill depth increases linearly down slope and rill depths are drawn from a normal distribution. The slope length was 10 m in the simulation and it was divided into 10 cells, each 1 m in length and width. The mean rill depth on the first cell (on the divide) is 2 cm and increases to 5 cm on the tenth cell (at the base of the slope). The discharge hydrographs (discharge is that from the base of the slope) from the simulations give results that are consistent with what we would expect. The surface with a single rill responds more rapidly to rainfall than that without a rill (Fig. 9.11a). Flow will concentrate in the rill and will be more rapid than on the surface without a rill. However, if we increase the number of rills on the surface to two (Fig. 9.11a) or three (Fig. 9.11b), the response to rainfall is slowed. This is because the same discharge is now divided between more rills, and flow in the rills is shallower. With shallower flow the hydraulic efficiency of the rills is reduced so the response to rainfall is slowed and dampened. This may, in part, explain the process of abstraction which takes place in early change on bare slopes and which was hypothesized by Horton (1945) and demonstrated by Schumm (1956). Abstraction would thereby give rise to a more effective evacuation of water especially for lower rainfall intensities. Of course, this effect is due partly to the length of slope considered. With a longer slope and an increase in discharge down slope, a surface with two or three rills could become more efficient, since one rill might not be able to accommodate all the downslope flow. As a result, the rill would overtop and the resulting excess water would flow more slowly on the interrill surface.

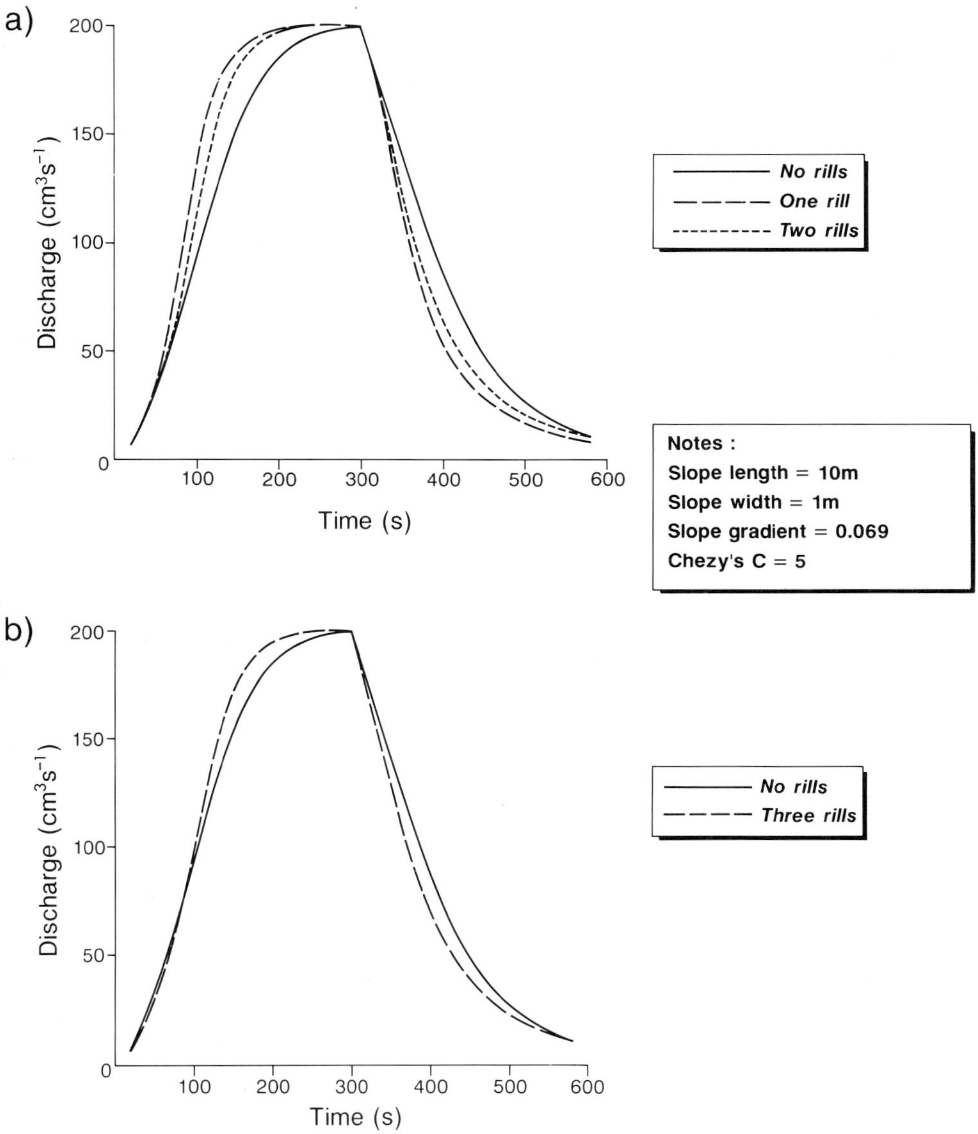

Figure 9.11 Example model output for rilled and non-rilled surfaces.

In the third situation we consider a surface with a single rill and a surface with two rills. The first rill on the surface with two rills is the same size as that on the surface with one rill. However, the second rill has a mean depth half that of the first rill. The slope length is 20 m and the larger rill increases in mean depth downslope from 2 cm at the divide to 8.3 cm at the slope base.

The hydrographs for both surfaces are given in Figure 9.12. Both hydrographs show the same rise until about 145 s into the simulation. The

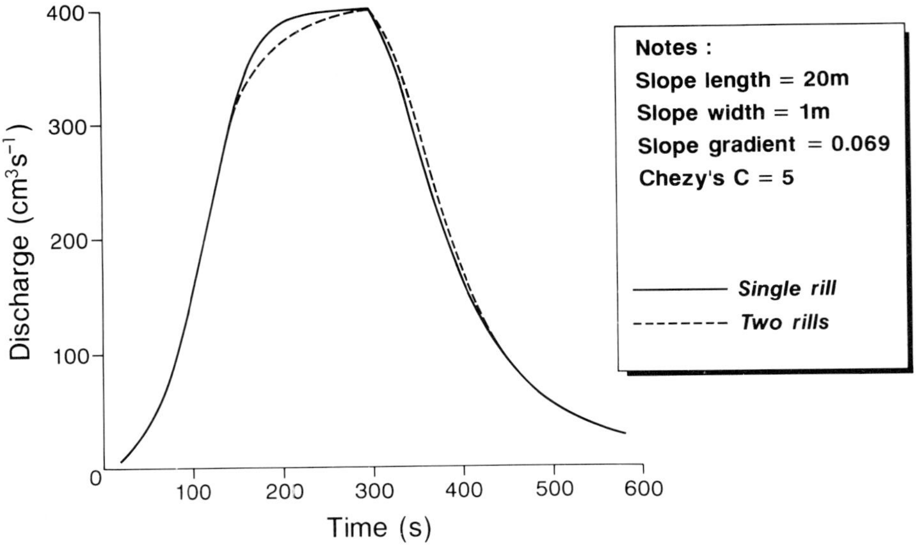

Figure 9.12 Example model output for rilled surfaces.

hydrograph from the surface with two rills then shows a delayed response. This is because some of the water arriving as active volume now starts to occupy the second rill instead of continuing to fill the larger rill, causing the hydraulic efficiency of the surface to be reduced. Although very easy to simulate, this type of behaviour is not accounted for in conventional flow routing models of rilled surfaces, where the proportion of flow reaching each rill is fixed throughout the simulation. We are currently undertaking experiments to compare the hydrographs generated with RETIC and those occurring on the SERFUB, and these will be published in due course.

Conclusions

In this chapter we have shown that use of the simple kinematic wave model in erosion studies does not go far enough. We adapted the model first in SAFMO by obtaining new approximations for the method of characteristics, by using a strip algorithm and by adopting a contour migration approach. Even with these modifications, the approach does not satisfy modelling needs in terms of erosion, though it may be satisfactory for hillslope hydrograph approximations, especially when roughness can be calibrated for particular slope problems.

However, the fundamental modification required is to accommodate the complex geometry of rills and gullies, which have an important control on erosion and feedback to the flow hydrograph itself.

We have therefore developed a new model (RETIC) which still depends on the core kinematic assumption, but allows flexibility in the determination of the hydraulic geometries and routes within the hillslope geometry. The model RETIC allows for and predicts the large variations in velocities expected on rilled and gullied slopes and is likely, therefore, to provide a better basis for erosion prediction. It should also be much better in dealing with other processes such as infiltration and organic matter transport.

We have briefly compared the predicted values of runoff using the conventional kinematic approach and the reticular flow approach and have shown that the hydrographs produced are quite different. We have also shown that the new model produces different hydrograph behaviour with different cross-slope rill configurations. These suggest that a single rill may be more efficient than a number of rills, especially for short-period hydrographs. This may partly explain the abstraction process. Since the model also allows dynamic transformation of the rill pattern and hence the hydrograph, the model may be better for exploring hillslope runoff on all types of slope surface for any type of storm event.

The RETIC model, unlike SAFMO, approaches the prediction problem through distribution functions, rather than trying to map detailed slope changes. This means, however, that we have a distribution of potential outputs, depending on how the actual depths are arranged, as we illustrated above. Clearly, the stochastic nature of the output needs more exploration. Where experiments are being conducted on field plots or in the laboratory, the actual original topography can be used in the model. However, the distribution function approach seems likely to be of much more value in the long term.

Clearly, both further development and rigorous testing are required. Experiments already carried out on the SERFUB indicate that the features produced by the model hydrographs can be observed in reality. A programme of more extensive testing will follow.

Acknowledgements

This work was partly undertaken with a grant from the Natural Environment Research Council, whose financial support is gratefully acknowledged. The authors also wish to thank the Department of Geography, University of Bristol for support during the research and completion of this chapter.

References

Abrahams, A. D., A. J. Parsons, S.-H. Luk 1989. Distribution of depth of overland flow on desert hillslopes and its implications for modelling soil erosion. *Journal of Hydrology* **106**, 177–84.

Abrahams, A. D., A. J. Parsons, S.-H. Luk 1990. Field experiments on the resistance to overland flow on desert hillslopes, southern Arizona. In *Erosion, transport and deposition processes*, 1–18. International Association of Hydrological Sciences, Publication Number 189. Wallingford, England: IAHS Press.

Croley, T. E. 1982. Unsteady overland sedimentation. *Journal of Hydrology* **56**, 325–46.

Cundy, T. W. & S. W. Tento 1985. Solution to the kinematic wave approach to overland flow routing with rainfall excess given by Philip's equation. *Water Resources Research* **21**, 1132–40.

Dietrich, W. E., C. J. Wilson, S. L. Reneau 1986. Hollows, colluvium and landslides in soil-weathered landscapes. In *Hillslope processes*, A. D. Abrahams (ed.), 361–88. Boston: Allen and Unwin.

Emmett, W. W. 1970. *The hydraulics of overland flow*. United States Geological Survey, Professional Paper 662A. Washington, DC: US Government Printing Office.

Foster, G. R. & L. D. Meyer 1975. Mathematical simulation of upland erosion by fundamental erosion mechanics. In *Present and prospective technology for predicting sediment yields and sources*. United States Department of Agriculture, Agricultural Research Service, Paper S-40.

Henderson, F. M. 1966. *Open channel flow*. New York: Macmillan.

Henderson, F. M. & R. A. Wooding 1964. Overland flow and ground-water flow from a steady rainfall of finite duration. *Journal of Geophysical Research* **69**, 1531–40.

Horton, R. E. 1945. Erosional development of streams and their drainage basins; hydrophysical approach to quantitative morphology. *Bulletin of the Geological Society of America* **56**, 275–330.

Howard, A. D. 1988. Groundwater sapping experiments and modeling. In *Sapping features of the Colorado plateau. A comparative planetry geology field guide*, A. D. Howard, R. C. Kochel, H. E. Holt (eds), 71–83. Washington: National Aeronautical Space Administration.

Ireland, H. A., C. F. S. Sharpe, D. H. Eargle 1939. *Principles of gully erosion in the Piedmont of South Carolina*. United States Department of Agriculture, Technical Bulletin 633.

Kemp (née Marchington), A. C. 1990. Towards the dynamics of gully growth. In *Erosion, transport and deposition processes*, 121–34. International Association of Hydrological Sciences, Publication Number 189. Wallingford, England: IAHS Press.

Kibler, D. F. & D. A. Woolhiser 1970. *The kinematic cascade as a hydrologic model*. Hydrology Paper No. 39, Colorado State University, Fort Collins, Colorado.

Kirkby, M. J. 1980. The stream head as a significant geomorphic threshold. In *Thresholds in geomorphology*, D. R. Coates & J. D. Vitek (eds), 53–73. London: Allen and Unwin.

Kirkby, M. J. 1985. Hillslope hydrology. In *Hydrological forecasting*, M. G. Anderson & T. P. Burt (eds), 37–76. Chichester, England: John Wiley.

Kirkby, M. J. 1990. A simulation model for desert runoff and erosion. In *Erosion, transport and deposition processes*, 87–104. International Association of Hydrological Sciences, Publication Number 189. Wallingford, England: IAHS Press.

Lighthill, M. J. & G. B. Whitham 1955. On kinematic waves. I: flood movement in long rivers. *Proceedings of the Royal Society of London*, **A229**, 281–316.
Marchington, A. C. 1987. Development of a dynamic model of gully growth. Unpublished Ph.D. thesis. University of Bristol, England.
Miller, J. E. 1984. *Basic concepts of kinematic wave models*. United States Geological Survey, Professional Paper 1302. Washington: Government Printing Office.
Moore, R. J. & R. T. Clarke 1981. A distribution function approach to run-off modeling. *Water Resources Research* **17**, 1367–83.
Philip, J. R. 1969. Theory of infiltration. *Advances in Hydroscience* **5**, 215–90.
Ponce, V. M. 1991. The kinematic wave controversy. *Journal of Hydraulic Engineering, Proceedings of the American Society of Civil Engineers* **117**, 511–25.
Ponce, V. M., R. M. Li, D. B. Simons 1978. Applicability of kinematic and diffusion models. *Journal of the Hydraulics Division, Proceedings of the American Society of Civil Engineers* **104**, 353–60.
Ponce, V. M., Y. H. Chen, D. B. Simons 1979. Unconditional stability in convection computations. *Journal of the Hydraulics Division, Proceedings of the American Society of Civil Engineers* **105**, 1079–86.
Rose, C. W., J.-Y. Parlange, G. C. Sander, S. Y. Campbell, D. A. Barry 1983. Kinematic flow approximation to runoff on a plane: an approximate analytic solution. *Journal of Hydrology* **62**, 363–9.
Schumm, S. A. 1956. Evolution of drainage systems and badland slopes at Perth Amboy, New Jersey. *Bulletin of the Geological Society of America* **67**, 597–646.
Thornes, J. B. 1984. Gully growth and bifurcation. In *Erosion control: man and nature*, 131–40. Denver: International Erosion Control Association.
Thornes, J. B., C. F. Francis, F. Lopez-Bermudez, A. Romero-Diaz 1990. Reticular overland flow with coarse particles and vegetation roughness under Mediterranean conditions. In *Strategies to control desertification in Mediterranean Europe*, J. L. Rubio & J. Rickson (eds), 228–43. Brussels: Commission of the European Community.
Willgoose, G., R. L. Bras, I. Rodriguez-Iturbe 1991. A coupled channel network growth and hillslope evolution model. *Water Resources Research* **27**, 1671–84.
Yair, A. & H. Lavee 1985. Runoff generation in arid and semi-arid zones. In *Hydrological forecasting*, M. G. Anderson & T. P. Burt (eds), 183–220. Chichester, England: John Wiley.

10

Process-oriented research on soil erosion and overland flow

B. T. Guy, R. P. Rudra, W. T. Dickinson

Abstract

Research into overland flow is being conducted at various scales at the University of Guelph. This chapter reviews some of the recent findings. Rainfall simulators have been developed, calibrated and validated to facilitate research. The single-nozzle version most commonly used produces rainfall at intensities from 17 to 215 mm h^{-1}, with characteristics similar to those of natural rainfall. In a study of infiltration and overland flow, the Green & Ampt equation was modified for layered soils. Laboratory experiments conducted in a deep-bed flume under simulated rainfall provided initial verification of the layered equation, and field data provided final verification. In an initial flume study of overland flow less than 1.5 mm deep over a sandy soil, rainfall impact was found to contribute 85% of the sediment-transport capacity. Reductions in flow competence were evident at small baseflows and low intensities. Based on results of a second study, an alternative formulation to the Shields curve was developed for transport inception in non-rainfall-impacted overland flow. Six fluvial transport equations provided poor matches with the measured transport capacities of overland flow, both with and without rain impact. A two-component model to describe the transport capacity was developed – one component due to flow and the other due to rainfall impact. Data were obtained to calibrate the model, and previous data were used for validation. Results of these and other studies are leading to significant modifications to models of runoff and soil erosion.

Introduction

In recent years there has been a rich array of research studies, undertaken by a wide variety of scientists, regarding processes related to the erosion of soil by water. Physical geographers have explored erosion processes primarily at a micro-scale and hillslope scale for the sake of improving our understanding of fundamental processes; soil scientists have observed and measured soil loss on a simple slope and a field scale in order to select soil and crop management options which minimize that loss; and agricultural and water resources engineers have monitored and modelled soil erosion and sediment transport on slopes, fields and watersheds in an attempt to predict what happens, when and where. The net result of this plethora of research work, as it is carefully integrated, is a more complete picture of erosion and transport mechanisms and an increased capacity to model both the mechanisms and the impact of climate and management changes on the mechanisms.

At Guelph, the approach to soil erosion being taken by researchers in the School of Engineering and Agriculture Canada spans a range of scales. The intent is to acknowledge explicitly differences among controlling processes, with results obtained at one scale providing input to or provoking questions to be resolved by studies at other scales. Micro-scale studies are being conducted on soil erosion flumes and in an environmental chamber in the laboratory, and on 1 m by 1 m field plots, with the aid of rainfall simulation. Hillslope monitoring and field sampling have also been undertaken. The laboratory facilities have provided opportunities to explore the mechanics of overland flow and infiltration on sloping and layered soils, and the transport capacity of overland flow. Data from the micro-scale field plots and hillslopes have been used to study and to validate infiltration rates, overland flow and suspended sediment concentrations and yields as functions of rainfall intensity, season of the year and soil and tillage practices. Field sampling of surface runoff, erosion and infiltration is yielding data regarding the stochastic nature of variables involved. The purpose of this chapter is to present an overview of some of the more process-related studies underway, and of the results obtained to date.

Development of rainfall simulation devices

In order to control and reproduce rainfall patterns for micro-scale field and laboratory studies, several simulation devices have been developed. Two generations of simulators have been developed, calibrated and validated: the GRS I, a fixed laboratory version (Pall et al. 1983); and the GRS II, a portable model for both field and laboratory use (Tossel 1987, Tossel et al. 1987). The GRS II has been used for many of the recent studies at Guelph regarding overland flow, infiltration and sediment transport capacity, as noted in following sections. The GRS II employs a continuous, downward-flow, wide-angle (120°), full-jet nozzle spray system. With a selection of low- to medium-rate nozzles with orifice diameters of 3.2 mm to 12.7 mm, the system can provide a range of rainfall intensities from 17 mm h^{-1} to 215 mm h^{-1}, exhibiting excellent spatial uniformity characteristics. The GRS II, its water supply system, generator and related facilities are readily transported to field sites in a small truck; the system is very simple to operate; and the construction cost is extremely low.

Calibration of the Guelph simulators has involved the development of relationships between rainfall intensity and its spatial uniformity on the one hand; and nozzle size, operating water pressure and height above the experimental facility or plot on the other (Pall et al. 1983, Tossel et al. 1987). In addition, for selected nozzles, photographs have been used as a basis for characterizing and mathematically describing the spatial distribution of angles of incidence of the simulated rainfall (Guy et al. 1990).

Validation has focused on evaluating the degree to which the simulated rainfall matches natural rainfall, the basis of comparison being drop size and drop velocity data. Several methodologies have been used for ascertaining the simulated drop size and velocity distributions, including (a) a low cost photographic technique, designed and developed to "freeze" multiple exposures of falling water droplets (Beals et al. 1983); (b) a laser-based optical array spectrometer (Kollenberg 1970, Tossel et al. 1990a); and (c) a distrometer, an impact sensing system which measures the momentum of falling droplets on a sensing cone (Joss & Waldvogel 1967, Tossel 1987). The modified Marshall-Palmer equation (Ulbrich 1983, Wong & Chidambaram 1985) has proven to be an effective predictor of the drop size distribution of the simulated rainfall; and this equation, used in conjunction with a velocity predictive equation (Gunn & Kinzer 1949), has provided reliable estimates of momentum flux density and kinetic energy flux density (Tossel et al. 1990a). Comparison of drop size distributions determined for simulated rainfall with limited data of natural storms has revealed that the GRS II simulated rainfall has drop size characteristics similar to natural rainfall through the mid to large drop size ranges (diameter > 2.00 mm). The simulated rainfall has greater counts of small drops (diameter < 1.50 mm); and the velocities of simulated drops are generally lower than the terminal velocities of natural rainfall, resulting in somewhat lower values of momentum and kinetic energy than those of natural rainfall (Tossel et al. 1990b). None the less, most nozzles produce precipitation which matches the characteristics of natural rainfall; and appropriate nozzles have been selected for laboratory and field research based on the specific requirements of each study.

Infiltration

The description and modelling of overland flow for erosion research and prediction must take into account the infiltration of water. In many climates, including those in much of Canada, the infiltration process is affected by the freezing and thawing of soils. More than 75% of the annual flow in Ontario streams occurs during late winter and early spring, when soils are experiencing freeze–thaw cycles; and it is during this same period that an even greater percentage of the annual sediment load in rivers is yielded by rural basins (Dickinson & Green 1988). The extent to which runoff potential during this period, and the associated erosion potential, depends on the characteristics of, and depth to, frost layers in the soil has been a topic of recent research at Guelph (Tan et al. 1987a, b, c).

The Green & Ampt infiltration equation (Green & Ampt 1911), modified for rainfall by Mein & Larson (1973), has been further modified to predict infiltration and runoff from layered soils (Tan et al. 1987a). Laboratory experiments were conducted in a soil erosion flume under simulated rainfall conditions for initial verification of the layered form of the equation (Rudra et al. 1986, Tan et al. 1987b); and data collected at field sites during natural freezing and thawing conditions were used for final validation (Tan et al. 1987c).

The Green & Ampt infiltration equation modified for unsteady rain and layered soil conditions can be presented in the following form:

For a thawed top soil layer,

$$f = K_{m_1}(1 + \frac{A_1}{F}) \ ; \tag{10.1}$$

for a frozen middle soil layer,

$$f = K_{m_2}(\frac{F + A_2}{F + B_2}) \ ; \tag{10.2}$$

and for an unfrozen bottom soil layer,

$$f = K_{m_3}(\frac{F + A_3}{F + B_3}) \tag{10.3}$$

where: A_1 = $M_1\ (S_{av,1} + D_p)$;
$A2$ = $M_2\ (S_{av,2} + D_p) + L_1\ (M_2 - M_1)$;
A_3 = $M_3\ (S_{av,3} + D_p) + L_2\ (M_3 - M_2) + L_1\ (\ M_3 - M_1)$;
B_2 = $L_1\ ((K_{m2}/K_{m1})\ M_2 - M_1)$;
B_3 = $L_2\ ((K_{m3}/K_{m2})\ M_3 - M_2) + L_1\ ((K_{m3}/K_{m1})\ M_3 + M_1)$;
F = cumulative infiltration;
M_i = initial soil water content deficit;
$S_{av,i}$ = effective suction head at the wetting front;
D_p = hydraulic head at the soil surface;

L_i = thickness of soil layer;
K_{mi} = saturated hydraulic conductivity; and
subscripts 1, 2 and 3 refer to the top, middle and bottom soil layers.

These equations were solved using the technique outlined by Tan et al. (1987b), and calibrated for results observed in laboratory experiments. The laboratory facility involved the GRS II rainfall simulator, and a 1.5 m long, 0.70 m wide and 0.15 m deep soil flume (Dickinson et al. 1982), with a thin perforated galvanized steel plate simulating a layered soil system (Tan et al. 1987a). The hydraulic conductivity of the plate was ascertained experimentally, and the effective hydraulic conductivity ratio, i.e. the ratio of effective hydraulic conductivity of the plate to the saturated hydraulic conductivity of the overlying soil, was determined to be 0.02. The laboratory approach adopted proved to be a very effective means for exploring the layered soil system under question, and the results revealed good agreement between predicted and observed infiltration patterns. It also became clear that the water content of the lower soil layer in this system is an important variable; and knowledge of that water content is critical to the reliable use of the modified Green & Ampt equation for runoff modelling from the layered soil.

To explore possible applicability of the infiltration equation modified for layered conditions to field situations, 0°C isotherm data, i.e. freeze and thaw penetration depths, were collected for two winter seasons at two sites in Ontario, for a loam and a silty clay loam soil (Tan et al. 1987c). The estimation of these freeze and thaw penetration depths is acknowledged to be difficult, the depth to, and thickness of, a frozen soil layer depending upon the soil water content at the time of first freeze, and the thickness of freeze and thaw penetration depths further depending upon air temperature, snow cover, soil management practices, the period of time from the day the air temperature first dropped below freezing in early winter and the period of time from the day the temperature rose above freezing in late winter (Benoit & Mostaghimi 1985, Pikul et al. 1986). Therefore, the following empirical equations were developed for estimating freeze and thaw depths at daily intervals:

$$L_{FZ} = -3.97\times10^{-2} - 3.71\times10^{-3}D_f + 1.92\times10^{-5}D_f^2 - 3.34\times10^{-4}T_a^2 - 0.276L_s \quad (10.4)$$

$$L_{TZ} = -4.27\times10^{-2} - 2.46\times10^{-3}D_{uf} + 1.33\times10^{-2}T_a - 6.93\times10^{-4}T_a^2 \quad (10.5)$$

$$L_{FP} = -1.63\times10^{-2} - 8.72\times10^{-3}D_f + 5.55\times10^{-5}D_f^2 - 3.03\times10^{-4}T_a^2 - 0.578 \tag{10.6}$$

$$L_{TP} = -5.96\times10^{-2} - 1.0\times10^{-2}D_{uf} + 2.43\times10^{-4}D_{uf}^2 - 1.55\times10^{-2}T_a - 8.23\times10^{-4}T_a^2 + 0.155L_s \tag{10.7}$$

where: L_{FZ} = freeze penetration depth under zero tillage, with positive upward direction, (m);
L_{TZ} = thaw penetration depth under zero tillage, with positive upward direction, (m);
L_{FP} = freeze penetration depth under conventional tillage, with positive upward direction, (m);
L_{TP} = thaw penetration depth under conventional tillage, with positive upward direction, (m);
D_{uf} = number of days from the day the mean daily air temperature first rose above freezing during the winter months;
D_f = number of days from the day the mean daily air temperature first dropped below freezing during the winter months;
T_a = mean daily air temperature, (°C); and
L_s = daily snow depth, (m).

When the equations predicted positive values for the winter field seasons, the freeze or thaw penetration depths were assumed to be zero, i.e. at the soil surface; and when the freeze penetration depth was predicted to be equal to or greater than the thaw penetration depth, the freeze–thaw cycle was assumed to have terminated for the season, i.e. these equations simulate only one freeze–thaw cycle and do not accommodate multiple cycles in one winter season. None the less, predicted freeze penetration depths have agreed well with observed results; and predicted thaw penetration depths have followed the pattern of such observed depths reasonably well (Tan et al. 1987c).

Six years of winter runoff data from plots at the University of Guelph were used for the application of the Green & Ampt infiltration equation modified by Tan et al. (1987c). Two field events were selected to have: (a) no snow cover on the soil surface; (b) a frozen soil layer overlain by a thawing soil layer; (c) an observed rainfall event; and (d) an observed runoff event. Runoff volumes predicted on the basis of the modified equation showed good agreement with observed values. Further, the saturated hydraulic conductivity for the frozen soil layer which produced the best fit for the runoff data was found to range from 6.4 x 10^{-9} to 3.8 x 10^{-7} m s^{-1}, within the range of published values; and the effective saturated hydraulic conductivity of the thawed Guelph loam soil, varying from 0.94 x 10^{-6} to 3.78 x 10^{-6} m s^{-1}, was also in an acceptable range.

The modified Green & Ampt equation proposed and validated by Tan et al. (1987b) has also been used as a basis for studying and simulating infiltration into soil profiles that experience surface crusting and/or sealing (Borah et al. 1989). An infiltration model has been hypothesized for a two- or three-layer soil system, consisting of a crusted or sealed surface layer and one or two non-homogeneous subsoil layers. Both laboratory and field experiments have been conducted to obtain model parameters; and simulated and observed values of infiltration rates, cumulative infiltration (and therefore runoff volumes) and times-to-ponding have been compared. Results to date have been very promising; and they have confirmed the utility of an integrated approach involving a hypothetical model, and laboratory and field observations.

In summary, it has been necessary to develop an expression for infiltration which can be applied to layered soil conditions, such as exist during freeze–thaw periods. It has been found that the Green & Ampt equation can be suitably modified for this purpose; that layered conditions of a freeze–thaw type can be simulated in the laboratory to explore infiltration; and that a modified Green & Ampt equation can provide useful estimates of infiltration and runoff amounts from thawing soil.

Hydraulics and transport capacity of overland flow

Two laboratory studies have recently been conducted to investigate and model the transport of sediment in overland flow. The first study was designed to investigate hydraulics and transport in interrill flow, both with and without rainfall, for a single soil type. This study prompted many questions, and led to a more comprehensive examination of hydraulics, transport inception, transport capacity, and the variation of transport capacity with rainfall and flow conditions. Results of these studies have been reported by Guy (1987, 1990), Guy et al. (1987, 1990, 1992), and Guy & Dickinson (1990). Two additional articles (Guy 1992a, b) are currently in preparation, based on the work reported in Guy (1990). The methods and the significant results of these studies are highlighted in this section.

Influence of rainfall on shallow-flow transport capacity
Research into soil erosion by overland flow has been constrained by difficulties in measurement of the relevant hydraulic and soil properties during rainfall–runoff events, and by the complexity of the processes involved. Nevertheless, the influence of rainfall, at least qualitatively, has been widely reported. Using a 0.69 m wide by 1.50 m long, roughened rigid boundary flume, a header tank and the GRS II rainfall simulator, the influence of rainfall on the transport capacity of very shallow flows was quantified. Two distinct flow types were examined – uniform baseflow, and spatially varied flow produced by the simulated rainfall, for a single soil type. Two replications were conducted at slopes of 0.02, 0.09 and 0.20 m m^{-1}; at three flowrates in the first case (1.2 x 10^{-5}, 7.2 x 10^{-5} and 24.0 x 10^{-5} m^2 s^{-1}), and at three rainfall intensities in the second (4.5, 14.0 and 18.0 cm h^{-1}). The rates of sediment delivery from the experimental flume were known to be constrained by the transport capacity, rather than by the rate of particle detachment from the bed.

Flow Reynolds numbers were less than 240, and flow depths were less than 1.5 mm. Sediment concentrations at the capacity transport rate ranged from 1.07 to 238 kg m^{-3}. Values of other hydraulic variables have been given by Guy et al. (1990). Since flows were within the laminar range, discharge per unit width and bedslope can be considered as fundamental hydraulic parameters. Knowing the values of these variables, along with fluid properties and a flow resistance coefficient, it is possible to determine values of most other hydraulic variables, such as boundary shear stress and stream power.

In spite of low values of relative submergence and relatively high viscosities in the sediment-laden uniform flows, velocities exceeded the predictions of the laminar velocity profile model. In the rainfall-generated flows, velocities were reduced compared to the uniform flows, in accordance with previous observations (e.g. Parsons 1949, Emmett 1970, Shen & Li 1973). In an important observation, the angle of incidence of the rainfall was found to significantly influence flow velocities within the flume. Over the upper portion, velocities were smaller than expected, whereas over the lower portion, velocities were larger than expected.

Both discharge and bedslope were found to influence transport capacity in the uniform flow runs, according to the relationship:

$$q_{sf} = aq^{b}s^{c} \tag{10.8}$$

where: q_{sf} = transport capacity of uniform flow;
q = discharge per unit width;
s = bedslope; and

a, b, c = coefficients.

For cases in which the flow was able to transport the full size distribution of the experimental soil, the exponents b and c were not significantly different from 2. However, at small slopes and small flow rates, transport became selective (i.e. only the smaller sizes of the soil mixture were transportable), which had the effect of reducing the exponent c. This observation indicates the importance of including a transport threshold term in an overland-flow transport equation. The exponents b and c for cases of non-selective transport in laminar flow are comparable to those of Meyer & Wischmeier (1969) for rough turbulent flow, who suggested that both exponents should be approximately 1.7. Although Aziz & Scott (1989) did not include a threshold term in their analysis, and arbitrarily forced the exponents b and c to be equal by selecting, a priori, a stream power function to describe their turbulent flow transport data, they found $b = c = 1.73$. In the rainfall-disturbed flows, the transport capacity enhancement caused by raindrop impact was determined by subtracting the flow contribution (as determined from results of the uniform flow runs) from the measured transport capacities. The transport capacities of these flows were 85% attributable to rainfall impact, and only 15% attributable to the flow itself. In addition, the competence of the rainfall-disturbed flows was greater than in comparable uniform flows. The rainfall enhancement of transport capacity was strongly related to both rainfall intensity and bedslope, as follows:

$$q_{sr} = dI^e s^f \tag{10.9}$$

where: q_{sr} = transport capacity enhancement attributable to rainfall;

I = rainfall intensity; and

d, e, f = coefficients.

The exponents e and f were not significantly different from 2 and 1, respectively. It is noteworthy that rainfall momentum and kinetic energy flux densities were better predictors for rainfall transport capacity enhancement than was rainfall intensity. This result is physically reasonable, since the micro-scale flow changes associated with raindrop impact can be more easily explained by exchanges of momentum or energy than by the simple addition of mass. Young & Wiersma (1973) have provided a clear demonstration of the importance of the rainfall kinetic energy. By placing a protective cloth layer over the experimental surface, the authors achieved an 89% reduction in the kinetic energy of the applied rainfall, without altering the intensity. The result was a 90% drop in the rate of soil loss. From a practical modelling standpoint, however, the rainfall intensity remains a superior predictor variable, because of its measurement ease and the widespread availability of data.

These experiments focused attention on the need to account separately for the different roles of rainfall and runoff in determining the transport capacity of shallow flow. They also identified the existence and importance of transport thresholds in such flows, and raised questions concerning their hydraulics. Other implications of the measurements include variation of the flow contribution to transport capacity with downslope distance, and variation of the rainfall contribution with flow depth.

In order to address these questions, a second and more comprehensive set of laboratory experiments was designed. These runs were conducted in a 1.50 m long by 0.253 m wide rectangular flume, with a glass bottom and Plexiglas sides. The GRS II rainfall simulator, centred over the flume, was used to provide simulated rainfall. Four very narrowly graded, non-cohesive materials were tested, each over a surface of natural roughness. The materials were selected to span a range of sizes and densities typified by natural soils, with median particle diameters ranging from 0.151 to 0.381 mm, and densities ranging from 1496 to 2650 kg m^{-3}. Capacity rate sediment transport for each material was observed in 207 runs. The hydraulic environments included uniform baseflow without rainfall, rainfall-generated flow without accompanying baseflow, and many combinations of rainfall and baseflow. Measurements included bedslope, baseflow rate, rainfall intensity, surface flow velocity (using a tracer dye) and flow depth (using a point gauge graduated in thousandths of an inch). Bedslopes ranged from 1% to 12%, baseflow rates ranged from 0 to 1090 mm^2 s^{-1} and rainfall intensity ranged from 0 to 13.9 cm h^{-1}. Flow depths did not exceed 5 mm. In addition to the transport capacity measurements, a separate set of 16 runs was performed in uniform baseflow, in which hydraulic conditions prevailing at the point of transport inception were determined. Because of the complexity of the experimental environment, the identification of precise transport thresholds under rainfall was not possible. Even with the lowest rain intensity, the smallest particles came into motion.

Transport inception

Data collected in both sets of experiments were applied to test the validity of the Shields transport inception criterion (e.g. Yalin & Karahan 1979). The Shields curve is used widely to determine the hydraulic conditions required to initiate motion of fluvial bed material. However, overland flow differs from riverine flow conditions in several important areas: characteristics of the underlying surface (including cohesion, and wide ranges in particle and aggregate size and density), steep slopes, high values of relative roughness and the influence of rainfall impact. These differences place overland flow beyond the range of hydraulic conditions for which the Shields criterion has been experimentally determined (Guy & Dickinson 1990). In addition, use of the curve requires knowledge of the appropriate frictional resistance law governing the flow. Flow resistance in shallow overland flow is not well understood, and Guy & Dickinson (1990) have shown with experimental data that different results can be obtained simply by varying the resistance law. Nevertheless, because of a lack of applicable research, the Shields curve has been applied to overland flow (Knisel 1980, Beasley et al. 1980).

Observations indicated clearly that overland flow was able to transport larger material when affected by rainfall. The Shields criterion significantly overpredicted the hydraulic conditions required to initiate motion in rain-impacted flows. In flows unaffected by rainfall, the transport threshold was observed to depend on the relative roughness, confirming the earlier work of Govers (1987). This dependence contributed to scatter in a plot of experimental data on the Shields diagram.

Data suggested that an alternative approach to transport inception is needed for overland flow, an approach that does not depend on knowledge of flow resistance, and which can be applied using relatively simple measurements of key hydraulic variables, and knowledge of fluid and particle properties. An approach similar to those of Schoklitsch (1962) and Bathurst et al. (1987) was adopted, with the following result:

$$q_{cr} = 1.062 \times 10^{-4} \varrho^{*\,2.36} d_{50}^{\;0.79} \sin\Theta^{-1.49} \qquad (10.10)$$

where:
- q_{cr} = unit discharge at transport inception (m^2/s);
- ϱ^* = $\varrho_s/\varrho_f - 1$;
- ϱ_s, ϱ_f = densities of sediment and fluid, respectively;
- d_{50} = median particle diameter (m); and
- Θ = angle of bed surface from horizontal (degrees).

The constant, *C*, incorporates the effects of gravity, flow resistance, and, for laminar flow, the kinematic viscosity. The equation is based on only 16 measurements, and must be refined before it can be applied in more general use. (The experimental range of key variables was as follows: $0.57 < \Theta < 6.80°$; $0.151 < d_{50} < 0.381$ mm; $0.50 < \varrho^* < 1.65$; $7 < q_{cr} < 600$ mm^2 s^{-1}.) Nevertheless, because of its advantages over the Shields criterion, this "critical discharge" based formulation has proven valuable for analysing the present experimental data.

Evaluation of fluvial transport equations for overland flow

Transport capacity and hydraulic measurements obtained in the 207 experimental runs were used to test the ability of six fluvial sediment transport equations to predict the transport capacity of overland flow. Equations tested were the Laursen (1958), Schoklitsch (1962), Yalin (1963), Bagnold (1966), Yang (1973) and a modified equation of the duBoys form. Most of these equations have either been recommended or used directly for modelling the transport capacity of rill, interrill or combined rill/interrill flow.

The transport capacity data obtained in uniform flow were considered separately from the data collected in rainfall-disturbed flow. Several statistics of the equation residuals were evaluated, including the mean error, the ratio of the mean prediction to the mean measurement, the root mean square error and the standard error of the residuals. In addition to the statistical evaluation, the performance of the equations in representing the trend of the data, the transport threshold, and material of different sizes and densities were all considered.

Whereas the general trend of the uniform flow data was reasonably well duplicated by the equations, that of the rainfall-disturbed data was not. For both uniform and rainfall-disturbed flow, most equations underpredicted transport rates, and showed significant variability. In confirmation of the transport threshold findings reported in the previous section, the threshold components of the fluvial equations provided poor predictions for the uniform flows, and extremely poor predictions for the rainfall-disturbed flows. Only the Laursen (1958) and Yalin (1963) equations were developed or have been modified for calculating the transport capacity of a sediment mixture, rather than for a bed composed of particles of equal size and density. The performance of the other equations suffered from this limitation.

None of the equations appears suitable for predicting the transport capacity of overland flow during rainfall. For flows without rain-impact, only the Schoklitsch (1962) equation provided a reasonable representation of the data. This result agrees with that of Bathurst et al. (1987), who reported that this equation is well suited to transport prediction in steep, rough flow. Since the data used to test the equations were obtained directly, the results are considered to be more conclusive than those of two previous model evaluation studies (Alonso et al. 1981, Julien & Simons 1985). The findings shed some doubt on the results of these two studies, in which some of the six equations have been recommended as appropriate for overland flow. The results also suggest possible weaknesses in several soil erosion models in common use.

Flow depth, velocity and resistance in overland flow

At least 1000 measurements of the surface velocity of sediment-laden laminar overland flow under rainfall were obtained during the transport experiments. The steep and rough boundary, particularly in combination with high viscosities and rainfall, would be expected to cause increased flow resistance and depth, and reduced velocities, compared to values over low gradient, smooth surfaces, without rainfall. However, differences between the measurements and the predictions of the smooth-surface, clear-water laminar velocity profile model could not be discerned by statistical analysis. This result supports our earlier finding based on more limited data. However, it does not lend support to the proposal of Savat (1980), who suggested a modification of the "constant" k_0 in the Darcy–Weisbach friction law to allow k_0 to increase as the surface becomes rougher and steeper, in non-sediment-laden flow. Neither does it support the conclusion of Shen & Li (1973) that the Darcy–Weisbach friction factor increases under rainfall. Flow depth measurements proved less conclusive, but also tended to support the applicability of the smooth-surface, clear-water profile model, when the thickness of the moving bed layer was accounted for.

It is suggested that reduced flow resistance due to high sediment concentrations may be partially responsible for these observations. Reductions in flow resistance due to high sediment concentrations have been confirmed in turbulent flows, but not yet in laminar flows. A second hypothesis is that in very shallow flow, the boundary assumes an increasingly significant role, compared to that of the flow, in completing the translation of the nearly vertical momentum of the rainfall to the flow direction. Raindrop size and velocity data suggested strongly that the flume surface, rather than the flow layer, performed most of the vertical to horizontal translation of momentum. Savat (1977) also found a relatively small rainfall effect on flow depths in very shallow flows (less than 3 mm deep).

Through these measurements and comparisons with previous data, a conceptual hydraulic picture of sediment-laden overland flow during rainfall has emerged. The velocity profile can be considered to comprise two components: a lower layer in which velocities are affected in an unknown way by the rough surface and by moving sediment, and an upper layer in which the time-averaged velocity distribution remains as expected over a smooth surface, in spite of momentary disruptions due to raindrop impact.

Overland flow transport capacity: model development and verification

Articles in which the significant results of Guy (1990) are reported are currently in preparation. A brief overview of the results is provided in this section.

A general model form, which is suitable for representing the transport capacity of overland flow, was developed. The model separates the two contributions to the transport capacity associated with flow and with rainfall-impact, as follows:

$$q_s = q_{sf} + q_{sr} \tag{10.11}$$

where: q_s = transport capacity;

q_{sf} = component of transport capacity due to the flow; and
q_{sr} = component of transport capacity due to rainfall-impact.

This model form is applicable to a broader range of conditions than other forms proposed earlier (e.g. Julien & Simons 1985), in which the two contributions are linked multiplicatively.

On the basis of theoretical and prior empirical information, key variables representing the driving hydraulic and rainfall forces, geometric characteristics of the boundary and properties of both fluid and solid were distilled from a large list of potentially relevant variables governing sediment transport. Relationships between these key variables and the experimental data were explored in the development of a transport capacity equation.

The optimal representation for transport capacity in non-rain-impacted flow was the following:

$$q_{sf} = a0(q-q_{cr})^{a1} \sin\Theta^{a2} \varrho^{*\,a3} {d_{50}}^{a4} \tag{10.12}$$

where q_{cr} is discharge per unit width at the point of transport inception, as defined above, and other variables are as previously defined (*a0, a1, a2, a3* and *a4* are coefficients). Flows under rainfall were consistently able to transport larger particles, and at higher rates, than equivalent undisturbed flows. The optimal representation for the transport capacity of rain-impacted flow was represented by the following model:

$$q_s = a0(q-q_{cr})^{a1} \sin\Theta^{a2} \varrho^{*\,a3} {d_{50}}^{a4} + b0 I^{b1} q^{b2} \sin\Theta^{b3} \varrho^{*\,b4} {d_{50}}^{b5} \tag{10.13}$$

where all variables are as previously defined, and *b0, b1, b2, b3* and *b4* are coefficients. It is noteworthy that the rainfall-impact effect (q_{sr}) is itself dependent on the flowrate (q). More detailed analyses with subsets of the data showed that the exponent *b2* became smaller as flowrate increased. These results show that, for very shallow flows, the rainfall effect increases with discharge, but at a diminishing rate as depths increase. Other studies have shown a similar effect (Palmer 1965, Mutchler & Larson 1971), whereby the raindrop impact effect on a surface first increases to a maximum value as water depth increases, then decreases and becomes insignificant. In terms of the relative contributions to transport, at the lower limit of the flowrates examined, below the transport threshold for undisturbed baseflows, the rainfall effect was entirely responsible for transport. As flows increased beyond the transport threshold, the flow contribution assumed an increasing share of the transport capacity. At the upper limit of the flows, the flow and rainfall contributions were about equally responsible for the transport capacity.

The data suggest that three ranges of sediment transport can be defined: a lower range, in which depths are shallow enough (less than approximately 5 mm) that the rainfall effect is directly related to flowrate, and dominates the transport capacity; an intermediate range, in which the rainfall effect is inversely related to flowrate, and reduces to insignificance; and an upper range, in which rainfall has no effect on sediment transport. The present study was limited to the lower region, whereas fluvial transport occurs within the upper region. The intermediate region, including its upper limit, remains undefined.

Both the flow and the rainfall-impact components of the transport capacity equation, and the full equation, were tested against independent data. The tests demonstrated that sufficient data were collected to fully account for the variation of the independent variables within the measured ranges. Comparison of model predictions with results of six previous studies confirmed the adequacy of the models, but also indicated limitations. In particular, additional work is needed to confirm the behaviour of transport capacity of material smaller than, and of material with densities intermediate between, the values examined in these experiments.

Concluding remarks

The erosion of soil by overland flow is extremely complex. However, our understanding of the processes involved, and our ability to make qualitative and quantitative predictions about those processes, continue to improve as we explore various hypotheses both in the laboratory and in the field. The findings from process-oriented research regarding soil erosion and overland flow are leading to significant modifications of field and watershed models relating to runoff, sediment yield and non-point-source pollution.

References

Alonso, C. V., W. H. Niebling, G. R. Foster 1981. Estimating sediment transport capacity in watershed modelling. *Transactions of the American Society of Agricultural Engineers* **24**, 1211–20, 1226.

Aziz, N. M. & D. E. Scott 1989. Experiments on sediment transport in shallow flows in high gradient channels. *Hydrological Sciences Journal* **34**, 465–78.

Bagnold, R. A. 1966. *An approach to the sediment transport problem from general physics.* US Geological Survey Professional Paper 422-I. Washington DC: US Government Printing Office.

Bathurst, J. C., W. H. Graf, H. H. Cao 1987. Bed load discharge equations for steep mountain rivers. In *Sediment transport in gravel-bed rivers*, C. R. Thorne, J. C. Bathurst, & R. D. Hey, (eds), 453–77. Chichester, England: John Wiley.

Beals, D., W. T. Dickinson, D. S. Helsdon, R. Pall 1983. An approach for characterizing falling water droplets. *Canadian Agricultural Engineering* **25**, 189–92.

Beasley, D. B., L. F. Huggins, E. J. Monke 1980. ANSWERS: A model for watershed planning. *Transactions of the American Society of Agricultural Engineers* **23**, 938–44.

Benoit, G. R. & S. Mostaghimi 1985. Modelling soil frost depth under three tillage systems. *Transactions of the American Society of Agricultural Engineers* **28**, 1499–505.

Borah, M. K., R. P. Rudra, W. T. Dickinson 1989. *Application of Green and Ampt infiltration equation in overland flow simulation.* American Society of Agricultural Engineers Paper No. 89-2169, American Society of Agricultural Engineers, St. Joseph, MI.

Dickinson, W. T., R. P. Rudra, G. J. Wall 1982. *Seasonal variations in soil erodibility.* American Society of Agricultural Engineers Paper No. 82-2573, American Society of Agricultural Engineers, St. Joseph, MI.

Dickinson, W. T. & D. Green 1988. Characteristics of sediment loads in Ontario streams. *Canadian Journal of Civil Engineering* **15**, 1067–79.

Emmett, W. W. 1970. *The hydraulics of overland flow on hillslopes.* US Geological Survey Professional Paper No. 662-A. Washington DC: US Government Printing Office.

Govers, G. 1987. Initiation of motion in overland flow. *Sedimentology* **34**, 1157–64.

Green, W. H. & G. A. Ampt. 1911. Studies on soil physics. 1. The flow of air and water through soils. *Journal of Agricultural Science* **4**, 1–24.

Gunn, R. & G. D. Kinzer 1949. Terminal velocity of water droplets in stagnant air. *Journal of Meteorology* **6**, 243–8.

Guy, B. T. 1987. Investigation of sediment transport at the capacity rate in interrill flow. Unpublished M.Sc. thesis. University of Guelph, Guelph, Ontario.

Guy, B. T. 1990. Sediment transport capacity of shallow overland flow. Unpublished Ph.D. thesis, University of Guelph, Guelph, Ontario.

Guy, B. T. 1992a. Overland flow transport capacity: model development. (in preparation).

Guy, B. T. 1992b. Overland flow transport capacity: model verification. (in preparation).

Guy, B. T. & W. T. Dickinson 1990. Inception of sediment transport in shallow overland flow. In *Soil erosion – experiments and models (Catena Supplement 17)*, R. B. Bryan (ed.) 91–109. Cremlingen: Springer-Verlag.

Guy, B. T., W. T. Dickinson, R. P. Rudra 1987. The roles of rainfall and runoff in the sediment transport capacity of interrill flow. *Transactions of the American Society of Agricultural Engineers* **30**, 1378–86.

Guy, B. T., W. T. Dickinson, R. P. Rudra, G. J. Wall 1990. Hydraulics of sediment-laden sheetflow and the influence of simulated rainfall. *Earth Surface Processes and Landforms* **15**, 101–18.

Guy, B. T., W. T. Dickinson, R. P. Rudra 1992. Evaluation of fluvial sediment transport equations for overland flow. *Transactions of the American Society of Agricultural Engineers* (in press).

Joss, J. V. & E. G. Waldvogel 1967. Ein Spectrograph für Niederschlagstropfen mit automatischer Auswertung. *Pure and Applied Geophysics* **68**, 240–6.

Julien P. Y. & D. B. Simons 1985. Sediment transport capacity of overland flow. *Transactions of the American Society of Agricultural Engineers* **28**, 755–62.

Knisel, W. G. (ed.) 1980. *CREAMS: a field-scale model for chemicals, runoff, and erosion from agricultural management systems.* US Department of Agriculture, Conservation Service Report No. 26.

Kollenberg, R. G. 1970. The optical array: an alternative to scattering or extinction for airborne particle size determination. *Journal of Applied Meteorology* **9**, 86–103.

Laursen, E. 1958. The total sediment load of streams. *Journal of the Hydraulics Division, Proceedings of the American Society of Civil Engineers* **84**, 1530-1–6.

Mein, R. G. & C. L. Larson 1973. Modelling infiltration during a steady rain. *Water Resources Research* **9**, 384–94.

Meyer, L. D. & W. H. Wischmeier 1969. Mathematical simulation of the process of soil erosion by water. *Transactions of the American Society of Agricultural Engineers* **12**, 754–8, 762.

Mutchler, C. K. & C. L. Larson 1971. Splash amounts from waterdrop impact on a smooth surface. *Water Resources Research* **7**, 195–200.

Pall, R., W. T. Dickinson, D. Beals, R. McGirr 1983. Development and calibration of a rainfall simulator. *Canadian Agricultural Engineering* **25**, 181–87.

Palmer, R. S. 1965. Waterdrop impact forces. *Transactions of the American Society of Agricultural Engineers* **8**, 69–70, 72.

Parsons, D. A. 1949. *Depths of overland flow.* Soil Conservation Service Technical Paper 82.

Pikul, Jr, J. L., J. F. Zuzel, R. N. Greenwalt 1986. Formation of soil frost as influenced by tillage and residue management. *Soil and Water Conservation Journal* **41**, 196–9.

Rudra, R. P., W. T. Dickinson, G. J. Wall, K. A. Tan 1986. Runoff response to frost layering. *Transactions of the American Society of Agricultural Engineers* **29**, 735–40.

Savat, J. 1977. The hydraulics of sheet flow on a smooth surface and the effect of simulated rainfall. *Earth Surface Processes* **2**, 125–40.

Savat, J. 1980. Resistance to flow in rough supercritical sheet flow. *Earth Surface Processes* **5**, 103–22.

Schoklitsch, A. 1962. *Handbuch des Wasserbues*, 3rd edn. Vienna: Springer-Verlag.

Shen, H. W. & R. M. Li 1973. Rainfall effect on sheet flow over smooth surface. *Journal of the Hydraulics Division, Proceedings of the American Society of Civil Engineers* **99**, 771–92.

Tan, K. A., R. P. Rudra, W. T. Dickinson, D. E. Elrick, G. J. Wall 1987a. *The experimental procedure for the verification of Green and Ampt equation.* Canadian Society of Agricultural Engineering Paper No. 87-304, Canadian Society of Agricultural Engineering, Ottawa.

Tan, K. A., R. P. Rudra, W. T. Dickinson, D. E. Elrick, G. J. Wall 1987b. *Verification of Green-Ampt infiltration equation for layered soils.* American Society of Agricultural Engineers Paper No. 87-2002, American Society of Agricultural Engineers, St. Joseph, MI.

Tan, K. A., R. P. Rudra, W. T. Dickinson, D. E. Elrick, G. J. Wall 1987c. *Modelling of infiltration under thawing soil conditions.* American Society of Agricultural Engineers Paper No. NAR87-209, American Society of Agricultural Engineers, St. Joseph, MI.

Tossel, R. W. 1987. A comparison of three methods used to characterize simulated rainfall properties. Unpublished M.Sc. thesis. University of Guelph, Guelph, Ontario.

Tossel, R. W., W. T. Dickinson, R. P. Rudra, G. J. Wall 1987. A portable rainfall simulator. *Canadian Agricultural Engineering* **29**, 155–62.

Tossel, R. W., G. J. Wall, W. T. Dickinson, R. P. Rudra, P. H. Groenevelt 1990a. The Guelph rainfall simulator II: Part I – Simulated rainfall characteristics. *Canadian Agricultural Engineering* **32**, 205–13.

Tossel, R. W., G. J. Wall, R. P. Rudra, W. T. Dickinson, P. H. Groenevelt 1990b. The Guelph rainfall simulator II: Part II – A comparison of natural and simulated rainfall characteristics. *Canadian Agricultural Engineering* **32**, 215–23.

Ulbrich, C. W. 1983. Natural variations in the analytical form of the raindrop size distribution. *Journal of Climatology and Applied Meteorology* **22**, 1764–75.

Yalin, M. S. 1963. An expression for bed-load transportation. *Journal of the Hydraulics Division, Proceedings of the American Society of Civil Engineers* **89**, 221–50.

Yalin, M. S. & E. Karahan 1979. Inception of sediment transport. *Journal of the Hydraulics Division, Proceedings of the American Society of Civil Engineers* **105**, 1433–43.

Yang, C. T. 1973. Incipient motion and sediment transport. *Journal of the Hydraulics Division, Proceedings of the American Society of Civil Engineers* **99**, 1679–1704.

Young, R. A. & J. L. Wiersma 1973. The role of rainfall impact in soil detachment and transport. *Water Resources Research* **9**, 1629–39.

Wong, R. K. W. & N. Chidambaram 1985. Gamma size distributions and stochastic sampling errors. *Journal of Climatology and Applied Meteorology* **24**, 568–79.

11

Evaluation of transporting capacity formulae for overland flow

Gerard Govers

Abstract

The prediction of the transporting capacity of overland flow is, in most models, obtained by implementing an existing formula developed from observations in channels and alluvial rivers. However, this approach is questionable, as the hydraulic conditions in overland flow are often totally different. Here, the performance of a number of formulae is tested using an experimental dataset obtained under laboratory conditions simulating rill flow. It is concluded that no existing formula performs well over the whole range of available data. The fundamental reason for this failure is that all existing formulae are to some extent empirical. Therefore, use of them outside the domain for which they were developed may lead to erroneous results. On the other hand, simple empirical equations based on shear stress, unit stream power and effective stream power, as well as the formula of Low (1989), can be used to predict the sediment transport capacity of overland flow, at least in some cases. These equations cannot only be used for our own data, but they also allow reasonable to good predictions of sediment loads measured by other authors. Further development of these equations is certainly possible. However, this requires that more experimental data be available, e.g. on the effects of sediment density and the effect of macroroughness on the transporting capacity of overland flow.

Introduction

Research, both experimental and theoretical, on the transporting capacity of overland flow has been stimulated by the development of so-called physically based erosion models during the past two decades, which describe sediment detachment and transport as distributed processes. A transporting capacity equation is necessarily a part of such a model, as the transporting capacity of the flow equals the upper limit of erosion. Furthermore, detachment is often modelled as a function of transporting capacity deficit.

At present, most deterministic erosion models incorporate existing transport formulae, the development of which is mainly based on experimental work in rivers or channels. Among the formulae most frequently used are the bedload formula of Yalin (1963) and the total load formula of Yang (1973). Although these and most other formulae are based on physical principles, they are also calibrated using experimental data: the sediment transport problem is so complex that, at present, it defies a completely deterministic prediction from basic physical properties alone (Yang 1973, Bagnold 1980). Calibration has proven to be a necessary step in the development of a sediment transport formula for channel flow.

In general, overland flow often has a very limited depth, and slopes are often one or two orders of magnitude higher than those encountered in alluvial channels: the possible consequence of this is that a semi-empirical formula developed to predict sediment transport in channels will give erratic predictions when used for overland flow. The problem is well illustrated by the work of Smart (1984), who found that the classic bedload formula of Meyer-Peter & Müller (1948) could not be used to predict sediment transport on steep slopes. Sediment transporting capacity was systematically underestimated on slopes steeper than 3%, and on a 20% slope transport rates were often an order of magnitude higher than the predicted values. Consequently, a new equation with an additional slope factor was developed (Smart 1984).

Similar problems may be expected when channel formulae are used in the overland-flow domain: thus, channel formulae should only be used after careful evaluation. Otherwise, predictions of sediment transporting capacity may become severely biased.

Until now, relatively few attempts have been made to evaluate the performance of channel formulae in overland-flow conditions. The use of the Yalin formula was proposed by Foster & Meyer (1972). Alonso et al. (1981) evaluated a wide range of sediment-transport formulae under various conditions. They concluded that for shallow flows, the Yalin equation gave the best predictions. In recent years, this equation has been incorporated in several models, including the erosion component of WEPP-project (Foster et al. 1989). Moore & Burch (1986) stressed the potential of the formula of Yang (1973) which is based on the concept of unit stream power. They obtained good results for sheet as well as rill flow conditions. Loch et al. (1989) found that the Yang formula gave satisfactory results for rill flow. Lu et al. (1989) also evaluated the capability of seven bedload and three total-load formulae to predict the sediment transporting capacity of flow over a deposition area using sand of different sizes. The best results were obtained using the formula of Engelund & Fredsoe (1976), while those of Meyer-Peter & Müller (1948), Einstein (1950) and Bagnold (1973) also gave good results. They also proposed a transport formula which was developed from their own experimental data.

The validation attempts reported above are all based on experimental data covering a rather limited range of parameter values. In the shallow flow experiments used in the evaluations reported by Alonso et al. (1981), slope never exceeded 0.07, while grain size was between 150 and 350 μm and specific density varied. Furthermore, flow was always in the laminar domain. The data used by Moore & Burch (1986) were all collected on slopes below 0.05, while median sediment diameters ranged from 100 μm to *c*. 450 μm and specific density of the wet sediment varied between 2650 and 1420 kg m^{-3}. Unit discharges used in the experiments of Lu et al. were low ($\leq 4 \times 10^{-4}$ m^3 (m s)$^{-1}$): equilibrium slopes are not reported but the reported dimensionless shear stress values suggest that they were relatively low as well. The evaluations of Foster & Meyer (1972) and Loch et al. (1989) were based on data covering an even more limited range.

A more thorough evaluation of the performance of transporting capacity formulae requires first of all that more data should be available. Govers (1990) collected data on a wider range of slopes, discharges and materials than the datasets that were used in the evaluations reported above, although not all possible conditions were covered. Unit discharges in this study were in the intermediate to high range (2–150 $\times 10^{-4}$ m^2 s^{-1}), so that the data were more representative for rill flow than for sheet or interrill flow. Slopes ranged from 0.017 to 0.21 and five well-sorted quartz materials with a median grain size ranging from 58 to 1100 μm and with a specific density near 2650 kg m^{-3} were tested (Table 11.1). In total, 465 experiments were carried out. In this chapter, these data are used to evaluate the performance of a number of transporting capacity equations. Also, some simple equations are proposed which can be used to predict sediment transporting capacity of overland flow

Table 11.1 Median grain size (D_{50}), grain roughness (k_s), critical shear stress – determined by initiation of motion experiments in laminar (τ_{crl}) and turbulent flow (τ_{crt}) as reported by Govers (1987) – and critical shear stress determined from solid discharge–shear stress plots (τ_{cr}) for the five materials used.

Material	D_{50} μm	k_s μm	τ_{crl} Pa	τ_{crt} Pa	τ_{cr} Pa
A	58	60	0.13	–	0.20
B	127	140	0.29	0.20	0.25
C	218	250	0.42	0.24	0.27
D	414	500	0.60	0.39	0.30
E	1098	1200	1.04	0.58	0.35

with sufficient accuracy: the performance of these equations is evaluated using independently collected data.

Materials and methods

Data on the transporting capacity of overland flow were collected using a non-recirculating flume 6 m long and 0.117 m wide. Details of the experimental procedure are given in Govers (1990). However, two points need further elaboration: the first concerns the calculation of the hydraulic characteristics of the flow, and the second concerns the elimination of edge effects.

As data were collected on plane beds, hydraulic characteristics of the flow were calculated using the algorithm of Savat (1980), which has been tested extensively for the prediction of friction factors and hence mean velocity and depth for clear water (Govers & Rauws 1986, Rauws 1988). However, while doing velocity measurements using dye in sediment-laden flows, Govers (1990) observed that the flow velocities which were measured using dye tracing were significantly higher than the flow velocities predicted using Savat's algorithm. Consequently, it was proposed to correct flow velocities and fluid depths using:

$$\bar{u}_s = \frac{\bar{u}_{cw}}{(1 - C_v)} \qquad (11.1)$$

where $\bar{u}_s$ = the velocity of the sediment-laden flow;
$\bar{u}_{cw}$ = the velocity of the clear-water flow with the same discharge over a bed with similar roughness;
C_v = the volumetric sediment concentration,
= $q_{s,v}/q$;
$q_{s,v}$ = the volumetric unit solid discharge; and
q = the unit discharge.

However, recent data concerning the effect of sediment load on mean flow velocity are ambiguous. Guy et al. (1990) found that mean surface velocities of sediment-laden sheetflow were 12% higher than those measured in clear water. On the other hand, Aziz & Scott (1989) consistently measured an increase in flow depth when sediment was present in the flow, even for the experiments without any noticeable development of bedforms, thus suggesting a negative effect of sediment load on flow velocity. When sediment load was present, Aziz & Scott measured total depth: thus, part of the increase in flow depth is due to the presence of sediment in the flow. The (fictitious) fluid depth may be calculated as:

$$d_f = d_t(1 - C_f) \tag{11.2}$$

where C_f = the volumetric sediment concentration in the fluid;
d_f = the fictitious fluid depth; and
d_t = the total flow depth.

C_f is not equal to the volumetric sediment concentration obtained by dividing the volumetric solid discharge by the volumetric fluid discharge: the average velocity of the sediment particles is generally lower than the mean fluid velocity. Assuming, as a first approximation, that:

$$C_f = 1.5\,C_v \tag{11.3}$$

fictitious fluid depths were calculated from the depths reported by Aziz & Scott (1989) for sediment-laden flows. Figure 11.1 shows that fictitious fluid depths are also higher in sediment-laden flows than in clear-water flows. It is rather surprising that both Guy et al. (1990) and Aziz & Scott (1989) find the effect of sediment load on velocity or depth to be independent of the intensity of sediment movement. This suggests that some of the differences observed might be due to bias on the depth and/or velocity measurements in sediment-laden flow or, in the study of Aziz & Scott, to bedform development. One may, indeed, expect that if sediment load has an influence on flow properties, the change in flow properties should show systematic variation with sediment load.

Perhaps the most detailed experiments on the influence of (suspended) load on flow velocity on steep slopes were carried out by Rickenmann (1990), who studied the influence of a suspension load, consisting of a silt-sized material, on flow hydraulics. Discharges in this study were relatively high, while the bed consisted of fixed gravel, resulting in flow depths of several centimetres. Considering the high energy level of the flow and the grain size distribution of the suspension material, it may be assumed that the sediment was moving with the same mean velocity as the fluid. Rickenmann's results indicate that the presence of suspension material does not have any significant influence on flow velocity for volumetric sediment concentrations below 0.20 when total volumetric discharge, i.e. the sum of the volumetric water discharge and the volumetric sediment discharge, is kept constant. A further increase in suspended sediment concentration causes the mean flow velocity to decrease. However, when data are compared on the basis of water discharge alone, it can be observed that mean flow velocity increases for a given fluid discharge with increasing sediment concentration up to a volumetric concentration of *c.* 0.20 (Fig. 11.2). At higher concentrations mean flow velocity decreases again. However, increases in flow velocity are generally lower than suggested by

Govers (1990) and, overall, the effect of sediment load on flow properties remains limited.

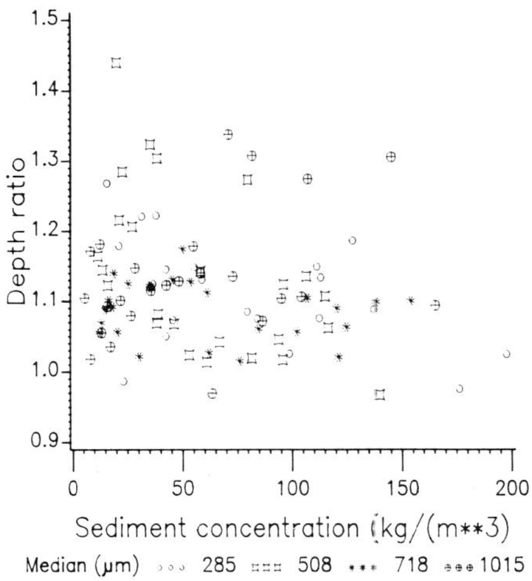

Figure 11.1 Ratio of fictitious fluid depth with sediment to fluid depth without sediment vs. sediment concentration (data from Aziz & Scott 1989).

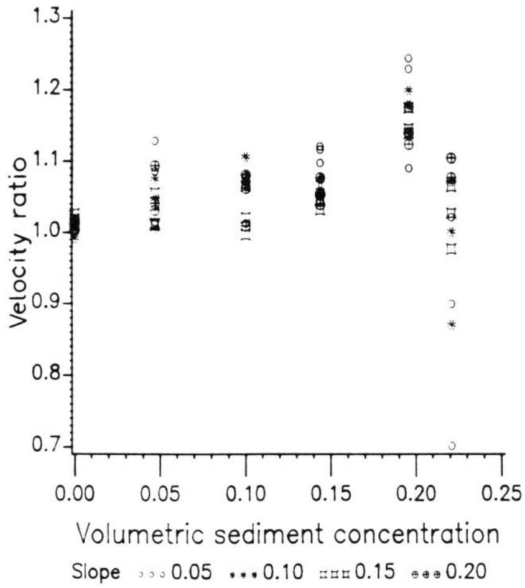

Figure 11.2 Measured flow velocity vs. flow velocity calculated for the same water discharge in the absence of sediment (data from Rickenmann 1990; flow velocities for clear water were calculated using a non-linear regression of flow velocity on slope and unit discharge derived from the measurements made by Rickenmann in clear water).

Considering the various contrasting experimental results, it was assumed, for this study, that sediment load does not influence flow properties. Therefore, fluid depth and flow velocity were calculated using the algorithm of Savat (1980), using the bed roughness values indicated in Table 11.1, and no correction for sediment load was made. It should be kept in mind that total depth may be significantly higher than fluid depth in sediment-laden flows, as volumetric sediment concentrations as high as 0.35 were measured. In order to calculated total depth, the speed of the sediment load must be known. Parsons (1972) and Govers (1989) measured velocities of individual grains in overland flow, but it is not clear to what extent these data are representative for flows with a high sediment load, as grain–grain interactions may become important (Leeder 1979).

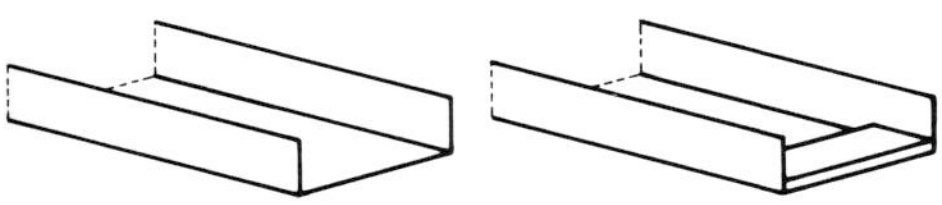

Figure 11.3 Sketch of the downslope end of the flume as used in the original experiments (a), and in the additional experiments (b).

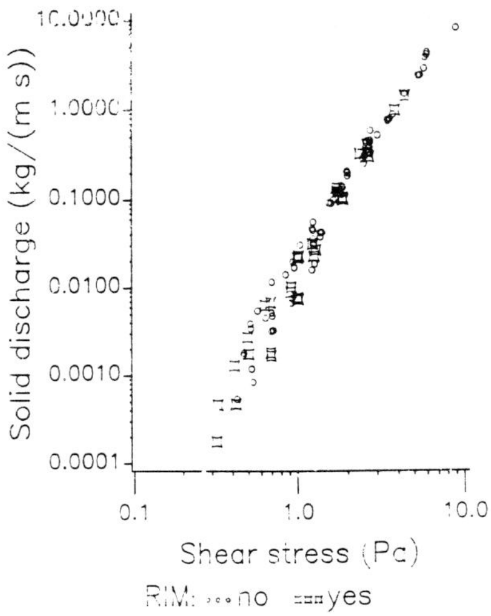

Figure 11.4 Solid discharge vs. fluid shear stress for material C without correction.

All experiments described by Govers were made in a flume with an open bottom end (Fig. 11.3). The sediment was put on top of the flume bed and it was assumed that the energy slope equalled the bedslope. While doing some additional experiments with coarse sand on low slopes to collect more information about sediment transport at near-critical conditions, it was observed that, on low slopes, an edge effect was present. The drawdown of the water surface near the flume outlet was such that sediment transport was intensified near the flume outlet, so that excessively high sediment transport rates were measured. Therefore, an additional series of experiments was carried out whereby a 0.01 m high stopper was placed on the flume bed near the flume outlet to contain the sediment (Fig. 11.3b). Plotting solid discharge versus shear stress reveals that this modification did not affect the results significantly, for fine sand (material C) but that a clear effect was present for coarse sand (Figs 11.4 & 5). For material E, the discrepancy between the two datasets could be eliminated by increasing the slope used in the calculations of hydraulic parameters by 0.015 (Fig. 11.5) for those experiments carried out in an open-ended flume. The original data were left unmodified for the fine materials (A, B and C), while for material D, slope was increased by 0.0075.

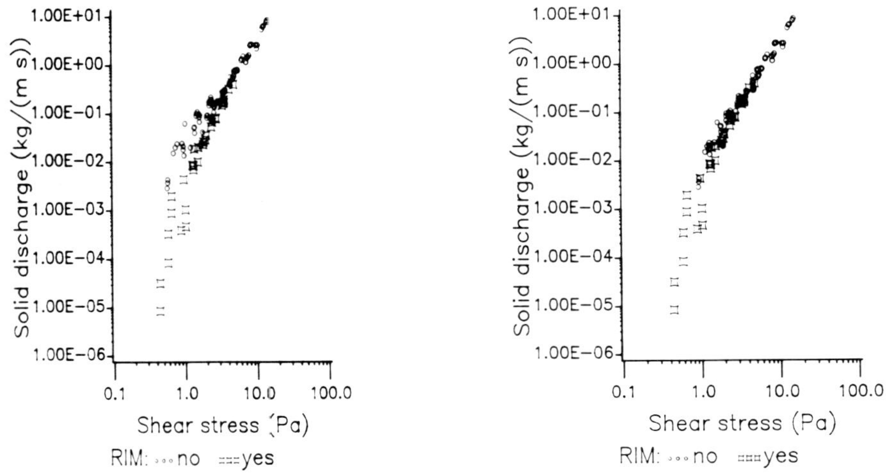

Figure 11.5 Solid discharge vs. fluid shear stress for material E (a) before and (b) after correction.

Evaluation of sediment transport formulae

Selected formulae

The following formulae were selected for evaluation. They are listed below in a format similar to the one suggested by Low (1989) so that a direct comparison of the parameters involved can be made:

Meyer-Peter & Müller (1948):

$$q_s = 8(Y - Y_{cr})D(gD(s - 1))^{0.5}\rho_s \tag{11.4}$$

Yalin (1963):

$$q_s = GY^{0.5}(Y - Y_{cr})D(gD(s - 1))^{0.5}\rho_s \tag{11.5}$$

$$G = \frac{0.635}{Y_{cr}} \frac{1 - \ln(1 + as^*)}{as^*} \tag{11.6}$$

$$as^* = \frac{2.45}{s^{0.4}} Y_{cr}^{\ 0.5} \left[\frac{Y}{Y_{cr}} - 1 \right] \tag{11.7}$$

Yang (1973):

$$(1)\quad \log C_t = 5.435 - 0.286\log\left(\frac{wD}{\nu}\right) - 0.457\log\left(\frac{u_*}{w}\right) + \left[1.799 - 0.409\log\left(\frac{wD}{\nu}\right) - 0.314\log\left(\frac{U_*}{w}\right)\right]\log\left(\frac{S\bar{u}}{w} - \frac{S\bar{u}_{cr}}{w}\right) \qquad (11.8)$$

$$q_s = \rho\frac{C_t}{1E6}\left[1 - \frac{C_t}{1E6}\right]^{-1} q \qquad (11.9)$$

$$(2)\quad \log C_t = 5.165 - 0.153\log\left(\frac{wD}{\nu}\right) - 0.297\log\left(\frac{u_*}{w}\right) + \left[1.780 - 0.360\log\left(\frac{wD}{\nu}\right) - 0.480\log\left(\frac{u_*}{w}\right)\right]\log\left(\frac{S\bar{u}}{w} - \frac{S\bar{u}_{cr}}{w}\right) \qquad (11.10)$$

$$q_s = \rho\frac{C_t}{1E6}\left[1 - \frac{C_t}{1E6}\right]^{-1} q \qquad (11.9)$$

Low (1989):

$$q_s = \frac{6.42}{(s-1)^{0.5}}(Y - Y_{cr})DS^{0.6}\bar{u}\rho_s \qquad (11.11)$$

Lu et al. (1989):

$$\phi = 5.09(Y - 0.05)^{1.25} \qquad (11.12)$$

$$q_s = \rho_s\phi\left[\frac{(\gamma_s - \gamma)gD^3}{\gamma}\right]^{0.5} \qquad (11.13)$$

where q_s = dry mass solid discharge/unit width;
Y = the dimensionless shear stress;
= $\tau/\gamma'_s D$;
τ = the shear stress,
= $\rho\ g\ R\ S$;
γ'_s = the submerged specific weight,
= $(\rho_s - \rho)g$;
D = the grain diameter;
ρ_s = the specific density of the sediment;
ρ = the specific density of the fluid;
g = the acceleration due to gravity;
R = the hydraulic radius;
S = the energy slope;
Y_{cr} = the critical dimensionless shear stress for sediment motion,
= $\tau_{cr}/\gamma'_s D$;
τ_{cr} = the critical shear stress;
s = ρ_s/ρ;
C_t = gravimetric sediment concentration (ppm);
w = the fall velocity of the sediment;
ν = the kinematic viscosity;
u_* = the shear velocity,
= $(g\ R\ S)^{0.5}$;
$S\bar{u}$ = the unit stream power;
$\bar{u}$ = the mean flow velocity;
$S\bar{u}_{cr}$ = the critical unit stream power for sediment motion, determined by the criterion of Yang (1973); and
ϕ = dimensionless transport rate.

The formula of Meyer-Peter & Müller was retained because it is one of the early formulae which has been most frequently used. The Yalin and Yang formulae were retained because they have been used frequently in the modelling of sediment transport by overland flow. The formula of Low is a modified version of the formula of Smart (1984) to account for density effects. This formula is interesting because it is the only formula that has been developed using experimental data obtained on steep slopes, although unit discharges were considerably higher and sediment sizes considerably coarser than those used during our experiments.

It should be noted that the calculation of the various parameters involved can cause some discrepancies in the predictions. For the calculation of shear stresses, we used the (fictitious) fluid depth, calculated using the algorithm of Savat (1980). In this way, the fluid shear stress, i.e. the shear stress exerted by the fluid alone, is calculated. However, Smart (1984) calculated shear stresses using total flow depth. In some cases this will be significantly different from the fluid shear stress as volumetric sediment concentrations became relatively high.

Critical shear stresses were derived from logarithmic plots of solid discharge vs. shear stress and are given in Table 11.1. For the coarse materials, critical shear stress values are lower than those derived from experiments on the initiation of motion in overland flow (Govers 1987). This is not uncommon. Similar findings were made by Bridge & Dominic (1984) and Govers (1987) with respect to grain velocities. Luque & Van Beek (1976) as well as Smart (1984) and Poesen & Torri (1989) suggest that critical shear stress should be slope-dependent, as sediment motion on steep slopes is facilitated by the downslope component of the weight force. However, Bayazit (1978), Bettes (1984) and Govers (1987) found that critical shear stresses tend to increase with decreasing relative depths. As relative depths are smaller on steep slopes, this effect may compensate for the effect of the weight force. Therefore, no correction of the critical shear stress was applied.

Fall velocities of the various grains were calculated using the procedure proposed by Dietrich (1982), assuming a specific density of 2650 kg m^{-3}, a shape factor of 0.6 for material A and 0.7 for all others, and a roundness index of 2.5 for material A and 3.5 for all others.

Not all available data were used in the evaluation of the sediment transport formulae, as it was observed that during experiments with high flow intensities a maximum sediment concentration of *c.* 1000–1200 kg m^{-3} was reached. A further increase in flow intensity did not lead to a further increase in sediment concentration. Therefore, experiments where the sediment concentration exceeded 1000 kg m^{-3} were eliminated. In total, 434 experiments were retained for analysis.

Performance

Performance of the various formulae was evaluated using logarithmic graphs of observed vs. predicted unit solid discharge. The use of logarithms is necessary as unit solid discharge varies over six orders of magnitude.

The formula of Meyer-Peter & Müller (1948) overpredicts solid discharge at low flow intensities (Fig. 11.6). The underprediction at high flow intensities could be expected from the results of Smart (1984), who obtained a similar result on steep slopes. Furthermore, there is a clear grain-size effect on the predicted solid discharge over the whole range of sediment tested.

Yalin's formula (1963) basically yields the same pattern, although scatter is even higher (Fig. 11.7). Observed values are near predictions only for materials C and D. Sediment transporting capacity is overpredicted for coarse and underpredicted for fine materials.

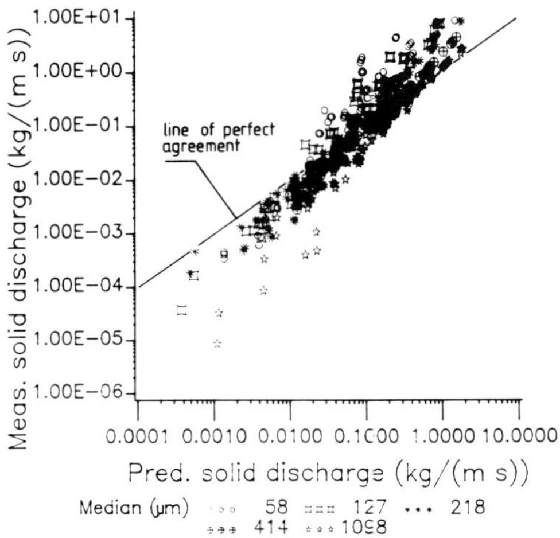

Figure 11.6 Observed vs. predicted solid discharge (formula of Meyer-Peter & Müller).

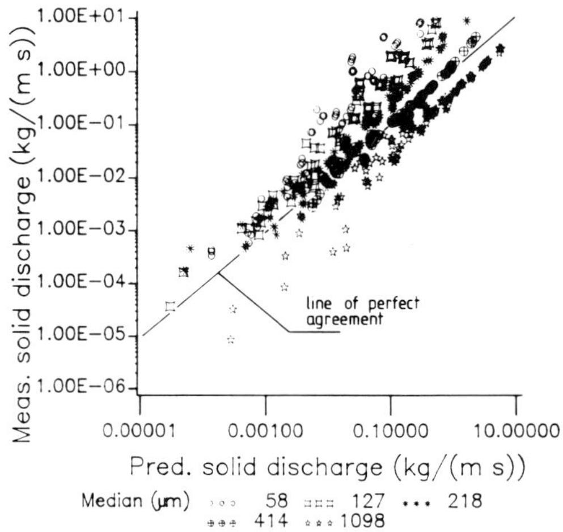

Figure 11.7 Observed vs. predicted solid discharge (formula of Yalin).

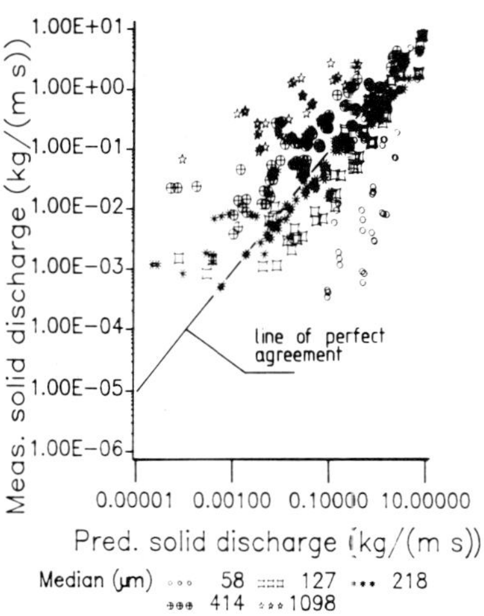

Figure 11.8 Observed vs. predicted solid discharge (formula 1 of Yang).

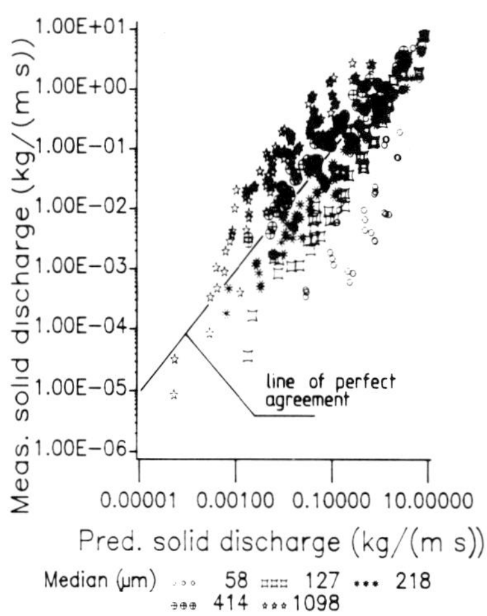

Figure 11.9 Observed vs. predicted solid discharge (formula 2 of Yang).

Yang's formulae (1973) have a specific problem, as in some cases predicted sediment concentrations exceed 1,000,000 p.p.m. For these cases a p.p.m of 500,000 was assumed. Even so, predictions based on Yang's first formula compare very poorly with observed values (Fig. 11.8). Furthermore, in a number of cases zero sediment transport was predicted as mean flow velocity was below the calculated critical value. Yang's second equation gives somewhat better results and predictions are near observed values for material D (Fig. 11.9). Contrary to the former equations, an overprediction of transporting capacity is observed for fine materials while it is underpredicted for coarse materials. This is apparently due to the sensitivity of predictions to the fall velocity of the sediment.

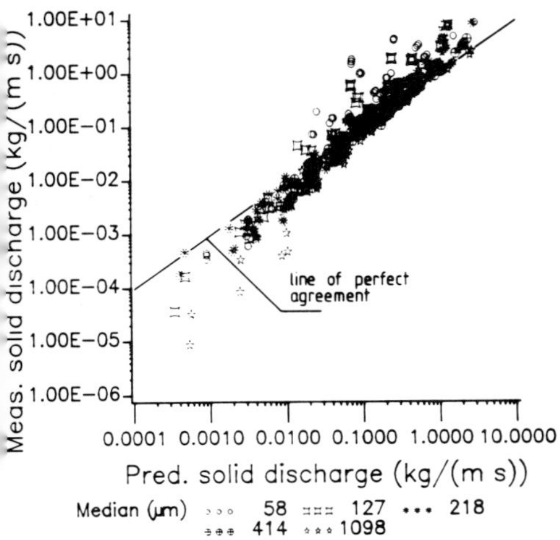

Figure 11.10 Observed vs. predicted solid discharge (formula of Low).

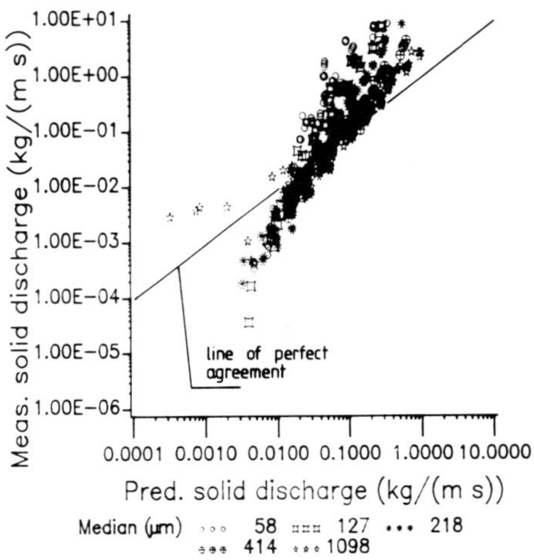

Figure 11.11 Observed vs. predicted solid discharge (formula of Lu et al.).

Agreement between predicted and observed values is excellent for materials D and E for the formula of Low at unit solid discharges exceeding *c.* 2E-3 kg $(\mathrm{m\ s})^{-1}$ (Fig. 11.10). Sediment transporting capacity is underpredicted for fine materials.

As could be expected from its formulation, the formula of Lu et al. (1989) performs in a very similar manner to the one of Meyer-Peter & Müller, except for the low sediment transport rates (Fig. 11.11). Predictions for the latter are influenced by the choice of a fixed critical dimensionless shear stress value of 0.05. Again, a systematic variation with grain size can be observed and predicted values are only near to observed ones at intermediate transport rates.

Discussion

Both the formula of Meyer-Peter & Müller and of Yalin belong to the family of classic excess-shear bedload formulae, which includes also those of Bagnold (1966) and Engelund & Fredsoe (1976). For this type of formulae it can be shown that they predict that:

$$q_s \sim \tau^{1.5} \tag{11.14}$$

for sediment transport at flow intensities well above the critical level (Yalin 1977). All formulae of this type may be expected to show a similar behaviour to those of Meyer-Peter & Müller, Yalin and Lu et al., and it can be concluded that they are not suitable for overland flow. Indeed, our data indicate that the shear stress exponent for overland flow must be higher. The fact that earlier evaluation attempts of the Yalin formula were successful is probably due to the limited range of the data that were used for testing.

Yang's formulae are based on a different concept, i.e. unit stream power. Several authors have indicated the potential of unit stream power for the prediction of sediment transport in overland flow (Govers & Rauws 1986, Moore & Burch 1986, Govers 1990, Loch et al. 1989). However, the original formulae of Yang were based on channel and river data and, as for the formula of Yalin, it can only be concluded that his formulae cannot be directly applied to overland-flow conditions. Earlier success in the validation of the Yang equation(s) is probably also due to the limited range of conditions tested.

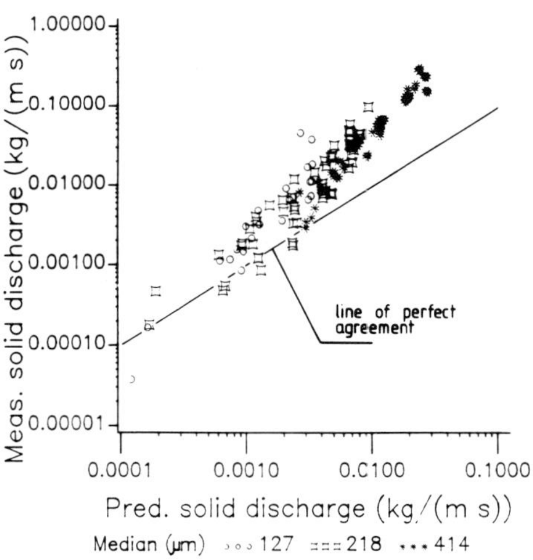

Figure 11.12 Observed vs. predicted solid discharge (corrected formula of Lu et al., Y ≤ 3E−1, materials B, C and D).

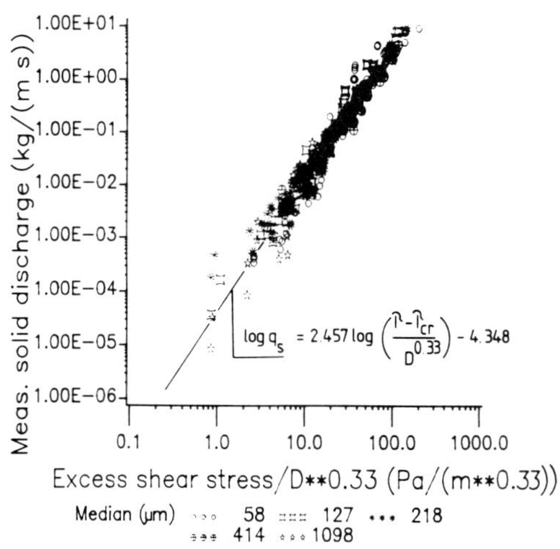

Figure 11.13 Solid discharge vs. $(\tau - \tau_{cr})/D^{0.33}$, all experiments.

Low's formula performs well for materials D and E, at least at shear stress levels well above the critical value. However, the formula underpredicts transporting capacity for finer materials. The latter is not surprising as Low's formula is developed to predict bedload. Govers (1987) showed, on the basis of the measurement of grain velocities, that only the latter two materials exhibit a bedload-type behaviour in overland flow. Finer materials can move at velocities that are significantly higher than the near-bed fluid velocity, which indicates that they are not only moving near the bed but also in suspension. It is not surprising that transporting capacity cannot be predicted accurately for these materials when a bedload formula is used.

The formula of Lu et al. does not perform as well as that of Low, despite the fact that it was developed from data in overland-flow conditions. Again, this must be attributed to the limited range of conditions tested in their experiments: in their experiments, Y never exceeded 3E−1. In our experiments, Y reached values up to 100. Furthermore, flow depth is calculated assuming laminar flow and:

$$f = \frac{16}{Re} \tag{11.15}$$

where f = the Darcy–Weisbach friction factor;
Re = the flow Reynolds number,
= $R\,\bar{u}/\nu$.

Equation 11.15 is incorrect: all studies on the hydraulics of laminar flow on rough surfaces indicate that the friction coefficient is considerably higher than the theoretical value of 24 (Woo & Brater 1961, Phelps 1975, Nittim 1977, Savat 1980, Rauws 1988). Equation 11.15 was first proposed by Davis et al (1983), who derived this relation from measured surface velocities. However, surface velocities in laminar flow are, in theory, 33% higher than the mean flow velocity. Furthermore, it has been shown that, with increasing roughness, mean flow velocity decreases much more rapidly than surface velocity (Emmett 1970, Phelps 1975, Govers 1986). For the grains studied by Lu et al. flow depths are 20–40% higher than those predicted when Equation 11.15 is used. If, from the present study, Y is calculated as $2/3\ \tau/(\gamma'_s D)$ and only those data are taken where $Y < 3 \times 10^{-1}$, while the critical shear stress is allowed to vary with grain size, it can be seen that agreement between predicted and observed values is acceptable (Figure 11.12).

From the analysis above, it can be concluded that the application of transport formulae developed from channel experiments and river observations to overland flow conditions is at least problematic. Only Low's formula shows real potential to predict the transporting capacity of overland flow, at least for coarser particles. The basic reason for the failure of the other formulae is that they were not developed and/or calibrated using experimental data on steep slopes. This confirms once more that any formula that is empirical in nature should not be used outside of the domain for which it was developed.

This does not mean that sediment transport in overland flow is controlled by fundamentally different variables when compared to channel flow. Figures 11.13, 11.14 & 11.16 show that, at least for part of the data, sediment transporting capacity can be well related to:

(a) the (excess) fluid shear stress :

$$\log q_s = 2.457 \log\left(\frac{\tau - \tau_{cr}}{D^{0.33}}\right) - 4.348 \qquad (11.16)$$

($r^2 = 0.95$, n = 434);

(b) the (excess) effective stream power, as defined by Bagnold (1980) for materials, B, C and D and solid discharges > 5E−4 kg $(\text{m s})^{-1}$:

$$\log q_s = 1.081 \log(\Omega - \Omega_{cr}) - 2.528 \qquad (11.17)$$

($r^2 = 0.97$, n = 284)

where Ω = the effective stream power,
= $(\tau\, \bar{u})^{1.5}/d^{2/3}$;
Ω_{cr} = $(\tau_{cr}\, \bar{u})^{1.5}/d^{2/3}$;
d = the flow depth; and

(c) the unit stream power as defined by Yang (1972) for materials A, B and C on slopes ≤ 0.14:

$$C = \frac{86.7(S\bar{u} - 0.005)}{\sqrt{D}} \qquad (11.18)$$

($r^2 = 0.94$, n = 226)

where C = the sediment concentration (kg m^{-3}).

A good relationship exists between solid discharge and excess shear stress divided by $D^{1/3}$ for all grain sizes (Fig. 11.13). However, it can be seen that, for material A, and to a lesser extent for material B, sediment transport is overpredicted at low shear stresses, while it is underpredicted for intermediate to high excess shear stresses. The more rapid increase of sediment transport with excess shear stress can be explained by the non-linear relationship between grain velocity and excess shear velocity for these fine materials (Govers 1987). Therefore, Equation 11.16 should not be extrapolated to finer grain sizes.

No significant grain-size effect is present when the excess effective stream power is used to predict sediment-transporting capacity for materials C, D and E (Fig. 11.14). In his original formulation, Bagnold (1980) assumed sediment transport rate to be inversely proportional to the square root of the grain size. Apparently, this is not the case in flows on steep slopes. The reduction in transporting capacity with increasing grain size is entirely accommodated by the reduction in flow velocity and increase in flow depth due to the increasing bed roughness. The robustness of the effective excess stream power–solid discharge relationship is well illustrated by plotting Smart's data together with our data for material E. Figure 11.15 shows that both datasets blend very well together, although Smart worked with much coarser grain sizes as well as with much higher discharges, and used a different experimental set-up and procedure.

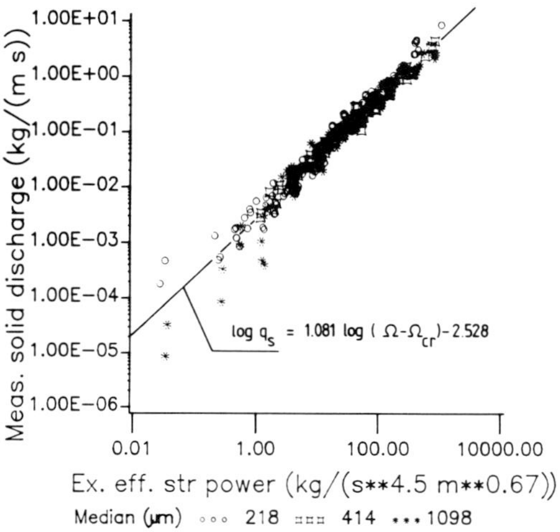

Figure 11.14 Solid discharge vs. excess effective stream power, materials C, D and E.

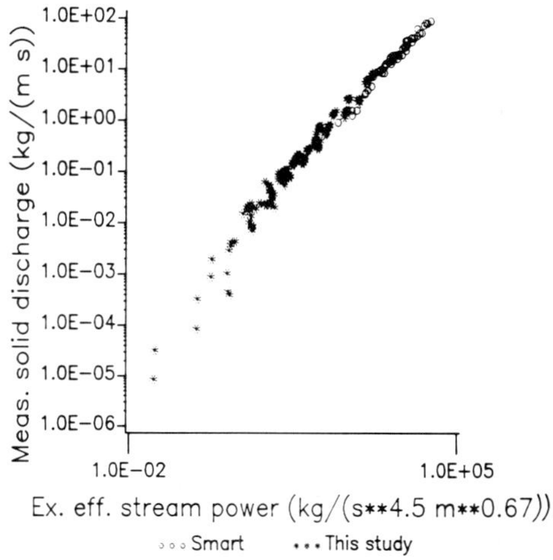

Figure 11.15 Solid discharge vs. excess effective stream power for materials E (D_{50} = 1098 μm) and data of Smart (1984) (D_{50} = 2000, 4200 and 4300 μm).

Unit stream power is well related to sediment concentration for materials A, B and C (Fig. 11.16). Yang (1972) proposed to incorporate grain size and density effects by using the dimensionless unit stream power, being the unit stream power divided by the fall velocity of the grain. This approach did not yield good results for our data: the reduction in sediment-transporting capacity with increasing grain size is far less important than is suggested by the increase in fall velocity. However, grain-size effects can be incorporated by dividing excess unit stream power by the square root of the grain diameter. The critical value of unit stream power (0.005 m s^{-1}) is determined by regression and has no further physical basis. Sediment transport does occur at lower unit stream power values, but concentrations are low.

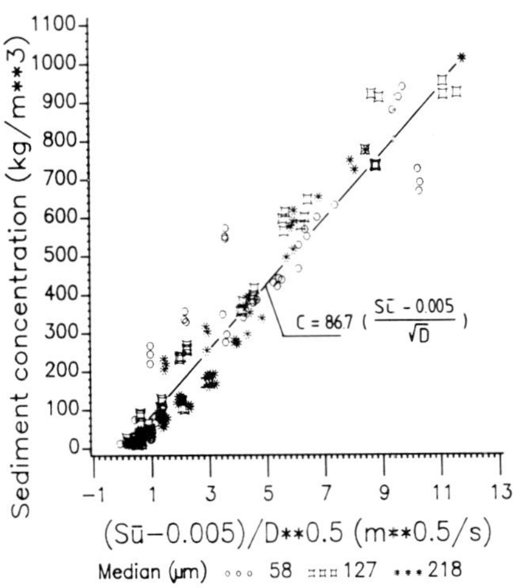

Figure 11.16 Sediment concentration vs. $(S\bar{u}-S\bar{u}_{cr})/D^{0.5}$, materials A, B and C, S < 0.14.

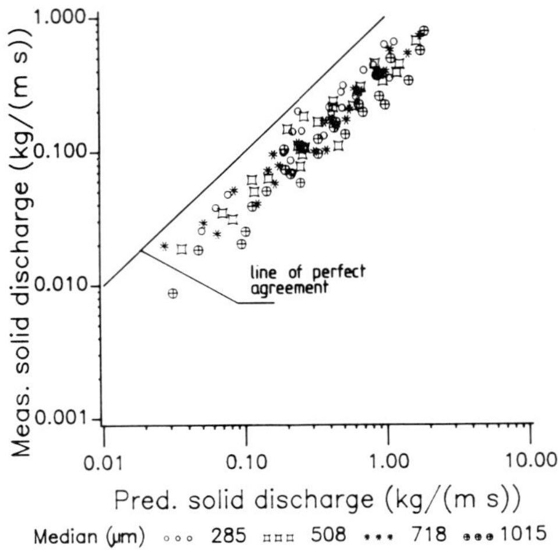

Figure 11.17 Observed vs. predicted solid discharges (data of Aziz & Scott 1989; Eq. 11.16).

Comparison with other datasets

Available data

During recent years, some data have become available on sediment transport in overland flow which can be used to evaluate the performance of the various formulae proposed above. Data are summarized in Table 11.2.

The most complete dataset has probably been collected by Aziz & Scott (1989) who measured sediment-transporting capacity using a set-up similar to the one used by Smart (1984). Sediment-transporting capacity was determined as the amount of sediment that could be transported over a fixed bed without any deposition occurring. Aziz & Scott reported grain size and density, flow discharge, sediment discharge and total flow depth for each experiment. Fictitious fluid depth was determined from total flow depth using Equation 11.2. Mean flow velocity was determined by dividing unit water discharge by the fictitious fluid depth.

Figures 11.17, 11.18 & 11.19 show that the sediment-transport rates they

Table 11.2 Overview of other datasets on sediment transport capacity in overland flow.

Author(s)	N	D_{50} μm	S	q $m^2\ s^{-1}$	q_s kg $m^{-1}\ s^{-1}$	C kg m^{-3}
Aziz & Scott (1989)	96	285–1015	0.03–0.10	0.0016–0.0047	0.088–0.78	5.32–197.6
Riley & Gore (1988)	35	470–660	0.007–0.062	0.00088–0.0015*	0.00038–0.08*	0.29–91.4
Meyer et al. (1983)	64	76–603	0.002–0.05	0.0017–0.0058	6.44×10^{-6}–0.78	0.0039–132.8
Rauws (1984)	16	110	0.03–0.087	0.00018–0.0062	0.00002–2.42	0.06–392
Kramer & Meyer (1969)	120	33–121	0.04–0.10	0.00012–0.00030*	0.00016–0.29	1.37–993

* Calculated assuming uniform flow over the whole flume width.

measured were generally lower than those predicted from Equations 11.16, 11.17 & 11.11. The excessive effective stream-power equation, as well as the formula of Low, perform better than the shear stress equation. On the other hand, the trend of the data is well predicted by all three equations. There are several possible reasons for the observed discrepancies:

(a) Measurements of flow depths in such thin flows loaded with sediment are very difficult to make and some systematic bias may be present. If flow depth is overestimated, then shear stress will be overestimated, which will lead to an overprediction of sediment-transporting capacity if Equation 11.16 is used. However, an overestimation of flow depth should lead to an underestimation of transporting capacity when Equation 11.17 is used. As this is not the case, errors in flow depth measurements are probably not the most important reason for the discrepancies.

(b) There is some variation in specific density for the materials used by Aziz & Scott. For the 718 and 1015 μm grains, specific density is 2900 and 3100 kg m^{-3}, respectively. Thus, for these grains, solid discharges will be overestimated when an equation is used which does not take into account variations in density.

(c) All equations have been developed from experiments on plane beds. Therefore, the shear stress equation (Eq. 11.16) will be valid only in the case of totally plane beds. If form roughness is present, part of the shear stress will be no longer available for sediment transport, so that transport rates will be overestimated when Equation 11.16 is applied. Some form resistance may have developed during the experiments of Aziz & Scott, although they report that no bedforms were visible during most of the experimental runs.

(d) A different method has been used to determine sediment-transporting capacity. The same flow may be capable of transporting more sediment over a mobile bed than over a fixed bed. On a mobile bed, several grain layers may be in motion simultaneously near the bed level (Wilson 1987). Such a situation is hardly possible on a fixed bed when no deposition is allowed. Therefore, lower sediment transport rates may be expected in the experiments of Aziz & Scott compared to our experiments. This would also explain why the formula of Low (1989) also overpredicts transport rates for the experiments of Aziz & Scott, despite the fact that density is taken into account in this formula (Fig. 11.19).

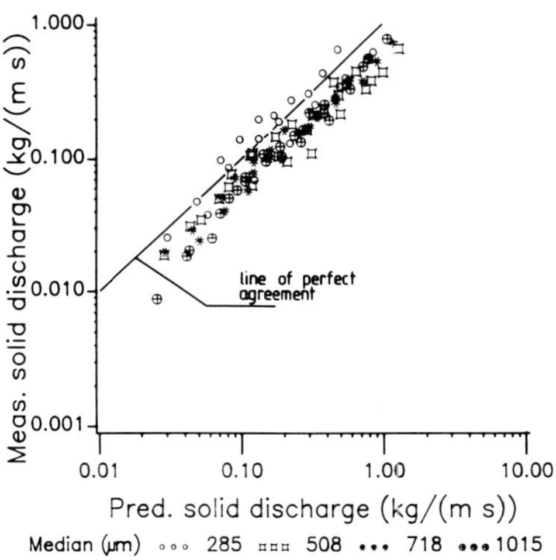

Figure 11.18 Observed vs. predicted solid discharge (data of Aziz & Scott 1989; Eq. 11.17).

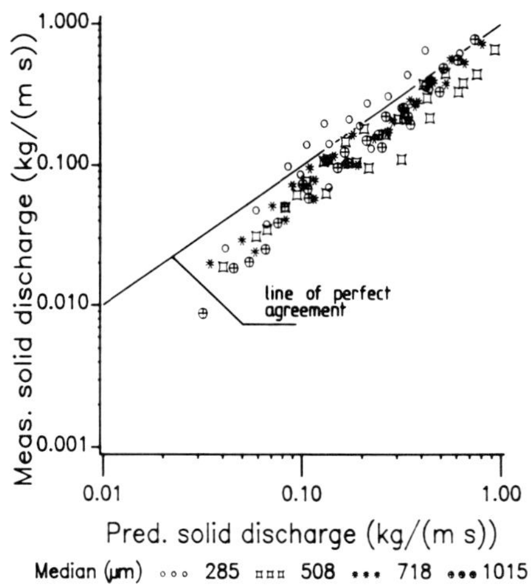

Figure 11.19 Observed vs. predicted solid discharge (data of Aziz & Scott 1989; formula of Low 1989).

Riley & Gore (1988) calibrated an experimental flume of 0.16 m wide and 0.5 m long using two sandy and two gravelly materials. Apart from fluid and sediment discharge, only crude information on flow depths during some of the runs is given. When shear stress is calculated from these depths, it can be seen that there is only a very poor correlation between predicted and measured sediment load for the sandy materials (Fig. 11.20). The most probable explanation for this is that during some of the experiments a large part of the

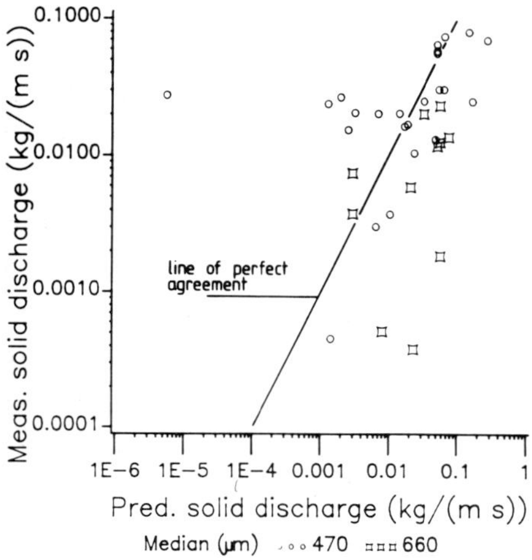

Figure 11.20 Observed vs. predicted solid discharge (data of Riley & Gore 1988; Eq. 11.16).

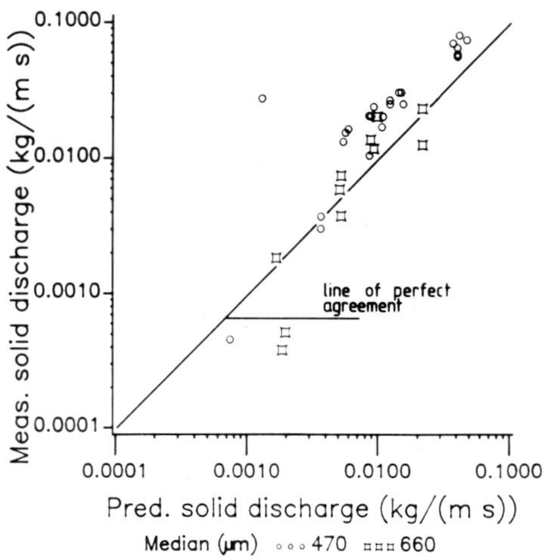

Figure 11.21 Observed vs. predicted solid discharge (data of Riley & Gore 1988; Eq. 11.17).

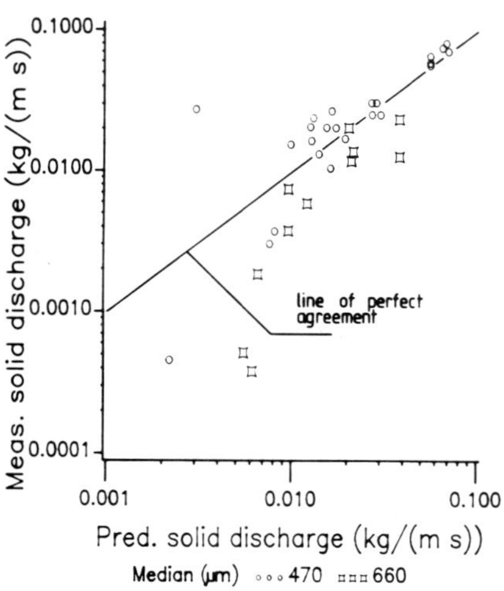

Figure 11.22 Observed vs. predicted solid discharge (data of Riley & Gore, formula of Low).

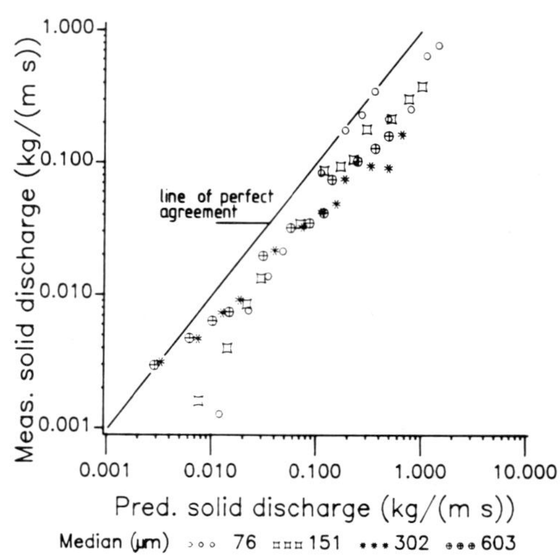

Figure 11.23 Observed vs. predicted solid discharge (data of Meyer et al., Eq. 11.16).

shear stress is due to form roughness and therefore ineffective in the transportof sediment. If the excess effective stream power is used, a good correlation between predicted and observed sediment discharge is obtained (Fig. 11.21). A similar result is obtained with the formula of Low (Fig. 11.22). The underestimation of sediment discharge for the 470 μm material is probably due to the presence of a considerable percentage of easily transportable fines in this material.

Meyer et al. (1983) measured sediment transport rates along a simulated row crop furrow, using a procedure similar to the one of Aziz & Scott (1989, Table 3). No hydraulic data were measured, so they can only be calculated approximately. As furrow geometry could be described by a parabola, friction factors were calculated using an iterative procedure based on a modified version of the algorithm of Savat (1980). Then, shear stress, effective stream power and unit stream power were calculated. Figure 11.23 shows that, when Equation 11.16 is used, there is a tendency to overpredict transporting capacity, but the trend of the data is well predicted. A similar pattern arises when predictions are based on Equation 11.17 (Fig. 11.24). Excellent predictions are

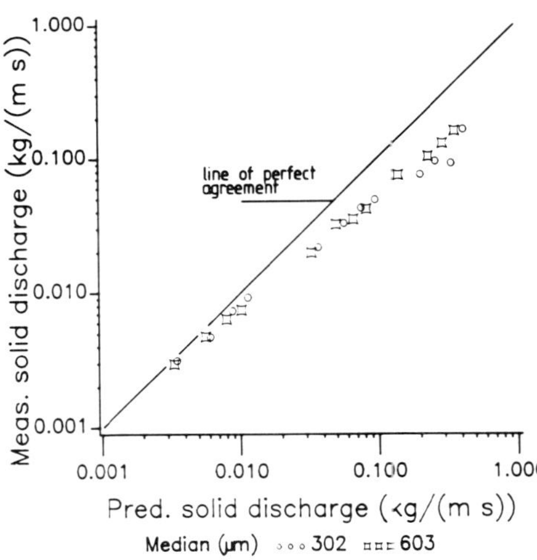

Figure 11.24 Observed vs. predicted solid discharge (data of Meyer et al., Eq. 11.17).

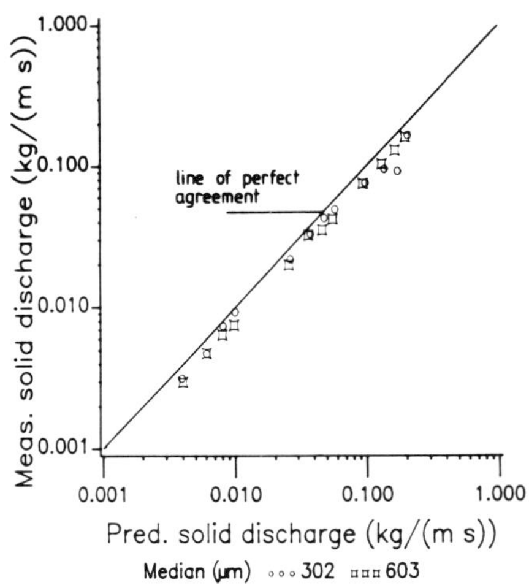

Figure 11.25 Observed vs. predicted solid discharge (data of Meyer et al., formula of Low).

obtained using the formula of Low for the coarse materials (Fig. 11.25). There is a good correlation between observed sediment concentrations and those predicted using the unit stream power equation, but predicted concentrations are considerably higher than observed ones, especially at high unit stream powers (Fig. 11.26).

Davis et al. (1983) measured transporting capacity of overland flow using the concave-slope technique, which was also used by Lu et al. (1989) in the development of their equation. In his publication, data are not tabulated, but plots of solid discharge vs. shear stress are given. For a 342 μm sand, Davis et al. propose the following equation:

$$q_s = 0.0482\tau(\tau - 0.27). \qquad (11.19)$$

Again, shear stress was calculated using Equation 11.15: therefore shear stress is underestimated and Equation 11.19 is given in $2/3\tau$ rather than τ:

$$q_s = 0.482(2/3)\tau\ ((2/3)\tau - 0.27). \qquad (11.20)$$

In Figure 11.27 this equation is compared with Equation 11.16 for the domain covered by the experiments of Davis et al. (assuming a critical shear stress of 0.3 Pa in Eq. 11.16). It can be seen that agreement between both equations is very good. However, Equation 11.19 tends to underpredict sediment loads measured in our experiments at higher shear stresses, which is due to its lower shear stress exponent. This illustrates again the problems associated with the extrapolation of empirically calibrated transport formulae.

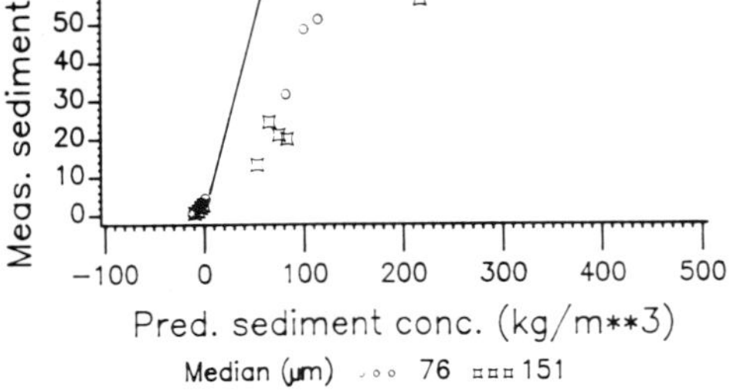

Figure 11.26 Observed vs. predicted sediment concentration (data of Meyer et al. 1983; Eq. 11.18).

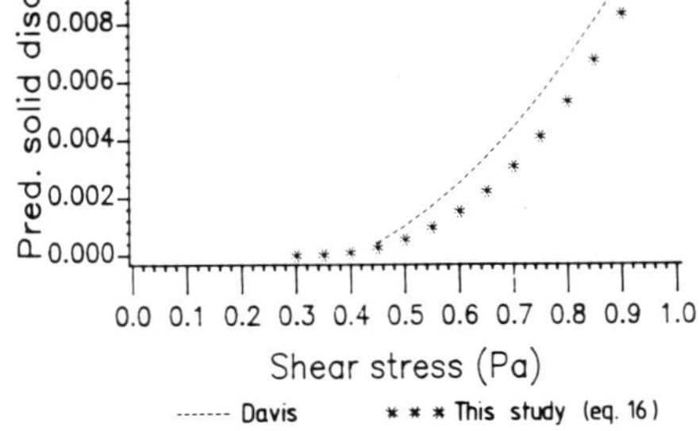

Figure 11.27 Comparison of the corrected equation of Davis et al. (1983) and Equation 11.16 for shear stress ≤ 1 Pa.

Rauws (1984) and Kramer & Meyer (1969) only report mean flow velocities and sediment discharges, so their data can only be used to evaluated Equation 11.18. Rauws measured sediment-transporting capacity in a single rill using a silty loam soil having a median grain diameter of 110 μm. Mean velocity was measured by dye tracing. Kramer & Meyer evaluated the effect of cover on sediment transport from a flume filled with 33 or 121 μm glass beads. During their experiments a braided pattern developed on the flume surface. They used the salt tracer technique to measure mean flow velocity. Figure 11.28 shows that for the experiments of Rauws, as well as for the experiments of Kramer & Meyer on bare surfaces, the agreement between observed and predicted values of sediment concentration is good, even for the 33 μm grains. Agreement is also very good for the experiments of Kramer & Meyer with cover and the 121 μm material (Fig. 11.29). However, the results are poorer with the 33 μm material when cover is present: sediment loads are systematically underpredicted for intermediate unit stream power values (Fig. 11.30). Underprediction increases with increasing mulch cover: this suggests that the phenomenon might be due to the fact that for such small grains sediment transport is enhanced due to local turbulence around the simulated vegetation elements. On the other hand, it is also possible that part of the disagreement is due to the extrapolation of Equation 11.18 to grain sizes for which it was not tested.

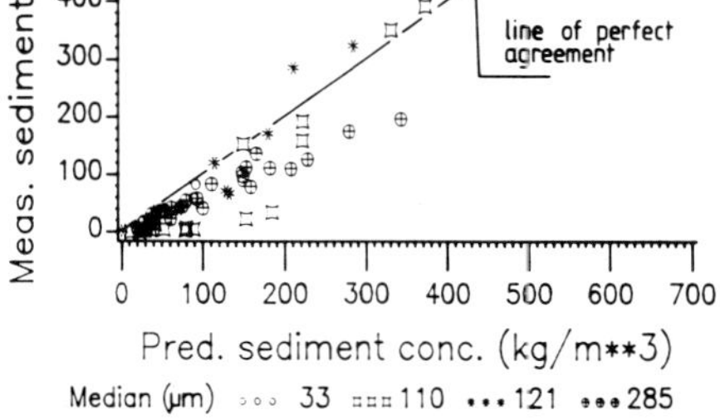

Figure 11.28 Observed vs. predicted sediment concentration (data of Kramer & Meyer 1969 (bare surfaces) and Rauws 1984; Eq. 11.18).

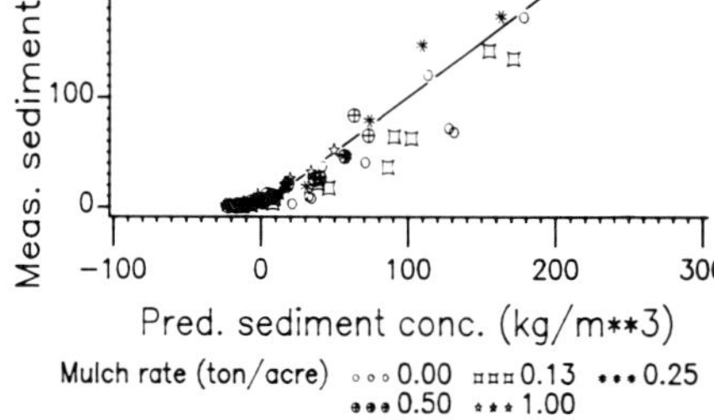

Figure 11.29 Observed vs. predicted sediment concentration (data of Kramer & Meyer 1969, bare and covered surfaces, 121 μm grains, Eq. 11.18).

Discussion

Both for the formula of Low as well as for the empirical equations proposed, the predictions generally compare favourably with measured sediment discharges. In some cases, predictions are systematically higher than the measured sediment discharges. This is probably at least partly due to differences in procedures and criteria used to measure and define the transporting capacity of overland flow.

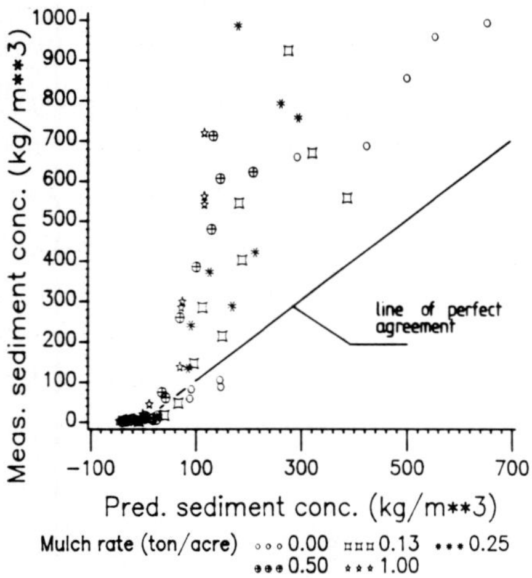

Figure 11.30 Observed vs. predicted sediment concentrations (data of Kramer & Meyer 1969, bare and covered surfaces, 33 μm grains, Eq. 11.18).

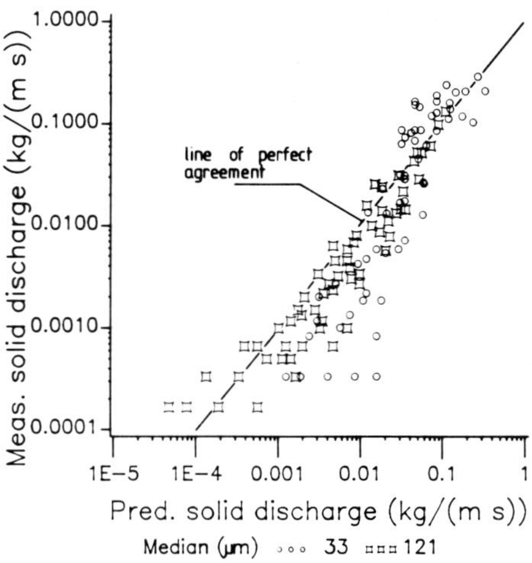

Figure 11.31 Observed vs. predicted solid discharge (data of Kramer & Meyer, 33 and 121 μm grains, bare and covered surfaces, Eq. 11.16; grain shear stress calculated using Laursen's procedure).

However, there remains a need for further experimental work. At the moment, there are almost no data on the effect of sediment density on the transporting capacity of overland flow. This is an important issue as, on cohesive soils, most of the material is transported in the form of aggregates (Loch & Donnollan 1983). Furthermore, no data on very fine grains exist.

A major problem is also the effect of bed roughness on the transporting capacity of overland flow. It is generally agreed that the increase in friction due to non-transportable flow obstacles like vegetation stems, stones, soil clods, etc. reduces the erosivity and transporting capacity of the flow (Foster et al.

1982, Govers & Rauws 1986). A transporting capacity equation for overland flow should be capable of predicting such reduction. From this point of view, the empirical equations based on unit stream power and effective stream power seem to be most promising as, in these equations, a reduction in flow velocity is directly translated into a reduction in transporting capacity. The comparison of predictions with the results obtained by Kramer & Meyer (1969) and Riley & Gore (1988) suggest that the predicted reduction in transporting capacity is at least of the correct order of magnitude. A reduction in flow velocity by macroroughness leads to an increase in predicted transporting capacity when Equation 11.16 is used, whereas predictions from the formula of Low basically remain unchanged: the form of this formula implies that a reduction in flow velocity is compensated by an increase in shear stress. Govers & Rauws (1986) proposed a method to calculate the grain shear stress, i.e. that fraction of total shear stress which can be used to transport sediment, for flows on irregular beds as:

$$\tau_g = \frac{1}{8} f_g \rho \bar{u}^2 \qquad (11.21)$$

where τ_g = the grain shear stress; and
f_g = the Darcy–Weisbach grain friction factor.

f_g was calculated using the algorithm of Savat (1980), assuming that the relation between f_g, Reynolds number, grain roughness and slope is unaffected by the presence of form roughness.

Although initial tests of the method were encouraging, it has to be kept in mind that the method is very sensitive to errors in grain roughness estimation as well as to errors in velocity measurements. Furthermore, as the reduction in shear stress is assumed to be proportional to the square of mean velocity, a reduction to 50% of the mean flow velocity by macroroughness would lead to a reduction of the transporting capacity to less than 10% of the original value if Equation 11.16 is used, as:

$$q_s \sim (\tau - \tau_{cr})^{2.5} \qquad (11.22)$$

Under the same assumptions, Equations 11.17 and 11.18 would predict a reduction in transporting capacity to *c.* 50%, which appears to be more realistic.

Another method of grain shear stress calculation is the one proposed by Laursen (1958, in Petit 1989), which was successfully used by Petit (1989) to calculate grain shear stresses in gravel bed rivers and flumes. Laursen's method derives the grain hydraulic radius from the Manning formula:

$$R_g = (n_g \frac{\bar{u}}{S^{0.5}})^{1.5} \tag{11.23}$$

where n_g = the Manning–Strickler roughness coefficient due solely to grain roughness.

Shear stress is then simply calculated as:

$$\tau_g = \rho g R_g S \tag{11.24}$$

Laursen's method essentially calculates the hydraulic radius corresponding to the measured mean velocity assuming that the bed is plane. This method has the advantage that no information about the Reynolds number is needed to calculate grain shear stresses. The method can also be transposed to overland

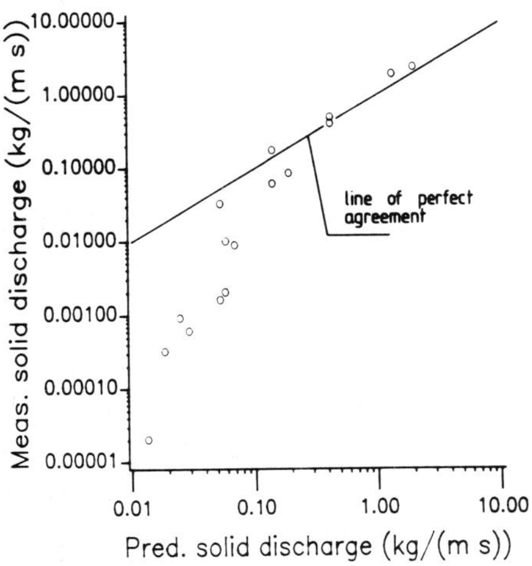

Figure 11.32 Observed vs. predicted solid discharge (data of Rauws 1984; Eq. 11.16; grain shear stress calculated using Laursen's procedure).

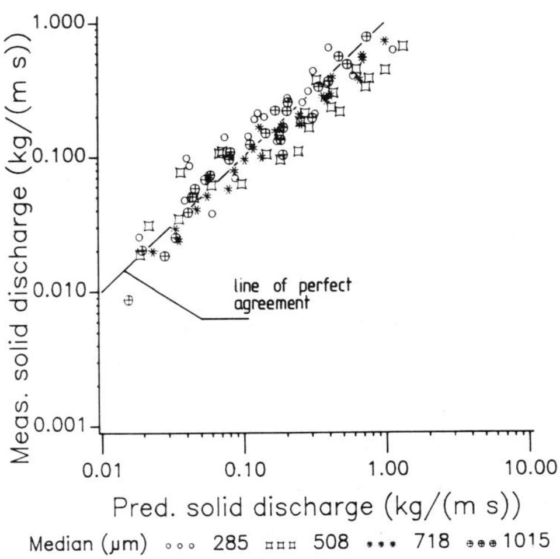

Figure 11.33 Observed vs. predicted solid discharge (data of Aziz & Scott 1989; Eq. 11.16; grain shear stress calculated using Laursen's procedure).

flow, as a minor modification of the algorithm of Savat allows determination of the grain hydraulic radius, equivalent to the measured mean velocity, by iteration. Figure 11.31 shows that there is reasonable agreement between predicted and measured sediment loads for the data of Kramer & Meyer, especially with 121 μm grains. The fact that sediment discharges are systematically overpredicted for low shear stresses and underpredicted for higher shear stresses, in the case of 33 μm grains, may be due to the already noted increase of the shear stress exponent with decreasing grain size. Another source of error is that the unit discharge is not exactly known, as flow concentration occurred during the experiments. For the data of Rauws, deviations are more important, especially at low shear stresses, while agreement is acceptable at high shear stresses (Fig. 11.32). For the data of Aziz & Scott, agreement between predicted and observed transport rates is also much improved (Fig. 11.33). Even if all these problems are solved, it might be necessary to use different equations under different circumstances, as the role of some variables may be dependent on the domain the formula is applied to. Our data indicate that solid discharge is independent of grain size if effective stream power is used as a predictor and if $D > 300$ μm. For finer grains, a grain-size effect is clearly present. Also, considering the findings on the initiation of motion, it is quite possible that the effect of relative depth may be of some importance only when relative depth is below some critical value. Therefore, this variable might be included in a equation for interrill flow, while it may be totally unimportant for rill flow.

Conclusions

The evaluation of existing transport formulae in overland flow shows that none of the equations yields good predictions over the whole range of conditions tested. Former evaluations gave promising results, merely because they were either based on a limited number of data or because the available data only covered a limited range of discharges, slopes and/or grain sizes. It is to be expected that, when a large number of equations is tested using data covering only a limited range, at least one will give satisfying results.

The use of these formulae in deterministic erosion models might therefore lead to severe systematic bias on the predictions of such models. Often, such bias will go undetected, as it is corrected for by optimizing one or more other parameters. However, this will severely affect the applicability range of such a model. As a higher level of generality is one of the major reasons for the development of physically based models, this should be avoided whenever possible.

Better predictions of the transporting capacity of overland flow are certainly possible. Empirical equations developed from our data, as well as the formula of Low, show reasonable to very good agreement with other datasets as long as data are in the same range. These equations may, therefore, be used in erosion models, more specifically for rill flow. However, before a more general (set of) equations can be proposed, experimental work is necessary to:

(a) evaluate the transportability of finer grains and grains of lower specific density;

(b) evaluate the influence of form roughness on the transporting capacity of overland flow; and

(c) evaluate transporting capacity at lower discharges on steep slopes.

References

Alonso, C. V., W. H. Neibling, G. R. Foster 1981. Estimating sediment transport capacity in watershed modeling. *Transactions of the American Society of Agricultural Engineers* **24**, 1211–26.

Aziz, N. M. & D. E. Scott 1989. Experiments on sediment transport in shallow flows in high gradient channels. *Hydrological Sciences Journal* **34**, 465–79.

Bagnold, R. A. 1966. *An approach to the sediment transport problem from general physics*. US Geological Survey Professional Paper 422-I, Washington DC: US Government Printing Office.

Bagnold, R. A. 1973. The nature of saltation and of bed load transport in water. *Proceedings of the Royal Society London Series* **A332**, 473–504.

Bagnold, R. A. 1980. An empirical correlation of bedload transport rates in flumes and natural rivers. *Proceedings of the Royal Society Series* **A372**, 453–73.

Bayazit, M. 1978. Scour of bed materials in very rough channels. *Proceedings of the American Society of Civil Engineers, Journal of the Hydraulics Division* **104**, 1345–9.

Bettes, R. 1984. Initiation of motion in gravel streams. *Proceedings of the Institute of Civil Engineers, Part 2* **77**, 79–88.

Bridge, S. J. & D. F. Dominic 1984. Bed load grain velocities and sediment transport rates. *Water Resources Research* **20**, 476–90.

Davis, S. S., G. R. Foster, L. F. Huggins 1983. Deposition of nonuniform sediment on concave slopes. *Transactions of the American Society of Agricultural Engineers* **26**, 1057–63.

Dietrich, W. E. 1982. Settling velocity of natural particles. *Water Resources Research* **18**, 1615–26.

Einstein, H. A. 1950. *The bed-load function for sediment transportation in open channel flows*. Technical Bulletin 1026. Washington DC: USDA Soil Conservation Service.

Emmett, W. W. 1970. *The hydraulics of overland flow on hillslopes*. Geological Survey Professional Paper 662-A. Washington DC: US Government Printing Office.

Engelund, F. & J. Fredsoe 1976. A sediment transport model for straight alluvial channels. *Nordic Hydrology* **7**, 293–306.

Foster, G. R. & L. D. Meyer 1972. Transport of soil particles by shallow flow. *Transactions of the American Society of Agricultural Engineers* **15**, 99–102.

Foster, G. R., C. B. Johnson, W. C. Moldenhauer 1982. Hydraulics of failure of unanchored cornstalk and wheat straw mulches for erosion control. *Transactions of the American Society of Agricultural Engineers* **25**, 940–7.

Foster, G. R., L. J. Lane, M. A. Nearing, S. C. Finkner, D. C. Flanagan 1989. Erosion component. In *USDA – Water Erosion Prediction Project: hillslope profile version,* L. J. Lane & M. A. Nearing (eds) 10.1–10.12. West Lafayette: USDA-ARS National Soil Erosion Laboratory.

Govers, G. 1986. *Mechanismen van akkerosie op lemige bodems.* Ph.D. thesis K.U. Leuven, Belgium.

Govers, G. 1987. Initiation of motion in overland flow. *Sedimentology* **34**, 1157–64.

Govers, G. 1989. Grain velocities in overland flow: a laboratory study. *Earth Surface Processes and Landforms* **14**, 481–98.

Govers, G. 1990. Empirical relationships for the transporting capacity of overland flow. In *Erosion, transport and deposition processes (proceedings of the Jerusalem workshop)* A. Yair, S. Berckowicz, D. E. Walling (eds), International Association of Hydrological Sciences, Publication Number 189, 45–63.

Govers, G. & G. Rauws 1986. Transporting capacity of overland flow on plane and on irregular beds. *Earth Surface Processes and Landforms* **11**, 515–24.

Guy, B. T., W. T. Dickinson, R. P. Rudra 1990. Hydraulics of sediment-laden sheetflow and the influence of simulated rainfall. *Earth Surface Processes and Landforms* **15**, 101–18.

Kramer, L. A. & L. D. Meyer 1969. Small amounts of surface mulch reduce runoff velocity and erosion. *Transactions of the American Society of Agricultural Engineers* **12**, 638–45.

Laursen, E. M. 1958. The total sediment load of streams. *Proceedings of the American Society of Civil Engineers, Journal of the Hydraulics Division* **84**, 1530-1-6.

Leeder, M. R. 1979. 'Bedload' dynamics: grain impacts, momentum transfer and derivation of a grain Froude number. *Earth Surface Processes* **4**, 291–95.

Loch, R. J. & T. E. Donnollan 1983. Field rainfall simulator studies of two clay soils of the Darling Downs, Queensland II. Aggregate breakdown, sediment properties and soil erodibility. *Australian Journal of Soil Research* **21**, 47–58.

Loch, R. J., J. C. Maroulis, D. M. Silburn 1989. Rill erosion of a self-mulching black earth. *Australian Journal of Soil Research* **27**, 535–42.

Low, H. S. 1989. Effect of sediment density on bed-load transport. *Journal of Hydraulic Engineering* **115**, 124–38.

Lu, J. Y., E. A. Cassol, W. C. Moldenhauer 1989. Sediment transport relationships for sand and silt loam soils. *Transactions of the American Society of Agricultural Engineers* **32**, 1923–31.

Luque, R. F & R. Van Beek 1976. Erosion and transport of bed-load sediment. *Journal of Hydraulic Research* **14**, 127–44.

Meyer, L. D., B. A. Zuhdi, N. L. Coleman, S. N. Prasad 1983. Transport of sand-sized sediment along crop-row furrows. *Transactions of the American Society of Agricultural Engineers* **26**, 106–11.

Meyer-Peter, E. & R. Müller 1948. Formulas for bedload transport. In *Proceedings of the 2nd Congress of the IAHR, Stockholm*, 39-63.

Moore I. D. & G. J. Burch 1986. Sediment transport capacity of sheet and rill flow: application of unit stream power theory. *Water Resources Research* **22**, 1350–60.

Nittim, R. 1977. *Overland flow on impervious surfaces.* University of New South Wales Water Resources Laboratory Report 151.

Parsons, D. A. 1972. The speeds of sand grains in laminar flows over a smooth bed. In *Sedimentation, symposium to honour Prof. H. A. Einstein*, H. W. Shen (ed.), 1.1–1.24. Fort Collins, Colorado.

Petit, F. 1989. The evaluation of grain shear stress from experiments in a pebble-bedded flume. *Earth Surface Processes and Landforms* **14**, 499–508.

Phelps, H. O. 1975. Shallow laminar flows over rough granular surfaces. *Proceedings of the American Society of Civil Engineers, Journal of the Hydraulics Division* **101**, 367–84.

Poesen, J. & D. Torri 1989. Mechanisms governing incipient motion of ellipsoidal rock fragments in concentrated overland flow. *Earth Surface Processes and Landforms* **14**, 469–80.

Rauws, G. 1984. *De transportcapaciteit van afterflow over een ruw oppervlak: laboratoriumexperimenten.* M.Sc. thesis. K.U. Leuven, Belgium.

Rauws, G. 1988. Laboratory experiments on resistance to overland flow due to composite roughness. *Journal of Hydrology* **103**, 37–52.

Rickenmann, D. 1990. Bedload transport capacity of slurry flows at steep slopes. *Mitteilungen Versuchsanstalt für Wasserbau, Hydrologie und Glaziologie der Eidgenössischen Technische Hochschule Zürich* **103**.

Riley, S. J. & D. B. Gore 1988. Aspects of the design and calibration of a portable flume. *Soil Technology* **1**, 297–312.

Savat, J. 1980. Resistance to flow in rough, supercritical sheet flow. *Earth Surface Processes* **5**, 103–22.

Smart, G. M. 1984. Sediment transport formula for steep channels. *Journal of Hydraulic Engineering* **110**, 267–77.

Wilson, K. C. 1987. Analysis of bed-load motion at high shear stresses. *Journal of Hydraulic Engineering* ***113***, 97–103.

Woo, D. C. & E. F. Brater 1961. Laminar flow in rough rectangular channels. *Journal of Geophysical Research* **66**, 4207–17.

Yalin, M. S. 1963. An expression for bed-load transportation. *Proceedings of the American Society of Civil Engineers, Journal of the Hydraulics Division* **89**, 221–48.

Yalin, M. S. 1977. *Mechanics of Sediment Transport*. Oxford: Pergamon Press.

Yang, C. T. 1972. Unit stream power and sediment transport. *Proceedings of the American Society of Civil Engineers, Journal of the Hydraulics Division* **98**, 1805–25.

Yang, C. T. 1973. Incipient motion and sediment transport. *Proceedings of the American Society of Civil Engineers, Journal of the Hydraulics Division* **99**, 1679–703.

12

Mechanisms of overland-flow generation and sediment production on loamy and sandy soils with and without rock fragments

J. W. A. Poesen

Abstract

This chapter reviews process research aiming at a better understanding of overland flow and sediment production mechanisms operating at the micro- (10^{-4}–10^{0} m^2) and the meso-scale (10^{0}–10^{4} m^2). Special attention is paid to surface sealing, Hortonian overland-flow generation, interrill and rill erosion on bare loamy and sandy soils, with and without rock fragments. The emphasis is put on effects of soil properties and surface slope on processes mentioned above. Surface sealing, a widespread process of physical degradation of bare topsoils, conditions Hortonian overland-flow production. Effects of soil textural composition, cover, position and size of rock fragments, as well as effects of soil surface slope and length, on the intensity of surface sealing and on overland-flow generation are discussed. On interrill areas, raindrop impact has been recognized to be the dominant erosive force influencing erosion rate. An empirical splash-erosion model has been developed to predict the total mass of sediment being detached at the surface of soils having different characteristics as well as slopes. Soil surface sealing promotes runoff and hence the potential for rilling. Effects of slope on rill formation and of soil texture on sediment production by rills is discussed. Hydraulic conditions for incipient motion of rock fragments in rill or gully flow have been established.

Introduction

Soil erosion by water is one of the major soil degradation processes in Europe (De Ploey 1989, Oldeman et al. 1990). Therefore, the Laboratory of Experimental Geomorphology has devoted much attention to the understanding and prediction of the spatial and temporal variations in overland flow and sediment production processes during the past two decades. Quantitative information on these processes is increasingly needed for the development of process-based simulation models such as the European Soil Erosion Model (EUROSEM: Morgan et al. 1990).

This chapter reports the results of a series of laboratory experiments and field measurements aiming at a better understanding of overland flow and sediment production mechanisms operating at the micro- (10^{-4}–10^0 m^2) and at the meso-scale (10^0–10^4 m^2). The effects of soil properties and surface slope on surface sealing and Horton overland-flow generation, interrill and rill erosion will be discussed. Particular attention is given to the occurrence of these processes on bare loamy and sandy soils without and with rock fragments in their surface layer. Hitherto, the latter have received relatively little attention despite the fact that they cover more than 60% of the land area in the Mediterranean belt (Poesen 1990).

Surface sealing and overland-flow generation

Surface sealing is a widespread process of physical soil degradation which influences Horton overland-flow generation and, hence, also interrill and rill erosion. Surface sealing has been defined as the orientation and packing of dispersed soil particles in the immediate surface layer of the soil, rendering it relatively impermeable to water (Soil Science Society of America 1984). Several mechanisms are recognized to be important in sealing, i.e. chemical dispersion, physical dispersion and filtration (Römkens et al. 1990).

A series of laboratory experiments has been conducted on loamy and sandy sediments in order to investigate the effects of textural composition, rock fragments and soil surface topography on sealing intensity and on overland-flow generation. Surface sealing on these sediments is essentially due to physical dispersion by raindrop impact and to the formation of a filtration pavement. As shown in Figure 12.1, the depth of the filtration pavement (i.e. the washed-in layer) below the soil surface increases with increasing geometric mean particle size of the sediment.

Soil texture

Soils in northern Europe often show a transition from sands over loamy sands and sandy loams to silt loams when moving towards the south. Hence, the effect of a changing sand content in a sand/silt loam soil mixture on the intensity of surface sealing was studied. Eight binary soil mixtures were obtained by mixing dune sand (100% sand) and a silt loam (3% sand, 80% silt and 17% clay) in different sand/silt loam ratios. The soil samples were put into a plot box and subjected to simulated rainfall with constant intensity. The percolation rate was plotted versus time and, from the curves obtained, a sealing index (SI) was deduced as the rate at which percolation rate through the soil sample decreases (Poesen 1986b, 1987a). More details on the experimental set-up and procedures are given in Poesen (1986a, 1986b, 1987a). The results

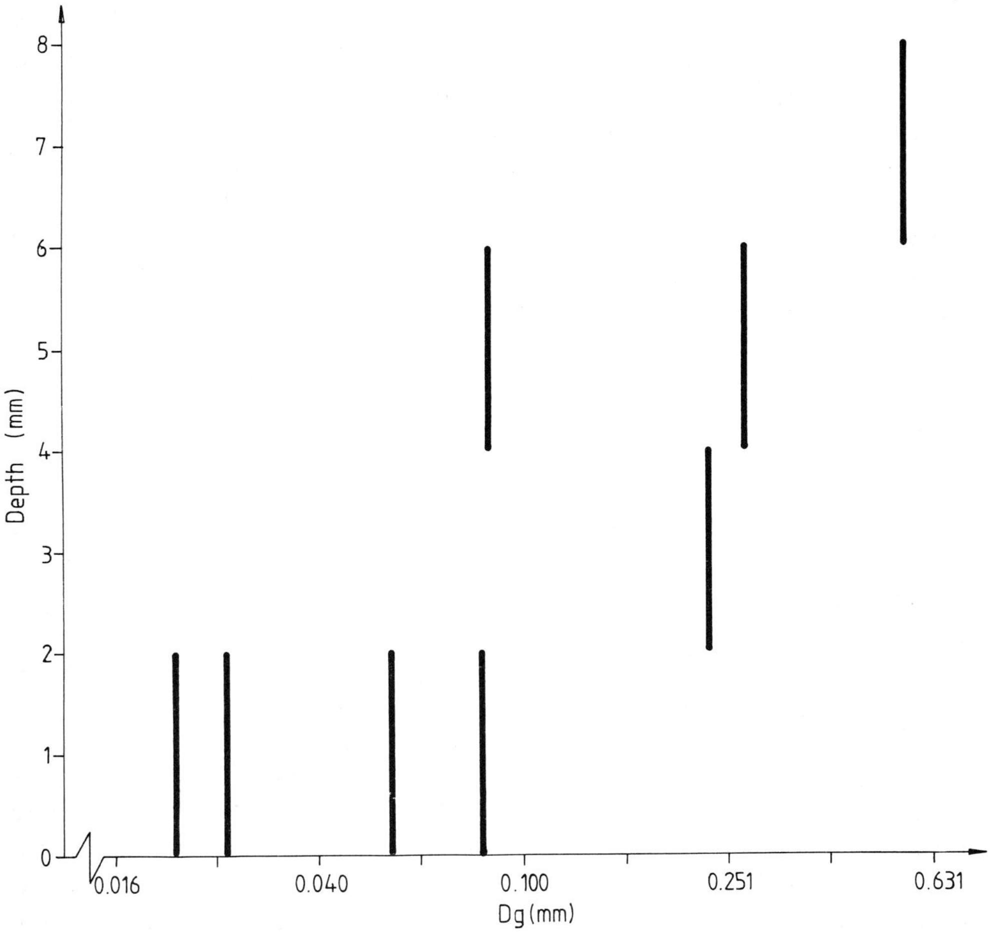

Figure 12.1 Relation between geometric mean particle size (*Dg*, Shirazi & Boersma 1984) of loose sediments and depth of filtration pavement (based on Poesen 1981).

are depicted in Table 12.1. This table shows that the sealing index is low for soils consisting of either 100% sand or less than 50% sand. A sediment consisting of 90% sand and 10% silt loam is most susceptible to surface sealing. Other field and laboratory results confirm that bare soils or sediments containing 80–94% sand and 20–6% silt and clay are extremely susceptible to sealing (Table 12.2). Soils and sediments with such a grain-size composition undergo a strong clogging of the (textural) pores by the formation of a filtration pavement. These observations are also in line with other research results: binary soil mixtures with similar proportions of "coarse" and "fine" particles are also the most susceptible to mechanical compaction in general (Bodman & Constantin 1965) and will yield the highest textural bulk densities (Fies & Stengel 1981). Because of their high susceptibility to surface sealing and

Table 12.1 Sealing index (SI) for binary soil mixtures with different sand/silt loam ratios (after Poesen 1987a).

Sand : silt loam ratio	SI $mm\ h^{-2}$
0 : 100	3.9
10 : 90	2.9
30 : 70	3.8
50 : 50	4.0
70 : 30	6.6
80 : 20	13.2
90 : 10	30.0
100 : 0	0.0

despite their considerable saturated hydraulic conductivity (i.e. > 50 $mm\ h^{-1}$), loamy sand soils will generate important overland-flow volumes even during low-intensity rainfall.

Rock fragments
A series of experiments has been conducted in order to find out how the vertical position, cover percentage and size of rock fragments affect overland-flow generation. Rainfall experiments conducted on a fine sandy soil (Poesen 1986a, b) having a 12% rock-fragment cover revealed that surface-sealing intensity was significantly controlled by the position of the rock fragments. The sealing index (SI) equalled 5.4 $mm\ h^{-2}$ when the rock fragments rested upon the soil surface, but equalled 9.0 $mm\ h^{-2}$ when the rock fragments were partly (for 50% of their volume) embedded in the top layer. In the first situation, the soil surface below the rock fragments, which was less sealed, could absorb part of the rock flow and part of the Horton overland flow created on the sealed soil surface. In such a situation, the rock fragments kept the sealing index low. In the second situation, the less permeable surface sealed tightly around the embedded rock fragments, resulting in a higher SI value.

In subsequent experiments conducted in an interrill flume, the effects of rock-fragment cover and the texture of the fine earth surrounding the rock fragments in the topsoil on overland-flow generation were investigated (Poesen et al. 1990, Poesen & Ingelmo-Sanchez 1992). The results of these experiments have been transformed into a structural model (Fig. 12.2). This model illustrates possible relations between rock-fragment cover (*Rc*) and relative interrill overland flow volume, as well as relative interrill sediment

Table 12.2 Textural composition sandy loam, loamy sand and sandy soils reported to be very susceptible to surface sealing (after Poesen 1988b).

Sand %	Silt %	Clay %	Soil identification	Location	Source
80	15	5	Loamy fine sand	Niono, Mali	Hoogmoed (1986)
83	2	15	Loamy sand	Coastal plain, Israel	Ben-Hur et al. (1985)
84	5	11	Ferralitic sandy soil	Adiopodoumé, Ivory Coast	Lafforgue & Naah (1976)
84	6	10	Loamy sand soil	Jodhpur, India	Sharma et al. (1983)
84	10	6	Loamy sand soil	Owerri, Nigeria	Boers et al. (1988)
85	2	13	Loamy sand soil	Sharon plain, Israel	Morin et al. (1981)
86	6	8	Sverdrup sandy loam	Minnesota, USA	Young & Onstad (1986)
89	2	9	Fine sandy soil	Tongeren, Belgium	Poesen (1984)
89	7	4	Princeton loamy fine sand	Indiana, USA	Mannering (1967)
90	6	5	Sand	Agadez, Niger	Valentin (1986)
92	5	3	Sand	Sadore, Niger	Hoogmoed (1986)
94	4	2	Sand	Indiana, USA	Mannering (1967)

yield for different (extreme) situations: that is, a positive relation for rock fragments partly incorporated into a surface seal developed on a top layer with essentially textural pore spaces, or on a top layer with structural pore spaces; and a negative relation for rock fragments embedded in a top layer with structural porosity, or for rock fragments resting on the surface of a soil top layer characterized by either essentially textural pore spaces or by structural pore spaces. Contrary to the current view, these results indicate that a cover of rock fragments can have an ambivalent effect on overland-flow volume and related sediment yield. Porosity of the topsoil near the rock fragments determines whether a positive or a negative relation between cover and runoff or sediment yield will occur, as well as the extent to which the positions of

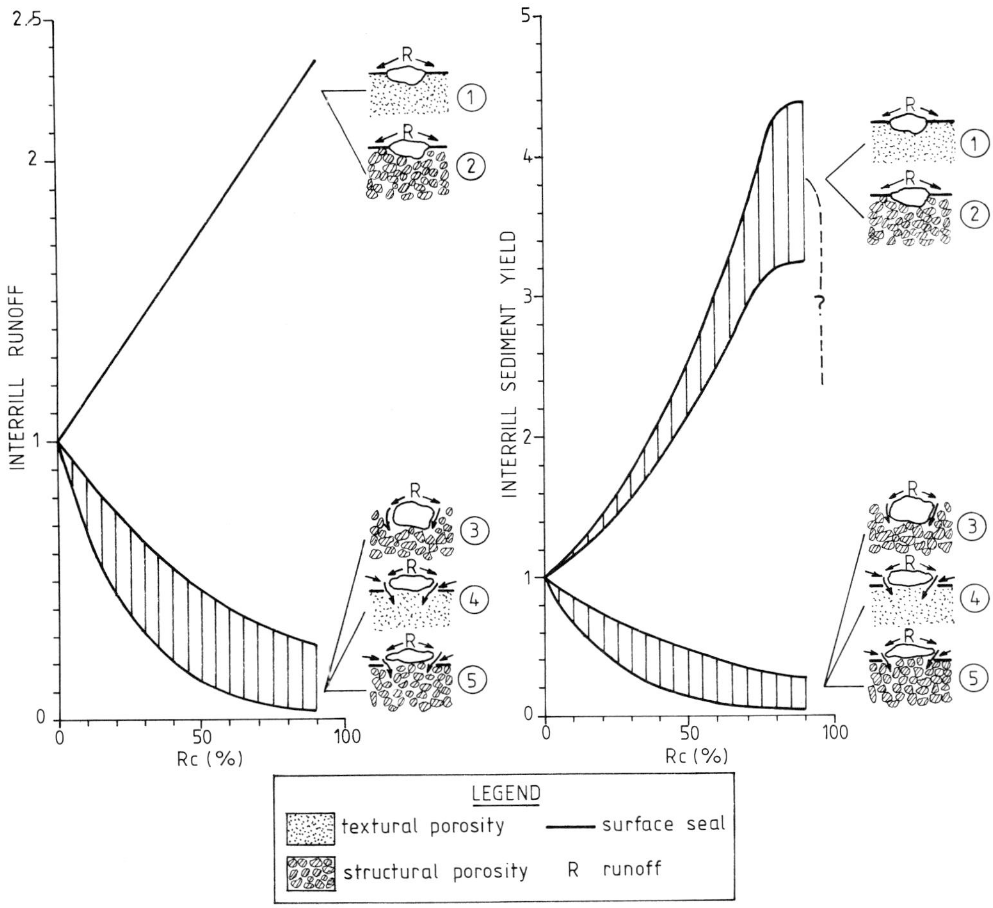

Figure 12.2 Structural model (after Poesen & Ingelmo-Sanchez 1992).

rock fragments affect this relation. Rock fragments increase overland flow volume (linear relation with cover) when they are well embedded in a surface seal (i.e. a top layer with essentially textural pore spaces; that is, pore spaces due only to the packing of primary particles). A negative linear or exponential relation occurs either where rock fragments are partly embedded within a top layer having a structural porosity or where the rock fragments rest upon the surface of a soil having either textural or structural pore spaces. The time at which macropores at the soil surface close is crucial, for then the effect of the cover of rock fragments on the generation of overland flow switches from negative to positive.

This structural model offers a possible explanation for seemingly contradictory field-plot data on the relation between rock-fragment cover and runoff yield (Poesen et al. 1990, Poesen & Ingelmo-Sanchez 1992). The opposite trends reported in the literature could be due to the fact that the topsoil structure surrounding the rock fragments in different field studies was of a different nature, e.g. cultivated soils with macropores versus non-cultivated compacted topsoils characterized by essentially textural porosity. As explained above, the effects of rock fragments on runoff and sediment yield will totally depend on the topsoil structure.

Table 12.3 Effects of rock-fragment size and cover on final runoff coefficient (in %; after Poesen & Lavee 1991, Lavee & Poesen 1991). Rock fragments rest on the soil surface.

	Rock fragment size (cm)			
Cover %	3	5.9	11.7	22.3
30	12.1	15.0	23.0	25.5
49	14.3	19.6	26.1	28.0
70	14.0	26.5	29.0	36.8
88	17.2	23.6	32.0	43.7
0		22.0		

Effects of rock-fragment size on overland-flow generation were studied by Poesen & Lavee (1991) and Lavee & Poesen (1991) using an interrill flume. Simulated rock fragments of four sizes ranging between 3 cm and 22.3 cm were applied at four cover percentages (30–88%) on a loamy sand, while rainfall was simulated at an intensity of 71 mm h^{-1}. Final runoff coefficients, i.e. runoff coefficients corresponding to steady-state conditions for the different treatments, are shown in Table 12.3. The experimental data clearly indicate that, for a given cover, rock-fragment size has a positive effect on overland-flow production. This finding is attributed to a positive effect of rock-fragment size on rock-flow discharge which flows onto the surrounding soil surface and which will continue as overland flow if the delivery intensity exceeds the infiltrability of the soil. It is interesting to note that, for the experimental conditions tested, the small rock fragments always reduce the runoff coefficient compared to the bare soil surface whereas the larger rock fragments, particularly at high covers, increase the runoff coefficient. Similar observations from the field were reported by Yair & Lavee (1976).

From these experiments it can be concluded that rock fragments at the soil surface have an ambivalent effect on overland-flow production. On the one hand, rocks prevent direct infiltration of (intercepted) raindrop water into the soil. On the other hand, rock fragments prevent soil-surface sealing by protecting the soil surface against raindrop impact forces and, therefore, they have a positive effect on water intake rate. Whether one or the other effect will dominate on infiltration rate and, hence, on overland flow production depends on various factors such as scale, position, size and cover of rock fragments, as well as on texture and structure of the fine earth. The trends indicated by the various experimental results might help in predicting rock-fragment effects on overland-flow generation.

Topography

Most studies on surface sealing and overland-flow generation have focused on the following factors: climate, soil, vegetation and surface treatments. Few studies have dealt with topographic effects on these processes. Therefore, a number of laboratory experiments (complemented by field measurements and observations) were conducted in order to investigate how micro- and macro-topography affects surface sealing and overland-flow production.

On rough, cultivated cloddy surfaces or surfaces with ridges, surface sealing often occurs discontinuously in space (Poesen 1988a). This is attributed to wind-driven rain, often falling with a mean rainfall obliquity (i.e. the angle between the vertical and the mean trajectory of the falling raindrops) ranging between 0° and 42° (Poesen 1986c). Consequently, a structural seal develops on the windward side of the clods on freshly cultivated fields during wind-driven rain. This surface seal continues as a depositional seal at the foot of the clods in the interclod depressions. At the leeward side of the clod, the initial microtopography remains unchanged. As long as a microtopography (due to tillage) exists, a continuous seal cannot form during oblique rain. Consequently, overland-flow generation will be discontinuous on such soil surfaces.

Effects of surface *slope steepness* on sealing intensity, infiltration rate and overland-flow generation were studied under simulated rainfall in the laboratory (Poesen 1984, 1986a, 1987a). The main results of these experiments are reproduced in Table 12.4 (Poesen 1984, 1986b). From this table it can be seen that sealing intensity is inversely proportional to slope steepness, i.e. the steeper the slope, the smaller the intensity of surface-seal formation, and the higher the final infiltration rate. This is also corroborated by a negative relation between slope steepness and topsoil shear strength (measured with a torvane), a measure of seal strength. The latter relation has also been observed in the field (Poesen 1984). The main explanation for this relation is given by a corresponding change in interrill and rill erosion rate. This is suggested by a strong negative correlation ($r^2 = 0.97$) between erosion rate and sealing index (Poesen 1987a). The high erosion rates on steep slopes prevent the development of a surface seal and thus the infiltration rate remains high. Recent experiments conducted by Agassi et al. (1989) and Bradford & Huang (1991) have confirmed this mechanism. The negative effect of surface slope on sealing intensity will only occur on soils that are very susceptible to sealing

Table 12.4 Effect of slope steepness on sealing intensity (SI), final infiltration rate (FI), erosion rate (ER) and topsoil shear strength (C) of a loamy sand at steady state (after Poesen 1984, 1986b).

Slope	SI mm h^{-2}	FI mm h^{-1}	ER g min^{-1}	C kPa
0.035	10.6	1.4	0.5	2.6
0.15	10.3	2.3	9.1	0.8
0.20	8.4	3.9	26.9	0.6
0.27	4.4	7.5	66.6	0.2
0.36	3.1	7.4	70.5	0.2

(Poesen 1984, Bradford & Huang 1991).

The laboratory-based results (Poesen 1984, 1986b) are confirmed by field measurements. Poesen (1987a) observed a linear negative relation between surface slope and overland-flow volume from bare interrill plots (0.5–0.7 m^2) covered with sandy loam soil and located on a convex–concave slope in central Belgium. Similar results were obtained more recently by Govers (1991a).

From these experiments it could be concluded that each factor causing increased erosion of a sealed topsoil will counteract the effects of surface sealing and, hence, result in an increase of infiltration rate and a decrease of overland-flow volume. Beside slope steepness, slope length could also be such a factor. Laboratory experiments conducted in a 17 m flume revealed that the relationship between slope length and Horton overland-flow volume per unit area was controlled by erosion and deposition patterns occurring within the plots (Bryan & Poesen 1989, Poesen & Bryan 1989). Slope segments where rills and headcuts developed were characterized by high infiltration and percolation rates. On the other hand, slope segments completely covered by sealed interrill surfaces or depositional seals, showed reduced infiltration rates.

Data from several authors show that overland-flow volume per unit area (*RE*) for bare soil surfaces decreases (exponentially) with increasing slope length for loamy soils which are susceptible to surface sealing (Fig. 12.3). In order to facilitate comparison, each *RE* value has been divided by the *RE* value for a 15 m long slope belonging to the same *RE* dataset (Poesen & Bryan 1989). The decrease of *RE* with slope length has been explained by a corresponding increase in the volume of infiltrated water. The latter has been explained by the facts that: (a) on longer plots there is more time for overland flow to be absorbed by the soil than on shorter plots (Duley & Ackerman 1934); and (b) that on longer slopes surface detention and, hence, overland-flow depths are larger compared to shorter plots, resulting in an extra hydraulic head (Mutchler & Greer 1980).

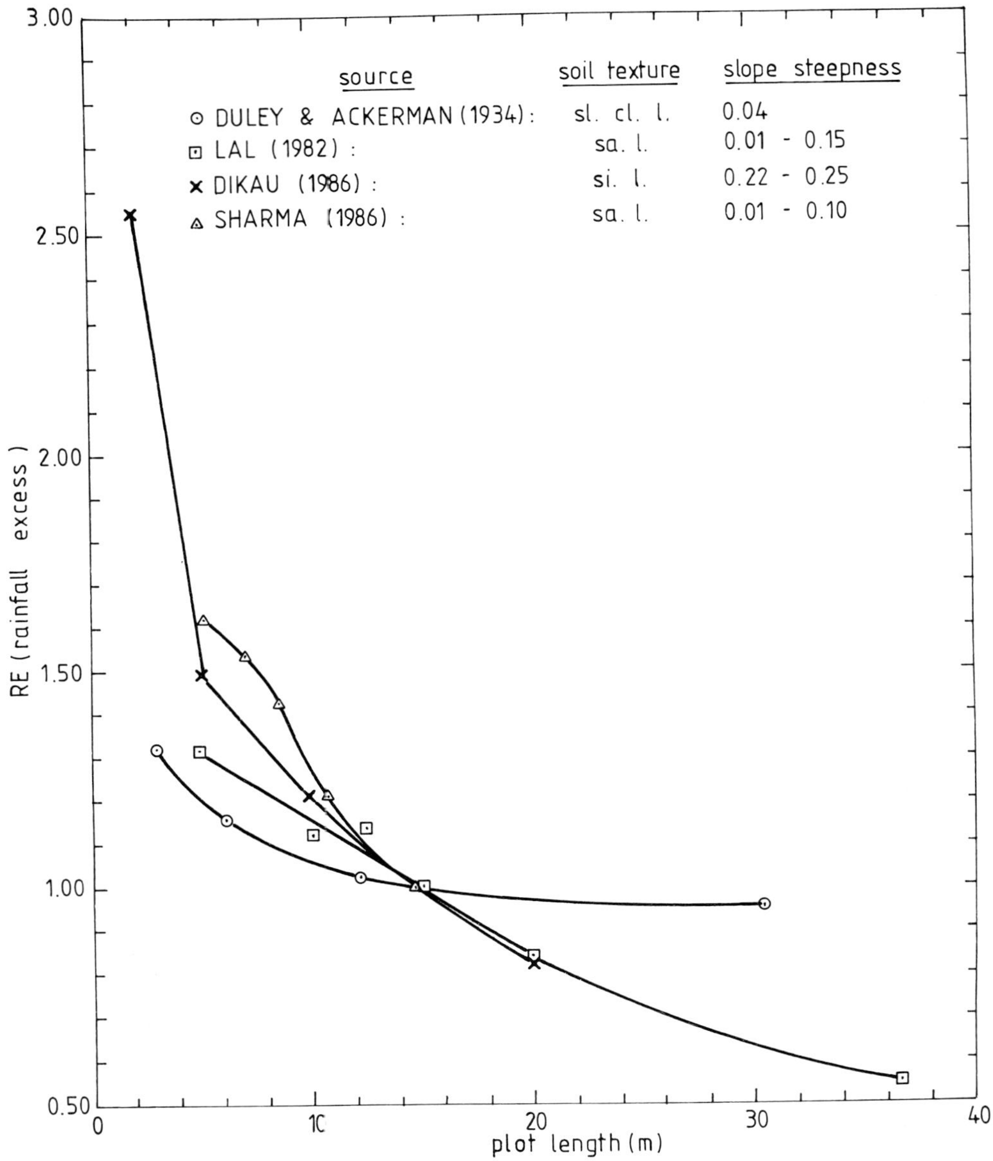

Figure 12.3 Effect of plot length on relative overland-flow volume per unit area (*RE*) for bare soil surfaces as reported by several investigators (after Poesen & Bryan 1989).

Our results indicate that the increased erosion rates of the sealed topsoil with increasing slope length might also help to explain the observed relationships between slope length and *RE*.

The experimental results on the effects of slope steepness and length on overland-flow generation reveal the existence of an important feedback from erosion processes occurring on hillslopes to the hydrological processes (infiltration, overland-flow generation). This is due to changing topsoil properties related to changing erosion and deposition rates, which can result in important compensations along a hillslope. Consequently, realistic hydrological models aiming at predicting overland-flow discharge along hillslopes should be (partly) driven by an erosion/deposition model.

Interrill erosion

Raindrop detachment

On interrill areas, raindrop impact has been recognized usually to be the dominant erosive force affecting erosion rates. Researchers aiming at predicting interrill erosion rates have often addressed two questions: (a) How much sediment will be detached by raindrop impact; and (b) What is the net loss of sediment from the interrill area? Potentially, sediment produced by raindrop detachment can be transported by air splash or by overland flow provided that the sediment size matches the interrill-flow competence (Parsons et al. 1991).

The amount of sediment detached by raindrop impact on a bare unit interrill area depends on two main factors: (a) the erosivity of the impacting raindrops expressed by, for instance, their kinetic energy; and (b) the resistance of the soil surface to raindrop detachment. At present, it is still difficult to allocate the kinetic energy of falling drops to the various component processes, such as soil matrix deformation (e.g. crater formation, surface sealing), breaking of physical and chemical bonds (detachment), transport of soil particles by air splash, as well as heating of the topsoil (including air and water) (Römkens et al. 1987). Experimental results reveal that the energy utilized to move the detached soil particles by splash is a remarkably small percentage of the incoming kinetic rain energy, i.e. it ranges between 0.2% and 20% depending on the soil surface state (Poesen 1983, Brandt & Thornes 1987). This is attributed to important frictional heat losses during impact.

Because of our limited quantitative knowledge about the expenditure of kinetic rainfall energy, a simple empirical model has been proposed in order to predict the rate of soil particle detachment on bare interrills (Poesen 1985):

$$SD = (1 - Rc) KE . RD^{-1} \qquad (12.1)$$

where SD = mass of detached sediment per unit area and per unit time (kg m^2 $time^{-1}$);
Rc = cover of soil surface by rock fragments (fraction);
KE = kinetic rainfall energy (J m^{-2} $time^{-1}$); and
RD = resistance of bare soil to raindrop detachment (J kg^{-1}).

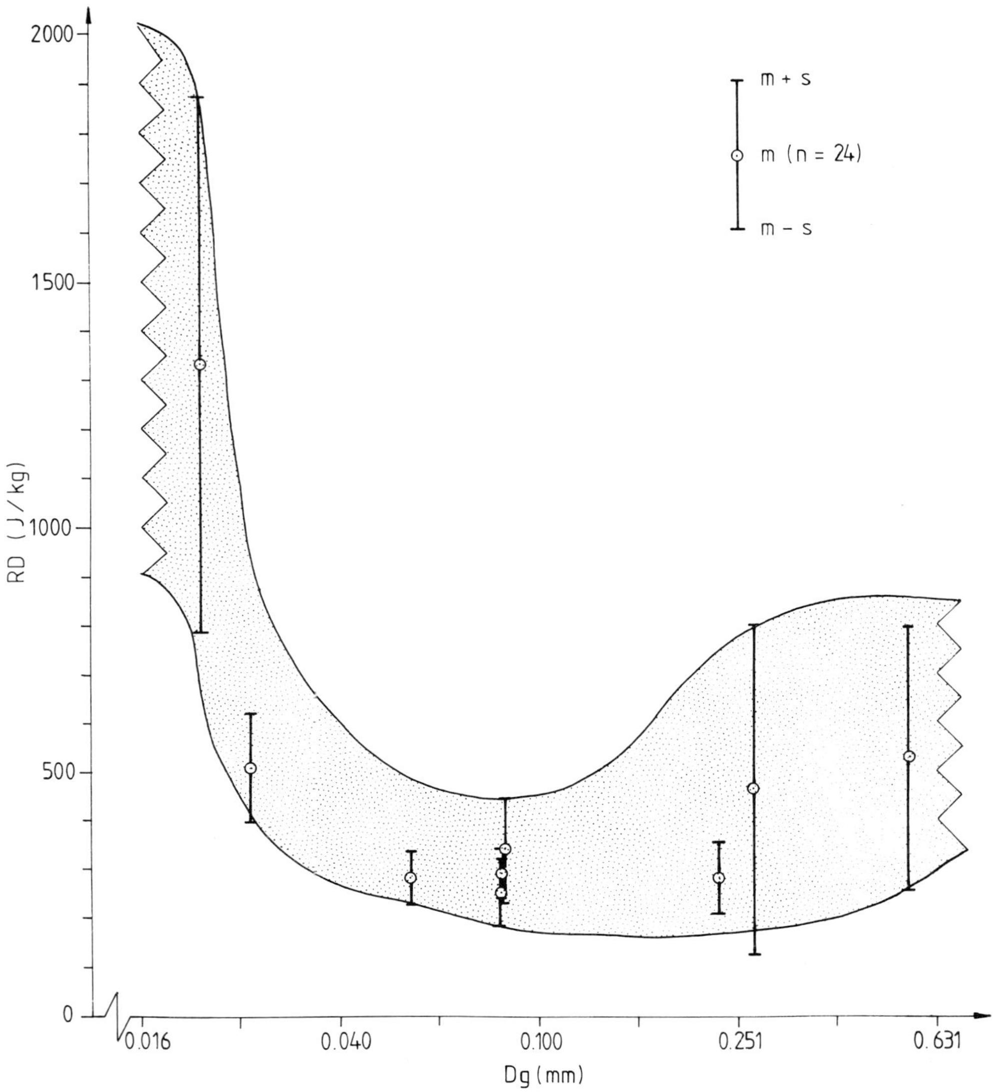

Figure 12.4 Relation between geometric mean particle size (*Dg*, Shirazi & Boersma 1984) and resistance to raindrop detachment (*RD*) of loose sediments. The standard deviation (s) of experimental values is shown with each mean (m) (after Poesen 1985).

RD indicates the kinetic rain energy required to detach 1 kg of sediment. Procedures for measuring *R* are discussed elsewhere (Poesen & Torri 1988, Torri & Poesen 1988a). For bare soil surfaces, *RD* depends on soil texture (Fig. 12.4, Poesen 1985, Poesen & Torri 1988), topsoil shear strength (Poesen 1981), soil-water content and the presence of a surface-water layer (Poesen 1985), and soil structure (Poesen 1985), as well as on soil surface slope (Fig. 12.5, Poesen 1985, Torri & Poesen 1992). The European Soil Erosion Model

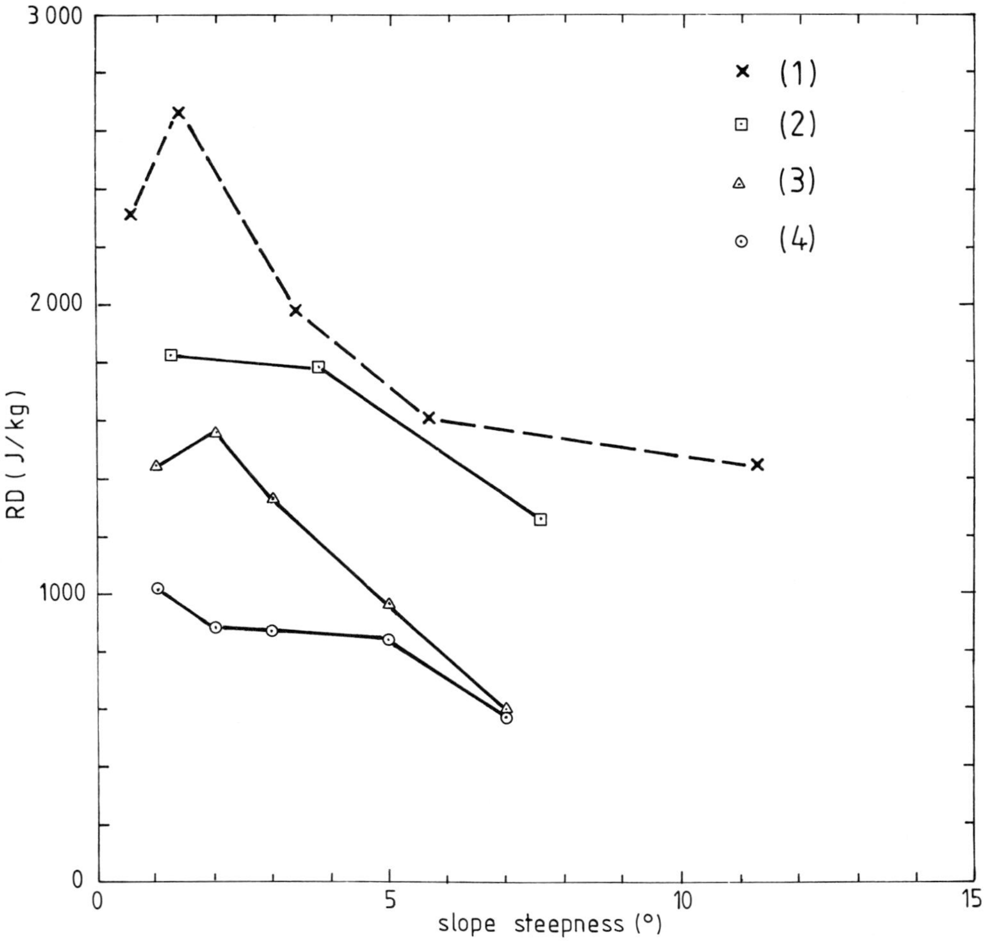

Figure 12.5 Relation between surface slope steepness and resistance to raindrop detachment (*RD*). Calculations are based on data published for several soils: (1) silt loam (Harmon & Meyer 1978), (2) silt loam (Lattanzi 1973), (3) silt loam sediment (Poesen 1985), (4) aggregated silt loam (Poesen 1985).

(EUROSEM; Morgan et al. 1990) uses this approach to predict the amount of sediment detached by raindrop impact.

On a sloping soil surface with vertical rain, or during oblique rains, net splash transport of sediment occurs in a downslope direction for vertical rain or in a downwind direction for wind-driven rain. Such transport rates can be calculated using a validated model proposed by Poesen (1985). In field conditions, however, these transport rates are relatively small compared to transport rates due to (interrill) overland flow. In EUROSEM, the latter is calculated for a unit interrill area using the equation (Styczen & Nielsen 1989):

$$DR = SDq(2v)^{-1} \qquad (12.2)$$

where DR = mass of detached sediment available for transport by overland flow (kg m^{-1} $time^{-1}$);
q = unit flow discharge (m^2 $time^{-1}$); and
v = settling velocity of sediment particles (m $time^{-1}$).

Rock fragments
Effects of position, cover and size of rock fragments, as well as of fine earth structure, on interrill sediment yield has been studied using an interrill flume (Poesen et al. 1990, Poesen & Lavee 1991, Poesen & Ingelmo-Sanchez 1992). Part of these results have been transformed into a structural model depicted in Figure 12.2.

For rock fragments partly embedded in a top layer with structural porosity, or for rock fragments resting on the surface of a soil top layer characterized by either essentially textural or by structural pore spaces, the following relation is obtained between rock-fragment cover (*Rc, %*) and relative sediment yield (*SY*) due to (interrill) overland flow:

$$SY = e^{-aRc} \qquad (12.3)$$

with *a* being a coefficient indicating the effectiveness of the rock cover in reducing soil loss. In these experiments *a* equalled 0.020 for the partly embedded fragments but 0.040 for fragments placed on the soil surface, indicating that rock fragments resting on the soil surface are more efficient in reducing interrill sediment yield than the partly embedded ones. This is attributed to a negative effect of rock-fragment cover on (a) overland-flow volume (Fig. 12.2) and, hence, on overland-flow velocity (Poesen & Ingelmo-Sanchez 1992), as well as on (b) the mass of raindrop-detached sediment (Eq. 12.1).

For rock fragments well embedded in a sealed top layer (i.e. a topsoil having essentially textural pores), however, a positive exponential relation between rock fragment cover (*Rc*, %) and relative sediment yield (*SY*) due to (interrill) overland flow was observed:

$$SY = e^{bRc} \qquad (12.4)$$

with *b* equal to 0.0164 (Poesen & Ingelmo-Sanchez 1992). This relation was obtained for cover percentages ranging between 0 and 83. The positive effect of cover on soil loss is attributed to both a corresponding increase in volume (Fig. 12.2) and in velocity of the reticular overland flow (Poesen & Ingelmo-Sanchez 1992). The rapidly increasing velocity of this flow with increasing cover is not only due to an increase in overland-flow volume (Fig. 12.2) but also to a corresponding reduction in the area between the fragments, i.e. the interrill flow zone. Hence, interrill overland flow is increasingly channelled between the fragments leading to larger unit flow discharges and, hence, to deeper and faster flow (Poesen & Ingelmo-Sanchez 1992). The latter phenomenon has also been reported by Thornes et al. (1990).

For very high covers of rock fragments the relation between cover and sediment yield should switch to a negative one, reaching a minimum sediment yield at a cover of 100% (Fig. 12.2). The increase in sediment yield due to an increase in cover will only last until more subsurface rock fragments are exposed and the cover exceeds a critical value beyond which cover has a negative effect on sediment yield. Finally, the ambivalent effect of cover on interrill sediment yield and its causes offers an explanation for reported contradictory field observations (Poesen & Ingelmo-Sanchez 1992).

Table 12.5 Effects of rock-fragment size and cover on relative solid discharge from interrills at steady state (after Poesen & Lavee 1991).

	Rock-fragment size (cm)			
Cover %	3	5.9	11.7	22.3
30	0.43	0.57	1.18	1.10
49	0.26	0.66	1.10	1.20
70	0.06	0.61	0.58	1.25
88	0.11	0.14	0.44	0.64
0		1.0		

The effects of rock-fragment size on steady-state sediment discharge from interrills were also studied by Poesen & Lavee (1991). The main results are represented in Table 12.5. These data indicate that for a given cover percentage the size of rock fragments has a positive effect on sediment yield. This is caused by a positive effect of rock-fragment size on overland-flow volume as well as on overland-flow velocity (Poesen & Lavee 1991). This also fits field observations made by Fletcher & Beutner (1941). These investigators concluded that the erodibility of soils containing rock fragments increased with the size of the fragments: rocky soils (i.e. soils with rock fragments > 40 mm in size) were on average 32% more susceptible to erosion by overland flow than gravelly soils (i.e. soils with rock fragments between 2 and 60 mm in size) within any one soil type.

Rill erosion

Effects of slope

Surface sealing of cultivated soils promotes overland-flow generation and, hence, the potential for rilling. On the other hand, however, a strongly developed seal or the presence of a compacted topsoil might hamper rill initiation (Poesen & Govers 1986). The genesis of a rill is controlled by the ratio of overland-flow forces to the resisting forces of the soil. In other words, incipient rilling on a bare, sealed soil surface occurs if the shear ratio (*SR*) exceeds a critical value (Torri et al. 1987):

$$SR = \tau C^{-1} > Z \qquad (12.5)$$

where τ = overland flow shear stress (Pa),
= (ρgRS) with ρ = fluid density,
g = acceleration due to gravity,
R = hydraulic radius, and
s = bed slope;

C = shear strength of the soil top layer (Pa), measured with a vane-test apparatus; and

Z = critical shear ratio for incipient rill formation ($0.0001 < Z < 0.0005$).

The apparent inconsistency in terms of the orders of magnitude of the differences between soil strength and flow shear stresses, i.e. a difference in magnitude of 2,000–10,000, can be explained by the fact that average shear strength overestimates local soil strength, whereas average flow shear stress underestimates local shear stresses associated with turbulent burst events (Torri et al. 1987, Nearing 1991).

These threshold conditions match with those described by Rauws & Govers (1988) for rill generation relatively well (Poesen 1988c, Poesen & Govers 1990):

$$U_{*cr} = 0.86 + 0.56C \qquad (12.6)$$

where U_{*cr} = critical effective shear velocity (cm s^{-1}) for rill generation:
= $(g\ R\ S)^{0.5}$ where R = hydraulic radius due to grain resistance; and

C = shear strength of the topsoil (5 mm thick) measured at saturation (kPa).

Several researchers have stressed the positive effect of slope steepness on rill generation (e.g. Savat & De Ploey 1982, Govers 1985). According to

Equations 12.5 and 12.6 an increase in slope steepness leads to an increase in shear stress or shear velocity of the flow and therefore to a higher probability of rill formation or to the development of deeper rills, as observed by several investigators (Poesen 1984, Parsons 1987, Govers 1991b). However, an increase of slope steepness often also leads to a decrease in the erosion resistance of the topsoil. Beside soil properties such as texture, structure and antecedent moisture content (Poesen 1981), slope steepness also determines shear strength (Poesen 1984, 1986a, Luk et al. 1989, Bradford & Huang 1991). Several field and laboratory observations revealed the following negative relation

$$C = C' e^{-(fSA)} \tag{12.7}$$

where C = shear strength of the 5 mm thick sealed top layer measured with a torvane (kPa);
C' = C value for $SA = 0°$;
SA = surface slope angle (°); and
f = a coefficient; the following values for f have been reported: 0.038 (Poesen 1984), 0.05–0.15 (Poesen 1986a), and 0.014–0.017 (Luk et al. 1989).

Consequently, for bare loamy and sandy soils, very susceptible to sealing, an increase in surface slope leads to an increased probability for rill formation not only because of an increase in overland flow erosivity but also because of a corresponding decrease in topsoil shear strength.

Effects of soil texture

The conceptual model expressing the potential for rill formation on a sealed topsoil (*SR* value, Eq. 12.5) is also useful when attempting to explain the relation between geometric mean particle size (*Dg*) of a soil and its susceptibility to rill (and interrill) flow erosion (*K*). This relation can be represented by the following equation (Römkens et al. 1987; Fig. 12.6):

$$K = 0.0035 + 0.0388 \exp\left[-0.5\left[\frac{(\log Dg + 1.519)}{0.7584}\right]^2\right] \tag{12.8}$$

where K = soil erodibility factor (Universal Soil Loss Equation, Wischmeier & Smith 1978) for soils containing less than 10% of rock fragments. K is soil loss rate due to overland-flow erosion per erosion index unit for a given soil as measured on a unit field plot, which has a length of 22.1 m, a slope of 9% and is in continuous clean-tilled fallow ($\text{t.ha.h.(ha.MJ.mm)}^{-1}$);

Dg = geometric mean particle size (mm; Shirazi & Boersma 1984):

$$Dg = \exp\left(0.01 \sum_{i=1}^{i=n} f_i(\ln m_i)\right) \qquad (12.9)$$

where f_i = particle size fraction (%);
m_i = arithmetic mean of the particle size limits (mm); and
n = number of particle size fractions.

Although K is calculated from total soil loss measured on field plots, this factor basically reflects the long-term average susceptibility to rill erosion for most soils (except for very sandy soils which are susceptible to liquefaction; De Ploey 1971). This is deduced from the fact that, on average, 75% of soil lost from field plots originates from rill areas, whereas interrill erosion only accounts in the mean for 25% of total soil loss (McCool & George 1983). A similar proportion between rill and interrill erosion was found by Govers & Poesen (1988).

Figure 12.6 illustrates that Equation 12.8, which is based on 249 published field-measured K values for soils having less than 10% rock fragments (Römkens et al. 1987), describes the relation between Dg and the average K value per Dg class rather well for 304 soils. K incorporates both the soil properties affecting infiltration as well as its resistance to overland flow detachment.

For clayey soils, i.e. soils having a $Dg < 0.010$ mm, the shear ratio (*SR*, Eq. 12.5) during rainfall is usually relatively low, mainly because of important topsoil shear strength values (Fig. 12.7). In addition, if the clay is not of a dispersive nature, these soils have good hydraulic properties so that they can absorb a high percentage of rainfall, keeping overland flow discharge and, hence also flow shear stresses, low. Furthermore, since these soils have a high percentage of water-stable clods (Kemper & Koch 1966) they remain rougher, on average, than soils with a coarser texture. This implies that more flow shear stress is spent on form roughness elements and less on grain roughness compared to soils with a coarser texture. Since flow shear stress acting to transport soil particles is related to grain roughness (Govers & Rauws 1986, Nearing et al. 1990), sediment transport rates should be low for clayey soils.

For loamy soils with a Dg ranging between 0.010 and 0.025 mm, soil erodibility increases with increasing geometric mean particle size because the shear ratio increases during rainfall. Topsoil shear strength for these loamy soils is lower than for clayey soils (Fig. 12.7). If these loamy soils have a low organic matter content, they are susceptible to sealing and compaction, which leads to important overland-flow discharges and thus also to important flow shear stresses.

A maximum rill (and interrill) erodibility is observed for soils having a Dg

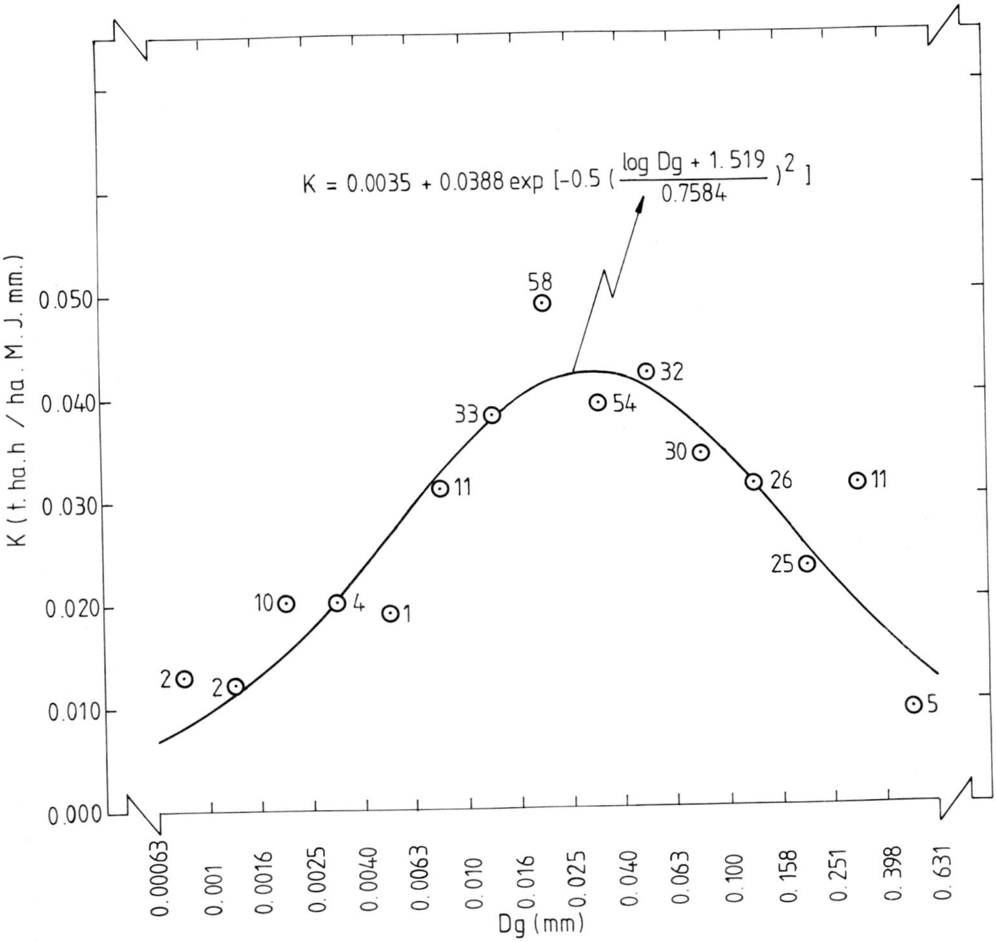

Figure 12.6 Relation between geometric mean particle size (*Dg*) of 304 soils containing less than 10% of rock fragments and their mean rill (and interrill) erodibility expressed by the *K* factor (Universal Soil Loss Equation). Sources of *K* data are given in Poesen (1988c). Number of soils from which the average was obtained for each *Dg* class is indicated with each datapoint.

in the range 0.016 – 0.063 mm (i.e. silt loams and very fine sandy loams). This peak does not correspond to the minimum resistance to raindrop detachment, which is situated at a *Dg* value between 0.040 and 0.250 mm (Fig. 12.4).

With coarsening texture (i.e. $Dg > 0.025$ mm) the soils become less erodible again although topsoil shear strength is rather low (Fig. 12.7). The high infiltration rate of sandy soils diminishes flow discharge and, hence, also flow shear stress. The sediment-transport capacity of the overland flow decreases with increasing grain size, adding to a decreasing erodibility in the grain size

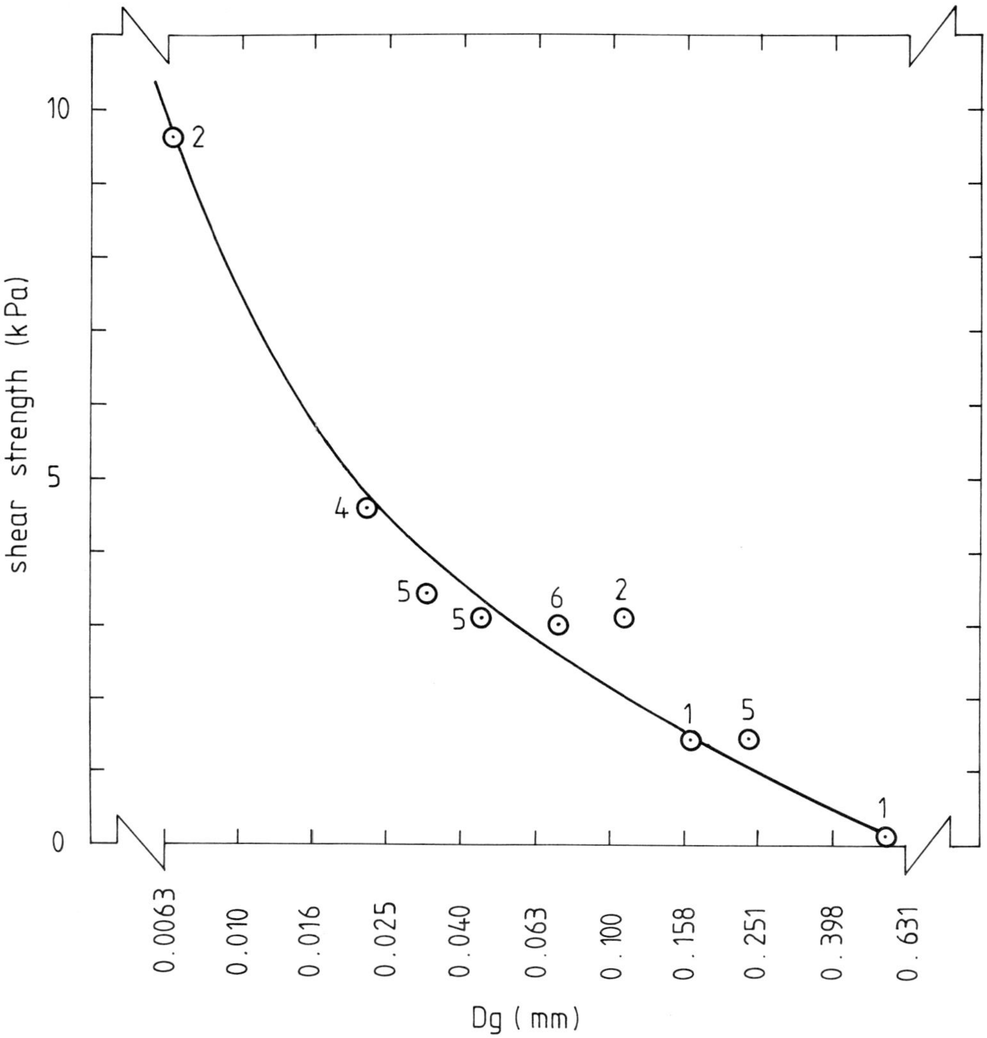

Figure 12.7 Relation between geometric mean particle size (*Dg*) for 31 soils and the corresponding average shear strength (*C*) of their saturated top layer (5 mm thick). *C* was measured with a torvane under steady-state conditions after the sieved soil received 30–60 mm of rain in 1 h. Numbers refer to number of soils for which shear strength was measured. (Data from Poesen 1981, 1983, Verhaeghen 1984, Torri et al. 1987.)

range of *Dg* between 0.024 mm to 0.631 mm (Govers 1990, Everaert 1991). Large *K* values in the *Dg* class 0.251–0.398 mm can probably be attributed to reduced infiltrability of these loamy sand soils due to their high susceptibility to surface sealing (see above).

It should be remarked that although the ratio between the maximum average *K* value (i.e. for silt loams) and minimum average *K* values (i.e. for clays and sands) is only 5 (Fig. 12.6), differences in rill and interrill erodibility between individual soils can be much more important. *K* reflects essentially a long-term average rill (and interrill) erodibility. On a very short term, however, rill erodibility for a given soil can exhibit even more important variations than is suggested by Figure 12.6 due to, for instance, moisture content variations (Govers et al. 1990).

Effects of rock fragments

For soils containing more than 10% of rock fragments, a negative relation between rock fragment cover at the soil surface and sediment yield due to erosion by rill and interrill flow has generally been observed (e.g. Box 1981, Collinet & Valentin 1984, Simanton et al. 1984, Martin 1988). The relations reported are similar to Equation 12.3, with values for the coefficient a ranging between 0.025 (Box 1981) and 0.058 (Collinet & Valentin 1984).

An exception to this trend was reported by Barnett et al. (1965) who observed that a cover of rock fragments (>5 cm) had a positive effect on the soils' rill and interrill erodibility. This can be explained by the fact that the rock fragments during their field experiments were well embedded in a sealed topsoil (see above). In addition, the effect of the reticular flow typical on such soil surfaces (Thornes et al. 1990), as opposed to sheet flow, is to concentrate erosive forces on the soil surface between rock fragments. It should be remarked, however, that if rock fragments at the soil surface increase erosion rates, this positive effect will only be temporary since the increased erosion rates will expose subsurface fragments leading to the formation of an erosion pavement (Shaw 1929) and, hence, to armouring.

The generally observed negative effect of a rock-fragment cover on sediment yield due to rill and interrill erosion can be explained by the combined effects of rock fragments on various subprocesses. First of all, a rock cover will often increase infiltration rates by reducing physical degradation of the topsoil and, hence, diminish overland-flow rate. Secondly, rock fragments at the soil surface increase surface roughness and therefore flow resistance expressed by a friction factor (see Chs 1 & 2, this volume). If friction is expressed by Manning's n, the following relation based on published field data can be established (Fig. 12.8):

$$n = n_c(n')^{-1} = e^{0.018Rc} \qquad (12.10)$$

where n = relative Manning's n value, i.e. the ratio between the friction factor for a given rock cover (n_c) and the friction factor for a bare soil surface (n'); and

Rc = cover of rock fragments resting on the soil surface (%).

Even though overland flow is often laminar at the upper end of a slope, the Manning's equation is often used in overland-flow computations. Combining Manning's equation with Equation 12.10 yields:

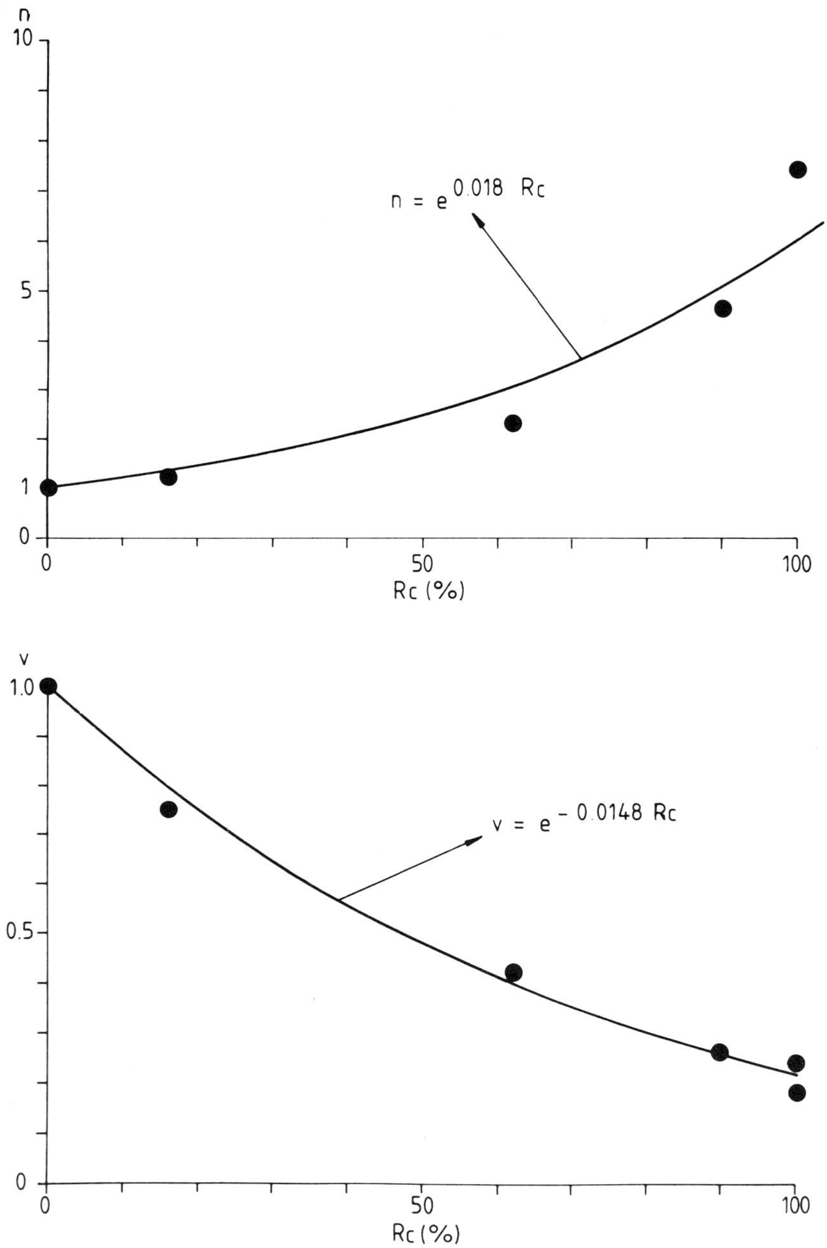

Figure 12.8 Relation between surface cover (*Rc*) by rock fragments (6.3–38.1 mm in size) and relative hydraulic roughness factor (Manning's *n*) and relative average overland flow velocity (*v*) at various distances down slope (4 to 53 m, slope = 0.20, rainfall intensity = 63.5 mm h^{-1}). All parameter values were divided by the corresponding parameter value for a bare soil surface. (Calculations are based on data published by Meyer et al. 1972 and Foster et al. 1980.)

$$v = (n')^{-1} e^{-0.018Rc} R^{0.67} S^{0.5} \qquad (12.11)$$

where v = average overland-flow velocity for a given rock cover (m s^{-1});
n' = Manning's n for a bare soil surface;
R = hydraulic radius (m); and
S = slope.

In the field, v has been observed to be proportional to $e^{-0.015\,Rc}$ (Fig. 12.8). Both detachment and transport of sediment are functions of flow velocity. Govers (1990) found that transport capacity (*TC*) of rill flow is proportional to unit stream power (*Sv*):

$$TC \approx Q(Sv - Sv_{cr})^x \qquad (12.12)$$

where TC = transporting capacity (unit solid discharge);
Q = unit flow discharge;
v = average flow velocity;
Sv_{cr} = critical unit stream power; and
x = a coefficient ranging between 0.6 and 1.0 depending on the sediment size (Govers 1990).

If one assumes interrill and rill sediment yield (*SY*) to be transport limited, a combination of Equations 12.11 & 12.12 yields the following proportionality:

$$SY \approx TC \approx e^{-zRc} \qquad (12.13)$$

where $0.011 < z < 0.018$. This indicates that part of the exponential decay of sediment yield due to rill and interrill erosion with increasing rock cover can be attributed to the effects of surface rock fragments on velocities of overland flow and, hence, on its transporting capacity.

Finally, with increasing rock cover the mass of sediment supplied to overland flow by raindrop impact will decrease linearly (Eq. 12.1), resulting in a smaller soil loss. The relative contribution of the various subprocesses to the overall reduction of rill and interrill sediment yield with increasing rock cover will vary depending on the site characteristics.

Rill flow competence

Field measurements revealed that rock fragments on upland areas are rather immobile in interrill flow while they can be readily entrained by rill flow (Poesen 1987b). Field experiments demonstrate that incipient motion conditions for rock fragments in concentrated flow are not well described by a Shields' entrainment parameter value (Θ) of 0.06 (Poesen 1987b, Abrahams et al. 1988). Therefore, laboratory experiments were set up to better describe these conditions (Torri & Poesen 1988b, Poesen & Torri 1989). Poesen & Torri (1989) found that rock fragments start moving in concentrated flow when

$$\Theta_m > 0.1Z^{-1} \qquad (12.14)$$

where Θ_m = a modified Shields entrainment parameter for concentrated overland flow on steep slopes; and
Z = relative flow depth.

$$\Theta_m = 1.5\rho g R \tan\alpha((\rho_s - \rho)gD_p(\cos\alpha\tan\beta - \sin\alpha))^{-1} \qquad (12.15)$$

where ρ, ρ_s = density of water, density of rock fragment;
g = acceleration due to gravity;
R = hydraulic radius;
α = slope angle of soil surface;
β = angle of repose; and
D_p = rock fragment diameter oriented parallel to the flow direction.

$$Z = R(D_u D_{50})^{-0.50} \qquad (12.16)$$

where D_u = rock-fragment diameter oriented perpendicular to the soil surface; and
D_{50} = median size of the particles at the soil surface.

Θ_m differs from the classical Shields entrainment parameter in that it incorporates effects of the downslope weight component of the rock fragments, their angle of repose (β) as well as their orientation with respect to the direction of the flow (D_p). Torri et al. (1990) extended Equation 12.15 to include cohesive forces (F_c) between the rock fragments and the rill bed:

$$\Theta_m = 1.5\rho g R \tan\alpha((\rho_s - \rho) g D_p(\cos\alpha\tan\beta - \sin\alpha) + F_c)^{-1} \qquad (12.15a)$$

Poesen & Torri (1989) demonstrated that the probability level of incipient motion of a single rock fragment in rill flow corresponded to the probability level of the angle of repose. Different β values can be assigned to a rock fragment resting on a rill bed, depending on the criterion (probability level of incipient motion) used to determine β. The relation between each β value and its corresponding probability of incipient motion can be well described by a normal distribution. Such a distribution reflects all possible local gradients a given rock fragment on a rill bed can assume. The steeper the local gradient the higher a rock fragment's probability of motion.

Flow competence predicted by Θ_m gives only a partial explanation of field observations. In the field, rock fragments having dimensions exceeding flow competence have been observed to move as well. The occurrence of collisions between rock fragments seems to be a likely mechanism explaining the observed discrepancy (Poesen & Torri 1989).

Conclusions

The study of various subprocesses involved in the generation of overland flow and its erosive activities has resulted in a number of conceptual models allowing the better understanding and prediction of spatial and temporal distribution of these subprocesses on bare soils. The most important conclusions are listed below.

(a) Loamy sands consisting of about 80–90% sand are extremely susceptible to surface sealing due to filtration pavement formation. This pavement will generate considerable overland flow during rainfall, despite the relatively high saturated conductivity of these loamy sands.

(b) Rock fragments can have a positive as well as a negative effect on overland flow and interrill sediment production, depending on the fine earth structure (i.e. porosity type), cover, position and size of rock fragments.

(c) Though erosion rates are relatively high on long and steep slopes this strong erosion can be counterbalanced by a self-regulating mechanism: the erosivity of overland flow often erases the surface seal, thus allowing infiltration rate to increase, which in turn leads to a decrease in the amount and erosive power of overland flow. Realistic hydrologic models aimed at predicting overland flow along hillslopes should be (partly) driven by an erosion/deposition model.

(d) The probability of rill formation on sealed topsoils increases with slope steepness. Such an increase is not only due to an increased erosive power of overland flow when slope steepens, but also to a corresponding decrease of topsoil shear strength and hence resistance to erosion.

(e) A maximum rill and interrill erodibility is observed for soils having a geometric mean particle size (*Dg*) in the range 0.016–0.063 mm. This maximum does not correspond to the peak of minimum resistance to raindrop detachment which is situated at a *Dg* ranging between 0.040–0.250 mm.

(f) The often reported negative relation between rock-fragment cover and rill (and interrill) erodibility is attributed to the combined effects of rock fragments on various subprocesses, i.e. rock fragments on top of the soil increase infiltration rates and increase friction, leading to a reduction in the flow's erosive power and reduced sediment supply by splash detachment. Little is known, however, about the relative contribution of the various subprocesses to the overall reduction in rill and interrill sediment yield with increasing rock cover for a given site. When well embedded in a sealed topsoil, rock fragments can increase rill and interrill sediment yield.

(g) A modified Shields entrainment parameter (Θ_m) has been developed to predict incipient motion of rock fragments in concentrated overland flow. Θ_m incorporates effects of the downslope weight component of the rock fragments, their angle of repose as well as their orientation with respect to the flow direction.

Acknowledgements

The author acknowledges all researchers who provided additional information on published *K* data, as well as Dr G. Govers and Dr K. Bunte for their comments on a previous draft.

References

Abrahams, A. D., S.-H. Luk, A. J. Parsons 1988. Threshold relations for the transport of sediment by overland flow on desert hillslopes. *Earth Surface Processes and Landforms* **13**, 407–19.

Agassi, M., I. Shainberg, D. Warrington, M. Ben-Hur 1989. Runoff and erosion control in potato fields. *Soil Science* **148**, 149–54.

Barnett, A., J. Rogers, J. Holladay, A. Dooley 1965. Soil erodibility factors for selected soils in Georgia and South-Carolina. *Transactions of the American Society of Agricultural Engineers* **8**, 393–5.

Ben-Hur, M., I. Shainberg, D. Bakker, R. Keren 1985. Effect of soil texture and $CaCO_3$ content on water infiltration in crusted soil as related to water salinity. *Irrigation Science* **6**, 281–94.

Bodman, G. B. & G. K. Constantin 1965. Influence of particle size distribution in soil compaction. *Hilgardia* **36**, 567–91.

Boers, T. M., H. O. Maduakor, D. P. Tee 1988. Sheet erosion from a bare sandy soil in south eastern Nigeria. In *Proceedings of the International Symposium on Erosion in S.E. Nigeria*, C. I. Ijioma, S. Anaba, T. M. Boers (eds), 23–34. Owerri, Nigeria: Federal University of Technology.

Box, J. E. 1981. The effects of surface slaty fragments on soil erosion by water. *Soil Science*

Society of America Journal **45**, 111–16.
Bradford, J. M. & C. Huang 1991. Comparison of interrill soil loss for laboratory and field procedures. *Soil Technology* (in press).
Brandt, C. J. & J. B. Thornes 1987. Erosional energetics. In *Energetics of physical environment*, K. J. Gregory (ed.), 51–87. Chichester, England: John Wiley.
Bryan, R. B. & J. W. Poesen 1989. Laboratory experiments on the influence of slope length on runoff, percolation and rill development. *Earth Surface Processes and Landforms* **14**, 211–31.
Collinet, J. & C. Valentin 1984. Evaluation of factors influencing water erosion in West Africa using rainfall simulation. *International Association of Hydrological Sciences Publication* **144**, 451–61.
De Ploey, J. 1971. Liquefaction and rainwash erosion. *Zeitschrift für Geomorphologie* **15**, 491–6.
De Ploey, J. 1989. *Soil erosion map of western Europe*. Cremlingen-Destedt, Germany: Catena Verlag.
Duley, F. & F. Ackerman 1934. Run-off and erosion from plots of different lengths. *Journal of Agricultural Research* **48**, 505–10.
Everaert, W. 1991. Empirical relations for the sediment transport capacity of interrill flow. *Earth Surface Processes and Landforms* **16**, 513–32.
Fies, J. C. & P. Stengel 1981. Densité texturale de sols naturels II.- Eléments d'interprétation. *Agronomie* **1**, 659–66.
Fletcher, J. E. & E. L. Beutner 1941. *Erodibility investigations on some soils of the Upper Gila Watershed*. United States Department of Agriculture Technical Bulletin 794, 1–31.
Foster, G. R., L. J. Lane, J. D. Nowlin 1980. A model to estimate sediment yield from field-sized areas: selection of parameter values. In *CREAMS: A field-scale model for chemicals, runoff, and erosion from agricultural management systems*, W. G. Knisel (ed.), 193–281. United States Department of Agriculture, Conservation Research Report 26.
Govers, G. 1985. Selectivity and transport capacity of thin flows in relation to rill erosion. *Catena* **12**, 35–49.
Govers, G. 1990. Empirical relationships for the transport capacity of overland flow. *International Association of Hydrological Sciences Publication* **189**, 45–63.
Govers, G. 1991a. A field study of topographical and topsoil effects on runoff generation. *Catena* **18**, 91–111.
Govers, G. 1991b. Rill erosion on arable land in Central Belgium. Rates, controls and predictability. *Catena* **18**, 133–55.
Govers, G. & J. Poesen 1988. Assessment of the interrill and rill contributions to total soil loss from an upland field plot. *Geomorphology* **1**, 343–54.
Govers, G. & G. Rauws 1986. Transporting capacity of overland flow on plane and on irregular beds. *Earth Surface Processes and Landforms* **11**, 515–24.
Govers, G., W. Everaert, J. Poesen, G. Rauws, J. De Ploey, J. P. Lautridou 1990. A long-flume study of the dynamic factors affecting the resistance of a loamy soil to concentrated flow erosion. *Earth Surface Processes and Landforms* **15**, 313–28.
Harmon, W. C. & L. D. Meyer 1978. Cover, slope, and rain intensity affect interrill erosion. *Proceedings Mississippi Water Resources Conference*, 9–16, Mississippi State University.
Hoogmoed, W. B. 1986. Crusting and sealing problems on West African soils. In *Assessment of soil surface sealing and crusting*, F. Callebaut, D. Gabriels, M. De Boodt (eds), 48–55. Belgium: State University of Gent.
Kemper, W. & E. Koch 1966. *Aggregate stability of soils from Western United States and*

Canada. United States Department of Agriculture Technical Bulletin 1355, 1–52.

Lafforgue, A. & E. Naah 1976. Exemple d'analyse expérimentale des facteurs de ruissellement sous pluies simulées. *Cahiers ORSTOM Série de Hydrologie* **13**, 195–237.

Lattanzi, A. R. 1973. Influence of straw-mulch rate and slope steepness on interrill detachment and transport of soil. Unpublished M.Sc. thesis, Purdue University, West Lafayette.

Lavee, H. & J. Poesen 1991. Overland flow generation and continuity on stone-covered soil surfaces. *Hydrological Processes* **5**, 345-60.

Luk, S. H., H. Chen, G. Q. Cai, Z. J. Jia 1989. Spatial and temporal variations in the strength of loess soils, Lishi, China. *Geoderma* **45**, 303–17.

McCool, D. K. & D. K. George 1983. *A second-generation adaptation of the Universal Soil Loss Equation for Pacific Northwest drylands*. American Society of Agricultural Engineers Paper 83-2066.

Mannering, J. V. 1967. The relationships of some physical and chemical properties of soils to surface sealing. Ph.D thesis. Purdue University, West Lafayette.

Martin, W. 1988. Die Erodierbarkeit von Böden unter simulierten und natürlichen Regen und ihre Abhängigkeit von Bodeneigenschaften. Ph.D. thesis, Technischen Universität München.

Meyer, L. D., C. B. Johnson, G. R. Foster 1972. Stone and woodchip mulches for erosion control on construction sites. *Journal of Soil and Water Conservation* **27**, 264–9.

Morgan, R. P. C., J. N. Quinton, R. J. Rickson 1990. Structure of the soil erosion prediction model for the European Community. In *Proceedings International Symposium on Water Erosion, Sedimentation and Resource Conservation, October 1990, Dehradun, India*, 49–59. Central Soil Water Conservation Institute.

Morin, J., Y. Benyamini, A. Michaeli 1981. The effect of raindrop impact on the dynamics of soil surface crusting and water movement in the profile. *Journal of Hydrology* **52**, 321–35.

Mutchler, C. & J. Greer 1980. Effect of slope length on erosion from low slopes. *Transactions of the American Society of Agricultural Engineers* **23**, 866–9, 876.

Nearing, M. A. 1991. A probabilistic model of soil detachment by shallow turbulent flow. *Transactions of the American Society of Agricultural Engineers* **34**, 81–5.

Nearing, M. A., L. J. Lane, E. E. Alberts, J. M. Laflen 1990. Prediction technology for soil erosion by water: status and research needs. *Soil Science Society of America Journal* **54**, 1702–11.

Oldeman, L. R., R. T. Hakkeling, W. G. Sombroek 1990. *World map of the status of human-induced soil degradation*. Wageningen International Soil Reference and Information Centre. Nairobi: United Nations Environment Programme I11.

Parsons, A. J. 1987. The role of slope and sediment characteristics in the initiation and development of rills. In *Processes and measurement of erosion*, A. Godard & A. Rapp (eds), 211–20. Paris: Editions du Centre National de la Recherche Scientifique.

Parsons, A. J., A. D. Abrahams, S.-H. Luk 1991. Size characteristics of sediment in interrill overland flow on a semiarid hillslope, southern Arizona. *Earth Surface Processes and Landforms* **16**, 143–52.

Poesen, J. 1981. Rainwash experiments on the erodibility of loose sediments. *Earth Surface Processes and Landforms* **6**, 285–307.

Poesen, J. 1983. Regenerosiemechanismen en bodemerosiegevoeligheid. Unpublished Ph.D. thesis. K. U. Leuven: Belgium.

Poesen, J. 1984. The influence of slope angle on infiltration rate and Hortonian overland flow. *Zeitschrift für Geomorphologie, Supplementband* **49**, 117–31.

Poesen, J. 1985. An improved splash transport model. *Zeitschrift für Geomorphologie* **29**, 193–211.

Poesen, J. 1986a. Surface sealing as influenced by slope angle and position of simulated stones in the top layer of loose sediments. *Earth Surface Processes and Landforms* **11**, 1–10.

Poesen, J. 1986b. Surface sealing on loose sediments: the role of texture, slope and position of stones in the top layer. In *Assessment of soil surface sealing and crusting*, F. Callebaut, D. Gabriels, M. De Boodt (eds), 354-62. Belgium: State University of Gent.

Poesen, J. 1986c. Field measurements to validate a splash transport model. *Zeitschrift für Geomorphologie, Supplementband* **58**, 81–91.

Poesen, J. 1987a. The role of slope angle in surface seal formation. In *International geomorphology II*, V. Gardiner (ed.), 437–48. Chichester, England: John Wiley.

Poesen, J. 1987b. Transport of rock fragments by rillflow: A field study. *Catena*, Supplement **8**, 35–54.

Poesen, J. 1988a. Surface sealing on sandy and loamy soils: some aspects of seal formation and the influence of sealing on water erosion subprocesses. *Quaderni di Scienza del Suolo* **1**, 9–19.

Poesen, J. 1988b. Surface seal formation and soil erosion subprocesses. In *Proceedings of the International Symposium on Erosion in S.E. Nigeria*, C. I. Ijioma, S. Anaba, T. M. Boers (eds), 60–7. Owerri, Nigeria: Federal University of Technology.

Poesen, J. 1988c. A review of the studies on the mechanisms of incipient rilling and gullying in the Belgian Loam Region. In *Proceedings of the International Symposium on Erosion in S.E. Nigeria*, C. I. Ijioma, S. Anaba, T. M. Boers (eds), 13–21. Owerri, Nigeria: Federal University of Technology.

Poesen, J. 1990. Erosion process research in relation to soil erodibility and some implications for improving soil quality. In *Soil degradation and regeneration in Mediterranean environmental conditions*, J. Albaladejo, M. A. Stocking, E. Diaz (eds), 159–170. Madrid: Consejo Superior de Investigaciones Cientificas.

Poesen, J. & R.B. Bryan 1989. Influence de la longueur de pente sur le ruissellement: rôle de la formation de rigoles et de croûtes de sédimentation. *Cahiers ORSTOM, Série de Pédologie* **25**, 71–80.

Poesen, J. & G. Govers 1986. A field-scale study on surface sealing and compaction on loam and sandy loam soils. Part II. Impact of soil surface sealing and compaction on water erosion processes. In *Assessment of soil surface sealing and crusting*, F. Callebaut, D. Gabriels, M. De Boodt (eds), 183–93. Belgium: State University of Gent.

Poesen, J. & G. Govers 1990. Gully erosion in the loam belt of Belgium: typology and control measures. In *Soil erosion on agricultural land*, J. Boardman, I. D. Foster, J. A. Dearing (eds.), 513–30. Chichester, England: John Wiley.

Poesen, J. & F. Ingelmo-Sanchez 1992. Runoff and sediment yield from topsoils with different porosity as affected by rock fragment cover and position. *Catena* **19** (in press).

Poesen, J. & H. Lavee 1991. Effects of size and incorporation of synthetic mulch on runoff and sediment yield from interrills in a laboratory study with simulated rainfall. *Soil & Tillage Research* **21**, 209–23.

Poesen, J. & D. Torri 1988. The effect of cup size on splash detachment and transport measurements Part I: Field measurements. *Catena, Supplement* **12**, 113–26.

Poesen, J. & D. Torri 1989. Mechanisms governing incipient motion of ellipsoidal rock fragments in concentrated overland flow. *Earth Surface Processes and Landforms* **14**, 469–80.

Poesen, J., F. Ingelmo-Sanchez, H. Mücher 1990. The hydrological response of soil surfaces to rainfall as affected by cover and position of rock fragments in the top layer. *Earth Surface Processes and Landforms* **15**, 653–71.

Rauws, G. & G. Govers 1988. Hydraulic and soil mechanical aspects of rill generation on agricultural soils. *Journal of Soil Science* **39**, 111–24.

Römkens, M. J., S. Prasad, J. Poesen 1987. Soil erodibility and properties. *Transactions of the 13th Congress of the International Society of Soil Science* **5**, 492–504.

Römkens, M. J., S. N. Prasad, F. D. Whisler 1990. Surface sealing and infiltration. In *Process Studies in hillslope hydrology*, M. G. Anderson & T. P. Burt (eds), 127–172. Chichester, England: John Wiley.

Savat, J. & J. De Ploey 1982. Sheetwash and rill development by surface flow. In *Badland geomorphology and piping*, R. Bryan & A. Yair (eds), 113–26. Norwich, England: Geo Books.

Sharma, K. D. 1986. Runoff behaviour of water harvesting microcatchments. *Agricultural Water Management* **11**, 137–44.

Sharma, K.D., H. Singh, O. Pareek 1983. Rainwater infiltration into a bare loamy sand. *Hydrological Sciences Journal* **28**, 417–24.

Shaw, C. 1929. Erosion pavement. *Geographical Review* **19**, 638–41.

Shirazi, M. & L. Boersma 1984. A unifying quantitative analysis of soil texture. *Soil Science Society of America Journal* **48**, 142–7.

Simanton, J. R., E. Rawitz, E. D. Shirley 1984. The effects of rock fragments on erosion of semiarid rangeland soils. In *Erosion and productivity of soils containing rock fragments*. Soil Science Society of America, Special Publication 13, 65–72.

Soil Science Society of America 1984. *Glossary of soil science terms*. Madison, Wisconsin.

Styczen, M. & S. A. Nielsen 1989. A view of soil erosion theory, process-research and model building: possible interactions and future developments. *Quaderni di Scienza del Suolo* **2**, 28–45.

Thornes, J. B., C. F. Francis, F. Lopez Bermudez, A. Romero Diaz 1990. Reticular overland flow with coarse particles and vegetation roughness under Mediterranean conditions. In *Strategies to combat desertification in Mediterranean Europe*, J. L. Rubio & R. J. Rickson (eds), 228–43. Commission of the European Communities, Directorate-General Agriculture, EUR 11175.

Torri, D. & J. Poesen 1988a. The effect of cup size on splash detachment and transport measurements Part II: Theoretical Approach. *Catena, Supplement* **12**, 127–37.

Torri, D. & J. Poesen 1988b. Incipient motion conditions for single rock fragments in simulated rill flow. *Earth Surface Processes and Landforms* **13**, 225–37.

Torri, D. & J. Poesen 1992. The effect of soil surface slope on raindrop detachment. *Catena* (in press).

Torri, D., R. Biancalani, J. Poesen 1990. Initiation of motion of gravels in concentrated overland flow: cohesive forces and probability of entrainment. *Catena, Supplement* **17**, 79–90.

Torri, D., M. Sfalanga, G. Chisci 1987. Threshold conditions for incipient rilling. *Catena, Supplement* **8**, 97–105.

Valentin, C. 1986. Surface crusting of arid sandy soils. In *Assessment of soil surface sealing and crusting*, F. Callebaut, D. Gabriels, M. De Boodt (eds), 40–47. Belgium: State University of Gent.

Verhaegen, T. 1984. Labo-experimenten en terreinwaarnemingen in verband met de erosiegevoeligheid van lemige bodems. Unpublished Ph.D. thesis. K. U. Leuven, Belgium.

Wischmeier, W. H. & D. D. Smith 1978. *Predicting rainfall erosion losses – A guide to conservation planning*. United States Department of Agriculture Agricultural Handbook

537.

Yair, A. & H. Lavee 1976. Runoff generative process and runoff yield from arid talus mantled slopes. *Earth Surface Processes* **1**, 235–47.

Young, R. A. & C. A. Onstad 1986. The effect of erosive forces on surface seal development. In *Assessment of soil surface sealing and crusting*, F. Callebaut, D. Gabriels, M. De Boodt (eds), 72–80. Belgium: State University of Gent.

13

Field investigations of sediment removal in interrill overland flow

Anthony J. Parsons & Athol D. Abrahams

Abstract

In this chapter we report on a series of field investigations of sediment removal in interrill overland flow conducted on runoff plots established at Walnut Gulch Experimental Watershed, Tombstone, Arizona. These investigations were undertaken to examine the roles of surface flow and rainfall in detaching sediment in interrill areas and to evaluate the limits to rates of sediment transport by interrill flow. Detachment by surface flow appears to be minimal, even though calculations indicate that natural rainstorms with a return period of less than 2 years generate surface flow capable of detaching sediment of size comparable to that being transported by the flow. This apparent paradox, we argue, is due to the dissipation of more than 90% of the power of the flow in overcoming form resistance. In contrast, detachment by raindrops is substantial but exhibits significant across-slope and downslope variation. This spatial variation has important implications for the rate of sediment transport. Away from the near-divide area, where it is presumed to be controlled by the transport capacity of the surface flow, the rate of sediment transport is less than the detachment rate. Only part of the detached sediment is removed by surface flow. The notion that, for much of the interrill area, the rate of sediment removal is equal to the detachment rate is an oversimplification that leads to an overestimate of the rate of erosion. For those parts of the interrill area where the rate of sediment transport is controlled by the capacity of the surface flow, our experiments show that this rate does not necessarily increase with gradient, as has been widely assumed. Relations between gradient and infiltration may confound any simple relation between gradient and transport capacity.

Introduction

In his study of soil erosion, Ellison (1947) identified four elements constituent to the process. Two agents, rainfall and surface flow, may each effect the two mechanisms of detachment and transportation. This conceptual framework has underpinned many models of sediment removal in interrill overland flow (e.g. Meyer & Wischmeier 1969, Rowlinson & Martin 1971). Investigations of the roles of the four elements (e.g. Young & Wiersma 1973, Quansah 1985), however, have yielded information on their relative importance, with the result that many models of interrill erosion now simplify the process to one in which rainfall is responsible for detachment (e.g. Rowlinson & Martin 1971, Foster & Meyer 1972) and surface flow alone is responsible for transportation (e.g. Hartley 1987).

In recent years we have undertaken a number of field investigations of sediment removal in interrill overland flow, the aim of which has been to elucidate the physical controls on detachment and transportation by rainfall and surface flow. Better understanding of these controls, we argue, will result in improved models for soil erosion by interrill overland flow. In some instances, our field investigations provide a physical explanation for some of the commonly accepted beliefs about the mechanics of sediment removal in overland flow; in others they suggest that commonly accepted beliefs are incorrect. This chapter summarizes and draws together several ideas that separately have been more fully expressed elsewhere (Abrahams et al. 1988, 1989, 1991, Parsons et al. 1990, 1991, 1992, Abrahams & Parsons 1991b, Parsons & Abrahams 1992). Our aim here is to present a synthesis of these ideas and to demonstrate their significance for modelling sediment removal in interrill overland flow. Specifically, we examine the roles of rainfall and overland flow in detaching sediment in interrill areas and evaluate limits to rates of sediment transport by interrill flow.

Study area and field methods

All the field experiments that are described in this paper were conducted on runoff plots established at Walnut Gulch Experimental Watershed, Tombstone, Arizona (31°43′N, 110°41′W). One of these plots, hereafter termed the large plot, was 18 m wide and 35 m long and encompassed an entire interrill portion of a hillslope, whereas the others, collectively termed small plots, were of two sizes; 1 m wide by 2.3 m long, and 1.8 m wide by 5.5 m long. On the large plot, gradient ranged from 0° at the divide to 4.5° at the outlet. The small plots, each more or less rectilinear, ranged in gradient from 1.3° to 33°.

The watershed has a warm semi-arid climate with a mean annual precipitation of 288 mm and a mean monthly temperature range of 8–27°C (Osborn 1983). Much of the watershed consists of dissected piedmont alluvium that is coarse, well cemented and of Quaternary age (Gilluly 1956, Gelderman 1970, Libby et al. 1970). Elsewhere, particularly around the margins, it is underlain by a mixture of metamorphic and igneous rocks. The large plot and most of the small plots were located within the area of dissected alluvium. Large areas of the watershed, including all locations of runoff plots, are vegetated by a sparse desert-shrub community. The shrub species vary from one location to another, but typically, in total, cover less than half the ground surface. Between these shrubs there is often a thin ground layer of forbs and grasses. The vegetation of the large plot included both shrubs and ground-layer species. The small plots, however, were deliberately sited between shrubs. Some of the plots were devoid of all vegetation but ground-layer forbs and grasses were present on others.

For all plots runoff was generated using simulated rainfall. For the large plot this rainfall was delivered from lines of sprinklers arrayed across the plot and supported 4 m above the ground on 31 vertical stand-pipes (Fig. 13.1). The rainfall had a near-uniform distribution over the plot, an intensity of 80 mm h^{-1}, a median drop size of 1.70 mm and provided 75% of the kinetic energy of natural rainfall at the same intensity. For the small plots the rainfall simulators were of the type described by Luk et al. (1986). In the configuration used on these plots, the simulators were designed to deliver rainfall at an intensity of 145 mm h^{-1}, with median drop size of 2.4 mm, and to provide 90% of the kinetic energy of natural rainfall at the same intensity.

The large plot

Within the interrill portion of the large plot, two measured sections (S1, S2) were established 12.5 m and 21 m, respectively, from the upper boundary (Fig. 13.1). At these sections a number of specially designed miniature flumes were installed which permitted the sampling of interrill overland flow with minimal disturbance to the flow. Details of the design and use of these flumes have been given elsewhere (Parsons & Abrahams 1989). Eight such miniature flumes were installed along S1, and 10 along S2. Along both sections the flumes were positioned so as to sample the range of overland-flow depths. Flumes of three widths were used (10, 20 and 40 cm), the size of flume installed at any location being determined by the anticipated overland flow discharge. The narrowest flumes were used where discharge was expected to be greatest so that the fill times of the sample bottles would not be too short. In addition to the flumes, S1 and S2 were equipped with thin rods spaced at 0.5 m intervals. Access to the rods and miniature flumes without stepping on the plot surface was gained via small metal gratings supported about 10 cm above the plot surface on narrow legs and placed just downslope of S1 and S2.

Down slope of S2, part of the overland flow became concentrated into a rill, through which it reached a supercritical flume located at the outlet of the plot (OF). The remaining interrill runoff was trapped in gutters that formed tapered lower boundaries of the plot (Fig. 13.1). Flow from these gutters was directed into calibrated flumes (LF and RF), to which automatic stage recorders were attached. However, as a result of local microtopography and the location of the boundary of the plot, much of the flow into the right hand gutter also

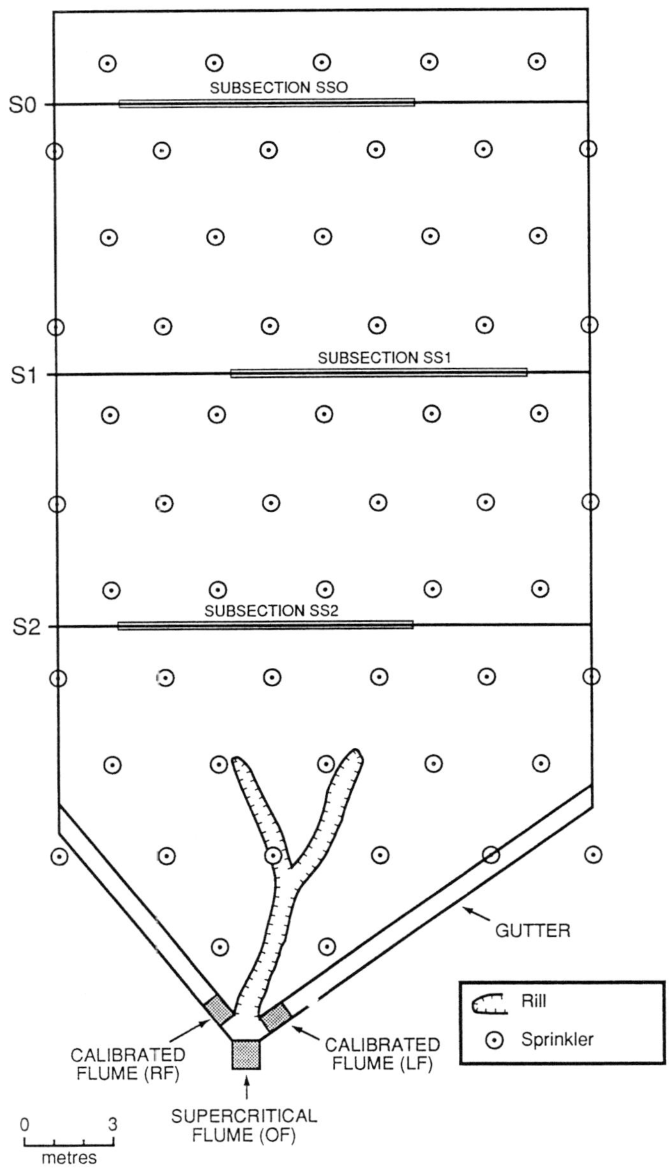

Figure 13.1 Plan of the large plot showing the locations of sprinklers, flumes, sections S0, S1 and S2, and subsections SS0, SS1 and SS2.

became sufficiently concentrated as to create a rill. Consequently, only the flow through LF can be truly regarded as interrill flow and comparable to that at S1 and S2.

Four rainfall simulation experiments (E1, ..., E4) were performed on the plot, lasting respectively for 10, 20, 10 and 30 min, the intervals between successive experiments being 48, 1 and 72 h. At the design rainfall intensity, these four experiments correspond to rainstorms with a return period of 3, 5, 3 and 10 years, respectively (Osborn & Renard 1988). During experiments flow depths adjacent to the rods at S1 and S2 were measured more-or-less simultaneously at 5 min intervals, beginning in minute 4. In the intervening periods, discharges passing over the lips of the miniature flumes, through the calibrated flumes and through the supercritical flume were sampled as often as possible by directing the water into 1 l or 2 l sample bottles. The hydrographs for S1, S2 and OF indicate that equilibrium was attained by minute 19 of experiments 2 and 4. Such conditions, therefore, have a return period of about 5 years at Walnut Gulch (Osborn & Renard 1988).

Estimation of sediment loads at S1 and S2 Sediment loads at S1 and S2 were estimated by multiplying the frequency distribution of flow depth at the section by a depth–sediment load rating equation. Frequency distributions of the flow depths measured at 5 min intervals at the rods along S1 and S2 (Fig. 13.2) reveal that all but five of the 28 distributions were adequately fitted by the negative exponential density, which was, therefore, adopted as a model for the distribution of flow depths. An attractive feature of this model is that it is characterized by a single parameter, which is the reciprocal of the mean (Abrahams et al. 1989). Thus it was possible to estimate, from the mean depth of flow measured at the rods, the distribution of flow depths across each section at 5 min intervals during the experiments.

Separate flow depth–sediment load rating equations for S1 and S2 were determined for each 5 min interval, centred on the 4th, 9th, 14th, 19th, 24th and 29th minutes of each experiment. Early efforts to compute these rating equations by least-squares regression yielded section sediment loads that were inconsistent and manifestly too large. This occurred because the equations usually overestimated the sediment loads in the shallowest flows. Inasmuch as the depth distributions were negative exponential, overestimation of the loads in the shallow flows often resulted in gross overestimation of the section sediment load. The most consistent section sediment loads were obtained when sediment load was plotted against flow depth on arithmetic graph paper, and the rating equation was forced to pass through the origin (Abrahams et al. 1991).

Estimation of sediment size characteristics at S1, S2 and LF Particles eroded during runoff events may be in the form of primary particles or aggregates. However, very sandy soils, such as those of the large plot (79.3% sand, 13.7% silt and 7.0% clay), are not well aggregated, and so tend to erode mostly as primary particles (Young 1980). Under these circumstances a laboratory determination of the dispersed sediment size distribution may be taken to be

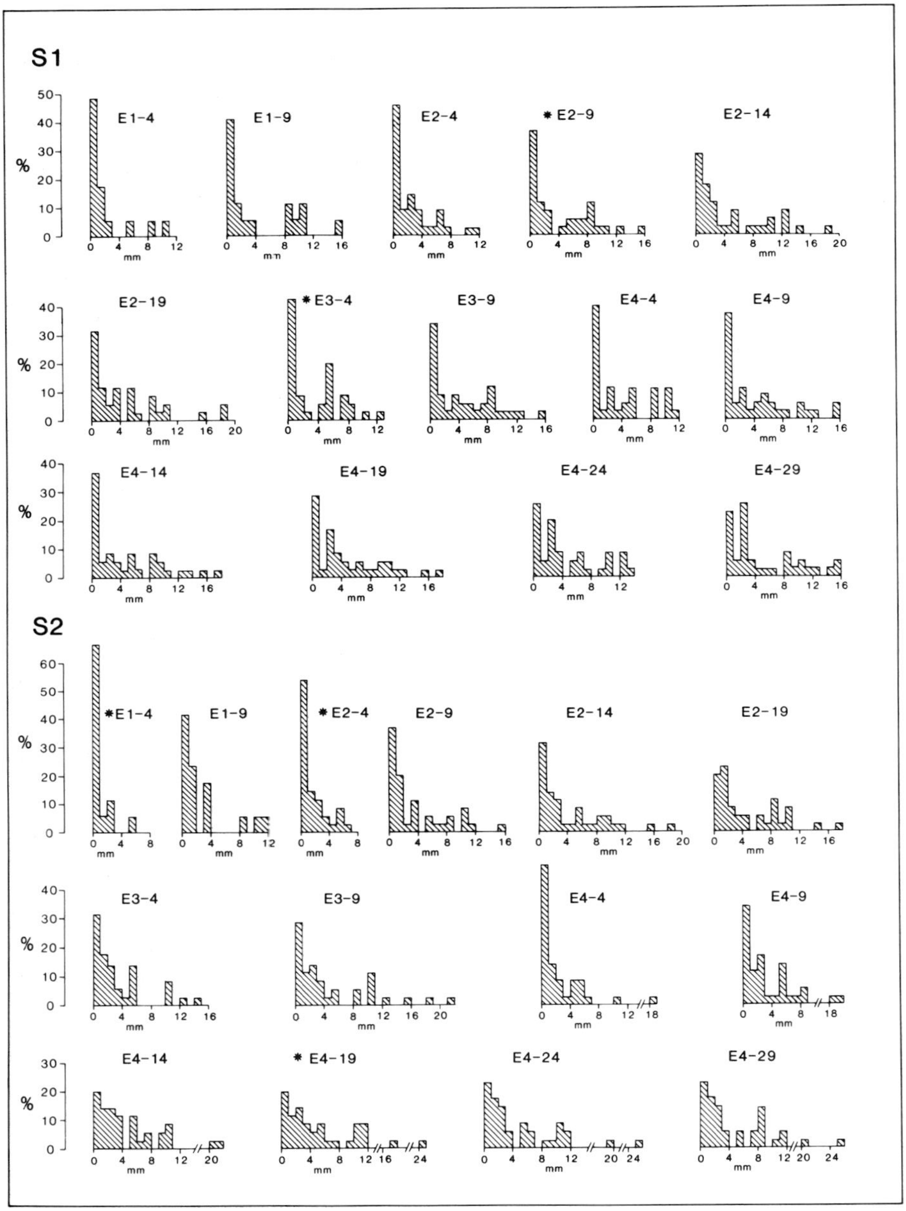

Figure 13.2 Distribution of depths of overland flow recorded at S1 and S2 on the large plot. Each distribution is labelled with the experiment number and the minute into the experiment when the depths were measured. Distributions that cannot be fitted by the negative exponential density are denoted by *.

equivalent to a field determination of the size distribution of the particles as they are eroded. Details of the laboratory determination used are given in

Parsons et al. (1991). It is sufficient to note here that because the samples collected from individual miniature flumes represent sediment carried in only part of the range of overland flow depths at S1 and S2, samples from all flumes at each of these sections were pooled to give an estimate of the average size characteristics of the sediment passing each of the sections. In addition, because no attempt was made to acquire samples from all miniature flumes and from LF simultaneously, individual samples were pooled to yield composite samples of sediment passing through S1, S2 and LF for each 5 min period during each of the four experiments. Finally, even after pooling the individual samples as described, some of the composite samples were too small to give a reliable estimate of percentages of sand, silt and clay, for which we deemed a minimum of 10 g to be necessary. In these cases, samples from contiguous 5 min periods within experiments were again pooled. In this way, with the exception of LF during E1 and minutes 5–9 of E2 for which the samples weighed only 8.0 and 9.1 g, respectively, samples were obtained for use in the analysis that were all in excess of 10 g.

The small plots

Thirty-minute rainfall simulation experiments were conducted on a total of 22 small plots. These plots may be divided into three types T1, ..., T3. Plots of T1 and T2 were 1.8 m by 5.5 m and those of T3 were 1 m by 2.3 m. T1 and T2 plots were underlain by Quaternary alluvium, whereas T3 plots were located in an area of metamorphosed, interbedded limestones and shales. With the exception of the latter parts of experiments on T1 plots, when the sampling intervals were longer, timed volumetric samples of surface runoff from the plots were obtained every minute. From these samples, sediment concentration was determined gravimetrically, and the runoff coefficient (i.e. total surface runoff expressed as a percentage of total rainfall) and sediment yield were computed for the 30 min simulated rainstorm. For the T1 plots the ground-layer forbs and grasses were clipped prior to the experiments, whereas for those of T2 similar vegetation was left undisturbed. Plots of T3 were devoid of vegetation.

Detachment by surface flow

Although the threshold conditions under which flowing water initiates movement of granular materials has been the subject of considerable study, for the purposes of predicting detachment by interrill overland flow much of this research has two significant shortcomings. First, it is largely empirical so that the capability to predict outside the range of existing experimental data is limited (James 1990). Secondly, most of the studies have been undertaken under conditions of turbulent flow and where the flow depth is large compared to the diameter of transported particles. In consequence, only limited information is applicable to interrill overland flow, which is generally shallow and may be laminar or turbulent. In a series of laboratory experiments, Govers (1987) demonstrated good agreement between his results for turbulent overland flow and Yang's (1972) unit stream power criterion of initiation of sediment movement (Fig. 13.3a). Furthermore, although Govers pointed to the impossibility of using a criterion based upon mean velocity of flow when the flow is laminar, he did obtain for laminar flow a relation between critical unit stream power and grain diameter (Fig. 13.3b).

On the large plot the median sizes of sediment transported through S1 and S2 averaged over all experiments are 75 μm and 98 μm, respectively. Using Govers' experimental results, coupled with the data from Table 13.1, it is evident that, regardless of whether overland flow is assumed to be laminar or turbulent, the minimum recorded flow is capable of initiating movement of the

Table 13.1 Flow hydraulics at S1 and S2 at minute 4 of E1 (minimum recorded values).

	S1	S2
Slope (sine)	0.0524	0.0612
Mean depth of flow (mm)	4.23	2.64
Mean velocity of flow (cm s^{-1})	3.84	4.77
Unit stream power (mm s^{-1})	2.01	2.92
Critical velocity to entrain sediment of median size (turbulent flow) (cm s^{-1})	2.24	1.83

median size of transported sediment. Inasmuch as a 4 min rainstorm at an intensity of 80 mm h^{-1} has a return period at Walnut Gulch of well under 2 years (Osborn & Renard 1988), these calculations, based upon the work of Govers (1987), seem to imply that flow detachment should be exceedingly common on this hillslope.

This implication receives further support from the analysis by Abrahams et al. (1989) of the significance of the distribution of overland-flow depths for flow detachment. Abrahams et al. showed that where overland flow has a negative exponential distribution, as is the case for S1 and S2 (Fig. 13.2), taking this distribution into account rather than employing mean depth will, under most circumstances, lead to higher estimates of the rate of flow detachment.

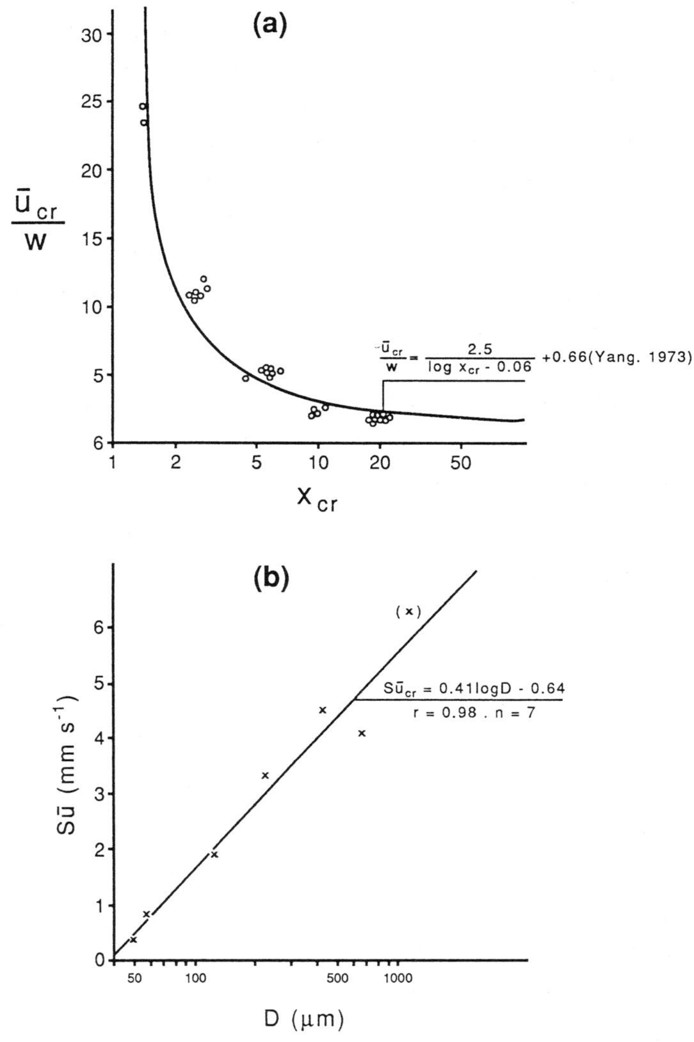

Figure 13.3 (a) Fit of experimental data for overland flow to Yang's equation relating critical dimensionless velocity ($\bar{u}_{cr}/w$) to shear velocity particle Reynolds number (X_{cr}). (b) Critical unit stream power ($S\bar{u}_{cr}$) as a function of grain diameter (D). (After Govers 1987.)

In contrast to the expectations of rates of flow detachment implied by Govers' experiments and the negative exponential depth distribution, are qualitative observations of the large plot which give little credence to the notion of significant detachment by overland flow. On this plot, because almost all of the overland flow occurs between the shrubs and because these shrubs are several decades in age, the overland flow lines must also have remained fixed for several decades. If significant flow detachment had occurred, then flowlines should have incised significantly over such a period of time. However, there are no signs of any such incision on the plot. These observations are supported by several field and laboratory studies which have concluded that in gentle interrill areas most sediment is detached mainly by raindrops and that flow detachment is a minor contributor to total sediment removal (e.g. Borst & Woodburn 1942, Ellison 1945, Young & Wiersma 1973, Walker et al. 1977, Quansah 1985). The discrepancy between the expected rate of flow detachment deduced from the experimental work of Govers and the observed distribution of overland-flow depths could be reconciled if it were assumed that the quantity of sediment removed by the runoff were limited by its transport capacity. However, such a possibility is unlikely, given the results of Gilley et al. (1985a, b). These authors used laboratory experiments coupled with a model of overland-flow processes to calculate that for a clay loam soil sediment transport capacity would be the limiting quantity over distances of 17 m and 11 m on 1% (0.6°) and 6% (3.4°) slopes, respectively. Given the size and gradient of the runoff plot and the locations of S1 and S2 within it, it seems improbable that sediment transport capacity limits erosion on all but the uppermost portion that does not include S1 and S2. In consequence, some other explanation is called for.

A probable explanation for the apparently low rates of flow detachment is that not all the power or shear stress of the flow may be available to mobilize particles on the ground surface. It is widely accepted (e.g. Einstein 1950, Singhal et al. 1980, Carson 1987) that in river flow sediment transport capacity is controlled not by total shear stress τ, but by τ':

$$\tau = \tau' + \tau'' \tag{13.1}$$

where τ' is grain shear stress and τ'' is form shear stress. Foster (1982) proposed that the same is true in overland flow, and Govers & Rauws (1986) and Govers (1988) presented evidence supporting this proposition. If this proposition is accepted, then because

$$\tau = \rho g d s \tag{13.2}$$

where ρ is fluid density, g is the acceleration due to gravity, d is flow depth and s is the energy slope, and because

$$gds = \frac{fv^2}{8} \tag{13.3}$$

where f is the Darcy–Weisbach friction factor and v is mean flow velocity, and assuming that grain resistance f' and form resistance f'' are additive (Einstein & Banks 1950, Einstein & Barbarossa 1952) and that there are no other sources of hydraulic roughness, that is

$$f = f' + f' \tag{13.4}$$

then

$$\tau = \tau' + \tau'' = \frac{\rho f' v^2}{8} + \frac{\rho f' v^2}{8}. \tag{13.5}$$

From Equation 13.5 it can be seen that grain shear stress expressed as a percentage of total shear stress $\%\tau'$ is equal to grain resistance expressed as a percentage of total resistance $\%f'$. Abrahams & Parsons (1991a) showed that for the intershrub areas of the large plot, where almost all of the runoff occurs, $\%f'$ is typically less than 10%, and so, therefore, is $\%\tau'$. The dissipation of a large part of the power of overland flow in overcoming form resistance therefore appears to provide an explanation for the apparently low rate of flow detachment, given the observed flow depths and velocities.

Although this result applies only to the special case of desert pavement surfaces where form resistance is largely due to coarse gravel mantling the surface, in all probability the result is qualitatively correct for humid hillslopes and agricultural lands where plant stems, litter and tilled surfaces generate form resistance and ensure that grain resistance remains relatively small. In consequence, the explanation presented here for the absence of flow detachment by interrill overland flow on the large plot most likely applies to many surfaces on which interrill overland flow occurs.

Detachment by rainfall

Soil detached by rainfall comprises that which is simply dislodged by the impact of falling raindrops as well as that which is carried in splash-crown droplets. In both cases, the rate of detachment depends upon the raindrop impact stress. Several studies have attempted to relate rates of detachment to characteristics of the rainfall, the soil and the depth of the surface water layer which either control raindrop impact stress or determine its effects. With the exception of Palmer (1964), most studies have reported that impact stress and, consequently, soil loss occur when the surface-water layer is very thin or absent (Mutchler & Young 1975, Ghadiri & Payne 1981, Torri et al. 1987). Inasmuch as all of these studies have been concerned solely with the effect of a water layer on the volume of soil carried in the splash-crown droplets rather than the total rate of detachment, which includes the soil that is simply dislodged by raindrop impact and is several times the splash rate, they do not actually demonstrate the significance of the surface-water layer for the rate of detachment. However, these findings are supported by Shultz et al. (1985), who have shown in laboratory experiments that the total detachment rate is, like the rate of splash, influenced by the depth of the surface-water layer.

Given the significance of the depth of the water layer for raindrop impact stress, it is clear that for a given rainfall intensity and soil type, spatial variation in detachment due to raindrops will occur in response to spatial variation in depths of overland flow. However, the effect of such spatial variation has, for the most part, been considered only in terms of downslope variation in detachment rate in response to a downslope increase in depth of overland flow (e.g. Gilley et al. 1985a, b). Given the evidence of across-slope variation in depth of overland flow (Fig. 13.2 and Abrahams et al. 1989), it is also apparent that significant across-slope variations in the rate of detachment must occur. The scale and importance of this variation was measured on the large plot in the following way.

Shortly after the completion of E4 the rainfall was restarted and, once equilibrium conditions had been attained, depth of flow was measured every 5 cm along 10 m subsections of S1 and S2, hereafter denoted as SS1 and SS2. In addition, the same measurements were obtained from a similar 10 m subsection of a section S0 located 3.5 m from the top of the plot, denoted SS0 (Fig. 13.1). Each of these depth measurements was taken as representative of a 5 cm length of the subsection centred on the point at which the measurement was made. Based upon the findings of Schultz et al. (1985, p.1883, Fig. 8 and Eqs 8 & 9) but modified for simplicity, we will assume that splash occurs equally over areas where the depth of the surface-water layer (d) $\leq$ 2 mm, but that there is zero splash where d > 2 mm; and that dislodgement occurs equally over areas where $d \leq$ 2mm, at 0.063 of this rate where d = 3 mm, at 0.028 of the rate where d = 4 mm, and is zero for greater values of d. Also, based upon the work of Schultz et al., we will assume that where $d \leq$ 2 mm the rate of dislodgement is four times the rate of splash. Using these values, the distribution of raindrop detachment at SS0, SS1 and SS2 is calculated to be as shown in Figure 13.4 and the cumulative amounts of detached soil per unit time (assuming SS0 to have a value of 100) are 100, 83.8 and 87.2 for SS0,

SS1 and SS2, respectively.

Many field studies have acknowledged that interrill overland flow is not a layer of water of uniform depth (e.g. Emmett 1970, Morgan 1980, Abrahams et al. 1986b). Models of soil erosion, however, have largely ignored this spatial variability in overland-flow depth and failed to consider its significance

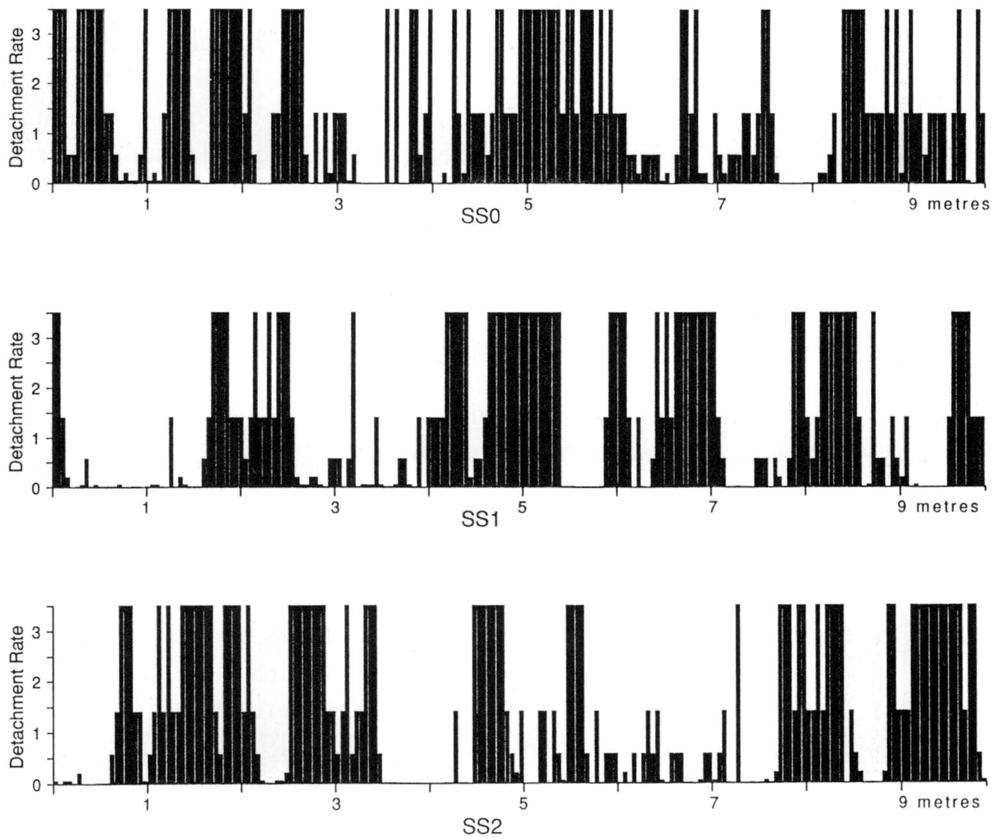

Figure 13.4 Calculated rates for detachment by raindrops along SS0, SS1 and SS2.

for rates of detachment. Our results for detachment on the large plot reveal both an across-slope variation in rates of detachment in response to across-slope variation in overland-flow depth and, due to the downslope concentration into threads of deeper flow (Parsons et al. 1990), a downslope variability in detachment rate. Detachment rate does not simply remain constant (as is generally assumed) nor does it systematically decrease down slope (as assumed by Gilley et al. 1985a, b). Inasmuch as the rate of detachment has often been considered to control the rate of interrill erosion, spatial variability in the former may imply complexity in the latter. This is discussed further below.

Transport

Inasmuch as rainsplash is a minor contributor to sediment transport, the rate of sediment transport is almost wholly a function of the rate at which overland flow transports sediment. This rate is thought to be limited either by the transport capacity of the flow – that is, it is transport-limited – or by the rate at which detached sediment is supplied to the flow – that is, it is detachment-limited. Because sediment-transport capacity is generally assumed to be zero at the divide and to increase down slope with discharge, whereas detachment capacity is assumed to be greater than zero at the divide and either to remain constant or to decrease down slope, it has been widely accepted that the interrill area can be divided into two portions. Close to the divide, erosion is controlled by the transport capacity, and further down slope the limit becomes the detachment capacity. We have examined rates of sediment transport on these two portions of the interrill zone using data on sediment removal from our small plots and our large plot, respectively.

The near-divide "transport-limited" section

Our investigation of the transport-limited portion of hillslopes focuses on the role of gradient as a determinant of the rate of erosion. The role of gradient was expounded by Gilbert (1877) who proposed that "in general erosion is most rapid where the slope is steepest; ... With increase in slope goes increase in the velocity of running water, and with that goes increase in its power to transport undissolved detritus" (p. 96). Gilbert's proposition has become virtually an axiom of geomorphology and is a standard feature of those modern process–response models of hillslope development by overland flow that assume erosion to be transport limited (e.g. Kirkby 1971, Band 1985), and is supported by the work of agricultural engineers who have sought to quantify the relation between the rate of overland-flow erosion and gradient using data collected from field and laboratory plots (e.g. Neal 1938, Zingg 1940, Musgrave 1947, Smith & Wischmeier 1957, Wischmeier & Smith 1978, McCool et al. 1987). The best known of these relations is that for the slope steepness factor in the Universal Soil Loss Equation (USLE). According to the USLE handbook (Wischmeier & Smith 1978), this relation may be applied throughout the United States to hillslopes with gradients up to 50% (26.6°). However, the relation is based in data collected from soil-covered hillslopes located exclusively in humid climates with gradients ≤ 25% (14°) (Smith & Wischmeier 1957).

The proposition is challenged by the findings of Yair & Klein (1973) who, in an investigation of natural runoff events on three debris-covered hillslopes with gradients of 15°, 19° and 25° in arid southern Israel, found an inverse relation between sediment yield and gradient. They attributed this relation to the fact that as gradient increases, debris size and surface roughness increase, causing infiltration to increase and runoff to decrease. As the decrease in runoff more than compensates for the increase in gradient, sediment yield decreases. Yair & Klein's work was significant for two reasons: (a) it showed that the relation between sediment yield and gradient for debris-covered arid hillslopes is different from the conventional one for soil-covered humid hillslopes, and (b) it suggested that the difference stems from different relations between gradient and runoff. As important as Yair & Klein's research was, it revealed only part of the sediment-yield–gradient relation for debris-covered hillslopes. Clearly, their inverse relation cannot be extrapolated to lower gradients, for to do so would imply that maximum sediment yield occurs at zero gradient, which is absurd!

We have investigated the relation between sediment yield and gradient for transport-limited portions of the interrill areas of debris-covered hillslopes, using data obtained from our small-plot experiments (Abrahams et al. 1988, Abrahams & Parsons 1991b). Although most of these plots were not physically located close to the divide, during the experiments they were rendered effectively so because no runoff was able to enter the plots from upslope of their upper boundaries. Graphs of the relations between sediment yield and gradient for T1, T2 and T3 are presented in Figure 13.5a. Each type shows essentially the same relation between sediment yield and gradient. For hillslopes up to about 12° there is an increase in sediment yield with gradient, but for gradients steeper than this there is a decrease in sediment yield with gradient, resulting in a convex-upward relation with a vertex at about 12°. Although these results are from only two substrates, they suggest that a convex-upward relation may apply to many debris-covered hillslopes on a variety of substrates.

The key to understanding this relation is the relation between runoff coefficient and gradient (Fig. 13.5b). For all the small plots, the runoff–gradient relation has basically the same shape – that is, there is a gradual increase in runoff as gradient increases up to about 12° and a rapid decrease in runoff as gradient increases beyond 12°. To identify the factors that control runoff and hence the shape of the runoff–gradient relation, stepwise multiple regression analyses were performed for the three types of plot, with runoff coefficient as the dependent variable, and mean surface particle size, % surface particles ≥ 2mm, % surface particles ≥ 5mm, % vegetation cover, % sand in the surface soil and % clay in the surface soil as the independent variables. In all cases mean surface particle size or % surface particles ≥ 5 mm entered the equation, explaining an average of 51.9% of the variance in runoff. These results imply that the shape of the runoff–gradient relation, and, in particular, the rapid decrease in runoff as gradient increases beyond 12°, are

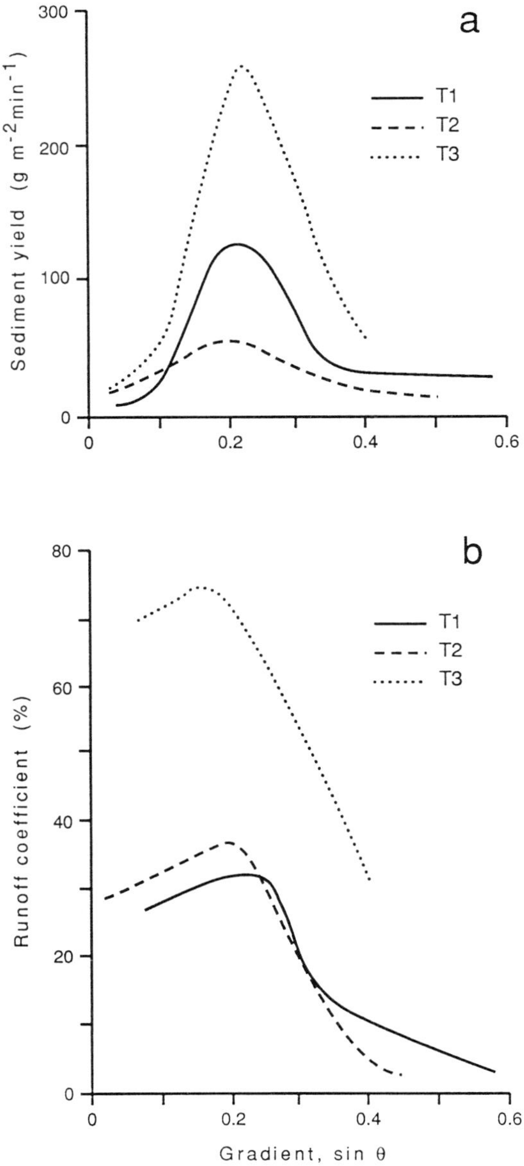

Figure 13.5 Graphs of (a) sediment yield, and (b) runoff coefficient against gradient for the three types of plot denoted by T1, T2 and T3, respectively.

contingent upon the presence of a coarse weathering mantle in which stoniness increases with gradient.

The results obtained in these experiments, taken in conjunction with those of Yair & Klein (1973), undermine the simple notion of an increase in sediment yield with gradient on transport-limited parts of interrill areas. Such a relation may only apply on hillslopes where there is no significant relation of infiltration, and hence runoff, with gradient. In these cases the increase in the downslope component of gravity with gradient will ensure that sediment yield increases with gradient. Where, however, infiltration and gradient are positively related, this relation may be sufficient to outweigh the relation between the downslope component of gravity and gradient.

The "detachment-limited" section

The notion that downslope of the "transport-limited" portion of hillslopes the rate of erosion is detachment limited was investigated by analyses of sediment loads at S1 and S2 and of sediment calibre at S1, S2 and LF on the large plot.

Downslope variation in section soil loss The average values of sediment load passing through S1 and S2 under equilibrium conditions are 8.18 and 17.98 g m^{-1} min^{-1} for S1 and S2, respectively (Abrahams et al. 1991). Under equilibrium conditions, water and sediment passing through S1 and S2 are the accumulated runoff and sediment contributed by the areas up slope of these cross sections. In order to determine the downslope pattern in the rate of erosion, it is necessary to determine the pattern of change of section soil loss, *G,* that gives rise to this pattern of accumulated sediment load. If it is assumed

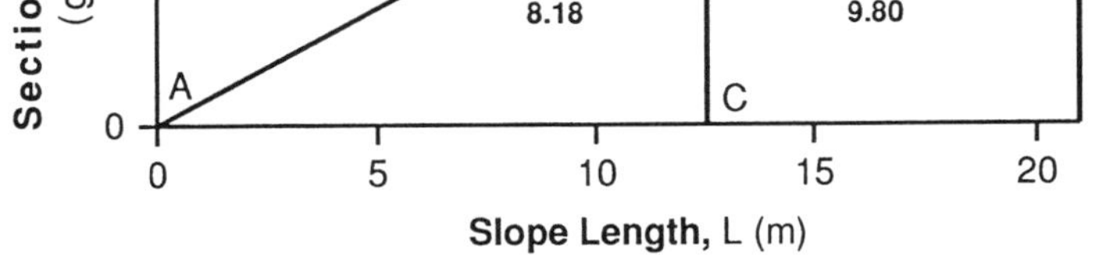

Figure 13.6 Graph showing the relation between computed average sediment loads at S1 and S2 under equilibrium runoff conditions and Equations 13.6 & 13.7.

that soil loss at the divide is zero, then the two values for the accumulated soil loss (8.18 and 17.98 g m^{-1} min^{-1}) can be represented by the areas ABC and ABDE in Figure 13.6. Consequently (Abrahams et al. 1991), the pattern of variation in section soil loss, *G*, with distance downslope, *L*, for the area between the divide and S1 is given by the equation

$$G = 0.105L \tag{13.6}$$

and between S1 and S2 it is given by

$$G = 1.89 - 0.44L. \tag{13.7}$$

Equations 13.6 & 13.7 indicate that although G increases down slope between the divide and S1, it decreases between S1 and S2. It is improbable that between S1 and S2 the erosion rate is controlled by the transport capacity of the flow because, as noted above, S1 and S2 are well downslope of any portion of the interrill area where this control might be expected to operate. Furthermore, the decrease in G between S1 and S2 is quite inconsistent with such a control. Equally, the decrease in G between S1 and S2 is inconsistent with the modelled rates of detachment at S1 and S2 which gave a higher value at S2 than at S1. Thus variation in the rate of detachment, as proposed in the model developed by Gilley et al. (1985a, b), also seems unable explain the observed pattern of soil loss.

Size characteristics of transported sediment The sediment collected at the miniature flumes at S1 and S2 and that passing through LF (average 52.4% sand, 28.0% silt and 19.6% clay) is, on average, finer than the matrix soil (average 79.3% sand, 13.7% silt and 7.0% clay). t-tests on the percentages of sand contained within the samples from which these average were obtained (Parsons et al. 1991) confirm that these percentages would not be expected in samples coming from the same population as the four samples of matrix soil ($p = 0.007$, $p = 0.01$ and $p = 0.004$ for S1, S2 and LF, respectively). That the interrill sediment is finer than the matrix soil implies, therefore, that either there is selective detachment by raindrops or selective transport by the interrill flow. Data on sediment carried in splash droplets (Parsons et al. 1991, 1992) show this sediment to be coarser than the matrix sediment found on the plot. These data, which show a tendency for sand-sized particles to be preferentially transported by rainsplash, accord with the results obtained by Ellison (1944) over a range of rainfall characteristics, and those of Poesen & Savat (1981), who investigated a range of soil types. Of particular importance is the fact that all these studies yield no evidence to indicate that the splashed sediment may be finer than the matrix soil of the runoff plot. Inasmuch as the greater part of the sediment detached by raindrops, namely that simply dislodged, will likely include even larger particles than splash droplets are competent to transport, the relative fineness of the sediment transported in the overland flow cannot be attributed to selective detachment by raindrops of the finer fraction of the matrix soil.

The alternative explanation is that the size characteristics of the transported sediment are related to overland-flow characteristics: either the flow is not competent to transport the coarser fraction of sediment supplied to it by raindrop detachment, or the flow selectively deposits the coarser part of its load, possibly in exchange for finer sediment. Inasmuch as sediment reaching S2 has a larger contributing area than that passing through S1, it might be anticipated that if selective deposition of coarser particles does occur, then in the latter part of the longer experiments, when runoff reaching both S1 and S2 has come from the entire portions of the plot upslope of these cross sections, finer sediment would be recorded at S2 than at S1. This hypothesis was tested by examining the calibre of sediment passing the two sections in minutes 14–19 of E2 and minutes 10–29 of E4. For these periods the mean sand percentage contained in the eroded sediment is the same at the two sections. By elimination, therefore, it appears that the relative fineness of the sediment in the interrill flow results from this flow not being competent to transport the coarser part of the load supplied to it by raindrop detachment.

Rates of detachment and erosion The disparity between the rates of detachment and erosion implied by the foregoing analyses was investigated further by means of an additional field experiment to determine the rate of raindrop detachment, a consideration of the competence of the overland flow and through the use of a simulation model. The field experiment (Parsons et al. 1991) yielded an estimate of the rate of sediment transport by rainsplash of 0.432 g m^{-2} min^{-1} that may be compared with the average rate of erosion of 0.856 g m^{-2} min^{-1}. Inasmuch as the rate of detachment has been shown in laboratory experiments by Schultz et al. (1985) to be 14–20 times greater than that of the soil carried in splash droplets, our measurement showing the rate of splash transport alone to be about 50% of the rate of transport implies that the total rate of detachment by raindrops is several times the rate of removal in overland flow.

An indication of the competence of the overland flow was obtained from data on the largest single particle collected in the samples obtained from S1 and S2 at any time in any of the experiments (9.0 mm), and the average sizes of the 10 largest particles collected during minutes 15–19 of E2 and minutes 19–29 of E4 (i.e. at or near equilibrium conditions) (3.7 mm and 5.0 mm for S1 and S2, respectively). In comparison, Moeyersons (1975) observed dislodged particles up to 20 mm in diameter in laboratory experiments using 4 mm diameter raindrops with lower kinetic energy than natural raindrops of the same size, and Kotarba (1980) indicates dislodgement of particles as large as 12 mm during low-intensity (12 mm h^{-1}) natural rainfall. Although these comparative data are rather limited, they do, none the less, support the proposition that overland flow lacks the competence to transport the largest detached particles. The evidence also indicates a difference in competence between S1 and S2, even though at both locations the mean depth of overland flow is the same (6.15 mm, see Abrahams et al. 1991, Table 2). A likely explanation for this difference is the difference in the maximum overland flow depth at the two locations, since it can reasonably be assumed that the largest particles are transported in the deepest threads of flow. At S1, the maximum equilibrium depth of any thread of flow is 20 mm, whereas at S2, because of the downslope concentration of overland flow into fewer, deeper threads (Parsons et al. 1990), it is 26 mm.

In revealing a difference between the maximum size of particles transported through S1 and S2 that is likely to be due to the difference in the maximum overland-flow depths at the two locations, the analysis of competence suggests that the spatial variability of overland flow not only leads to spatial variation in rates of detachment (Fig. 13.4) but also in the ability of overland flow to transport all detached particles. Just as the flow is not competent to transport the largest detached particles, so it is unable to transport all the detached sediment. To investigate this notion further, a simulation model was employed. Details of this model are given in Abrahams et al. (1991). It is sufficient to note here that:

(a) The model uses as input detailed measurements of overland flow through the 10 m subsections of S1 and S2, SS1 and SS2, described earlier. These subsections are considered to be representative of the flow conditions through the longer sections of which they are part. In addition to measuring depth at 5 cm intervals along these subsections, they were also divided into zones of relatively uniform flow characteristics, termed partial sections, for which surface velocity of the flow was measured by timing injected dye over a distance of 1 m. These surface velocities were then corrected to give mean velocities (Abrahams et al. 1986a). For the purposes of the simulation model, each subsection was divided into 5 cm elements centred on the points where flow depth was measured, and each element was assigned the flow depth measured at the point and the velocity measured for the encompassing partial section.

(b) The model has two adjustable parameters, D_L and v_c. The former is the rate of soil dislodgement in comparison to the rate of soil splash. The latter is the critical mean velocity required to sustain in transport an already detached particle of a given size. Initial values for these parameters were based upon experimental results obtained by Schultz et al. (1985, Fig. 8) as above, for D_L, and by Sundborg (1967) and Govers (1987), for v_c.

The simulation model was run for SS1 and SS2 using both the initial values

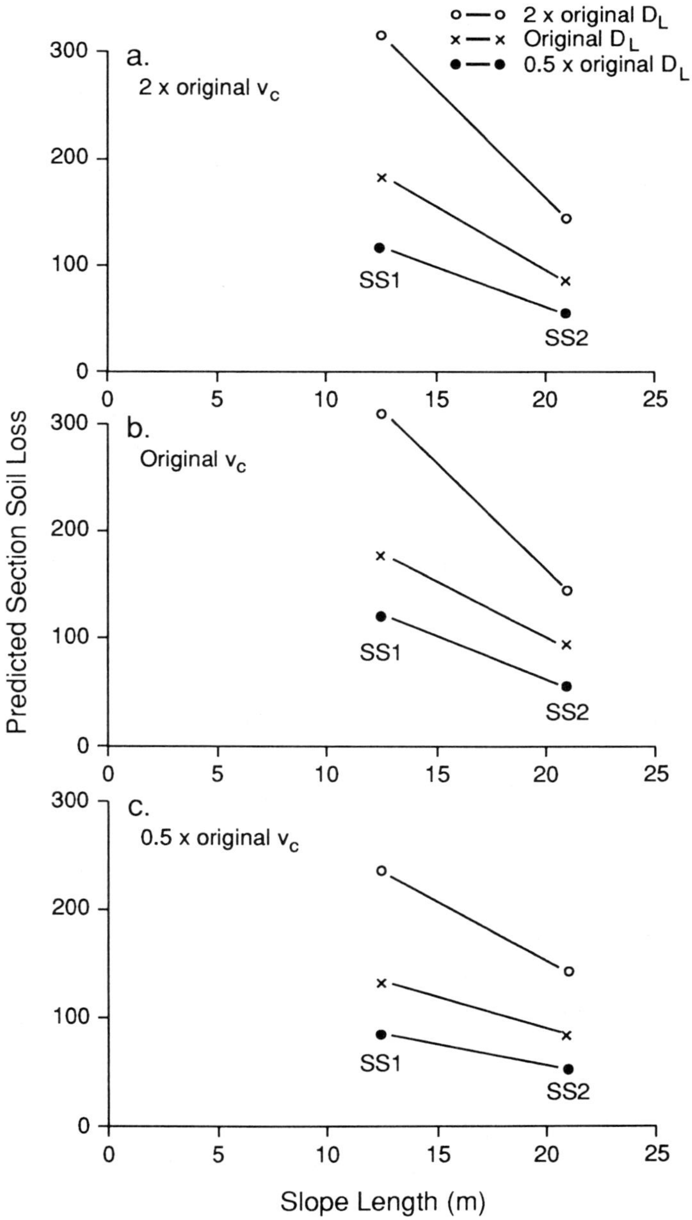

Figure 13.7 Output from the simulation model.

of the adjustable parameters and values 0.5 and 2 times these values. The latter values were employed to evaluate the sensitivity to the adjustable parameters of the downslope pattern in predicted section soil loss, G'. As can be seen in Figure 13.7, for all the tested values of the adjustable parameters, G' decreases between SS1 and SS2 just as G does. Moreover, G' is relatively insensitive to the values of the adjustable parameters. These findings lend support to the proposition that the detachment and removal mechanisms in the model closely resemble those operating on the runoff plot and suggest that the observed pattern of variation in G may be explained in terms of the across-slope distributions of overland-flow depth and velocity.

An examination of soil loss predicted by the simulation model (Fig. 13.8) reveals that soil loss along SS1 and SS2 tends to be greatest (a) where adjacent areas have shallow or no flow and, hence, are sources of splashed sediment; (b) where flow depths are sufficiently small to permit splash and dislodgement; and (c) where flow velocities are sufficiently high to remove particles splashed into or dislodged within the flow. These observations of predicted across-slope variations in soil loss provide a test of the model. Specifically, if the model is correct, the proportion of the surface covered by shallow, competent overland flow on the runoff plot might be expected to vary down slope in much the same manner as G.

A comparison of Figures 13.7 and 13.9 reveals this to be the case. For all tested values of v_c, the proportion of the surface covered by shallow ($\leq$ 2 mm), competent overland flow decreases between SS1 and SS2, just as G does. This decrease is a result of the progressive concentration of flow down slope into fewer and larger threads (Parsons et al. 1990). It is significant that the pattern of soil loss revealed by this simulation model is not the same as the pattern of detachment obtained using the same model (Fig. 13.8). Not all sediment that is detached is transported, either because it is detached in areas where there is no significant overland flow, or because what flow there is, lacks the competence to transport the coarser fraction of detached sediment.

As noted above, interrill overland flow, both on naturally sparsely vegetated hillslopes and on bare agricultural fields, has been reported as consisting typically of a shallow film of water with threads of deeper, faster flow. It can be anticipated, therefore, that the disparity between sediment detachment and transport rates identified on the present runoff plot is a common phenomenon. For these environments the notion that interrill erosion is controlled by the rate of detachment appears to be both an oversimplification and one that leads to an overestimate of the rate of erosion. In reality, the rate of erosion is a function not only of the rate of detachment but also of the spatial variability in depth and velocity of overland flow.

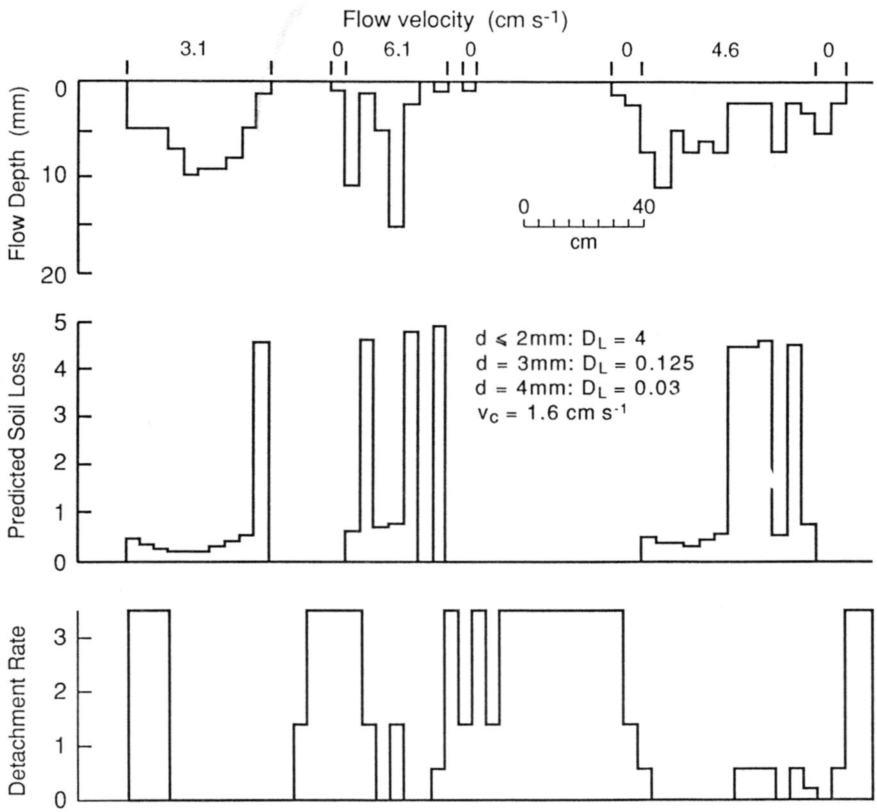

Figure 13.8 Relationship of spatial variation in overland-flow depth to soil loss predicted by the simulation model for part of SS1 and to calculated rates of detachment along the same part of SS1.

Conclusion

For models of sediment removal in interrill overland flow developed within the framework of the mechanisms of detachment and transportation by both rainfall and surface flow, the field studies reported in this paper both point to weaknesses in existing models and provide directions for future model development.

The analysis of flow detachment indicates that, whereas in the past, based upon a few experimental studies (e.g. Borst & Woodburn 1942, Quansah 1985), many models have simply assumed flow detachment to be zero, it may be possible now to examine more realistic estimates of the likelihood of flow detachment. Under conditions of considerable form resistance, such as those that obtain on desert debris slopes and recently ploughed fields, grain resistance

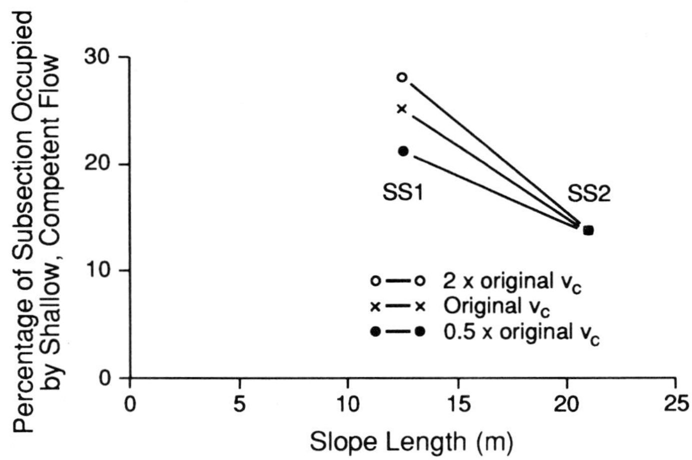

Figure 13.9 Graphs of percentage of subsection occupied by shallow ($d \geq 4$ mm), competent overland flow against slope length between SS1 and SS2.

will be a small part of total flow resistance, and flow detachment will be minimal. On smoother surfaces, such as less recently ploughed agricultural fields, grain resistance may constitute a higher proportion of total flow resistance and flow detachment may be greater.

Detachment by raindrops is strongly influenced by the depth of the surface-water layer. Because this layer typically varies in depth both across slope and down slope so, too, does the rate of raindrop detachment. Hitherto, models have assumed that raindrop detachment is spatially uniform or decreases down slope. Our experiments indicate that neither of these cases is true for our large plot. In so far as the character of overland flow on this plot is typical of interrill flow, it is likely that raindrop detachment seldom exhibits one of the simple spatial patterns that have been assumed in models.

Detachment, however, is only one element determining sediment removal, although, apart from the portion of the interrill zone close to the divide, it has generally been assumed to be paramount. Our studies, however, indicate that this is not the case. Of the material that is detached, only a part is removed in the surface flow. Models of soil erosion need to take into account discrepancy between soil detachment and sediment removal that results from the spatial variability in depth and velocity of overland flow.

For the near-divide portion of the interrill zone, where transport capacity of the surface flow is presumed to control the rate of sediment removal, our studies show that the assumption that this rate always increases monotonically with gradient is incorrect. For the specific case of desert debris slopes it appears that, due to the presence of a coarse weathering mantle and the relationship between mantle stoniness and gradient, the relationship between sediment removal and gradient is curvilinear. It seems unlikely that the curvilinear relation with a vertex at about 12°, obtained in our investigation, and the monotonically positive relation, hitherto widely assumed to be universally valid, together represent the only possible relations between hillslope gradient and sediment yield. More probably, our investigation suggests that a range of relations exists. If this is the case, then this range needs to be determined so that it may subsequently be incorporated into models of interrill sediment removal.

Acknowledgements

We acknowledge the assistance and co-operation of the USDA/ARS Aridland Watershed Management Research Unit, Tucson, in conducting the field experiments at Walnut Gulch. We thank the many friends and colleagues who helped with these experiments, especially Shiu-Hung Luk who collaborated with us in our early work. The research was supported by NATO grant RG. 85/0066 and NSF grant SES-8812587.

References

Abrahams, A. D. & A. J. Parsons 1991a. Resistance to overland flow on desert pavement and its implications for sediment transport modeling. *Water Resources Research* **27**, 1827–36.

Abrahams, A. D. & A. J. Parsons 1991b. Relation between sediment yield and gradient on debris-covered hillslopes, Walnut Gulch, Arizona. *Geological Society of America, Bulletin* **103**, 1109–13.

Abrahams, A. D., A. J. Parsons, S.-H. Luk 1986a. Field measurement of the velocity of overland flow using dye tracing. *Earth Surface Processes and Landforms* **11**, 653–7.

Abrahams, A. D., A. J. Parsons, S.-H. Luk 1986b. Resistance to overland flow on desert hillslopes. *Journal of Hydrology* **88**, 343–63.

Abrahams, A. D., A. J. Parsons, S.-H. Luk 1988. Hydrologic and sediment responses to simulated rainfall on desert hillslopes in southern Arizona. *Catena* **15**, 103–17.

Abrahams, A. D., A. J. Parsons, S.-H. Luk 1989. Distribution of depth of overland flow on desert hillslopes and its implications for modeling soil erosion. *Journal of Hydrology* **106**, 177–84.

Abrahams, A. D., A. J. Parsons, S.-H. Luk 1991. The effects of spatial variability in overland flow on the downslope pattern of soil loss on a semiarid hillslope, southern Arizona. *Catena* **18**, 255-70.

Band, L. E. 1985. Simulation of slope development and the magnitude and frequency of overland flow erosion in an abandoned gold mine. In *Models in geomorphology*, M.J. Woldenberg (ed.), 191–211. Boston: Allen and Unwin.

Borst, H. L. & R. Woodburn 1942. The effect of mulching and methods of cultivation on

runoff and erosion from Muskingham silt loam. *Agricultural Engineering* **23**, 19–22.
Carson, M. A. 1987. Measures of flow intensity as predictors of bed load. *Journal of Hydraulic Engineering* **113**, 1402–21.
Einstein, H. A. 1950. *The bed-load function for sediment transportation in open channel flows.* Technical Bulletin 1026, US Department of Agriculture, Washington DC.
Einstein, H. A. & R. B. Banks 1950. Fluid resistance of composite roughness. *Transactions of the American Geophysical Union* **31**, 603–10.
Einstein, H. A. & N. L. Barbarossa 1952. River channel roughness. *Transactions of the American Society of Civil Engineers* **117**, 1121–32.
Ellison, W. D. 1944. Studies of raindrop erosion. *Agricultural Engineering* **25**, 131–6, 181–2.
Ellison, W. D. 1945. Some effects of raindrops and surface flow on soil erosion and infiltration. *Transactions of the American Geophysical Union* **26**, 415–29.
Ellison, W. D. 1947. Soil erosion studies. *Agricultural Engineering* **28**, 145–6, 197–201, 245–8, 297–300, 349–51, 402–5, 422–4.
Emmett, W. W. 1970. *The hydraulics of overland flow on hillslopes.* US Geological Survey Professional Paper 662-A. Washington DC: US Government Printing Office.
Foster, G. R. 1982. Modeling the erosion process. In *Hydrologic modeling of small watersheds*, C. T. Haan (ed.), 297–300. St. Joseph, Michigan: American Society of Agricultural Engineers.
Foster, G. R. & L. D. Meyer 1972. A closed-form soil erosion equation for upland areas. In *Sedimentation: Symposium to honor Professor H. A. Einstein*, H. W. Shen (ed.), 12.1–12.19. Fort Collins, Colorado: Colorado State University.
Gelderman, F. W. 1970. *Soil survey of Walnut Gulch Experimental Watershed: a special report.* Portland, Oregon: US Department of Agriculture, Soil Conservation Service and Agricultural Research Service.
Ghadiri, H. & D. Payne 1981. Raindrop impact stress. *Journal of Soil Science* **32**, 41–9.
Gilbert, G. K. 1877. *The geology of the Henry Mountains.* Washington DC: US Geographical and Geological Survey.
Gilley, J. E., D. A. Woolhiser, D. B. McWhorter 1985a. Interrill soil erosion – Part I: development of model equations. *Transactions of the American Society of Agricultural Engineers* **28**, 147–53, 159.
Gilley, J. E., D. A. Woolhiser, D. B. McWhorter 1985b. Interrill soil erosion – Part II: testing and use of model equations. *Transactions of the American Society of Agricultural Engineers* **28**, 154–9.
Gilluly, J. 1956. *General geology of central Cochise County, Arizona.* US Geological Survey Professional Paper 281. Washington DC: US Government Printing Office.
Govers, G. 1987. Initiation of motion in overland flow. *Sedimentology* **34**, 1157–64.
Govers, G. 1988. Discussion of "Measures of flow intensity and predictors of bed load". *Journal of Hydraulic Engineering* **114**, 853–6.
Govers, G. & G. Rauws 1986. Transporting capacity of overland flow in plane and irregular beds. *Earth Surface Processes and Landforms* **11**, 515–24.
Hartley, D. M. 1987. Simplified process model for water sediment yield from single storms: Part I – Model formulation. *Transactions of the American Society of Agricultural Engineers* **30**, 710–17.
James, C. S. 1990. Prediction of entrainment conditions for nonuniform, noncohesive sediments. *Journal of Hydraulic Research* **28**, 25–41.

Kirkby, M. J. 1971. Hillslope process-response models based on the continuity equation. In *Slopes: form and process*. Institute of British Geographers Special Publication No. 3, D. Brunsden (ed.), 15–30.

Kotarba, A. 1980. Splash transport in the steppe zone of Mongolia. *Zeitschrift für Geomorphologie Supplementband* **35**, 92–102.

Libby, F., D. E. Wallace, D. P. Spangler 1970. *Seismic refraction studies of the subsurface geology of Walnut Gulch Experimental Watershed, Arizona, ARS 41-164*. Beltsville, Maryland: US Department of Agriculture, Agricultural Research Service.

Luk, S.-H., A. D. Abrahams, A. J. Parsons 1986. A simple rainfall simulator and trickle system for hydro-geomorphological experiments. *Physical Geography* **7**, 344–56.

McCool, D. K., L. C. Brown, G. R. Foster, C. K. Mutchler, L. D. Meyer 1987. Revised slope steepness factor for the Universal Soil Loss Equation. *Transactions of the American Society of Agricultural Engineers* **30**, 1387–96.

Meyer, L. D. & W. H. Wischmeier. 1969. Mathematical simulation of the process of soil erosion by water. *Transactions of the American Society of Agricultural Engineers* **12**, 754–8, 762.

Moeyersons, J. 1975. An experimental study of pluvial processes on granite grus. *Catena* **2**, 289–308.

Morgan, R. P. C. 1980. Field studies of sediment transport by overland flow. *Earth Surface Processes* **5**, 307–16.

Musgrave, G. W. 1947. The qualitative evaluation of factors in water erosion – a first approximation. *Journal of Soil and Water Conservation* **2**, 133–8, 170.

Mutchler, C. K. & R. A. Young 1975. Soil detachment by raindrops. In *Present and prospective technology for predicting sediment yields and sources. Proceedings of the Sediment Yield Workshop USDA-ARS-40*, 113–17. Oxford Mississippi: US Department of Agriculture.

Neal, J. H. 1938. The effect of degree of slope and rainfall characteristics on runoff and soil erosion. *Agricultural Engineering* **19**, 213–17.

Osborn, H. B. 1983. *Precipitation characteristics affecting hydrologic response of southwestern rangelands, ARM-W-34*. Oakland, California: US Department of Agriculture, Agricultural Research Service.

Osborn, H. B. & K. G. Renard 1988. Rainfall intensities for southern Arizona. *Journal of Irrigation and Drainage Engineering* **114**, 195–9.

Palmer, R. S. 1964. *The influence of a thin water layer on waterdrop impact forces*. International Association of Scientific Hydrology, Publication 65, 141–8.

Parsons, A. J. & A. D. Abrahams 1989. A miniature flume for sampling interrill overland flow. *Physical Geography* **10**, 96–105.

Parsons, A. J. & A. D. Abrahams 1992. Controls on sediment removal by interrill overland flow on semi-arid hillslopes. *Israel Journal of Earth Sciences* (in press).

Parsons, A. J., A. D. Abrahams, S.-H. Luk 1990. Hydraulics of interrill overland flow on a semi-arid hillslope, southern Arizona. *Journal of Hydrology* **117**, 255–73.

Parsons, A. J., A. D. Abrahams, S.-H. Luk 1991. Size characteristics of sediment in interrill overland flow on a semi-arid hillslope, southern Arizona. *Earth Surface Processes and Landforms* **16**, 143–52.

Parsons, A. J., A. D. Abrahams, J. R. Simanton 1992. Microtopography and soil-surface materials on semi-arid, piedmont hillslopes, southern Arizona. *Journal of Arid Environments* **22**, 107–15.

Poesen, J. & J. Savat 1981. Detachment and transportation of loose sediment by raindrop splash. *Catena* **8**, 19–41.

Quansah, C. 1985. Rate of soil detachment by overland flow, with and without rain, and its relationship with discharge, slope steepness and soil type. In *Soil erosion and conservation,* S. A. El-Swaify, W. C. Moldenhauer, A. Lo (eds), 406–23. Ankeny, Iowa: Soil Conservation Society of America.

Rowlinson, D. L. & G. L. Martin 1971. Rational model describing slope erosion. *Journal of the Irrigation and Drainage Division, Proceedings of the American Society of Civil Engineers* **97**, 39–50.

Schultz, J. P., A. R. Jarrett, J. R. Hoover 1985. Detachment and splash of a cohesive soil by rainfall. *Transactions of the American Society of Agricultural Engineers* **28**, 1878–84.

Singhal, M. K., J. Mohan, A. K. Agrawal 1980. Role of grain shear stress in sediment transport. *Irrigation Power* **37**, 105–8.

Smith, D. D. & W. H. Wischmeier 1957. Factors affecting sheet and rill erosion. *Transactions of the American Geophysical Union* **38**, 889–96.

Sundborg, A. 1967. Some aspects on fluvial sediments and fluvial morphology. I. General views and graphic methods. *Geografiska Annaler* **49A**, 333–43.

Torri, D., M. Sfalanga, M. Del Sette 1987. Splash detachment: runoff depth and soil cohesion. *Catena* **14**, 149–55.

Walker, P. H., J. Jutka, A. J. Moss, P. I. A. Kinnell 1977. Use of a versatile experimental system for soil erosion studies. *Soil Science Society of America, Journal* **41**, 610–12.

Wischmeier, W. H. & D. D. Smith 1978. *Predicting rainfall erosion losses – a guide to conservation planning*, US Department of Agriculture Handbook No. 537.

Yair. A. & M. Klein 1973. The influence of surface properties on flow and erosion processes on debris covered slopes in an arid area. *Catena* **1**, 1–18.

Yang, C. T. 1972. Unit stream power and sediment transport. *Journal of the Hydraulics Division, Proceedings of the American Society of Civil Engineers* **98**, HY 1805–25.

Young, R. A. 1980. Characteristics of eroded sediment. *Transactions of the American Society of Agricultural Engineers* **23**, 1139–42, 1146.

Young, R. A. & J. L. Wiersma 1973. The role of rainfall impact in soil detachment and transport. *Water Resources Research* **9**, 1629–36.

Zingg, A. W. 1940. Degree and length of land slope as it affects soil loss in runoff. *Agricultural Engineering* **21**, 59–64.

14

Upland erosion research on rangeland

J. R. Simanton & K. G. Renard

Abstract

Rangeland upland erosion research using rainfall simulation techniques has been conducted on a wide range of ecosystems in the western United States. The initial research began in 1981 and was to parameterize rangeland conditions for use in the Universal Soil Loss Equation (USLE) and later the Water Erosion Prediction Project (WEPP) erosion model. The field experimental design has progressed from simple replicated plots needed to parameterize the empirically based USLE to multi-intensity rainfall rates and overland-flow injections needed for the dynamic, process-based WEPP erosion model. Results from more than 10 years of research have shown that erosion pavement (surface-rock fragments) plays a major role in reducing soil erosion; there are temporal changes in rangeland soil erodibility; and vegetative canopy cover has very little direct effect on soil erosion rates. The Revised Universal Soil Loss Equation (RUSLE) has been developed and includes results of these rangeland erosion studies. The WEPP erosion model is nearing completion and results of 2 years of intensive rangeland experiments are incorporated into the rangeland, infiltration and soil modules of the model.

Introduction

Upland erosion from rangelands may seem relatively small compared to the very high rates reported for cropland areas. However, when viewed in terms of erosion rate per unit area and the vastness of rangelands worldwide, the magnitude of the situation can be appreciated. Furthermore, rangeland erosion rates are very critical because of the limited soil resource associated with many of the ecosystems involved. Rangelands cover extensive areas of the world and are an important land resource of the western and southwestern United States (Simanton 1991). Rangelands are usually used for livestock grazing, wildlife habitats, recreation areas and water-producing catchments. Precipitation is generally less than potential evapotranspiration, erratic and, in areas where intense thunderstorms dominate, upland erosion can be significant (Branson et al. 1981). This chapter traces the history of our efforts to develop technology for improved erosion prediction on rangelands of the western United States.

Universal soil loss equation

Efforts to estimate or predict soil erosion from rangelands have centred on the Universal Soil Loss Equation (USLE) (Wischmeier & Smith 1978), an equation developed for cropland situations but applied to range and forest land conditions (Dissmeyer & Foster 1980, Simanton et al. 1980).

The USLE estimates average annual soil loss using the equation:

$$A = RKLSCP \tag{14.1}$$

where: A = estimated soil loss (tons ha^{-1} yr^{-1});
R = rainfall erosivity factor (EI units yr^{-1}) (EI = MJ · mm (ha · h)$^{-1}$);
K = soil erodibility factor (tons ha^{-1} EI^{-1});
LS = slope steepness-length factor;
C = cover and management factor; and
P = erosion control practice factor.

These factors reflect the major variables influencing soil erosion by rainfall and resultant overland flow. The equation is based on plot data collected mainly from cropland areas in the eastern United States. The cropland rainfall simulation erosion research used relatively large plots, a standard plot tilled up–down slope, in fallow condition, with 9% plot slope and standard sequences of rainfall input (Wischmeier & Mannering 1969).

Rainfall simulation

Rainfall simulation is a useful tool for evaluating the hydrologic and erosional responses of the natural environment. The merits and shortcomings of rainfall simulation have been well documented (Neff 1979). The major objections to rainfall simulators are that they do not produce natural rainfall energies or variable intensities, and, in the case of larger simulators, the water used in the simulation can have a different water quality from that of natural rainfall. However, the major advantage of simulators, especially in arid and semi-arid environments, is that not only can maximum control be achieved over where, when and how data are collected, but there is also no need to wait for natural storms which are usually very sporadic. Plot runoff and erosion responses can be compared easily among ecosystems because the same rainfall sequence, intensity and amount can be applied and antecedent conditions can be controlled.

J. R. Simanton & K. G. Renard

USLE rangeland experimental procedures

Rangeland experiments
As part of an effort to improve the application of the USLE to a wide range of land types and uses, the United States Department of Agriculture's (USDA) Agricultural Research Service (ARS) initiated a programme to update the technology. The Southwest Watershed Research Center's (SWRC) staff in Tucson, Arizona participated in this effort by developing rainfall simulation procedures to evaluate and quantify soil and management parameters for conditions in various western United States rangeland ecosystems. Standards similar to those used for cropland studies were used in these simulator studies, so direct comparison could be made to other USLE research.

Figure 14.1 Operation of the rotating boom rainfall simulator used on rangeland erosion plots.

Rainfall simulator The SWRC began rangeland erosion plot studies in 1981 to develop rangeland soil loss factors for the USLE. These studies were conducted using a rotating boom rainfall simulator (Swanson 1965) on 3.05 m by 10.7 m plots (Fig. 14.1). The simulator is trailer-mounted, has 10 booms (each 7.6 m long) radiating from a central stem, and rotates at about 4 r.p.m. The arms support 30 V-Jet 80100 nozzles positioned at various distances from the stem. (Trade names are included for information only, and do not constitute endorsement by the authors or the US Department of Agriculture.) The nozzles spray downward from an average height of 2.4 m, apply rainfall intensities of about 65 or 130 mm h^{-1} and produce drop-size distributions similar to natural rainfall. Simulator energies are about 77% of those of natural rainfall and the simulator produces intermittent rainfall impulses at the plot surface as the booms pass over the plot. Rainfall spatial distribution over each plot has a coefficient of variation of less than 10%. Changes in rainfall intensities are produced by increasing or decreasing the number of open nozzles; 15 nozzles for 65 mm h^{-1} and 30 nozzles for 130 mm h^{-1}. Because of the simple design and portability of the simulator, and because two plots are covered during one run, many plots can be evaluated in a relatively short time (Fig. 14.2).

Rainfall/runoff/sediment Three rainfall simulation runs were made on each plot pair in the following sequence: dry run, initial 60 min rainfall on dry soil conditions; wet run, 30 min rainfall approximately 24 h after the dry run; and very wet run, 30 min rainfall 30 min after the completion of the wet run. Rainfall application rate was measured with a recording raingauge and rainfall distribution on each plot was measured with six non-recording raingauges. Plot runoff was measured by specially designed precalibrated flumes (4 l s^{-1} maximum capacity) equipped with water-level recorders that measured instantaneous flow depth. Continuous hydrographs were produced using the flume's depth/discharge rating table. During a run, times of ponding (0.5 of the plot surface had standing water), runoff initiation, sediment samples and end of runoff were recorded on field notes for later comparisons to recorder charts. Plot sediment yield was calculated from periodic sediment samples taken throughout the hydrograph. Sampling intervals were dependent on changes in the runoff rate with more frequent sampling (1–2 min intervals) when discharge was changing rapidly. Sediment samples were analysed for total concentration and particle size distribution. All rainfall, runoff and sediment data were used in computer programs developed especially for the simulator studies.

Plot characteristics Vegetation composition, foliar canopy cover and ground surface characteristics of each of the large plots were measured before and after treatment. Surface cover characteristics included: soil, gravel (5–20 mm), rock (>20 mm), litter, cryptogams and basal plant cover. A 49-pin point-meter, which was 3.05 m long with pin-holes spaced every 60 mm, was placed perpendicular to the plot slope and rested on the metal plot border at 10

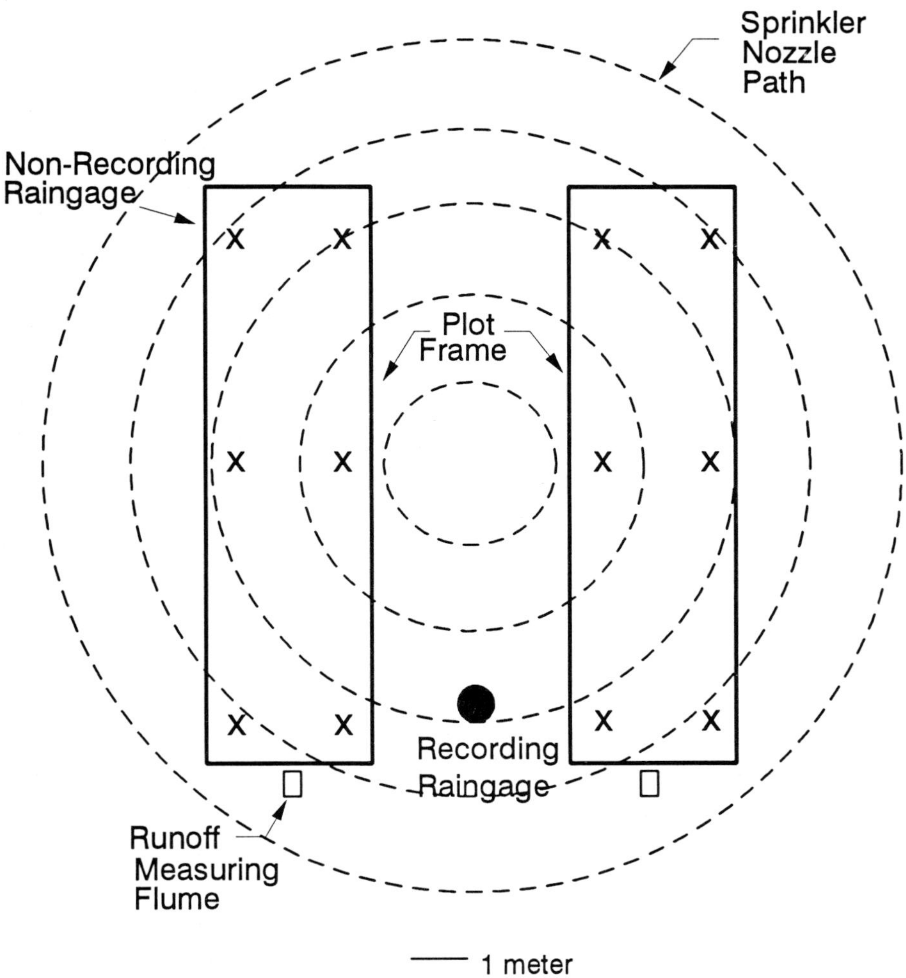

Figure 14.2 Plot layout of large plots used in rangeland field experiments.

positions evenly spaced along the plot. At each position, 49 pin-point surface and canopy measurements were made by dropping a pin through each pin-hole. Ten permanent transects across each plot produced 490 point readings to describe each plot's surface and vegetation canopy cover.

USLE rangeland study sites and treatments

The initial rangeland USLE experiments began in 1981 on plots in southeastern Arizona and two years later on sites on the Nevada Test Site in southern Nevada. The general procedure included spring and fall rainfall application on at least two replications of three or four treatments on one or more soil types in each ecosystem studied (Simanton & Renard 1982).

Arizona plots

The Arizona rangeland rainfall simulator USLE plots were located on the Walnut Gulch Experimental Watershed in southeastern Arizona (Fig. 14.3). The watershed is representative of millions of hectares of brush and grass rangeland found throughout the semi-arid Southwest and is considered a transition zone between the Chihuahuan and Sonoran Deserts. Average annual precipitation on the watershed is about 300 mm and is bimodally distributed, with 70% occurring during the summer thunderstorm season of July to mid-September (Osborn et al. 1979).

Soils Soils are generally well-drained, calcareous, gravelly loams with large percentages of rock and gravel on the soil surface (Gelderman 1970). Three soil series selected were: Bernardino (a thermic Ustollic Haplargid), Cave (thermic, shallow Typic Paleorthid) and Hathaway (thermic Aridic Calciustoll). The Bernardino series is a deep, well-drained, fine-textured soil formed in old calcareous alluvium and has 50%, by volume, gravel and cobbles in the surface 10 cm and usually less than 35% gravel in the remaining profile. The Cave series is a shallow, well-drained, medium-textured soil with indurated lime hardpans that have developed at less than 45 cm depth in old gravelly and cobbly calcareous alluvium. This soil has up to 60%, by volume, gravel and cobbles in the surface 10 cm and usually less than 40% gravel in the remaining profile. The Hathaway series is a deep, well-drained, gravelly medium and moderately coarse-textured soil over very gravelly, coarse-textured materials of moderate depths. This soil, formed from gravelly or very gravelly calcareous old alluvium, has up to 70%, by volume, gravel and occasional cobbles in the surface 10 cm and usually less than 50% in the remainder of the profile.

Vegetation Major vegetation of the watershed includes: creosotebush (*Larrea tridentata*), white-thorn (*Acacia constricta*), tarbush (*Flourensia cernua*), snakeweed (*Gutierrezia sarothrae*), burroweed (*Aplopappus tenuisectus*), black grama (*Bouteloua eriopoda*), blue grama (*B. gracilis*), sideoats grama (*B. curtipendula*) and bush muhly (*Muhlenbergia porteri*).

Plot treatment Treatments were initially imposed in the spring of 1981 and then reapplied, except for the tilled treatment, prior to subsequent season's rainfall simulations. These treatments were: natural cover or no treatment (both grass and shrub), clipped (vegetation clipped to a 20 mm height and

Figure 14.3 Rangeland erosion study sites in the western United States.

clippings removed), bare (vegetation clipped to the soil surface and all surface litter and rock fragments greater than 5 mm removed) and tilled (up- and down-slope moldboard ploughing and disking). The tilled treatment was intended to represent the standard USLE treatment for determination of the soil erodibility factor (*K*). The clipped treatment, not intended to represent grazing effects, was used to determine vegetation effects on erosion and the bare plot was to define the role of rock fragments (erosion pavement) on soil erosion.

Nevada Test Site plots

The Nevada Test Site (NTS) plots were established in 1983 at Area 11, which is located in a transition zone between the Great Basin and Mojave Desert; and at Mercury, Nevada, in the northern Mojave Desert (Fig. 14.3). Annual precipitation generally varies from 125 to 175 mm of which about 75% occurs between mid-September and late-March and the remainder falls during the summer season as scattered thundershowers (Romney et al. 1973).

Soils The soils do not have official series names but are Typic Durorthid (shallow, mixed thermic). The primary differences between the soils at the two study sites are in textural class and parent material. The soil at Area 11 is coarse-loamy and formed in material weathered from tuff, basalt and limestone. The soil at the Mercury site is loamy with randomly dispersed clay pockets, and formed in material weathered from limestone, quartz and tuff. Both study sites are underlain by a silica-lime hardpan; the soils are well drained with medium to rapid runoff, and both have moderate permeability. The soil at Mercury has higher water-holding capacity because of the higher clay content and less coarse sand through the profile.

Vegetation The major vegetation of Area 11 includes: Mormon tea (*Ephedra nevadensis*), spiney hopsage (*Artemesia spinescens*), shadescale (*Atriplex confertifolia*), boxthorn (*Lycium andersonii*) and Indian ricegrass (*Oryzopsis hymenoides*). Vegetation at Mercury includes spiney menodora (*Menodora spinescens*), shadescale, Mormon tea, desert alyssum (*Lepidium fremontii*), and creosotebush.

Plot treatments Plot treatments were the same as the Arizona plots except that the tilled treatment was not made. Other than this difference, all aspects of the study were identical to the Arizona erosion plot study.

USLE results and discussion

Arizona plots

Eight seasonal rainfall simulations (over 4 years) were made on the 24 Arizona erosion plots at Walnut Gulch. Summaries of runoff and erosion rates are given in Table 14.1. The tilled treatment proved to be of little value in evaluating all but the factor P in the USLE. Runoff and subsequent erosion from the tilled plots were practically non-existent except for the very wet runs.

Table 14.1 Average spring (Sp) and fall (Fa) runoff rate (mm EI^{-1}) for three treatments on the Bernardino, Hathaway and Cave soils.

		Runoff rate (mm EI^{-1})					
		Natural		Clipped		Bare	
Soil	Moisture	Sp	Fa	Sp	Fa	Sp	Fa
Bernardino	dry	0.017	0.012	0.021	0.038	0.052	0.066
	wet	0.016	0.018	0.029	0.052	0.055	0.060
	v. wet	0.025	0.026	0.040	0.057	0.060	0.068
Cave	dry	0.022	0.043	0.028	0.062	0.053	0.069
	wet	0.025	0.044	0.044	0.062	0.055	0.068
	v. wet	0.033	0.049	0.052	0.070	0.059	0.076
Hathaway	dry	0.020	0.046	0.028	0.060	0.042	0.064
	wet	0.018	0.035	0.032	0.056	0.039	0.059
	v. wet	0.024	0.042	0.040	0.066	0.056	0.069

		Erosion rate (kg ha^{-1} EI^{-1})					
		Natural		Clipped		Bare	
Soil	Moisture	Sp	Fa	Sp	Fa	Sp	Fa
Bernardino	dry	0.248	0.067	0.379	0.592	9.717	8.468
	wet	0.194	0.102	0.433	0.741	10.489	8.986
	v. wet	0.266	0.125	0.669	0.830	10.061	8.612
Cave	dry	0.367	0.486	1.013	1.737	9.547	11.074
	wet	0.414	0.436	1.253	1.543	8.099	9.316
	v. wet	0.449	0.437	1.637	1.684	7.371	8.985
Hathaway	dry	0.347	0.389	0.743	1.273	9.882	10.948
	wet	0.244	0.299	0.685	1.260	8.387	9.147
	v. wet	0.331	0.360	0.884	1.464	9.705	10.112

Because of the unexpected response from the tilled plots, only one replication on each soil was retilled after the first year of runs. This deviation from the original plan was designed so that the recovery rate and response of the tilled plot could be determined. The data for the tilled plots have not been summarized because of complications associated with sequences of retreatment and invasion of plants.

Seasonal differences Seasonal (spring–fall) runoff and erosion differences were found throughout the 4 year study period. The magnitude of these differences appears to be both treatment and soil variable but the trend was toward more runoff and erosion from the fall simulations on the clipped and bare treated plots (Fig. 14.4a,b). These vegetative cover-free plots would be influenced by the soil-surface compacting effects produced by the summer thunderstorm rainfall; an effect dissipated by the winter freeze–thaw process that tends to loosen the soil surface before the spring simulator runs. The natural-cover plots had more runoff but lower sediment concentration in the fall (Fig. 14.5a,b).

Soil differences Runoff rates among the three soils did not vary as greatly as the erosion rates. The Bernardino soil had lower erosion rates than either the Cave or Hathaway, regardless of the plot treatment. Erosion rate differences between the Cave and Hathaway soils were very small except for the clipped treatment, under which the Cave soil had higher erosion rates (possibly showing a higher soil erodibility and/or more exposed surface soil).

Antecedent moisture effects Runoff rates generally increased as soil moisture increased (dry surface to wet to very wet) on all soils, treatments and seasons.

Treatments Runoff rates varied between treatments and were affected by both soil and season. The bare-soil plot always had the greatest runoff and erosion rates regardless of the soil but were larger in the fall. The natural-cover plot had the lowest runoff rate but the spring rates showed the larger treatment differences. Treatment effects on erosion rates were very obvious, with the bare soil treatment on the Bernardino soil having an erosion rate nearly 90 times greater than the natural-cover treatment. The bare-soil treatment erosion rates of the other two soils were nearly 30 times greater than the natural-cover treatment.

Effects of surface and vegetation characteristics Results from the treatment comparisons of erosion rates separated the effects of various surface and canopy characteristics. A negative exponential relationship was found in the analysis of the effect of erosion pavement (surface rock fragments > 5 mm) on erosion rates, given by the equation $ER = a\exp(EPb)$, where ER = erosion rate in t

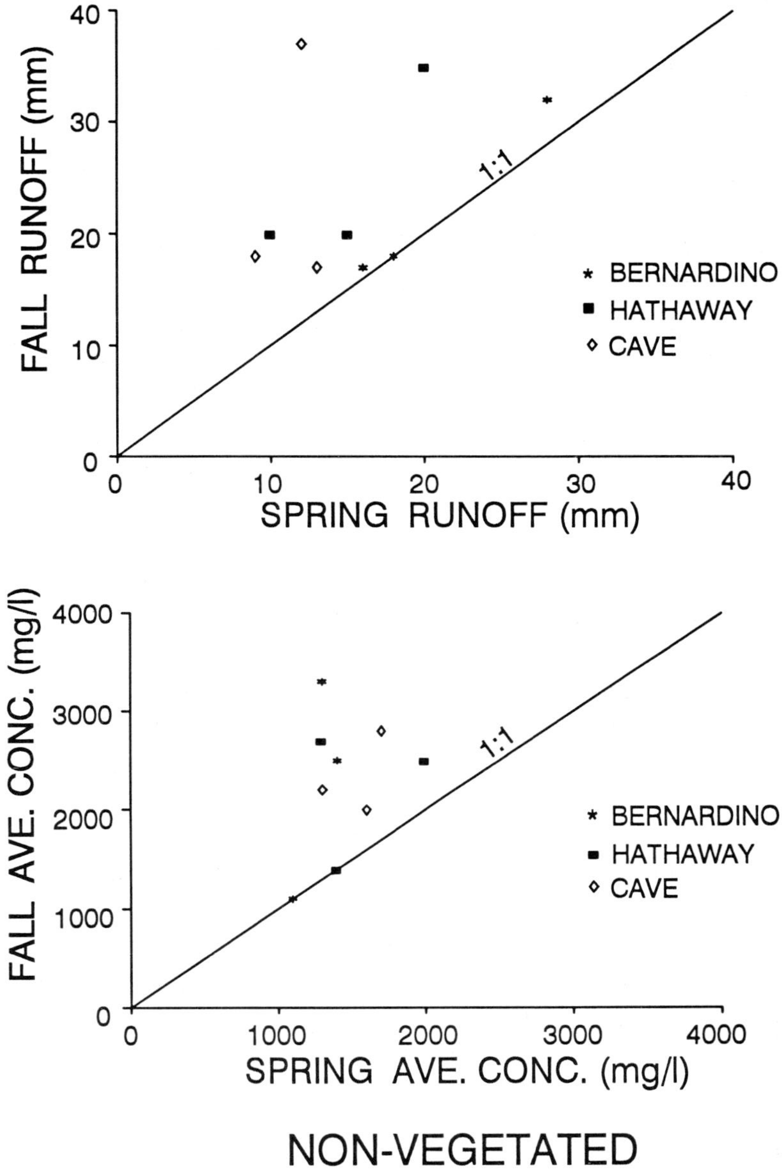

Figure 14.4 (a) Relation between runoff volumes (mm) measured in the spring and fall. Bare and clipped plots for 1981 on the Bernardino, Hathaway and Cave soils. (b) Relation between average sediment concentrations (mg l^{-1}) measured in the spring and fall. Bare and clipped plots for 1981 on the Bernardino, Hathaway and Cave soils.

ha^{-1} EI^{-1}, EP = percentage erosion pavement, and $b = -0.044$ (Fig. 14.6). The results of this analysis, based on data from the three soils for all runs over the 4 year period, are very similar to those reported by Simanton et al. (1984)

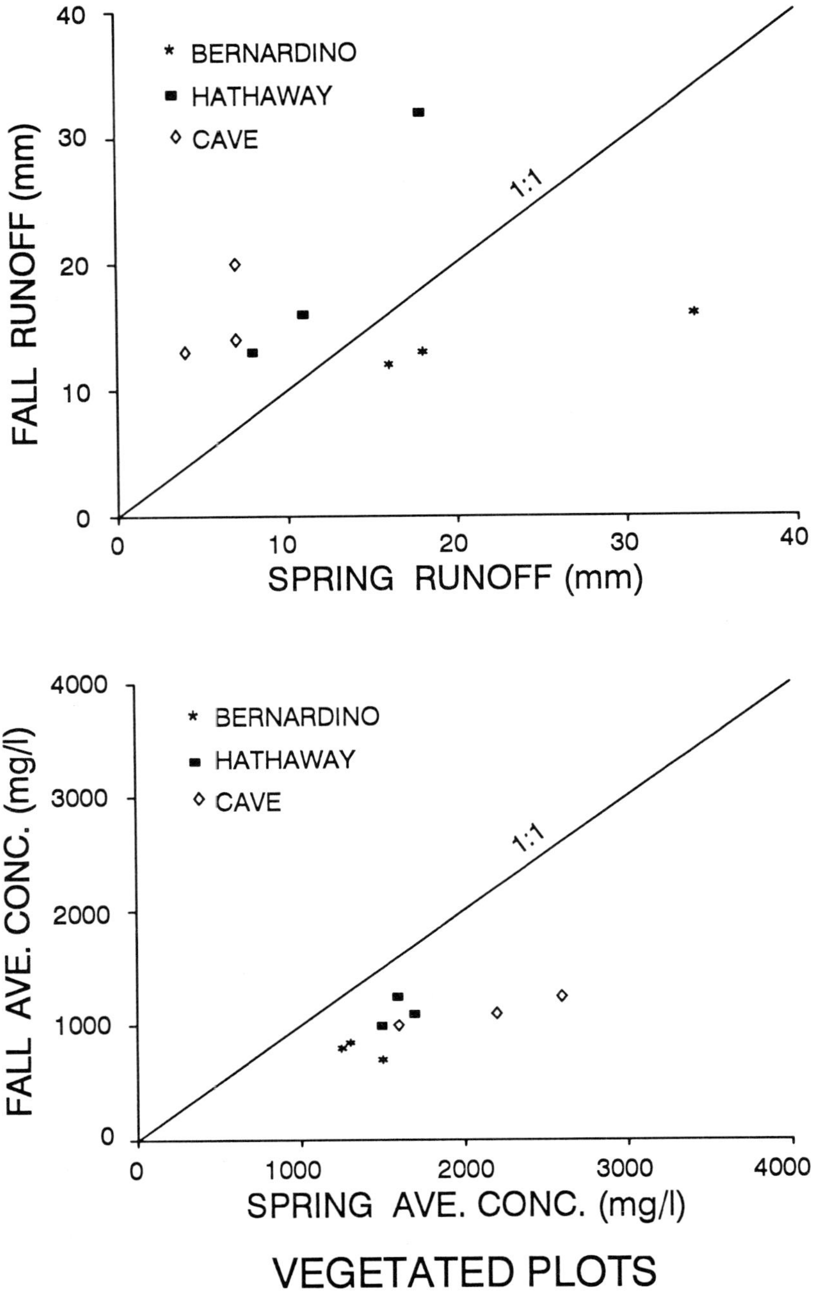

Figure 14.5 (a) Relation between runoff volumes (mm) measured in the spring and fall. Natural plots for 1981 on the Bernardino, Hathaway and Cave soils. (b) Relation between average sediment concentrations (mg l^{-1}) measured in the spring and fall. Natural plots for the 1981 on the Bernardino, Hathaway and Cave soils.

who used only 1 year of simulator data to develop the erosion pavement–erosion rate relationship.

Vegetation effects Vegetation canopy during the first two seasons of rainfall simulations did not affect runoff or erosion rates (Figs 14.7a, b & 14.8a, b). However, as vegetative cover became more dominant on the ungrazed, natural-cover plots, the erosion rate difference between the vegetated and clipped plots began to increase until there was almost a tenfold difference in erosion rates between the two. Interrelated to this cover increase on the natural plots was a corresponding decrease in organic matter, root mass and other indirect effects of vegetation removal on the clipped plots. Vegetation type differences did not affect average erosion rates; similar rates were found for both grass, grass/shrub and shrub-dominated canopies.

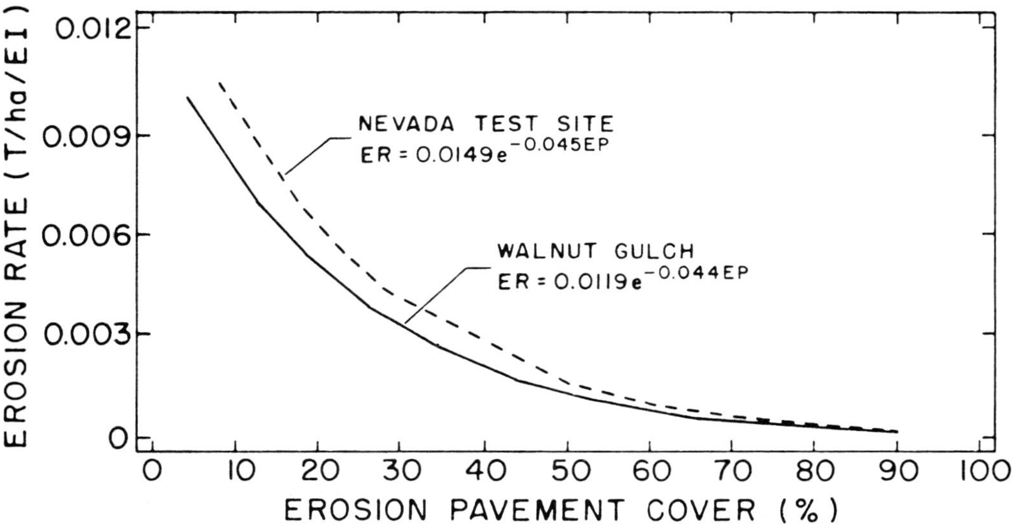

Figure 14.6 Exponential fit of relation between erosion rate (t ha^{-1} EI^{-1}) and erosion pavement for the Arizona and Nevada Test Site rangeland erosion plots (from Simanton et al. 1985).

USLE factor values Because the tilled plots did not have runoff or erosion rates much different from those of the natural plots, the bare plot was used as the rangeland "standard plot" to determine K values ($t \cdot ha \cdot hr \cdot (ha \cdot MJ \cdot mm)^{-1}$) for the three soils used in the study ($C = 1$ for the bare condition). Measured K values, as reflected in the erosion rate from the bare plots, did not vary greatly between spring and fall simulations but did change over the 4 year study period (Simanton & Renard 1986). Soil K values for the Bernardino and Cave soils levelled out after about 2 years (Fig. 14.9). The calculation to determine K from measured soil loss from the bare plot on each

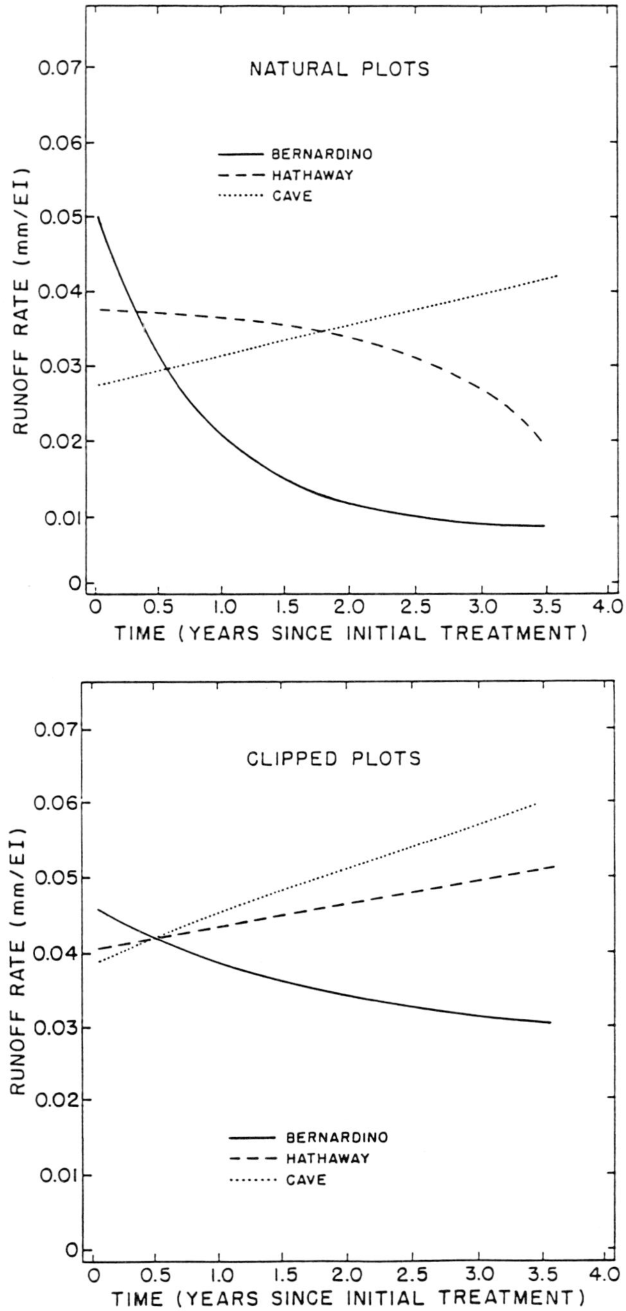

Figure 14.7 Non-linear least-squares fit of change with time in the measured runoff rate (mm EI^{-1}) on (a) natural plots and (b) clipped plots for the replicated run average of spring and fall runs, with time zero equal to spring 1981, for the Bernardino, Hathaway and Cave soils.

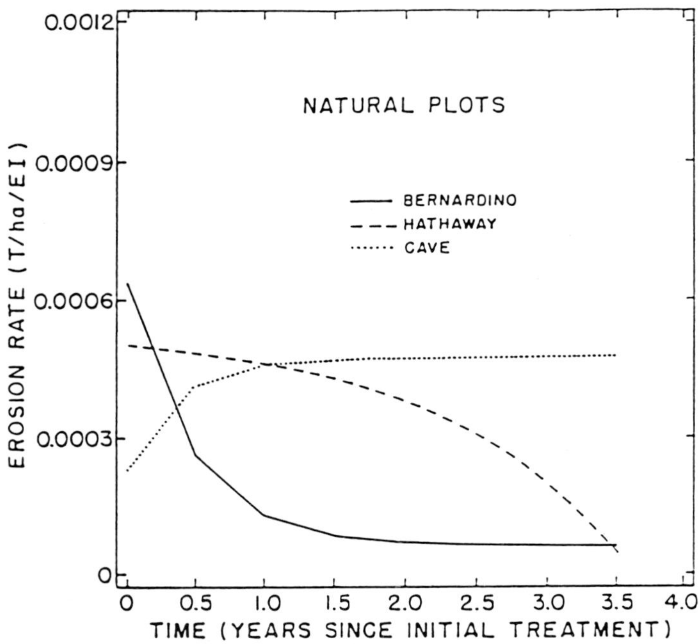

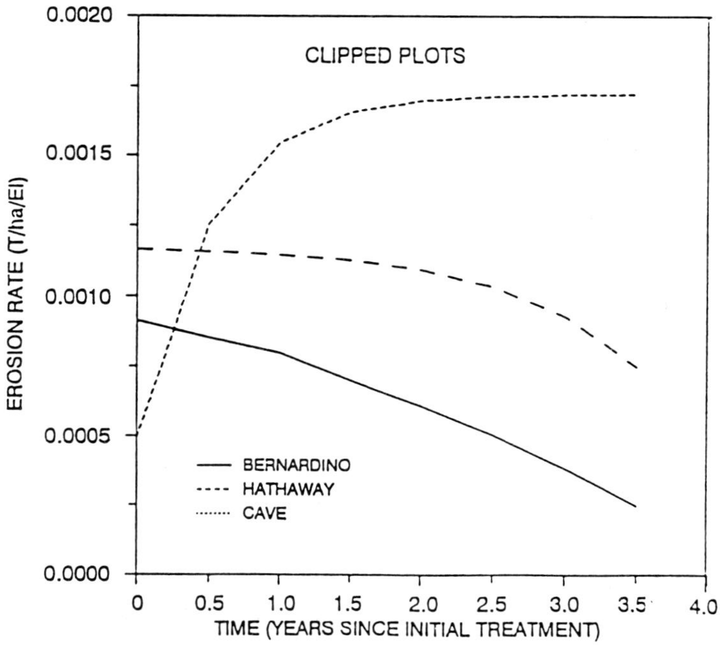

Figure 14.8 Non-linear least-squares fit of change with time in the measured erosion rate (t ha^{-1} EI^{-1}) on (a) natural plots and (b) clipped plots for the replicated run average of spring and fall runs, with time zero equal to spring 1981, for the Bernardino, Hathaway and Cave soils.

soil is:

$$K = \frac{A}{RCLSP} \qquad (14.2)$$

where: A = measured soil loss from the bare plot;
R = rainfall energy to produce the soil loss;
LS = slope and length correction for each plot;
P = 1 for rangeland conditions; and
C = 1 for the bare plot.

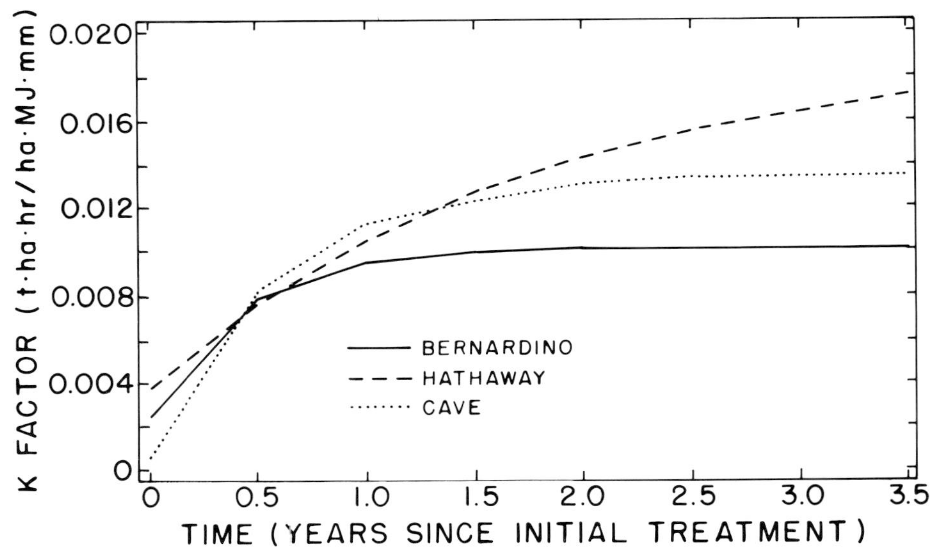

Figure 14.9 Non-linear least-squares fit of the change with time in the USLE *K* factor for the bare plots on the Bernardino, Hathaway and Cave soils. Time zero equals spring 1981 and *C* was assumed to be 1 for the bare surface condition.

When $C = 1$ for the bare plot, the simulator-derived K values were 0.009, 0.011 and 0.011 $(\mathrm{t \cdot ha \cdot h (ha \cdot MJ \cdot mm)^{-1}})$ for the Bernardino, Hathaway and Cave soils, respectively. The nomograph (Wischmeier et al. 1971) values for these same three soils were 0.021, 0.028 and 0.036 $(\mathrm{t \cdot ha \cdot h (ha \cdot MJ \cdot mm)^{-1}})$, respectively. If the bare-plot C value is assumed to be 0.45 (as given in Table 10 of Wischmeier & Smith 1978), and used to calculate K from the simulator bare- plot data, the K values would be 0.020, 0.027 and 0.024 $(\mathrm{t \cdot ha \cdot h (ha \cdot MJ \cdot mm)^{-1}})$, respectively. These are fairly consistent with the nomograph K values for the three soils. However, the 0.45 maximum C value (as mentioned above) was determined from an agricultural soil and represents the ratio of soil loss from a 7 year reconsolidated tilled soil to the 2 year average soil loss just after tillage (i.e. soil loss from the tilled soil was 2.2 times greater than the soil loss from the same soil 7 years after its last tillage). Results from the rangeland tilled treatment indicated that both runoff and erosion were reduced just after tillage, as compared to the natural condition, and that erosion increased with time as the soil reconsolidated. If the measured *K*-values from the bare plot are used to calculate *C* in the USLE, *C*-factor evaluation can be made from measured erosion and related to surface and vegetation characteristics of the erosion plots.

Temporal changes

Of the five factors in the USLE (the topographic effect (*LS*) is usually considered one term), the rainfall factor is the only one expected to significantly change annually on rangeland. This is also the factor over which man has no control. The cover-management factor can change naturally from year to year, but not as drastically as that for the rainfall factor. Perennial vegetation cover changes are difficult to perceive on a year-to-year basis and tend to leave the impression that a static cover condition exists on rangeland. Management changes and their effect on surface and canopy cover are usually not immediately evident on rangelands. Erosion studies in cropland areas have indicated that erosion rates vary temporally during the year for various cover-management situations and that these changes are often related to soil erodibility changes (Dissmeyer & Foster 1981, Van Doren et al. 1984). Thus, some of the cover-management parameter temporal changes observed in field data are possibly reflecting soil erodibility changes. Limited information is available concerning erosion rate change within natural rangeland conditions. Though the main objective of our rainfall simulations was to quantify USLE factors for rangelands, interesting temporal changes in erosion rates were found over the 4 year study at Walnut Gulch.

Tilled plots The concept of using a tilled fallow plot as the reference for the simulator studies on rangeland was abandoned after a short time because: (a) tillage is not a common practice on semi-arid rangelands; (b) the tilled rangeland plots did not yield appreciable runoff and subsequent erosion in contrast to erosion associated with tillage in agronomic cropped areas; and (c) the tilled plots remained artificially rough with large surface depression storage because of boulders, cobbles and gravel material brought to the soil surface by the treatment. Thus, after two seasons, only the clipped and bare treatments were left to compare to the natural plots.

Bare plots The bare-soil treatment produced the largest erosion rates (t ha^{-1} EI^{-1}) of all treatments and the rates increased with time for about 2 years before reaching an "equilibrium" with the energy input for both the Bernardino and Cave soils (Fig. 14.10). After 4 years, the Hathaway soil-erosion rate was still increasing. The increase in erosion rate for the bare-soil treatment closely emulated runoff changes which may be attributed to the decrease in root and residue material in the soil which, in turn, decreased the soil macropore structure (Dixon & Simanton 1979). Another reason for an erosion increase could be attributed to the formation of a rill network that developed after the vegetation and rock fragments were removed. Also, removing rock fragments would cause a decrease in surface roughness and a corresponding increase in runoff, erosion and runoff response-time to the simulated rainfall. Most likely, the increase in runoff and erosion rates is a combination of these factors. If the erosion rate increase was a function of plant and litter removal, the effect should be found in the clipped plot results.

Clipped plots The erosion rate for the clipped plot changed with time but not as drastically as that for the bare plot (Fig. 14.10). As with the bare plot, the change was associated with a similar change in runoff rate. This suggests that the influence of removal of plant canopy cover on erosion rate was small and that the formation of a rill network and the loss of surface rock fragments were, in fact, dominating the response of the bare plots. In addition, the erosion pavement may be effective in maintaining a high infiltration capacity by preventing soil-surface crusting or sealing (Lane et al. 1987).

Natural plots Erosion and runoff rates of the natural plots decreased for the Bernardino and Hathaway soils and increased on the Cave soil for about the first 2 years (Figs. 14.7a & 8a). The different shapes of the erosion and runoff rate curves are probably reflecting vegetation differences. The Bernardino natural plots were dominated by perennial grasses, the Cave natural plots were shrub and forb dominated, and the Hathaway natural plots had both grass and shrub-canopy cover.

Vegetation effects Vegetation effects on erosion rates were determined from differences in erosion rate between the clipped and natural treatments on all soils. By the end of the 4 year study, the clipped plots had an average equilibrium erosion rate almost five times greater than the average erosion rate of the natural plots (Fig. 14.8a, b). However, the bare plots had an average equilibrium erosion rate of more than 25 times the average rate of the natural plots. Even though the clipped plots did not have vegetation after the first year of treatment, the erosion rate changed very little with time, suggesting that the erosion-reducing effect of vegetation was not as significant as the effect of

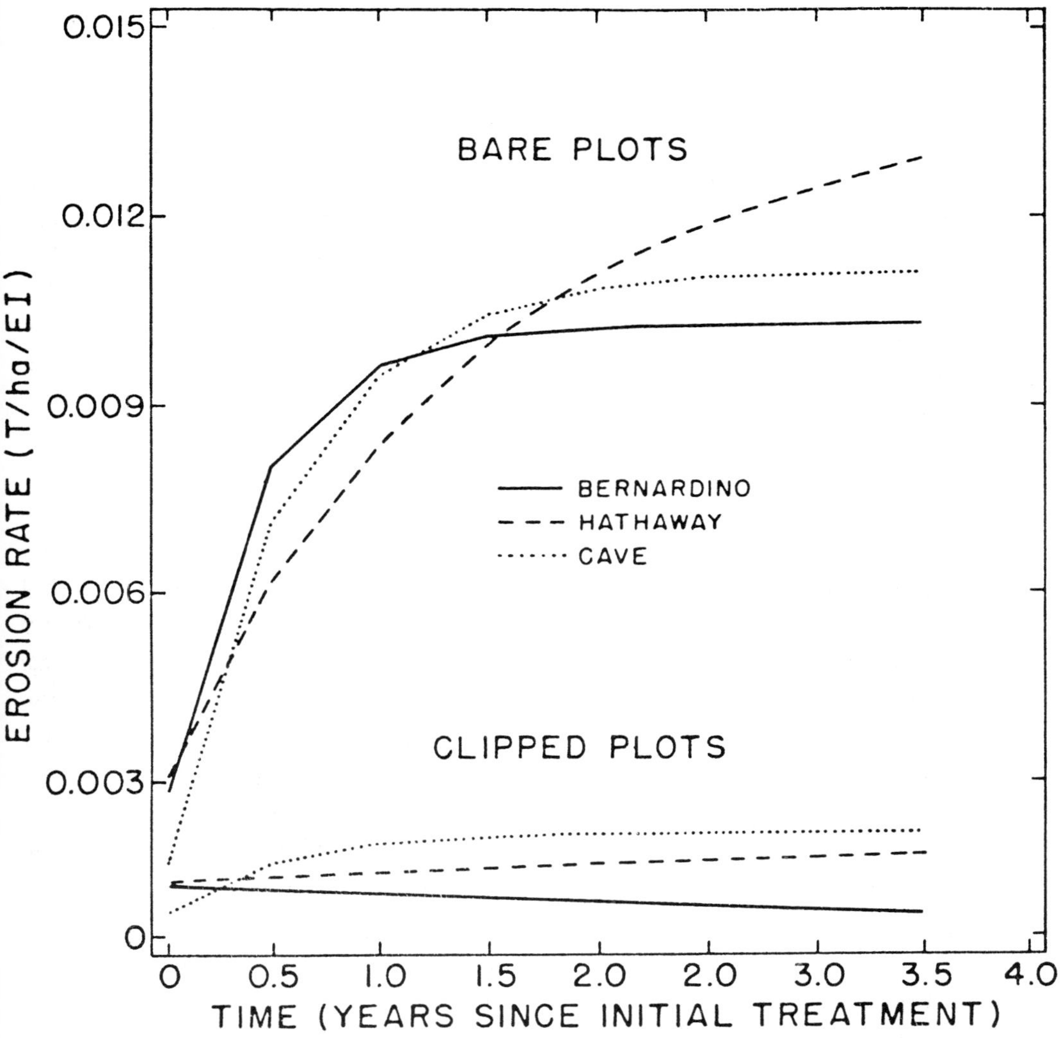

Figure 14.10 Non-linear least-squares fit of change with time in the measured erosion rate (t ha^{-1} EI^{-1}) on the bare and clipped plots for the replicated run average of spring and fall runs, with time zero equal to spring 1981, for the Bernardino, Hathaway and Cave soils.

surface rock fragments, as was shown by Simanton et al. (1984). Canopy cover of the natural plots tripled on the Bernardino and Hathaway plots and nearly doubled on the Cave plots over the 4 year study (Fig. 14.11). This increase was undoubtedly a result of the increased water applied, which stimulated vegetative growth, but also reflects the plot response to no livestock grazing. Litter cover on the natural plot's soil surface decreased with increasing

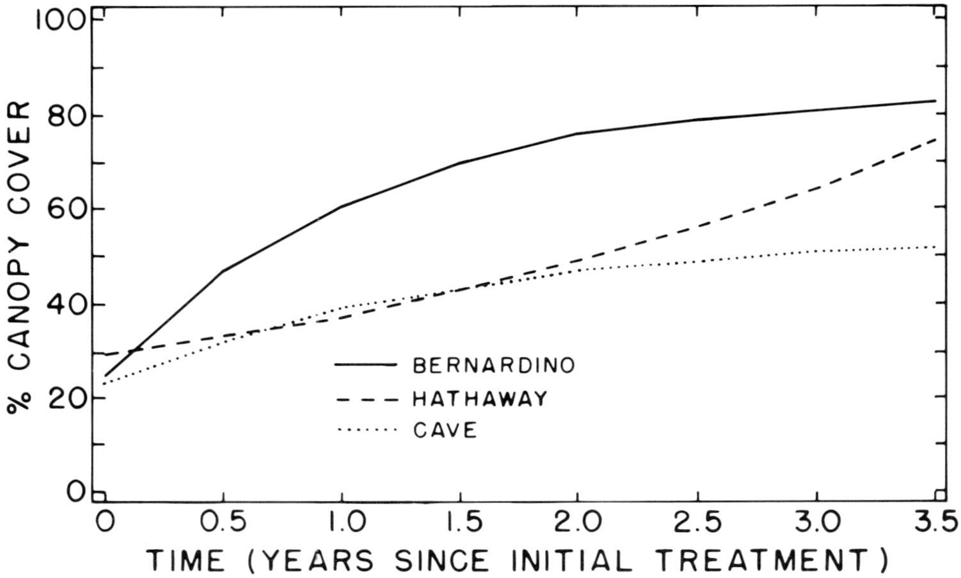

Figure 14.11 Non-linear least-squares fit of the change with time of the percent canopy cover on the natural plots, with time zero equal to spring 1981, for the Bernardino, Hathaway and Cave soils.

vegetation canopy but the amount of bare soil more than doubled over the 4 year study period (Fig. 14.12). The increase in percentage bare soil on the natural plots could be caused by vegetation trapping of wind-blown soil or, as evidenced by the almost complete lack of litter cover, termite activity. Termites bring soil to the surface and use it to coat litter particles so they can utilize the litter during the day, out of the direct rays of the sun (Whitford et al. 1982). With weathering, these termite casts break down and the soil remains on the surface.

USLE C *factor* The USLE cover-management factor (*C*) was calculated for the natural plots, assuming that the bare-plot *C* value was unity and that the calculated *K*, or soil erodibility factor, of the bare plot was valid for each of the soils. Because of the method of calculation, the *C* and *K* factors are not independent, and a decrease in one will produce an increase in the other. The *C* value decreased with time but at different rates for each soil–vegetation complex (Fig. 14.13). The rate of decrease for *C* on the Bernardino soil natural plot (grass vegetation) was over twice the increasing rate of *K* during the first year of study (Figs 14.9 & 13). The decrease in the *C* value of the Hathaway natural plot (shrub and grass vegetation) was about the same as the increase in the *K* value. The Cave natural plot (shrub and forbs) had a *C* value change that was six times less than the corresponding increase in the *K* value during the first year of study. The *C* value of the Bernardino and Cave soils reached

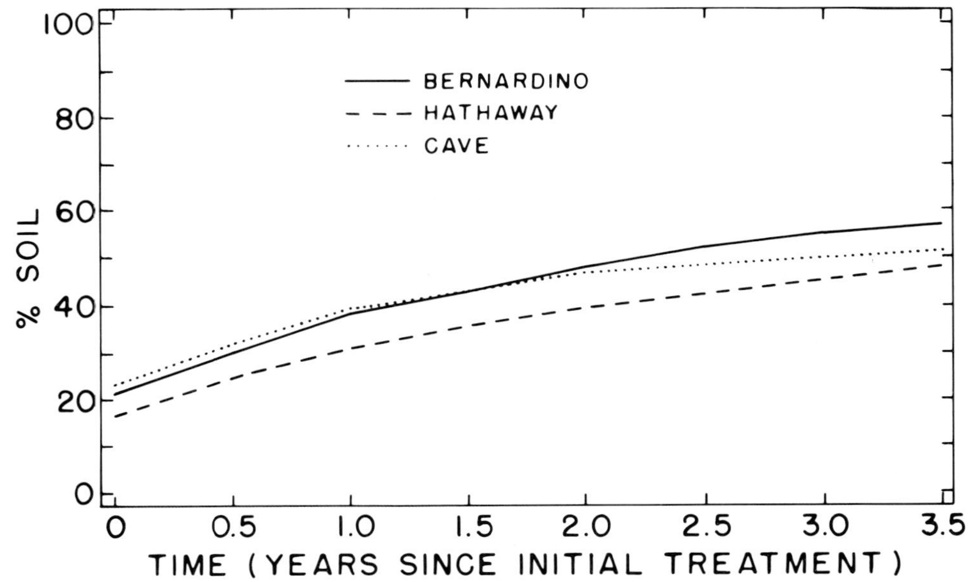

Figure 14.12 Non-linear least-squares fit of the change in percentage bare soil with time on the natural plots, with time zero equal to spring 1981, for the Bernardino, Hathaway and Cave soils.

equilibrium around 2 years after the start of the simulation study, whereas the natural plot on the Hathaway soil still had a slight downward trend after 4 years. The *C* factor response reflects vegetation canopy types. A grass canopy is more important in erosion control than a shrub canopy, but the vegetation type also influences runoff, which is interrelated with erosion.

Nevada Test Site

Two years, or four seasons, of rainfall simulations were made on the NTS erosion plots. Data summaries of the runoff and erosion rates are given in Table 14.2.

Seasonal differences As at the Arizona plots, seasonal (spring–fall) runoff and erosion rate (per EI unit) differences were found throughout the 2 year study period. Fall runoff rates were higher than those in the spring, regardless of the soil. However, on the Mercury soil, the season of higher runoff rates (fall) was not the season of higher erosion rates (spring), indicating that some factor, such as soil erodibility, was influencing the erosion rate.

Soil differences Runoff and erosion rate differences were found between the two soils. The Mercury soil had higher runoff and erosion rates than the Area-11 soil, regardless of the plot treatment.

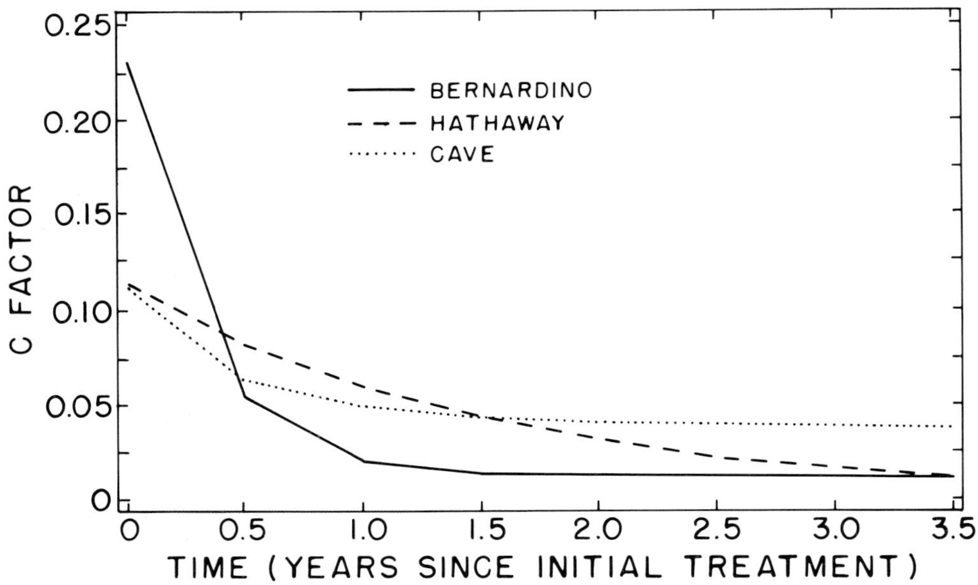

Figure 14.13 Non-linear least-squares fit of the change with time of the USLE *C* factor for the natural plots on the Bernardino, Hathaway and Cave soils. Time zero equals spring 1981. The *C* factor was calculated using the simulator-derived *K* value from the bare plot, whose *C* value was assumed to be 1.

Antecedent moisture effects Runoff rates increased as soil moisture increased (dry surface to wet to very wet) on both soils under all treatments. Erosion rates were more variable and decreased on the wet surface runs on the natural treatment.

Treatments Runoff rates varied among treatments and were affected by both soil and season. The bare plot always had the greatest runoff rate regardless of the soil or season; the rate was greater in the fall. The plot with natural cover had the lowest runoff rate and again, the fall rates showed the larger treatment differences. Bare-treatment erosion rates on the Mercury soil were about 20 times greater than the rates from the natural treatment and 10 times greater than the clipped treatment. The Area-11 bare-treatment erosion rates were about 45 times greater than the natural treatment and about 25 times greater than the clipped treatment. These differences in erosion rate between treatments are not as great as those found on the Arizona plots, but the general trend of treatment effect was the same.

Effects of surface and vegetation characteristics Erosion pavement appeared

Table 14.2 Average spring (Sp) and fall (Fa) runoff and erosion rates for three treatments on the Mercury and Area 11 soils.

		Runoff rate (mm EI^{-1})					
		Natural		Clipped		Bare	
Soil	Moisture	Sp	Fa	Sp	Fa	Sp	Fa
Mercury	dry	0.027	0.048	0.048	0.065	0.062	0.078
	wet	0.037	0.049	0.059	0.075	0.068	0.076
	v. wet	0.045	0.060	0.066	0.088	0.077	0.094
Area 11	dry	0.003	0.018	0.001	0.026	0.021	0.057
	wet	0.011	0.016	0.011	0.026	0.043	0.053
	v. wet	0.022	0.033	0.026	0.038	0.054	0.062

		Erosion rate (kg ha^{-1} EI^{-1})					
		Natural		Clipped		Bare	
Soil	Moisture	Sp	Fa	Sp	Fa	Sp	Fa
Mercury	dry	0.514	0.518	1.003	0.871	11.943	9.642
	wet	0.471	0.445	1.069	0.984	14.125	9.708
	v. wet	0.581	0.575	1.218	1.341	12.197	11.180
Area 11	dry	0.050	0.167	0.038	0.678	3.531	10.031
	wet	0.189	0.125	0.249	0.302	8.060	10.017
	v. wet	0.400	0.211	0.484	0.420	10.682	11.336

to be an important factor in the erosion process but not as dominant as was found on the Arizona plots. Vegetation was more effective in reducing erosion rates at the NTS than on the Arizona plots. Analysis of the effect of erosion pavement on erosion rates indicated that the relationship was exponential, similar to the Arizona relationship, and, based on data from the two soils for all runs over the 2 year period, had an exponent of −0.045 (Fig. 14.6).

USLE parameter values Assuming that the bare plot represented the USLE "unit plot" (corrected for *LS*) condition as the most erodible condition possible ($C = 1$), the *K*-factor values from the simulator results were 0.016 and 0.010 for the Mercury and Area-11 soils, respectively. These measured *K* values are

38% and 16% of the *K* values derived from the soil erodibility nomograph developed by Wischmeier et al. (1971). If $C = 0.45$ (the maximum allowed for rangeland in Wischmeier & Smith 1978) the *K* values would have been 84% and 35% of the nomograph values.

Revised universal soil loss equation

Major changes to the USLE
The USLE is in the process of being revised to be applicable to a wide range of land uses. Major improvements to the USLE incorporated into the Revised Universal Soil Loss Equation (RUSLE) include: expanded erosivity (*R*) map for the western United States; seasonal variability of the soil erodibility (*K*) determination; varying topographic (*LS*) factors dependent on soil's susceptibility to rill erosion; a subfactor approach to determine the cover-management (*C*) factor; detailed procedures for estimating the conservation support practice (*P*) factor for cropland, including values for contouring, terracing and stripcropping, and management practices on rangelands; and implementation of RUSLE using a computer program that will run on either DOS or UNIX computers (Renard et al. 1991).

R *Factor* Rainfall information from over 1,000 locations in the western United States was analysed to produce the improved erosivity (*R*) map for the western United States as well as EI distribution maps for the West and Hawaii. Because ponded water on the soil surface reduces the erosivity of the rain, the *R* factor in RUSLE is reduced where flat slopes occur in regions of intense rain storms (i.e. high *R*-factor areas). Finally, an *R* equivalent is used in the Pacific Northwest region to reflect the combined effect of freezing soil, and rain on snow on partially frozen soil.

K *factor* Erodibility data from around the world have been reviewed, and an equation was developed that estimates soil erodibility (*K*) as a function of "average" soil particle diameter. This then gives an estimate of *K* for soils where the nomograph does not apply. Use of this function is recommended only where the nomograph or no other procedure apply. RUSLE varies *K* seasonally. Experimental data showed that *K* is not temporally constant and varies with season, being greatest in early spring and lowest in mid-fall. The seasonal variability is determined for 15-day intervals by weighting the instantaneous estimate of *K* in proportion to the *EI* for the same period. Instantaneous estimates of *K* are predicted from equations relating *K* to the frost-free period and annual *R*-factor. RUSLE *K* is adjusted for rock fragments in the soil to account for their effect on infiltration. Rock fragments on the soil surface are treated as mulch or ground cover in the C factor. RUSLE also provides a procedure for identifying those soils that are highly, moderately or slightly susceptible to rill erosion relative to their susceptibility to interrill erosion.

L *and* S *factors* RUSLE slope length (L) utilizes three separate slope-length relationships that are functions of slope steepness and the susceptibility of the soil to rill erosion relative to interrill erosion. A slope-length relationship specifically for the Palouse region of the Pacific Northwest area of the United States is also included in the RUSLE. Experimental data and field observations, especially on rangelands, do not support the current USLE quadratic relationship for slope steepness effects when extended to steep slopes. RUSLE slope- steepness (S) relationships are much more linear than in the USLE, with the biggest differences being found on slopes 50% or greater. Slope segments estimated as a single plane in the USLE were usually a poor representation of the topography. Complex slopes are represented in RUSLE to provide a closer approximation of the topography effect.

C *factor* RUSLE cover-management (C) is determined using a subfactor approach to compute soil-loss ratios as a function of five subfactors; prior land use, canopy cover, ground cover and within-soil effects.

The subfactor relationship is given by:

$$C = PLU \cdot CC \cdot SC \cdot SR \cdot SM \tag{14.3}$$

where:
- PLU = a prior land-use subfactor;
- CC = a canopy subfactor;
- SC = a surface-cover subfactor;
- SR = a surface-roughness subfactor; and
- SM = a soil-moisture subfactor.

The surface-cover (SC) term of the subfactors has the greatest effect on erosion estimates. The erosion reduction effect of surface cover can vary from 65 to 95% for a 50% surface cover. In the RUSLE this variable effectiveness is represented by the equation:

$$SC = \exp(-bM) \tag{14.4}$$

where SC is mulch or ground-cover subfactor value; M is percentage ground cover; and the b coefficient can be 0.025 (the value in the USLE), 0.035 (the RUSLE "typical" value), 0.045 for rangelands, or 0.05 for conditions where rill erosion dominates. The value used is based on the ratio of rill to interrill erosion. Ground cover in the USLE consisted of only vegetation and litter cover, whereas in RUSLE, ground cover (*M*) includes rock fragments with the vegetation and litter. Subfactor values for the within-soil effect are calculated from amount of biomass in the soil, which accumulates from roots and incorporated residue. Rangeland options for this within-soil component include default values, for 14 rangeland ecotypes, which can be used to estimate root mass (Weltz et al. 1987). Grazing effects on rangeland, pasture and meadows are reflected in the effect of canopy height, ground cover and root biomass.

P *factor* RUSLE *P* factors for rangeland are based on estimates of roughness and runoff reduction associated with the conservation practice and, for certain combinations of roughness and runoff reduction, the slope upon which the practice was implemented.

RUSLE measured vs. predicted Measured soil losses from rainfall simulation erosion plots from throughout the western United States were compared to soil losses estimated by both the USLE and the RUSLE (Renard & Simanton 1990). Correlations between measured and estimated values varied among the 17 sites tested. The agreement of the RUSLE-estimated and measured values was better than that for USLE estimates, for the natural and clipped treatments.

Water erosion prediction project

Process-based erosion prediction technology

The USLE, RUSLE and other current procedures for predicting erosion have been criticized as inadequately representing rangeland erosion processes. The USLE and RUSLE models are not process based and do not simulate interactions of water, soil, plant and management responses in a realistic manner to assess soil-erosion responses to rangeland management actions. Emerging and current technology, coupled with faster, larger and more readily available personal computers, have focused the need for a new process-based technology to predict and assess erosion and sedimentation rates on rangelands (Lane et al. 1988). In 1985 USDA and ARS identified the development of new erosion prediction technology as one of its top research goals. The erosion prediction technology was to be developed in three stages; development of a hillslope version that then could be incorporated into a watershed version, with the ultimate version being a grid or multi-watershed version (Foster & Lane 1987, Stone et al. 1990). The Water Erosion Prediction Project (WEPP) was initiated in 1985 to meet this goal and was designed to collect experimental field data from both crop and rangeland soil and vegetation complexes (Foster & Lane 1987, Lane & Nearing 1989, Nearing et al. 1990). Data collected from rainfall simulator plots during this field effort would be used to parameterize the WEPP model through development of relationships among soil properties, vegetation, cover, erosion, runoff and infiltration. Because of the many ecosystems and land uses to be evaluated, the field experiments would allow the definition of management impacts on rangeland productivity and conservation.

Soil property–erodibility relationships

One objective of the WEPP rainfall simulation experiments was to determine soil erodibility values for a wide range of soil types and conditions. Developing an equation that predicts soil erodibility from easily measured soil properties makes the WEPP model widely applicable. The soil erodibility nomograph used to estimate the *K* factor of the USLE illustrated that such relationships could be developed and applied to a wide range of soils. Uniform treatment of plots for erodibility measurements is imperative for valid soil property–erosion relationships to be developed. Many soils in arid and semi-arid regions have thin horizons (2–5 cm) that would be mixed during tillage, and the resulting mixture may not be representative of the soil surface subject to erosion. Additionally, large rocks in the soil may cause tillage to be impractical or may unduly alter surface roughness and depressional storage. Because of these and other potential problems with tillage, soil erodibility was determined from the bare-plot treatment described previously.

Rangeland site characteristics

Vegetation and soil-surface characteristics may have a greater influence on erosion and runoff rates from rangelands than basic soil properties (bulk density, soil texture, soil strength, etc.). The WEPP rangeland field experiments were designed to separate the effects of vegetation canopy cover on runoff and erosion from the effects of surface cover. Algorithms expressing infiltration rates as functions of total foliar and ground cover are currently being evaluated and will be incorporated into a more complex infiltration routine in the WEPP model. Time to peak discharge, concentrated flow paths, overland-flow velocities and associated shear stresses on the soil surface are all affected by the type, quantity and distribution of vegetation and surface cover. Root mass, standing biomass, litter, random roughness, ground surface cover and shrub density are important components of the plant community structure that may affect overland-flow routing and sediment yields.

WEPP procedures

Modifications
The Southwest Watershed Research Center in Tucson, Arizona was given the responsibility for development of rangeland erosion parameters to be incorporated in the WEPP erosion model. The WEPP field procedures used to evaluate the diversity of rangeland ecosystems were modifications of the rangeland USLE procedures previously described in this chapter (Simanton et al. 1987). These modifications were necessary because WEPP was to produce a process-based dynamic erosion model. Co-operating US Government agencies, Department of Energy laboratories, and universities in Nevada, Idaho, New Mexico, Washington and Utah also use these new procedures for their WEPP-related rangeland field experiments.

Equipment and procedures Electric solenoid valves were attached to the 130 mm h^{-1} nozzles on the rainfall simulator so that instantaneous changes in rainfall intensities could be made to better define infiltration and soil-erodibility parameters. The very wet simulation had varying rainfall intensity (65 and 130 mm h^{-1}) and an addition of overland flow for variable time periods. An example of the rainfall and overland-flow application sequences for the very wet run is presented in Figure 14.14. Each water application rate remained constant until adequate sediment samples were taken at each runoff equilibrium rate. This very-wet-run sequence provided soil infiltration and erosion data needed to define the WEPP model infiltration process and define the interrill and rill soil-erodibility parameters. Depending on soil erodibility, three or four rates of clear-water flow were applied at the upper end of the bare plots during the final 65 mm h^{-1} rainfall application of the very wet run (Fig. 14.14). Flow rates ranged from 45 to 200 mm h^{-1}, with the duration of application dependent on time to reach runoff equilibrium at each overland flow rate. Soil rill erodibility and critical shear were determined from erosion rates associated with these overland flow additions.

Large and small plots Two large plots (3.05 by 10.7 m) of each of the natural, clipped and bare-soil treatments were installed at each rangeland site. All plots at a site were grouped within a 50 m by 50 m area that was determined by the USDA Soil Conservation Service (SCS) to be in the same soil and vegetation type. As with the USLE plots, this plot size was necessary because it reduced the ratio of plot border effects to total plot area, and the long length allowed evaluation of rill erosion and sediment transport and deposition associated with sheet and concentrated flow. Metal sheets (2 mm thick by 15 cm wide and 3 m long) were used to form the sides and upper end

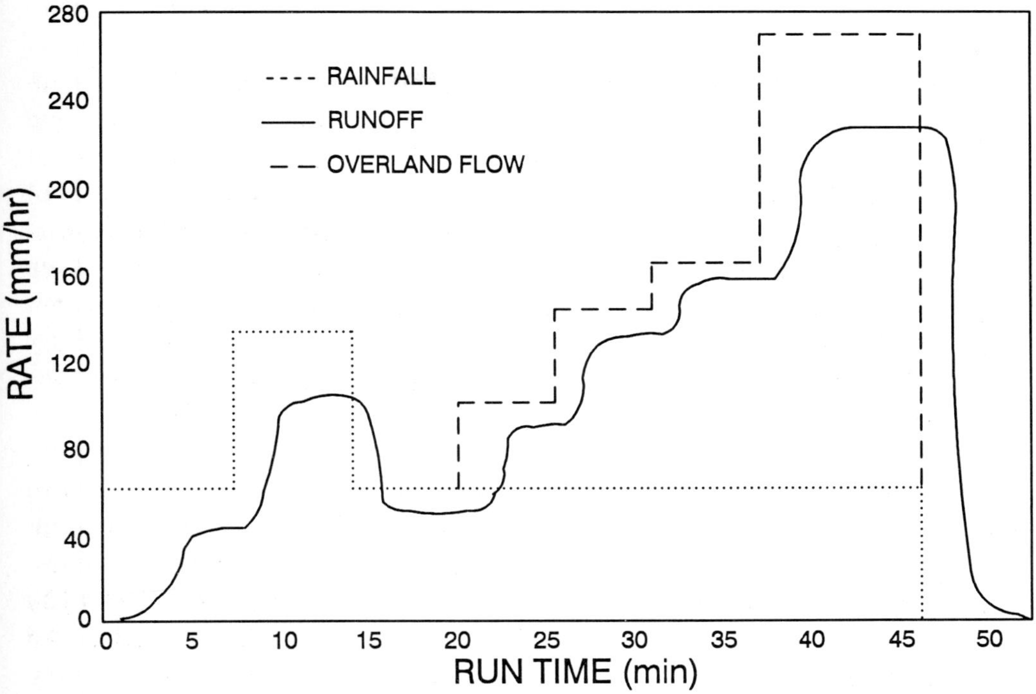

Figure 14.14 Sequence of rainfall application and overland-flow additions used during the very wet run of the WEPP rangeland field experiments.

of each plot. These sheets were inserted 3 cm into the soil so that a 12 cm high border delineated each plot. The downslope end of the plot had a 20 cm wide metal sheet, with a sill plate formed on the upper edge, inserted into the soil so that the sill plate was flush with the soil surface. Runoff and sediment from the plot was diverted into the runoff-measuring flume by troughs mounted below the sill plate. Interrill plots (0.6 by 1.2 m) were used to determine rates of erosion due to raindrops (soil interrill erodibility) as compared to the erosion rates due to the combination of raindrops and overland flow, produced on the longer, large plots. Also, effects of raindrop impact on soil crusting and infiltration were determined from comparisons between the two treatments on the interrill plots. Two interrill plots were installed next to each of the large plots of the bare-soil treatment. The interrill plots were treated in the same way as the large bare plot, with one of the interrill plots covered with window screen to dissipate raindrop impact and reduce soil-surface crusting. Interrill runoff hydrographs and sediment yields were determined from periodic (every 2 min during the rising hydrograph and at 5 min intervals during runoff equilibrium) volumetric samples manually collected during the rainfall simulations.

Biomass Vegetation-canopy and plot-surface characterizations were made with the same 49-pin point-meter described for the USLE experiments. Total above-ground herbaceous biomass was determined by clipping three 0.5 m by 1.0 m quadrats from the clipped and bare plots before they were treated. Above-ground woody biomass was determined by dimensional analysis using relationships between plant volume and weight. The relationships between leaf area and leaf weight were established from measurements taken at the time of simulation for the dominant plant species at each rangeland site. Below-ground biomass (excluding fauna) at each site was determined from soil cores taken after the wet runs. Microtopography (random roughness) of each plot was determined with a roughness meter (Kincaid & Williams 1966) and by photogrammetric methods.

Soils Detailed soil-pedon description, sampling and laboratory analysis were made by the SCS at each of the rangeland sites. Pedon analysis included particle-size distribution, soil moisture release curves, organic carbon, cation-exchange capacity, clay mineralogy and other physical and chemical properties. Soil surface (top 5 cm) bulk density was determined using the compliant cavity method before the dry and after the very wet runs. Soil surface (0–5 cm) and subsurface (5–20 cm) moisture contents were gravimetrically determined before the dry and wet runs and after the dry and very wet runs. Indices of soil strength were measured with the Torr Vane and pocket penetrometer after the dry and very wet runs. Bulk surface-soil samples collected prior to the dry run were sent to various laboratories for storage and subsequent testing. Undisturbed soil core samples taken after the very wet run were used for detailed morphological descriptions of the soil surface horizon and surface crust characteristics.

Rangeland sites A 2 year field programme began in 1987 to evaluate a wide range of rangeland soil/vegetation complexes in the western United States (Fig. 14.3, Table 14.3). Soils at the sites are in the orders of Mollisols, Alfisols, Entisols, and Inceptisols (Table 14.4). Moisture regimes are ustic, xeric and aridic. Surface textures range from loamy sand to clay, and many of the soils have appreciable contents of coarse rock fragments.

WEPP rangeland field data

Results

Preliminary results from these WEPP field studies include the determination of rangeland rill and interrill soil erodibility values for the WEPP model (Nearing

Table 14.3 Code, location and plant community of the Water Erosion Prediction Project's (WEPP) rangeland erosion plots.

WEPP code	Location	Plant community
A1	Walnut Gulch, AZ	Chihuahuan Desert shrub
A2	Walnut Gulch, AZ	Chihuahuan Desert grass
B1	Nevada Test Site, NV	Great Basin shrub
B2	Nevada Test Site, NV	Mojave Desert shrub
C1	Sonora, TX	Oak Savanna grass
D1	Chickasha, OK	Tallgrass prairie
D2	Chickasha, OK	Mixed grass prairie (reverted)
E1	Ft. Supply, OK	Mixed grass prairie
E2	Woodward, OK	Mixed grass prairie (cont. graze)
E3	Ft. Supply, OK	Mixed grass prairie
E4	Freedom, OK	Tall grass prairie (no graze)
E5	Freedom, OK	Mixed grass prairie (heavy
F1	Sidney, MT	Mixed grass prairie (club moss)
G1	Meeker, CO	Salt desert brush
H1	Cottonwood, SD	Mixed grass prairie (light graze)
H2	Cottonwood, SD	Shortgrass prairie (heavy graze)
I1	Los Alamos, NM	Pinyon–Juniper interspace
J1	Cuba, NM	Shortgrass desert grassland
K1	Susanville, CA	Great Basin shrub steppe
L1	Fresno, CA	Annual grassland

et al. 1989, Laflen et al. 1991), and the development of a crust factor for a Green Ampt infiltration model (Rawls et al. 1990).

Vegetation effects Because such a diverse range of soil/vegetation complexes were evaluated, the WEPP rangeland database provides an excellent opportunity to evaluate the direct effects of vegetation canopy cover on runoff and erosion (Simanton et al. 1991). Direct physical effects include interception losses, raindrop energy dissipation and surface roughness. Indirect effects include desirable levels of soil structure, organic matter, macro-porosity and litter cover. Separation of the direct and indirect effects of vegetation on runoff and erosion is possible by comparing responses of the natural and clipped plots within a site. To normalize differences in soil moisture content that could be present for the dry run, only runoff and erosion results from the wet run (field capacity) were analysed. Although run times and rainfall intensities were to be the same for each rainfall simulation and plot, water supply and wind problems sometimes caused different rainfall volumes to be applied. To account for these application differences in comparing plot runoff responses, a runoff ratio (*Q/P*) was determined for each plot by dividing total runoff volume (*Q*) by total rainfall volume (*P*) applied during the simulation.

Erosion rates (kg ha^{-1} Qmm^{-1}) were calculated by dividing total plot sediment yield by total plot runoff volume. Final infiltration rate (mm h^{-1}) was the difference between the rainfall rate and equilibrium runoff rate. Initial rainfall abstraction (mm) was calculated as the rainfall volume applied to the plot before runoff occurred. Initial infiltration rate was calculated as the difference between the rainfall rate (mm h^{-1}) and the runoff rate 5 min after runoff began. Comparisons of runoff ratios, erosion rates, initial and final infiltration rates, initial rainfall abstraction, soil moisture content and plot surface characteristics between paired natural and clipped plots were made using linear regression analysis and the corresponding 95% confidence interval of the regression line slope and intercept.

Except for comparisons of random roughness, soil moisture and plot slope, 21 natural/clipped pairs were available for comparison. Random roughness was measured only in 1988 so there were only seven pairs available. Soil moisture data were lost for one, thus giving only 19 pairs for comparison. Plot slopes did not change between 1987 and 1988 so the plots re-evaluated in 1988 were not included in the slope comparisons.

Five rangeland sites were selected in Oklahoma and represent grass prairies and shrub steppes of the Great Plains. The site near Chickasha (D1) is typical of native tallgrass prairie and had been lightly grazed prior to the 1987 evaluation and moderately grazed prior to the 1988 evaluation. The site is located on ARS watershed R-5 which has been used for extensive hydrologic and erosion studies (Sharma et al. 1980). The two sites at Ft. Supply (E1 and E3) represent different range conditions on mixed grass prairie that had intermixed brush. The sites are less than 1 km apart and both had been grazed. Two sites at Freedom (E4 and E5) were adjacent to one another, separated only by a fence. E4 had not been grazed in over 10 years, whereas E5 had been continuously heavily grazed. The site near Sidney, Montana (F1) had been lightly grazed and represents rangelands whose soil surface cover includes large amounts of club mosses. The salt desert shrub site near Meeker, Colorado (G1) had not been grazed within a year of the evaluation. This site represents a rangeland soil that is susceptible to rill formation. The grassland site near Cottonwood, South Dakota (H1) was lightly grazed prior to 1987 and moderately grazed prior to the 1988 evaluations. The brush site (K1) near

Table 14.4 Code, series, classification and texture of soils of the Water Erosion Prediction Project's (WEPP) rangeland erosion plots.

WEPP code	Soil series	Soil classification	Soil texture
A1	Stronghold	Coarse, loamy, mixed, thermic Ustochreptic Calciorthid	Gravelly sandy loam
A2	Forrest	Fine, mixed, thermic, Ustolloc Haplargid	Sandy clay loam
B1	NA	Shallow, mixed, tyhermic, Typic Durorthid	Gravelly fine sandy loam
B2	NA	Clayey-skeletal, montmorillonitic, thermic Haplic Nadurargid	Fine sandy loam
C1	Purves	Clayey, montmorillonitic, thermic Calciustoll	Silty clay loam
D1	Grant	Fine-silty, mixed thermic, Udic Argiustoll	Loam
D2	Grant, eroded	Fine-silty, mixed thermic, Udic Argiustoll	Very fine sandy loam
E1	Pratt	Sandy, mixed, thermic, Psammentic Haplustalf	Loamy fine sand
E2	Quinlan	Loamy, mixed, thermic, shallow, Typic Ustochrept	Loam
E3	Tivoli	Mixed, thermic, Typic Ustipsamment	Fine sand
E4	Woodward	Coarse-silty, mixed, thermic, Typic Ustochrept	Very fine sandy loam
E5	Woodward	Coarse-silty, mixed, thermic, Typic Ustochrept	Very fine sandy loam
F1	Vida	Fine-loamy, mixed Typic Argiboroll	Loam
G1	Degater	Fine, mixed, mesic, Typic Camborthid	Silty clay
H1	Pierre	Very-fine, montmorillonitic, mesic, Typic Torrent	Clay
H2	Pierre	Very-fine, montmorillonitic, mesic, Typic Torrent	Clay
I1	Hackroy	Loamy, mixed, mesic, shallow Aridic Haplustalf	Fine sandy loam
J1	Querencia	Coarse-silty, mixed, mesic Ustollic Camborthid	Fine sandy loam
K1	Jauriga	Fine-loamy, mixed, mesic, Typic Argixeroll	Gravelly sandy loam
L1	Apollo	Fine-loamy, mixed, thermic, Calcic Haploxeroll	Loam

Susanville, California had not been grazed for 1 year prior to evaluation.

Within-site effects Even though there was considerable effort in the site selection process to ensure homogeneity in soil and vegetation properties, variability within the site can occur (Devaurs & Gifford 1984). Comparisons of natural and clipped-plot characteristics of percentage litter, exposed soil, slope, and random roughness were made to strengthen the assumption that each site was homogeneous.

Plot characteristic comparisons of litter, bare soil, random roughness and slope are shown in Figure 14.15. The regressions of these comparisons have line slope coefficients of nearly 1.0 and relatively small intercepts. Confidence-interval tests at the 95% level showed that in all plot characteristic comparisons the regression intercept was not different from zero and the regression slope was not different from one; indicating that measured plot characteristics were not significantly different between the two treatments.

Natural vs. clipped Regression analysis indicated no difference between the natural and clipped plots' runoff ratios, final infiltration rates, soil moisture contents, initial abstractions and initial infiltration rates (Fig. 14.16). Erosion rates between the natural and clipped plots were different (Fig. 14.16). Though the intercept of the erosion rate comparison was not different from zero, the regression line slope was different from one. The regression line slope of less than one and the relatively small intercept indicate that the clipped plots had an erosion rate less than the natural plots.

The graph of initial abstraction (Fig. 14.16) shows two distinct groupings of points. The high initial abstractions were associated with sites with either very porous soil (E1 and E3) or sites that had been lightly or ungrazed 2 or more years prior to our evaluations (E4 and 1987 H1). The two extreme points on the erosion-rate graph in Figure 14.16 represent erosion rates from the G1 site. These relatively high erosion rates were ascribed to rill erosion. This site was the only one with noticeable rills on the plots before and after the rainfall simulations. Meyer et al. (1975) found, in rainfall simulation studies on tilled 6% sloped plots, that rilled plots produced about 3.4 times the soil loss of non-rilled plots. The G1 plots, sloped at 9–11%, produced 3.3 times the average erosion rate of comparably sloped plots at other sites evaluated (sites F1 and K1).

WEPP discussion

Site homogeneity is demonstrated by comparisons made of litter, exposed soil,

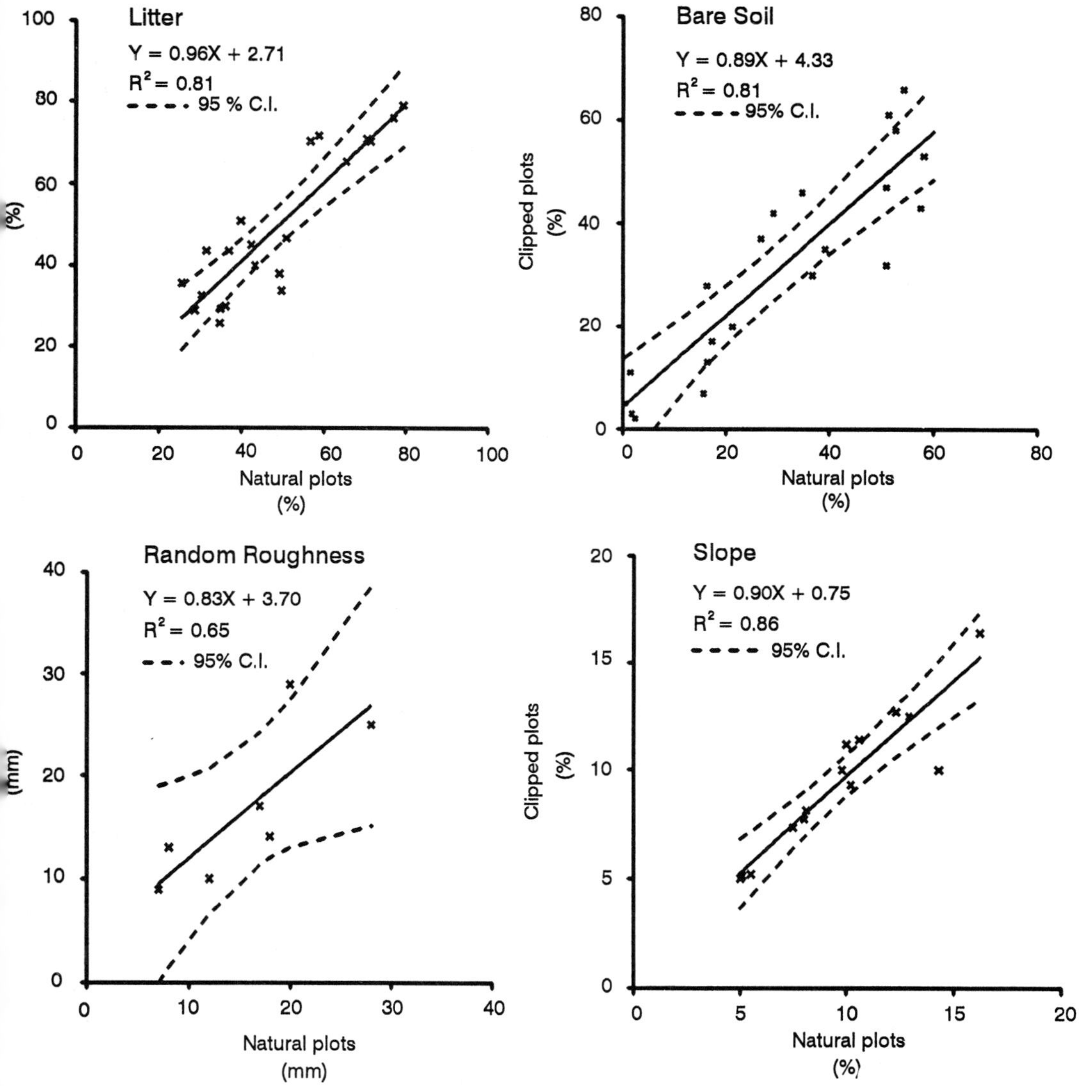

Figure 14.15 Comparison of natural and clipped-plot litter and bare-soil cover, random roughness and slope with regression equation, coefficient of determination and 95% confidence interval (CI) (from Simanton et al. 1991).

slope and random roughness. The similarity in soil moisture eliminates antecedent moisture effects and helps isolate the canopy-cover effects. Factors not quantified, such as distribution of vegetation and ground-surface cover on each plot, could have an effect on plot runoff and erosion response. Because plot treatments were randomly imposed at each site, these effects should also be random and not be biased for either treatment.

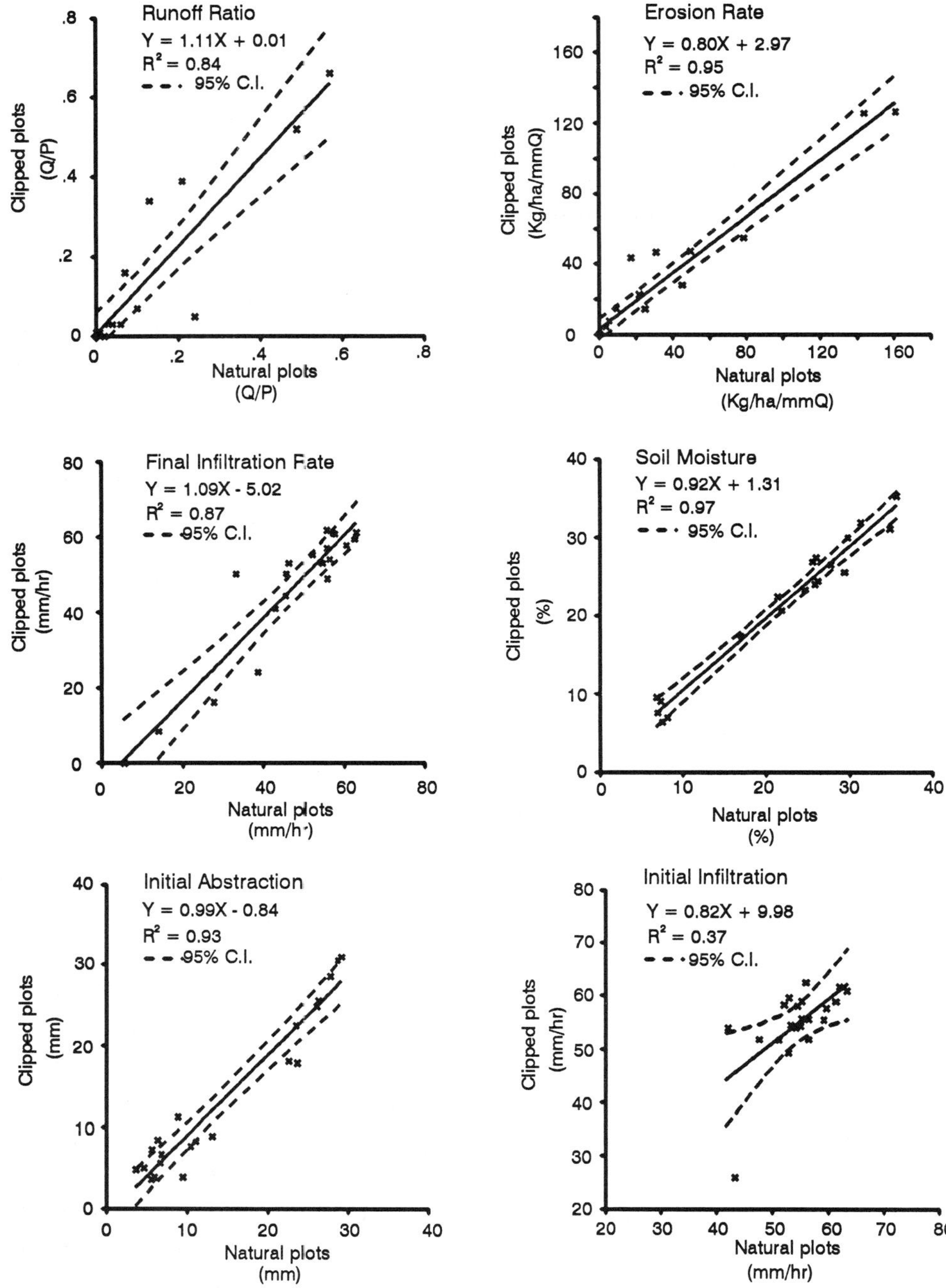

Figure 14.16 Comparison of natural and clipped plot runoff ratios, erosion rates, final infiltration rate, soil moisture, initial rainfall abstractions and initial infiltration with regression equation, coefficient of determination and 95% confidence interval (CI) (from Simanton, et al. 1991).

Final infiltration rate is a function of soil properties, ground-surface cover and rainfall application rate. If, in a rainfall simulation under conditions of field capacity soil moisture, application rate does not substantially exceed the infiltration rate, runoff equilibrium may not occur, or may occur late in the rainfall simulation. This indicates that the final infiltration rate was limited by the application rate rather than inherent soil or cover properties. This was the situation at sites E1 and E3 where runoff did not occur during the simulation and the infiltration rates for these soils have been reported to range from 55 to 270 mm h^{-1} (Rhoades et al. 1964). Runoff equilibrium was usually reached 10–15 min into our 30 min wet run. This relatively short period to reach equilibrium indicates that soil properties were controlling the final infiltration rate and that rainfall application rate was not limiting.

Initial rainfall abstractions are a function of the initial soil infiltration rate, surface roughness and storage, and interception losses to vegetation canopy (USDA, SCS 1972). These abstractions were nearly identical for the natural and clipped plots, as indicated by the very low intercept (−0.84) and 0.99 regression slope. Infiltration rate affects initial abstractions, but comparisons of initial infiltration rate indicated that there were no differences in the rate between the natural and clipped plots. There was no difference in litter, bare soil, random roughness or initial abstractions between treatments. As a result, it can be inferred that interception loss by canopy cover was not a significant component to initial abstraction. Interception losses are more significant under conditions of lower rainfall intensities and amounts than those used in this study (Thurow et al. 1987). The significance of canopy cover in erosion prediction models is small compared to ground surface cover (Table 10 in Wischmeier & Smith 1978). Canopy cover's direct influence on erosion is through its dissipation of raindrop energy, with the magnitude of this influence dependent on the ground surface condition found under the canopy. The more bare soil under the canopy, the larger the canopy cover effect (Wischmeier & Smith 1978). Bare soil on the rangeland natural and clipped plots averaged 30.6% and, of this, only 8% was under canopy. Under this condition, and assuming that only bare soil would be detached by raindrops, the clipped-plot sediment yield would not be significantly affected by the soil eroded from the now canopy-free bare soil. Khan et al. (1988) found that as canopy-cover height increased from 0.25 m to 1.0 m the erosion rate also increased. The clipped plots had a canopy height of 0.02 m and the natural plots canopy height could be as great as 1 m. This may explain the slightly greater erosion rates from our natural plots.

The clipped treatment in the WEPP rangeland experiments was designed to separate, into direct and indirect effects, the interacting relationships among vegetation canopy cover, runoff, infiltration and erosion. Because the same procedures, plot size and rainfall simulator were used in the WEPP rangeland experiments, the unique dataset allows direct comparisons of runoff and erosion responses over a wide variety of rangeland soils and vegetation types. Comparisons indicate that, under the simulated rainfall conditions and soil and vegetation types evaluated, canopy cover had little direct effect on runoff, infiltration, initial abstractions and erosion rate. Canopy cover's direct contributions to interception losses and soil-surface protection from raindrop impact are not large in rangeland runoff and erosion responses.

Summary

Ten years of rainfall simulation studies on rangeland erosion plots have produced a large database used to parameterize models for predicting soil erosion. These studies, conducted over a very wide range of rangeland ecosystems represent a unique database that will be very difficult to duplicate. Rangeland erosion studies are a relatively new research area and the results from our studies have only begun to answer some of the basic questions regarding techniques for estimating erosion on rangelands. Additional studies, research approaches, and analyses are still needed to fully understand the rangeland erosion processes.

USLE Four years of seasonal rainfall simulation studies on rangeland USLE-type plots have indicated that erosion and runoff rates, per unit of *EI*, change with time for the first 1–2 years and then tend to reach an equilibrium rate. Associated with these changes were changes in the USLE *K* (bare plots) and *C* factors, vegetation canopy, and amount of bare-soil accumulation on the plot surface (natural plot). If data are to be used in erosion models that estimate long-term management effects on erosion, the rainfall simulation database needs to extend for more than 1 year. The importance of erosion pavement on the erosion process of western rangelands has been demonstrated and appears to be more dominant than vegetation canopy.

RUSLE New algorithms have been developed to reflect rangeland response to the rainfall (*R*), topography (*LS*), cover-management (*C*) and conservation practice (*P*) factors of the RUSLE.

WEPP Data from the WEPP rangeland experiments represent a wide range of soils and vegetation types and can provide a better understanding of the processes involved in the interactions among vegetation, soil, runoff and erosion.

References

Branson, F. A., G. F. Gifford, K. G. Renard, R. F. Hadley 1981. *Rangeland hydrology*. Denver: Society for Range Management

Devaurs, M. & G. F. Gifford 1984. Variability of infiltration within large runoff plots on rangelands. *Journal of Range Management* **37**, 523–8.

Dissmeyer, G. E. & G. R. Foster 1980. *A guide for predicting sheet and rill erosion on forest land*. Technical Publication SA-TP 11, USDA-FS.

Dissmeyer, G. E. & G. R. Foster 1981. Estimating the cover-management factor (C) in the Universal Soil Loss Equation. *Journal of Soil and Water Conservation* **36**, 235–40.

Dixon, R. M. & J. R. Simanton 1979. Water infiltration processes and air–earth interface conditions. In *Surface and subsurface hydrology*, H. J. Morel-Seytoux et al. (eds), 314–30. Littleton: Water Resources Publication.

Foster, G. R. & L. J. Lane 1987. *User requirements: USDA-Water Erosion Prediction Project (WEPP)*. NSERL Report No.1. West Lafayette: USDA-ARS National Soil Erosion Research Laboratory.

Gelderman, F. W. 1970. *Soil survey, Walnut Gulch experimental watershed*. Arizona Special Report, USDA-SCS.

Khan, M. J., E. J. Monke, G. R. Foster 1988. Mulch cover and canopy effect on soil loss. *Transactions of the American Society of Agricultural Engineers* **31**, 706–11.

Kincaid, D. R. & G. Williams 1966. Rainfall effects on soil surface characteristics following range improvement treatments. *Journal of Range Management* **19**, 346–51.

Laflen, J. M., W. J. Elliot, J. R. Simanton, C. S. Holzhey, K. D. Kohl 1991. WEPP: Soil erodibility experiments for rangeland and cropland soils. *Journal of Soil and Water Conservation* **46**, 39–44.

Lane, L. J. & M. A. Nearing (eds) 1989. *USDA-Water erosion prediction project: Profile model documentation*, NSERL Report No. 2. West Lafayette: National Soil Erosion Research Laboratory.

Lane, L. J., J. R. Simanton, T. E. Hakonson, E. M. Romney 1987. Large-plot infiltration studies in desert and semiarid rangeland areas of the Southwestern USA. In *International Conference on infiltration development and application*, Jan. 1987, 365–76. University of Hawaii at Manoa.

Lane, L. J., E. D. Shirley, V. P. Singh 1988. Modeling erosion on hillslopes. In *Modeling geomorphological systems*, M. G. Anderson (ed.), 287–308. Chichester, England: John Wiley.

Meyer L. D., G. R. Foster, M. J. M. Römkens 1975. Source of soil eroded by water from upland slopes. In *Present and prospective technology for predicting sediment yields and sources, Proceedings of sediment-yield workshop*, ARS-S-40, 77–89. Oxford, MS: USDA Sediment Laboratory.

Nearing, M. A., D. I. Page, J. R. Simanton, L. J. Lane 1989. Determining erodibility parameters from rangeland field data for a process-based erosion model. *Transactions of American Society of Agricultural Engineers* **32**, 919–24.

Nearing, M. A., G. R. Foster, L. J. Lane, S. C. Finkner 1990. A process-based soil erosion model for USDA Water Erosion Prediction Project technology. *Transactions of American Society of Agricultural Engineers* **32**, 1587–93.

Neff, E. L. (ed.) 1979. *Rainfall simulator workshop*, ARM-W10. Tucson: USDA-SEA.

Osborn, H. B., K. G. Renard, J. R. Simanton 1979. Dense networks to measure convective rainfall in the Southwestern United States. *Water Resources Research* **15**, 1701–11.

Rawls, W. J., D. L. Brakensiek, J. R. Simanton, K. D. Kohl 1990. Development of a crust factor for a Green Ampt model. *Transactions of American Society of Agricultural Engineers* **33**, 1224–8.

Renard, K. G. & J. R. Simanton 1990. Application of RUSLE to rangelands. In *Proceedings of Conference on Watershed Planning and Analysis in Action*, 164–73. American Society of Civil Engineers, Irrigation and Drainage Division.

Renard, K. G., G. R. Foster, G. A. Weesies, J. P. Porter 1991. RUSLE: Revised universal soil loss equation. *Journal of Soil and Water Conservation* **46**, 30–3.

Rhoades, E. D., L. F. Locke, H. M. Taylor, E. H. McIlvain 1964. Water intake on a sandy range as affected by 20 years of differential cattle stocking rates. *Journal of Range Management* **17**, 185–90.

Romney, E. M., V. Q. Hale, A. Wallace, J. Childress, T. L. Ackerman 1973. *Some characteristics of soil and perennial vegetation in northern Mojave Desert areas of the Nevada Test Site*. University of California, Los Angeles, Laboratory of Nuclear Medicine and Radiation Biology Report, UCLA 12-916.

Sharma, M. L., G. A. Gander, C. G. Hunt 1980. Spatial variability of infiltration in a watershed. *Journal of Hydrology* **45**, 101–22.

Simanton, J. R. 1991. Revegetation of semiarid rangelands: problems, procedures, probabilities. *Rangelands* **13**, 129–32.

Simanton, J. R. & K. G. Renard 1982. Seasonal change in infiltration and erosion from USLE plots in southeastern Arizona. *Hydrology and Water Resources in Arizona and the Southwest* **12**, 37–46.

Simanton, J. R. & K. G. Renard 1986. Time related changes in rangeland erosion. In *Erosion on rangelands: emerging technology and data base: Proceedings of the rainfall simulator workshop*, 14–15 January 1985, L. J. Lane (ed.), 18–22. Denver: Society for Range Management.

Simanton, J. R., H. B. Osborn, K. G. Renard 1980. Application of the USLE to Southwestern rangelands. *Hydrology and Water Resources in Arizona and the Southwest* **10**, 213–20.

Simanton, J. R., E. Rawitz, E. D. Shirley 1984. The effects of rock fragments on erosion of semiarid rangeland soils, In *Erosion and productivity of soils containing rock fragments*. Soil Science Society of America Special Pub. 13. 65–72. Madison: Soil Science Society of America.

Simanton, J. R., C. W. Johnson, J. W. Nyhan, E. M. Romney 1985. Rainfall simulation on rangeland erosion plots. In *Erosion on rangelands: emerging technology and data base: Proceedings of the rainfall simulator workshop*, 14–15 January 1985, L. J. Lane (ed.), 11–17. Denver: Society for Range Management.

Simanton, J. R., L. T. West, M. A. Weltz, G. D. Wingate 1987. *Rangeland experiments for water erosion prediction project*. American Society of Agricultural Engineers, Paper No. 87-2545. St. Joseph, MI: American Society of Agricultural Engineers.

Simanton, J. R., M. A. Weltz, H. D. Larsen 1991. Rangeland experiments to parameterize the water erosion prediction project model: vegetation canopy cover effects. *Journal of Range Management* **44**, 276–82.

Stone, J. J, V. L. Lopes, J. J. Stone 1990. Water Erosion Prediction Project (WEPP) watershed model: Hydrologic and erosion calculations. In *Proceedings of watershed planning and analysis in action*, 184–90. American Society of Civil Engineering, Committee on Watershed Management & Irrigation & Drainage Division.

Swanson, N. P. 1965. Rotating-boom rainfall simulator. *Transactions of American Society of Agricultural Engineers* **8**, 71–2.

Thurow, T. L., W. H. Blackburn, S. D. Warren, C. A. Taylor, Jr 1987. Rainfall interception by midgrass, shortgrass, and live oak mottes. *Journal of Range Management* **40**, 455–60.

US Department of Agriculture, Soil Conservation Service 1972. Section 4, *Hydrology, national engineering handbook*. Washington, DC: US Government Printing Office.

Van Doren, D. M., W. C. Moldenhauer, G. B. Triplett, Jr 1984. Influence of long-term tillage and crop rotation on water erosion. *Soil Science Society of America Journal* **48**, 636–40.

Weltz, M. A., K. G. Renard, J. R. Simanton 1987. Revised universal soil loss equation for Western rangelands. In *US/Mexico Symposium on Strategies for Classification and Management of Native Vegetation for Food Production in Arid Zones*, 104–11. USDA-FS General Technical Report RM-150.

Whitford, W. G., Y. Steinberger, G. Ettershank 1982. Contributions of subterranean termites to the "economy" of Chihuahuan desert ecosystems. *Ecologia* **55**, 298–302.

Wischmeier, W. H. & J. V. Mannering 1969. Relation of soil properties to its erodibility. *Soil Science Society of America Proceedings* **33**, 131–7.

Wischmeier, W. H. & D. D. Smith 1978. *Predicting rainfall erosion losses – a guide to conservation planning*. USDA Agriculture Handbook No. 537. Washington, DC: US Government Printing Office.

Wischmeier, W. H., C. B. Johnson, B. V. Cross 1971. A soil erodibility nomograph for farmland and construction sites. *Journal of Soil and Water Conservation* **26**, 189–92.

15

Description of the US Department of Agriculture water erosion prediction project (WEPP) model

L. J. Lane, M. A. Nearing, J. M. Laflen, G. R. Foster, M. H. Nichols

Abstract

The USDA Water Erosion Prediction Project (WEPP) was initiated in 1985 to develop new generation water erosion prediction technology for use in soil and water conservation and in environmental planning and assessment. The WEPP computer models represent erosion technology based on fundamentals of infiltration, surface runoff, plant growth, residue decomposition, hydraulics, tillage, management, soil consolidation and erosion mechanics. Process-based erosion models provide several major advantages over empirically based erosion prediction technology, including most notably: (a) capabilities for estimating spatial and temporal distributions of net soil loss; and (b) being process-based, the model can be extrapolated to a broad range of conditions which may not be practical or economical to field test. Soil detachment, transport and deposition processes are represented in the models using a steady-state sediment continuity equation which represents rill and interrill processes. Rill detachment rate is dependent upon the ratio of sediment load to transport capacity, rill erodibility, hydraulic shear stress, surface cover, below-ground residue and consolidation. Rill hydraulics are used to calculate shear stresses in rills. Net deposition is calculated when sediment load is greater than transport capacity. Interrill erosion is represented as a function of rainfall intensity, ground cover, canopy cover and interrill soil erodibility. The models are designed to accommodate spatial and temporal variability in topography, surface roughness, soil properties, hydrology and land-use conditions on hillslopes. A process-based erosion model used with a process-based hydrologic model, a daily water-balance model, a plant growth and residue-decomposition model, a climate generator and a soil-consolidation model represents a potentially powerful tool for estimating soil loss and selecting agricultural management practices for soil conservation.

Introduction

The United States Department of Agriculture (USDA) Water Erosion Prediction

Project (WEPP) was initiated in 1985 with the stated objective: "To develop new generation water erosion prediction technology for use by the USDA–Soil Conservation Service (SCS), USDA–Forest Service (FS), United States Department of Interior (USDI)–Bureau of Land Management (BLM), and other organizations involved in soil and water conservation and environmental planning and assessment." The new erosion prediction technology is based on modern hydrologic and erosion science and is process oriented. The first version of the technology was delivered to user agencies in August 1989. The technology will undergo extensive testing and evaluation by user groups while research continues to refine the relationships in the model. Delivery of the version intended for general use is expected in 1993. It is anticipated that the WEPP technology will eventually replace the Universal Soil Loss Equation for routine assessment of soil erosion and planning purposes in the United States.

The WEPP technology consists of three computer models: a profile version, a watershed version and a grid version. The profile version computes soil detachment and deposition on a hillslope profile and provides the basis for the other two versions. The profile version applies to hillslopes similar to those for the USLE, except that the WEPP model computes both detachment and deposition on the hillslope, as well as the net total soil loss from the end of the slope. The watershed and grid versions can estimate net soil loss or gain over a small watershed or field-sized area at all points including channels. The models are intended to incorporate the influence of climate, soils, topography, management and supporting practices on erosion, deposition, sediment yield and sediment size distributions over the area of interest. The models are based on continuous simulation, and output from the models include predictions of the net soil loss or gain at each point on the hillslope for all times of the year. Detailed goals for the project were formulated with specific input by expected users of the technology and those involved with the technical development of the model (Foster & Lane 1987). The objective of this chapter is to present a summary of the WEPP profile version erosion-prediction technology with emphasis on the erosion calculations within the model.

Model summary

The profile version of the WEPP model will be executed primarily as a continuous simulation model, although it can be run on a single-storm basis. Continuous simulation means that the processes that influence erosion, including management practices and climate, are modelled as a function of time. For example, surface residue may influence the amount of soil lost during a given rainfall event. A plant-growth and residue-decay model within the WEPP model estimates the amount of crop residue on the soil surface for each day of the year. The model adjusts surface cover as a function of leaf drop during senescence and residue remaining after harvesting. The amount of residue buried during tilling is also used by the plant-growth and residue-decay model. Most calculations in the WEPP model are made on a daily time step. Soil parameters, residue amounts, crop growth, soil water content, surface roughness and other adjustments to model parameters are also made on the daily time step.

Because the model inputs are in terms that the general user understands: planting dates, tillage dates, harvest dates, yields, implement types, etc. the WEPP model is user friendly. Various sources are available to provide technical information that is required to run the WEPP model. Climatic information, for instance, can be generated by the CLIGEN model, which is a stochastic weather generator. Crop-specific information, such as growth parameters, will be provided to the model user by Agricultural Research Service (ARS) and SCS technical experts. Soils information required by the model is available from an SCS soil characterization database which contains information routinely collected for soil surveys. Required topographic information is compatible with current methods of measuring slope profiles in the field.

Model structure

The WEPP profile computer model includes six major components: climate, infiltration, water balance, plant growth and residue decomposition, surface runoff and erosion. A brief description of each major component is given below.

The climate component, CLIGEN (Nicks 1985), is run separately from WEPP. The CLIGEN model generates rainfall amount, duration, maximum intensity, time to peak intensity, maximum and minimum temperature, solar radiation and wind speed and direction for the on-site location. Output from CLIGEN is stored in a file which is read by the WEPP model. Temperature determines whether precipitation takes the form of rain or snow, and wind speed and direction are used to determine the redistribution of snow on the slope profile. Runoff and erosion caused by snowmelt are also calculated.

The number and distribution of precipitation events are generated using a two-state Markov chain model. Given the initial condition that the previous day was wet or dry, the model determines stochastically if precipitation occurs on the current day. A random number (0–1) is generated and compared with the appropriate wet–dry probability. If the random number is less than or equal to the wet–dry probability, precipitation occurs on that day. Random numbers greater than the wet–dry probability give no precipitation. When a precipitation event occurs, the amount of precipitation is determined from a skewed normal distribution function. The rainfall duration for individual events is generated from an exponential distribution using the monthly mean durations. The amount of daily precipitation is partitioned between rainfall and snowfall using daily air temperature. If the average daily air temperature is 0°C or below, precipitation is considered to be snowfall. Daily maximum and minimum temperatures and solar radiation are generated from normal distribution functions. A disaggregation model is included to generate time–rainfall intensity data or breakpoint data from daily rainfall amounts. Given a rainfall amount and rainfall duration, the disaggregation model derives a rainfall intensity pattern with properties similar to those obtained from analysis of breakpoint data. The breakpoint rainfall data are required by the infiltration component to compute rainfall excess rates and runoff.

The infiltration component of the hillslope model is based on the Green and Ampt equation, as modified by Mein & Larson (1973), with the ponding-time calculation for an unsteady rainfall (Chu 1978). The infiltration process is divided into two distinct stages: a stage in which the ground surface is ponded with water and a stage without surface ponding. During an unsteady rainfall the infiltration process may change from one stage to another. Under a ponded surface the infiltration process is independent of the effect of the time distribution of rainfall. When the infiltration rate reaches its maximum capacity it is referred to as the infiltration capacity. At this stage, rainfall excess is computed as the difference between rainfall rate and infiltration capacity. Without surface ponding, all the rainfall infiltrates into the soil. Under these conditions, infiltration rate equals the rainfall intensity, which is less than the infiltration capacity, and rainfall excess is zero.

The water-balance and percolation component of the profile model is based on the water-balance component of SWRRB (Simulator for Water Resources in Rural Basins) (Williams & Nicks 1985), with some modifications for improving estimation of percolation and soil-evaporation parameters. The water-balance component will estimate daily snowpack evaporation and melt, potential evapotranspiration, soil and plant evaporation, soil-water content in the root zone, and percolation throughout the simulation period. The water-balance component has been designed to use input from the climate (daily precipitation, temperature and solar radiation), infiltration (infiltrated water volume) and plant-growth (daily leaf-area index, root depth and residue cover) components.

The plant-growth component of the WEPP model simulates plant growth and residue decomposition for cropland and rangeland conditions. The purpose of this component is to simulate temporal changes in plant variables that influence the runoff and erosion processes. Crop-growth variables computed in the cropland model include growing degree-days, mass of vegetative dry matter, canopy cover and height, root growth, leaf-area index, plant basal area, etc. (Alberts et al. 1989). The effect of tillage on residue and soil properties is also included in the model. The rangeland plant-growth model estimates the initiation and growth of above- and below-ground biomass for range plant communities by using a unimodal or a bimodal potential growth curve. Range plant variables computed in the rangeland model include plant height, litter cover, foliar canopy cover, ground surface cover, exposed bare soil and leaf-area index (Weltz & Arslan 1990). The cropland plant-growth and decomposition models will accommodate mono-, double, rotation and strip cropping practices. The user is asked to select the desired cropping practice option. In the current model, crop choices in double cropping, rotation and strip cropping systems are limited to annual crops specified in the WEPP User Requirements (Foster & Lane 1987) plus perennial crops of alfalfa and grasses. A challenge for the next few years is to develop a method that would allow parameterization of any crop for the WEPP model using standard reproducible techniques.

Many of the soil parameters that are used in the hydrology and erosion calculations change with time as a result of crop growth stage, tillage operations, soil freezing and thawing, compaction, weathering or past history of precipitation. The soils component makes adjustments to soil properties on a daily time step. Examples of temporally varying factors include soil bulk density, saturated conductivity, surface roughness and erodibility parameters.

Erosion from areas irrigated using stationary sprinkler or furrow irrigation systems can be estimated using the irrigation component of the WEPP model. The stationary sprinkler systems include solid set, sideroll, and handmove systems. Stationary irrigation systems provide water to all locations within an area simultaneously and thus simulate natural rainfall of uniform intensity. Furrow irrigation systems supply water to the upper end of a furrow with channel hydraulics determining advance and infiltration along the length of the furrow. Either natural precipitation or irrigation events may cause erosion. The relative contribution of these processes to runoff and soil loss from an irrigated area can be identified by the irrigation component of the profile model. If irrigation is available, the user can choose one of three scheduling options. The first option uses available soil moisture depletion criteria. This option requires a data file of irrigation periods when irrigation is allowed. The model determines when the irrigation will occur and the depth applied. A second option uses a database of irrigation dates and depths. The final option allows a combination of the first two options.

Surface runoff is represented in two ways in the WEPP model:

(a) broad, uniform sheet flow is assumed for the overland-flow routing to calculate the overland flow hydrograph, with hydraulic roughness terms being weighted averages of the rill and interrill areas;

(b) flow is partitioned into broad sheet-flow for interrill erosion calculations and concentrated flow for rill erosion calculations.

The proportion of the area in rills is represented by a rill density statistic (equivalent to a mean number of rills per unit area) and an estimated rill width. Representative rill cross sections are based on the channel calculations derived from extensive field experimentation. Depth of flow, velocity, and shear stress in the rills are calculated assuming rectangular channel cross sections. The erosion calculations are then made for a constant rate over a characteristic time to produce estimates of erosion for the entire runoff event.

The erosion component of the model uses a steady-state sediment continuity equation which calculates net values of detachment or deposition rates along the hillslope profile. The erosion process is divided into interrill and rill components, with the interrill areas supplying sediment to the rills or small channels. Within the rills, the sediment may be carried down slope or deposited in the rill. Scour by rill flow is calculated for the case when flow shear exceeds critical shear of the soil and when sediment load is less than calculated sediment capacity. The erosion component of the model is discussed in more detail in a later section.

Model inputs and outputs

Four input data files are required to execute the WEPP profile model: (a) a climate file; (b) a slope profile file; (c) a soil file; and (d) a management file. For the case of irrigation, additional input files are required.

CLIGEN is used to generate the climate file for the continuous simulation option of the WEPP model. Model use and climate at the location where the model is to be applied determine the most appropriate number of years of simulated climatic data. Three years of simulation are normally adequate (given the current set-up of CLIGEN) for comparing various management practices for making soil-conservation decisions. More than 3 years will be required for climates which are semi-arid or arid, or if more accurate long-term predictions of soil loss are desired. The model will not run partial years of simulation. It will not normally be feasible for the user to generate climate files without the aid of CLIGEN for the continuous simulation option of the model.

The slope profile is described by length-slope pairs starting at the upper end of the hillslope. Breakpoints for the end of input segments should be made at the locations on the hillslope of the most obvious changes in slope. A typical S-shaped profile, for instance, might best be described by three input segments: a relatively flat segment at the upper end of the hillslope, a steeper mid-segment and a flatter end-segment at the toe of the slope. Slope length does not end where deposition begins. The slope profile must be described to the end of the field, or to a concentrated flow channel, grass waterway or terrace. The point where detachment ends and deposition begins is calculated by the model and given as output. Representative slope profiles must be chosen by the user for building the slope input file for the field.

Downslope variability is accommodated in the model by dividing the slope profile into overland-flow elements. An overland-flow element is defined as a section of the hillslope which is homogeneous in terms of cropping, management and soil properties. Erosion on the slope profile is calculated for each of 100 increments on each overland-flow element. Each overland-flow element is described topographically by the user with one or more slope input segments, which are described below in the section on model inputs. The model can accommodate up to 10 overland-flow elements on the profile.

The soil profile may be represented by up to 10 layers. The first line of the soil file contains general information about the soil, including soil name, texture class, soil albedo, initial saturation, rill and interrill erodibilities (if available) and critical shear stress (if available). The remainder of the file contains information for each soil layer, including bulk density, saturated conductivity (if available), field capacity (if available), 15-bar water content (if available), percentage sand, silt and clay, organic matter content, cation exchange capacity and percentage rock fragments.

Differences in soil type down the slope profile may be described using the overland-flow element for each soil type. An overland-flow element is defined as a section of the hillslope which is homogeneous in terms of cropping, management and soil properties. The user should be aware, however, that each additional overland-flow element significantly increases computational time. If soil properties, for example, are not greatly different down the slope (i.e. if soils do not vary in texture classes), the improvement in erosion prediction on the hillslope may not be significant enough to warrant multiple overland-flow elements for the downslope soil-texture variation.

The structure of the management file will depend on the land use. At present, croplands and rangelands are the two land uses supported by the WEPP model. Disturbed forest lands will be added. The management file for croplands includes crop-growth and residue-decay parameters for the crop-growth model, tillage dates, tillage implements, information on contour farming (if any), planting, harvesting and grazing dates, data on weed cover and data on the size of equipment used. The rangeland management file contains plant information for the ecological range community, dates of grazing and number and type of animals grazed.

Up to three irrigation input files may be required to run the model for the case of irrigation, depending upon the irrigation scheduling option specified in the management data file. These files may be: (a) a depletion-level scheduling file; (b) a fixed-date scheduling file; and (c) a sprinkler irrigation control file. The control file includes a description of the irrigation system used and dates on which irrigation may be active. The depletion-level file is used if irrigation is to be based on water content of the soil as calculated by the water-balance component of the model. A combination of depletion-level and fixed-date scheduling may be chosen. Details of the input requirements for irrigation are presented in the WEPP Profile Model Documentation (Lane & Nearing 1989).

The output of the continuous simulation model represents time-integrated estimates of erosion. In nature, as well as in the model predictions, a large percentage of erosion occurs due to a small percentage of rainfall events. The model simulates yearly erosion and sums the total soil loss over those years for each point on the hillslope to obtain average annual values along the hillslope. The model calculates both detachment and deposition. It predicts where deposition begins and/or ends on a hillslope, which may vary from storm to storm. Certain points on the hillslope may experience detachment during some rainfall events and deposition during other events. The output of the continuous simulation model represents an average of the erosion events.

The model output includes two sections, one for onsite effects of erosion and one for offsite effects. These two sections are clearly delineated in the output. Onsite effects of erosion include a section on time-integrated (average annual) soil loss over the areas of net soil loss. This quantity is the one which is most analogous to USLE estimates. It is the soil loss estimate which is most closely tied to onsite loss of productivity. The section for onsite effects also includes estimates of average deposition over the areas of net deposition. Lastly, it provides a table of soil loss at each of a minimum of 100 points down the slope. The second section of the output is for offsite effects of erosion. It includes estimates of sediment loads leaving the profile. This is the sediment which is a potential problem in terms of delivery to waterways, as well as the offsite delivery of agricultural pollutants which may be bound to soil particles. This section also includes sediment particle-size information. Since agricultural pollutants are preferentially bound to certain size classes of sediment, this information can have significance in assessing offsite pollution problems.

The output options also include the potential for obtaining monthly or daily (storm-by-storm) estimates of onsite and offsite effects of erosion. The output as a whole provides a potentially powerful tool for conservation planning. The model estimates explicitly where and when soil loss problems are occurring on a particular hillslope for a given management option on a selected field. It also provides a quick and inexpensive method for evaluating conservation methods.

The model may also be executed in the single-storm mode. For this case, all of the parameters used to drive the hydrology and erosion components of the model must be input by the user, including soil conditions for the day of the rainfall event, crop canopy, surface residue, days since last disturbance, surface random roughness, oriented roughness, etc. In the continuous simulation mode the influence of these user inputs, which represent the initial conditions for the simulation, is small since the model adjusts each of these variables internally. In the single-storm mode, user inputs have a major influence on the output. The single-storm option of the model requires a great deal more knowledge on the part of the user to interpret and use the output for planning, evaluation and design for conservation purposes. The single-storm model helps in understanding and evaluating the factors which influence erosion on a hillslope; it is of limited value in evaluating conservation systems where conditions change as a function of time.

Erosion equations

In this section the erosion component of the WEPP profile model is described briefly. The fundamental equations for sediment continuity, detachment, deposition, shear stress in rills, and transport capacity are presented. Relationships describing temporal modifications to baseline erodibility parameters as a function of above- and below-ground residue, plant canopy and soil consolidation are also presented.

Sediment continuity equation

The WEPP erosion model computes estimates of net detachment and deposition using a steady-state sediment continuity equation:

$$\frac{dG}{dx} = D_f + D_i \qquad (15.1)$$

where x (m) is distance downslope, G (kg s^{-1} m^{-1}) is sediment load, D_i (kg s^{-1} m^{-2}) is interrill erosion rate and D_f (kg s^{-1} m^{-2}) is rill erosion rate. Interrill erosion, D_i, is considered to be independent of x. Rill erosion, D_f, is positive for detachment and negative for deposition.

Interrill erosion in the model is represented as a process of sediment detachment and delivery to concentrated flow channels, or rills, whereby the interrill sediment is then either carried off the hillslope by the flow in the rill or deposited in the rill. Sediment delivery from the interrill areas is considered to be proportional to the square of rainfall intensity, and the constant of proportionality is the interrill erodibility parameter. The function for interrill sediment delivery also includes terms to account for the effects of ground and canopy cover.

Net soil detachment in rills is calculated when hydraulic shear stress exceeds the critical shear stress of the soil and when sediment load is less than sediment transport capacity. For rill detachment

$$D_f = D_c \left[1 - \frac{G}{T_c} \right] \tag{15.2}$$

where D_c (kg s^{-1} m^{-2}) is detachment capacity by flow and T_c (kg s^{-1} m^{-1}) is sediment-transport capacity in the rill. When shear stress exceeds critical shear, detachment capacity, D_c, is expressed as:

$$D_c = K_r \, (\tau_f - \tau_c) \tag{15.3}$$

where K_r (s m^{-1}) is a rill soil erodibility parameter, τ_f (Pa) is flow shear stress acting on the soil particles, and τ_c (Pa) is the rill detachment threshold parameter, or critical shear stress, of the soil.

Net deposition is computed when sediment load, G, is greater than sediment transport capacity, T_c. For the case of deposition

$$D_f = \left[\frac{V_f}{q} \right] [T_c - G] \tag{15.4}$$

where V_f (m s^{-1}) is effective fall velocity for the sediment, and q (m^2 s^{-1}) is flow discharge per unit width.

Hydrologic inputs

Three hydrologic variables are required to drive the erosion model. They are (a) effective rainfall intensity, I_e (m s^{-1}); (b) peak runoff per unit area, P_r (m s^{-1}); and (c) effective runoff duration, t_r (s). Rainfall intensity is generated by the CLIGEN climate generator and the runoff peak and duration are computed by the hydrologic component of the WEPP model. The simplest method of transposing the dynamic hydrologic information into steady-state terms for the erosion equations is to assign the value of P_r to the peak value of runoff on the hydrograph. The effective duration of runoff, t_r, is then calculated as the time required to produce a total runoff volume equal to that given by the hydrograph with a constant runoff rate of P_r. Thus, t_r is given as:

$$t_r = \frac{V_t}{P_r} \tag{15.5}$$

where V_t (m) is the total runoff volume for the rainfall event. Effective rainfall intensity, I_e, which is used to estimate interrill soil loss, is obtained from the equation

$$I_e = \left[\frac{\int I^2 dt}{t_e} \right]^{1/2} \tag{15.6}$$

where I is rainfall intensity (m s^{-1}), t is time (s) and t_e is the total time (s) during which rainfall rate exceeds infiltration rate.

Flow shear stress

Shear stress of channel flow is computed at the end of an average uniform profile length by assuming a rectangular channel geometry. The uniform profile is assumed to have a constant or uniform gradient, $\bar{S}$, that passes through the endpoints of the profile. The shear stress from the uniform profile is used as the normalization term for hydraulic shear along the profile. Width, w (m), of the channel at the end of the rill is calculated using the relationship

$$w = c\, Q_e^d \tag{15.7}$$

where Q_e (m^3 s^{-1}) is flow discharge at the end of the slope and c and d are coefficients derived from data on the effect of rill geometry on flow rate and discharge from the study of Gilley et al. (1990). Discharge rate is given by

$$Q_e = P_r\, L\, R_s \tag{15.8}$$

where L (m) is slope length, and R_s (m) is the average distance between flow channels on the hillslope.

The sensitivity of the model to rill spacing, R_s, and channel width, w, was investigated by Page (1988). Estimates of predicted sediment load were sensitive to rill spacing when an increase in flow shear from increased rill spacing (hence discharge) caused flow shear to exceed the threshold of critical shear of the soil and initiate rilling. The effect of rill spacing on average sediment loss per unit area was minimal for the condition that shear stress was always greater than critical shear stress. Increased rill spacing causes a greater flow volume in the rill, a higher shear stress acting on the soil, and increased sediment load. However, the loss of soil must then be averaged over the larger contributing area to the rill, resulting in the relative insensitivity of average soil loss per unit area to rill spacings.

A similar effect was observed for rill width. Decreased rill width causes increased flow depth and shear. However, the area of scour in the rill is less and hence average soil loss is not greatly affected. A large effect was seen only when increase in flow shear crossed the threshold of critical shear of the soil. Since most sediment is lost for large runoff events where critical shear of the soil is greatly exceeded, the effect of rill spacing and width on predicted soil loss was not considered to be great in terms of overall model sensitivity.

Depth of flow is computed using the friction factor of the rill, the channel width and the average slope gradient. Hydraulic radius, R (m), is then computed from the flow width and depth of the rectangular channel. Shear stress acting on the soil at the end of the uniform slope, τ_{fe} (Pa), is calculated using the equation

$$\tau_{fe} = \gamma \bar{S} R \left[\frac{f_s}{f_t} \right] \tag{15.9}$$

where γ is the weight density of water (kg m^{-2} s^{-2}), $\bar{S}$ is average slope gradient, f_s is friction factor for the soil, and f_t is total rill friction factor. The ratio of f_s/f_t represents the partitioning of the shear stress between that acting on the soil and the total hydraulic shear stress, which includes the shear stress acting on surface cover (Foster 1982). Shear stress along the rill is then calculated as a function of distance, x, and shear stress at the end of the hillslope.

Sediment transport capacity

Sediment transport capacity and sediment load are calculated on a unit channel width basis within the erosion component. Sediment load is converted to a unit field width basis at the end of the calculations. Transport capacity is calculated as a function of x, using a simplified form of the Yalin sediment transport equation of the form

$$\tau_c = k_t \, \tau_f^{3/2} \tag{15.10}$$

where τ_f is hydraulic shear acting on the soil (Pa), and k_t is a transport coefficient ($m^{1/2}\ s^2\ kg^{-1/2}$). Transport capacity at the end of the slope is computed using the Yalin equation as modified by Foster & Meyer (1972) for non-uniform sediment. The coefficient k_t is calibrated from the transport capacity at the end of the slope, T_{ce}, using the method outlined by Finkner et al. (1989). A representative shear stress is determined as the average of the shear stress at the end of the representative uniform average slope profile and the shear stress at the end of the actual profile. The representative shear stress is used to compute T_{ce} using the Yalin equation and k_t is then determined from the relationship given in Equation 15.10. Differences in sediment transport capacity between the simplified Yalin and the original Yalin equations, using the calibration technique, are minimal (Finkner et al. 1989).

Limits of application

The erosion predictions from the WEPP profile model are applicable to "field-sized" areas or conservation treatment units. The maximum size "field" is about a section (640 acres or 260 hectares), although an area as much as three times this large may be needed for some rangeland applications. As topographic complexity increases, the field size to which the model output applies decreases. On some very complex areas, the "field" may be much smaller. The WEPP model cannot be applied to areas where permanent channels, such as classical gullies and stream channels, are found.

The profile model cannot be applied to areas with channels which are farmed over and known as concentrated flow or "cropland ephemeral gullies". However, the watershed version of the WEPP model specifically addresses areas with ephemeral gullies. The watershed version should also be used for estimating erosion in terrace channels or grassed waterways on cropland and in rangeland and forestland applications where "fields" may contain large concentrated flow channels.

Summary

The USDA/WEPP profile computer model represents a new generation of technology for estimating soil erosion caused by rainfall and overland flow on hillslopes and is an alternative to currently used erosion prediction technology in the US. The model is based on hydrologic and erosion processes, including major components for climate, infiltration, water balance, crop growth and residue decomposition, surface runoff and erosion. It calculates spatial and temporal distributions of soil loss. The model has been designed to include a user interface which is easily useable by soil conservation planners in the field. The model structure is modular to facilitate replacement of components as new research provides refinement and improvement of existing prediction procedures. A steady-state sediment continuity equation is used as the basis for the erosion computations of net detachment and deposition. Similar to other erosion models, such as the one used in CREAMS (Foster et al. 1981), the WEPP erosion model calculates erosion from rill and interrill areas and uses the concept that detachment and deposition rates in rills are a function of the portion of the transport capacity which is filled by sediment. However, unlike other recent models, the WEPP erosion model partitions runoff between rill and interrill areas and calculates shear stresses based on rill flow and rill hydraulics rather than sheet flow (Page 1988).

Erodibility parameters are based on the extensive field studies of Laflen et al. (1987) and Simanton et al. (1987), which were specifically designed and interpreted for the erosion model. Temporal variations of erodibility are based on the consolidation model of Nearing et al. (1988). Cropping-management effects are directly represented in the model by terms for plant canopy, surface cover and buried-residue effects on soil detachment and transport. Because the WEPP erosion routines use daily water balance and infiltration routines which are spatially varied, the model can calculate erosion for the case of non-uniform hydrology on hillslopes, resulting in estimates of spatially varied erosion and sediment yield.

References

Alberts, E. E., M. A. Weltz, F. Ghidey 1989. Plant growth component. In *USDA-Water Erosion Prediction Project: Hillslope profile version,* L. J. Lane & M. A. Nearing (eds) NSERL Report No.2. West Lafayette: USDA-ARS National Soil Erosion Laboratory.

Chu, S. T. 1978. Infiltration during an unsteady rain. *Water Resources Research* **14**, 461–6.

Finkner, S. C., M. A. Nearing, G. R. Foster, J. E. Gilley 1989. A simplified equation for modeling sediment transport capacity. *Transactions of American Society of Agricultural*

Engineers **32**, 1545–50.
Foster, G. R. 1982. Modeling the erosion process. In *Hydrologic modeling of small watersheds,* C. T. Haan (ed.), 297–360. American Society of Agricultural Engineers Monograph No. 5. St. Joseph, Michigan: American Society of Agricultural Engineers.
Foster, G. R. & L. J. Lane 1987. *User requirements, USDA-water erosion prediction project (WEPP).* National Soil Erosion Research Laboratory Report No. 1. West Lafayette, Indiana: National Soil Erosion Research Laboratory.
Foster, G. R., L. J. Lane, J. D. Nowlin, J. M. Laflen, R. A. Young 1981. Estimating erosion and sediment yield on field-sized areas. *Transactions of American Society of Agricultural Engineers* **24**, 1253–62.
Foster, G. R. & L. D. Meyer 1972. A closed-form soil erosion equation for upland areas. In *Sedimentation: Symposium to honor H. A. Einstein,* H. W. Shen (ed.). Ft. Collins, Colorado: University of Colorado.
Laflen, J. M., A. Thomas, R. Welch 1987. *Cropland experiments for the WEPP project.* American Society of Agricultural Engineers Paper No. 87- 2544. St. Joseph, Mi: American Society of Agricultural Engineers.
Gilley, J. E., E. R. Kottwitz, J. R. Simanton 1990. Hydraulic characteristics of rills. *Transactions of American Society of Agricultural Engineers* **33**, 1900–6.
Lane, L. J. & M. Nearing (eds) 1989. *USDA-Water Erosion Prediction Project: Hillslope Profile Version, NSERL Report No.2.* West Lafayette: USDA-ARS National Soil Erosion Laboratory.
Mein, R. G. & C. L. Larson 1973. Modeling infiltration during a steady rain. *Water Resources Research* **9**, 384–94.
Nearing, M. A., L. T. West, L. C. Brown 1988. A consolidation model for estimating changes in rill erodibility. *Transactions of American Society of Agricultural Engineers* **31**, 696–700.
Nicks, A. D. 1985. Generation of climate data. In *Proceedings of the Natural Resources Modeling Symposium, 16–21 October 1983,* D. G. DeCoursey (ed.). USDA-ARS ARS-30.
Page, D. I. 1988. Overland flow partitioning for rill and interrill erosion modeling. Unpublished M.S. thesis. University of Arizona, Tucson, AZ.
Simanton, J. R., L. T. West, M. A. Weltz, G. D. Wingate 1987. *Rangeland experiments for water erosion prediction project.* American Society of Agricultural Engineers Paper No. 87-2545. St. Joseph, Mi: American Society of Agricultural Engineers.
Weltz, M. A. & A. B. Arslan 1990. Water Erosion Prediction Project (WEPP) estimating plant components that are used to simulate soil erosion on rangelands. In *Proceedings of the International Symposium on Water Erosion, Sedimentation and Resource Conservation, Dehradun, India, 9–13 October 1990,* Y. K. Arora (ed.).
Williams, J. R. & A. D. Nicks 1985. SWRRB, a simulator for water resources in rural basins: an overview. In *Proceedings of the Natural Resources Modeling Symposium, 16–21 October 1983,* D. G. DeCoursey (ed.), 17–22. USDA-ARS, ARS-30.

16

An aspect-driven kinematic routing algorithm

Nicholas J. Lea

Abstract

This chapter presents a procedure for predicting the path taken by overland flow as it moves across digitally modelled terrain. It argues that "nearest neighbour" algorithms provide insufficiently accurate predictions, and that they cannot be used for the construction of subcatchment boundaries or to predict the location of principal streams.

The algorithm presented is based around calculation of the aspect vector. It is argued that flow moves across a planar surface in the direction defined by this vector, and that the heterogeneous terrain surface can be approximated by a grid of planes. The calculation of "best fit" planes is discussed and a method presented for the calculation of the aspect vector. The routing algorithm models flow moving kinematically as a point source from the centre of the source pixel until it reaches a perimeter point. Once at the perimeter, flow is transferred to the coincident perimeter point on a neighbouring pixel.

The use of this algorithm is demonstrated in the derivation of the principal stream network of a large river basin in the Philippines, and also in defining the areas of subwatersheds. Its application to sediment-yield modelling is discussed in the estimation of a distributed delivery ratio, used to model sediment redeposition alongside empirical erosion estimators.

Introduction

In hydrology and erosion mechanics, the use of distributed models is becoming more commonplace. Instead of correlative models that use large-scale measurements of catchment characteristics, spatially distributed data are used by models that attempt to represent the interaction of large numbers of small-scale physical processes (Abbot et al. 1986, Beven et al. 1987). A component of all such models is a procedure to predict the route taken by water or sediment as it flows from upland sources toward the catchment outlet. Presented here is one such routing algorithm and a demonstration of its use in determining the locations of principal streams and their drainage areas within a river basin. It was developed as part of a larger task to model the erosion processes within a river basin for use in conservation planning and the prediction of reservoir sedimentation. It was found necessary to design a routing algorithm that could accurately predict drainage paths for the construction of sediment delivery ratios and that also could predict the location of subwatersheds to enable local calibration of erosion estimates to be made. (The term "watershed" is used throughout this chapter following the British usage, to indicate the line dividing one catchment (drainage basin) from another.) Existing routing algorithms were found to be insufficiently accurate for both these tasks and especially so in complex terrain.

The algorithm described requires a three-dimensional representation of the basin topography. This is implemented by dividing the basin area, as if by a large regular grid, into small squares or pixels. A spot height at the centre of each pixel is measured or estimated, thus forming a two-dimensional array of elevation values. This array constitutes the basis of a digital terrain model (DTM) that enables estimates of the surface elevation to be made at all points within the catchment (see Petrie & Kennie 1986). The case of triangular pixels is not considered, although common in DTMs, for two reasons. First, the use of triangular pixels in distributed models greatly increases the necessary processing but gives little return in accuracy. Secondly, the use of remotely sensed data for the land-use component of erosion prediction immediately lends itself to the use of square rather than triangular elements.

Lowest neighbour routing

The most widely used routing algorithms that model flow across a grid of pixels, could be described as "lowest neighbour" algorithms. A pixel has eight neighbours: north, northeast, east, southeast, etc. All the flow from a source pixel is assumed to transfer to the neighbouring pixel with the lowest elevation below that of the source pixel. When no such neighbour can be found, flow stops. In the case of two or more neighbouring pixels having equally low elevations, a convention is adopted such as flow being divided equally between them. The problem with such an approach is that flow is constrained to travel in eight (or fewer) directions. This results in significant oversimplification in the estimation of flow paths and, consequently, invalid predictions of watershed locations. To illustrate the potential error when using an algorithm of this sort, consider the following example.

Suppose a catchment is modelled using grid squares defined with north–south and east–west sides. Consider a large area of hillslope within this catchment that faces south-southeast, i.e. a bearing of 157.5° (see Fig. 16.1). Since flow proceeds downhill in the direction of steepest slope, water will flow south-

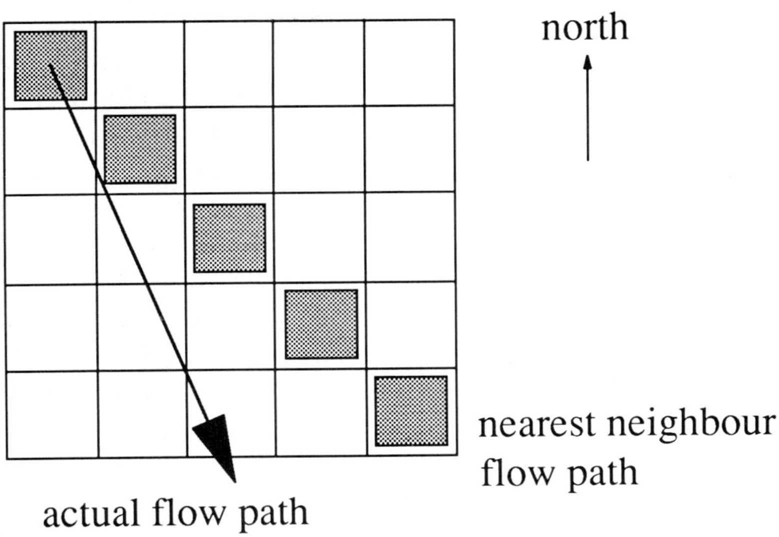

Figure 16.1 Error in the bearing of a nearest neighbour flow path.

southeast. Using a lowest neighbour algorithm, any flow on a slope with this bearing will always move into the southeast pixel. Since the area of hillslope is large, this process will be repeated many times, resulting in an overall flow path that runs directly southeast over many pixels. The error in the bearing of this flow path is 22.5°. Thus, if it has a length of d pixels, the distance from the true flow path will be (1-tan(22.5°))d pixels, which is equivalent to a spatial error of 58% of the distance. Because a subcatchment can be defined to be the area drained by a specified stream, or segment of a stream, routing algorithms may be used to predict subcatchment areas. Pixels with flow that routes to the specified stream are included in the subcatchment area, and those that route elsewhere are excluded. The divergent errors in flow paths produced by "lowest neighbour" algorithms make them unsuitable for the prediction of subcatchment areas.

Problems also occur when routing is attempted on shallow slopes. It is possible on shallow slopes that the difference in elevation between upslope and downslope neighbouring pixels is too small to be differentiated by the computer. This is especially likely when the model contains a very large number of pixels and economies have to be made in the use of computer memory. Using a "lowest neighbour" algorithm, flow must stop as if in a hollow, unless a more satisfactory solution can be found.

Aspect-driven routing

Overland flow on a planar surface

In considering a replacement algorithm, we should first investigate the behaviour of overland flow on a planar surface. Consider a point source on a uniform hillslope and the direction in which water will flow from it. The hillslope can be modelled as a plane in three-dimensional cartesian coordinates with the z-ordinate representing elevation. The equation of an arbitrary plane intersecting the origin is:

$$z = Ax + By. \tag{16.1}$$

Consider a unit path beginning at the origin in the XY plane and described by the angle θ. The projection onto the plane of the end of the unit path is:

$$z = A\cos\theta + B\sin\theta. \tag{16.2}$$

Water will flow along the path leading to the lowest point on the plane (i.e. along the steepest downhill gradient). This path will have direction θ when Equation 16.2 above is at a minimum.

We can rewrite Equation 16.2 in the form:

$$z = R\cos(\theta + \alpha) \tag{16.3}$$

where α = -arctan(B/A) and $R = \surd(A^2 + B^2)$. Since R is positive, the minimum value of z occurs when:

$$\theta + \alpha = \pi \tag{16.4}$$

giving:

$$\theta = \pi + \arctan(\frac{B}{A}). \tag{16.5}$$

The cartesian unit vector in the direction of θ in the XY plane ($\cos\theta$,$\sin\theta$) is equal to $(-A/\surd(A^2+B^2),\ -B/\surd(A^2+B^2))$. Thus flow proceeds across a plane defined by the equation $z = Ax+By$, in the direction of the vector $(-A, -B)$.

The *aspect vector* is defined as the projection of the normal to the surface onto the XY plane. The normal to a surface is the three-dimensional vector defined in terms of the directional derivatives $(-\delta z/\delta x, -\delta z/\delta y, 1)$ and its projection onto the XY plane is the two-dimensional vector $(-\delta z/\delta x, -\delta z/\delta y)$.

Since *A* and *B* are the coefficients of *x* and *y*, respectively, they are also the directional derivatives, $\delta z/\delta x$ and $\delta z/\delta y$. Thus the vector describing the direction of flow is equal to the aspect vector.

Given that flow proceeds in the direction of the aspect vector, the routing algorithm presented in this paper determines flow paths using only aspect information. Elevation data is used solely in the initial determination of aspect. The basis of the algorithm developed below is to model the entry and exit points of flow on the perimeter of each pixel and to consider flow within a pixel as a point source moving in the direction of the aspect vector. This overcomes the inaccuracies encountered with "lowest neighbour" algorithms and has proved to be suitable for the prediction of subcatchment boundaries.

Aspect calculation

Before defining the flow algorithm, it is necessary to consider the problem of deriving aspect values from the matrix of elevation points that comprise the digital terrain model. Since the aspect vector is derived from a surface and the DTM specifies spot heights at the centre of pixels, the first task is to fit planes representing the surface of each pixel.

Suppose the DTM uses a three-dimensional cartesian coordinate system with the *z*-ordinate representing elevation. Without loss of generalization it is possible to shift onto a coordinate system where the southwest corner of the pixel is at (0,0) in the XY plane, and the northeast corner has coordinates (1,1). A plane is then required of the form:

$$z = \alpha x + \beta y + \gamma. \tag{16.6}$$

For the purpose of calculating the aspect vector (which only requires α and β), we neglect the spot height defined at the centre point and fit a plane to the co-ordinates of the four corners. Since three linearly independent points in three dimensions uniquely define a plane, a "best fit" plane must be found to fit the four points given.

Using the principle of least squares, the best-fit plane is obtained when:

$$\alpha = \frac{(c_{ne} - c_{nw}) + (c_{se} - c_{sw})}{2} \tag{16.7}$$

$$\beta = \frac{(c_{ne} - c_{se}) + (c_{nw} - c_{sw})}{2} \tag{16.8}$$

where c_{ne}, c_{se}, c_{nw}, c_{sw} are the estimated elevations of the four corners of the pixel. Since α and ß are the estimated coefficients of x and y in the equation of the plane, the aspect vector, as already discussed, is simply $(-\alpha,-ß)$. Equivalently, it can be defined as the angle θ which is equal to $\pi+\arctan(ß/\alpha)$.

For the purposes of overland-flow routing, it is necessary for the aspect to be defined at every pixel. It is quite possible that α and ß are both computed to be zero, in which case the aspect, θ, would remain undefined. Since this would never happen on real terrain – there is always a direction of greatest slope no matter how shallow it is – an estimate must be made of the aspect under these conditions.

Suppose the aspect vector is undefined, i.e. $\alpha = 0$ and $ß = 0$. By solving Equations 16.6 and 16.7 simultaneously, it can be seen that the situation only occurs when:

$$c_{se} = c_{nw} \tag{16.9}$$

and,

$$c_{ne} = c_{sw} \tag{16.10}$$

that is, when both pairs of opposite corners have equal elevations. In such cases, the estimated elevations of the corner pixels can be recalculated using a wider consideration of the terrain data. Initially the estimated elevation of each corner is calculated as the mean of the four surrounding spot heights. This process can be expanded and the mean of a three by three square of spot heights with its corner on the centre pixel may be used for the new estimate. If these estimates cause the same problem, a four by four square is used and so on (see Fig. 16.2). By this means, the angle of the aspect vector may be

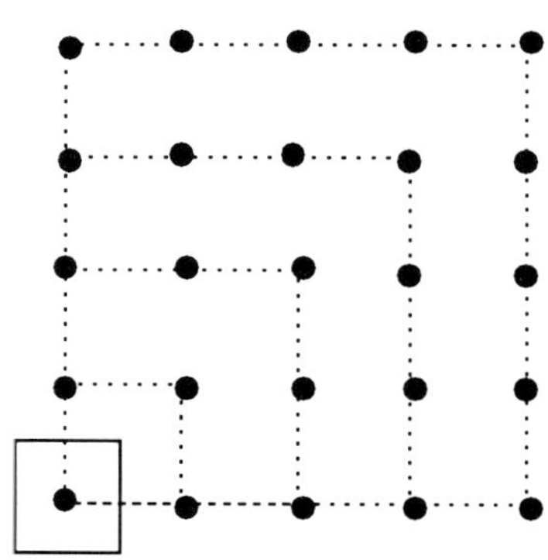

Figure 16.2 Iterative estimation of corner elevations.

calculated for each pixel. The routing algorithm defined in the next section makes use of only this parameter since it contains all the information needed to derive flow paths.

Flow routing

Flow is assumed to originate at the centre of the source pixel and travel kinematically as a point source. In the following analysis the coordinate system defined above is used with the southwest corner of the source pixel having coordinates (0,0) and the northeast corner having coordinates (1,1). Consider the source pixel (Fig. 16.3). Flow originates at the point (0.5,0.5) and

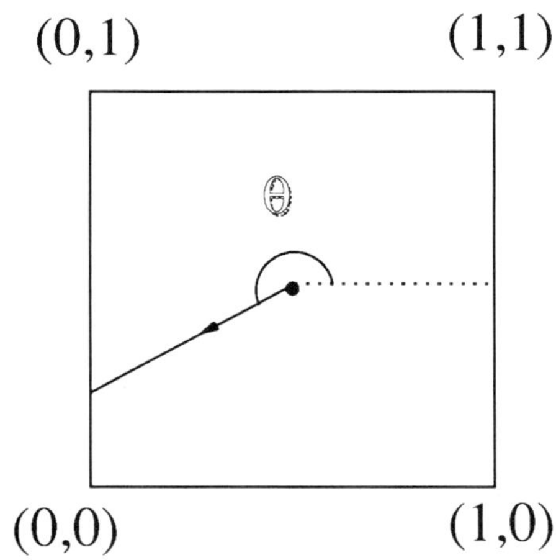

Figure 16.3 Initial flow path from the centre point of a pixel.

proceeds in the direction of the aspect, θ, until it reaches the perimeter. The coordinates (x_o,y_o) are defined to be the "outlet point", i.e. the intersection of the pixel's perimeter with the flow path. The relationship between the outlet point and the aspect is derived by simple trigonometry and is given by:

$$\begin{aligned}
(x_o,y_o) &= (\,1,\ (1+\tan\theta)/2\,) \quad \text{for } -\pi/4 < \theta \leq \pi/4 && \text{eastern exit}\\
&= (\,(1+\cot\theta)/2,\ 1\,) \quad \text{for } \pi/4 < \theta \leq 3\pi/4 && \text{northern exit}\\
&= (\,0,\ (1-\tan\theta)/2\,) \quad \text{for } 3\pi/4 < \theta \leq 5\pi/4 && \text{western exit}\\
&= (\,(1+\cot\theta)/2,\ 0\,) \quad \text{for } 5\pi/4 < \theta \leq 7\pi/4 && \text{southern exit}
\end{aligned}$$

The outlet point (x_o,y_o) translates to an inlet point (x_i,y_i) on the perimeter of the neighbouring pixel that shares the edge on which the outlet point lies (Fig. 16.4). The inlet point is also defined with respect to coordinates whose origin lies on the southwest corner of the new pixel. The relationship between the outlet and inlet points is thus:

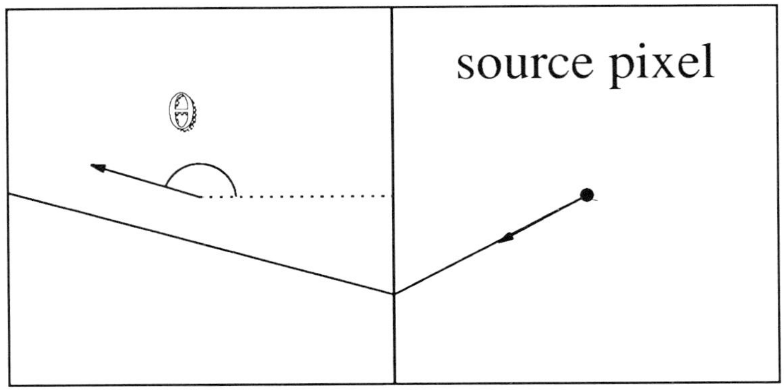

Figure 16.4 The flow path across adjacent pixels.

Outlet	$(1,t)$	flows to eastern pixel with inlet	$(0,t)$
Outlet	$(t,1)$	flows to northern pixel with inlet	$(t,0)$
Outlet	$(0,t)$	flows to western pixel with inlet	$(1,t)$
Outlet	$(t,0)$	flows to southern pixel with inlet	$(t,1)$

where t is an arbitrary parameter defining position along an edge and $0 \leq t < 1$.

Unlike in the previous example where flow originates at the centre of the source pixel, there is a chance that flow starting at the perimeter of the new pixel will not be routed across the pixel at all. Flow will cross a pixel when the aspect vector has a negative component in the direction of the inlet edge (i.e. points away from it). For example, if the inlet point is on the eastern edge, as in the diagram above, flow will cross to another edge if $\pi/2 < \theta < 3\pi/2$. In the cases when this does not happen, the slopes of the two pixels face each other and effectively form a valley. Since the inlet edge forms the valley floor, flow is routed along the inlet edge to neighbouring pixels.

When flow is known fully to cross a pixel, it travels along the line from the inlet point in the direction θ. Arbitrary points on this line of the form (x,y) satisfy the equation:

$$(y - y_i)\cos\theta = (x - x_i)\sin\theta. \qquad (16.11)$$

The outlet point is defined as the intersection of this line with the edges of the pixel except the inlet edge. Since a line can only intersect the perimeter of a square twice, and it is known that this line intersects the inlet edge, it must intersect precisely one of the other three edges. The edges of the pixel lie on the lines: $x = 0$, $y = 0$, $x = 1$, $y = 1$. Intersecting Equation 16.11 with these lines gives the possible outlet points:

$x = 1$ (eastern edge) $\quad (1 , y_i + (1-x_i)\tan\theta)$
$y = 1$ (northern edge) $\quad (x_i + (1-y_i)\cot\theta, 1)$
$x = 0$ (western edge) $\quad (0 , y_i - x_i\tan\theta)$
$y = 0$ (southern edge) $\quad (x_i - y_i\cot\theta, 0)$

Unless $\theta = 0$ or $\theta = \pi/2$, each one of these intersection points will be defined, but an intersection with an edge requires the point (x_o,y_o) to be such that $0 \leq x_o, y_o \leq 1$. Intersection points outside of this range are discarded, leaving one legitimate outlet point. As before, this outlet point is translated to the inlet point of the appropriate neighbouring pixel and the process repeated.

In the case where the aspect has positive component in the direction of the inlet edge and a full crossing is not possible, then a prediction can still be made

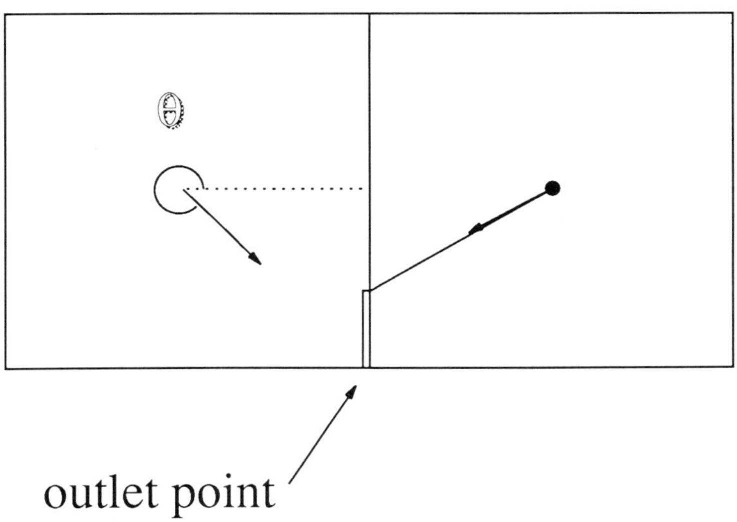

Figure 16.5 An example where full crossing is not possible.

of the flow path. Consider the example illustrated by Figure 16.5. The adjacent pixels have slopes facing each other. From the inlet point of the left pixel, flow is neither possible eastwards nor westwards, but since the aspect has a positive southern component, flow is routed to the most easterly point on the southern edge. In the computer implementation of this algorithm, this point is represented as (0.999999,0). The procedure described here can be generalized to hold for the whole perimeter in the following way. Flow unable to cross a pixel is routed to the edge adjacent to the inlet in the direction with positive aspect component. The outlet point is the point on this edge nearest to the inlet edge.

The algorithm outlined above describes the routing of flow across individual

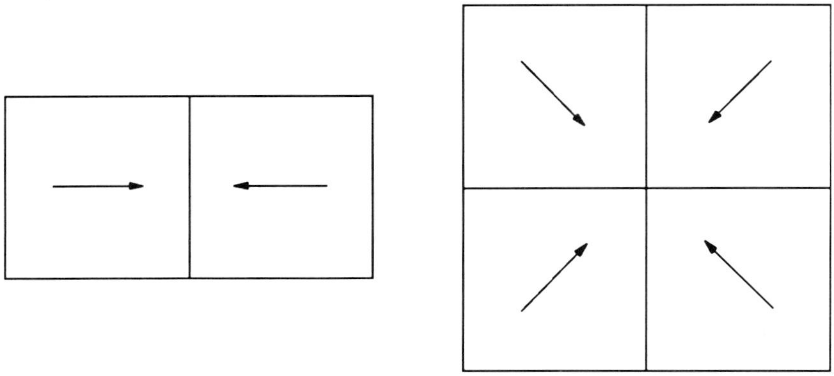

Figure 16.6 Aspect vectors that represent typical hollows.

pixels. Flow paths are constructed by the repeated application of the algorithm until the catchment outlet is reached, or a hollow prevents the continued progress of flow. Hollows are represented by groups of pixels whose aspects prevent flow from escaping. This is illustrated by the two examples in Figure 16.6. When constructing flow paths it is therefore necessary to record pixels already in the flow path so that when hollows are reached, flow is prevented from sloshing indefinitely backwards and forwards or round and round.

Applications of the routing algorithm

Estimation of the principal stream network

A comprehensive prediction of river locations within a river basin would require a distributed hydrological model (Abbot et al. 1986). Among other things, this would need to take into account varying infiltration rates, hydraulic roughness of vegetation, the effect of slope, groundwater recharge, and backwater effects on shallow slopes before such a prediction could be reached. Without such sophistication, however, acceptable predictions of river locations are possible using the routing algorithm described.

Instead of using hydrological modelling, drainage paths may be constructed from each pixel and their convergence used to identify streams. The number of paths converging on a particular pixel can be used as an index of the overland discharge at that point. The resultant array of values provides sharp distinction between river locations and their drainage areas. An arbitrary threshold dictates the number of flow paths needed to converge on a pixel before it is classified as a stream. It is useful therefore to refer to published maps of stream locations with which to calibrate these predictions.

Figure 16.7 shows a convergent flow-paths map of the Magat river basin of the Philippines. The reservoir in the northeastern corner drains a catchment of approximately 4000 km^2 and was modelled using a 500 by 700 grid of 2 hectare pixels. The map is coloured such that the dark areas represent the highest convergence of flow paths (i.e. the river system) and the lighter areas represent the lowest (uplands and watershed boundaries). By further calibration against cartographic maps, a reliable estimate of stream network location may be obtained.

Derivation of subcatchment boundaries

A subcatchment may be defined as the drainage area of a stream segment. As in the discussion of the previous section, by constructing flow paths from each pixel in the basin, it is possible to derive the drainage area of every pixel that lies on the stream network. By use of the routing algorithm, the problem of deriving subcatchment boundaries is thus reduced to the problem of identifying segments of the stream network that generate distinct sub-catchments.

In the example presented, using the data from the Philippines, river-links spanning confluences, junctions or source nodes were used to generate subcatchments. The stream network derived in the previous section by reference to convergent flow paths needs to be refined before it is suitable for this calculation. To provide accurate distinction between stream links, it is necessary for streams to be represented as sparsely as possible. An algorithm is used that constructs the sparsest possible representation that maintains an eight-connected stream network (i.e. pixels are connected if they are N, NE, E, SE, S, SW, W or NW neighbours).

The algorithm chooses an eight-connected path that follows the centre pixels of the stream network. A system of weighting is used assigning each stream pixel with the value of the number of neighbouring stream pixels and an eight-connected path is then traced, following the pixels with greatest weighting. When problems are encountered, the user directly specifies the drainage path, and the final sparsely eight-connected network is corrected to ensure that it drains the whole basin.

A recursive routine is used to "tree-walk" this stream network and assign a different value to each river-link. The routing path from each pixel in the basin is calculated and the source pixel assigned the same value as the first stream pixel reached. By this method, all pixels in the basin are assigned the value of their drainage river-link and sub-watershed boundaries are thus derived.

Figure 16.8 shows the results of the calculation for the Philippino example with the eight-connected stream network superimposed. The map is shaded to differentiate neighbouring subcatchments.

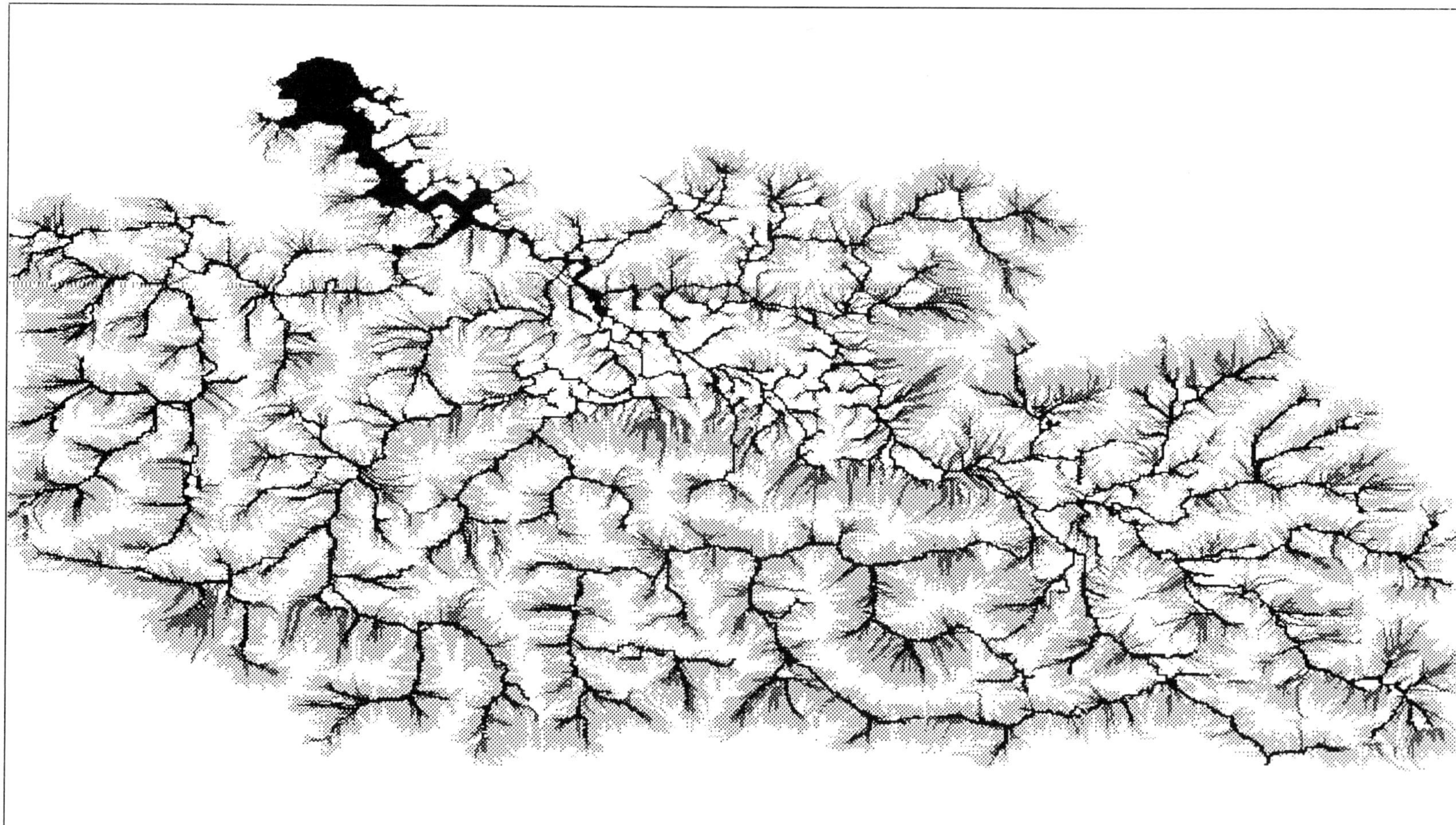

Figure 16.7 Number of convergent flowline paths at a pixel.

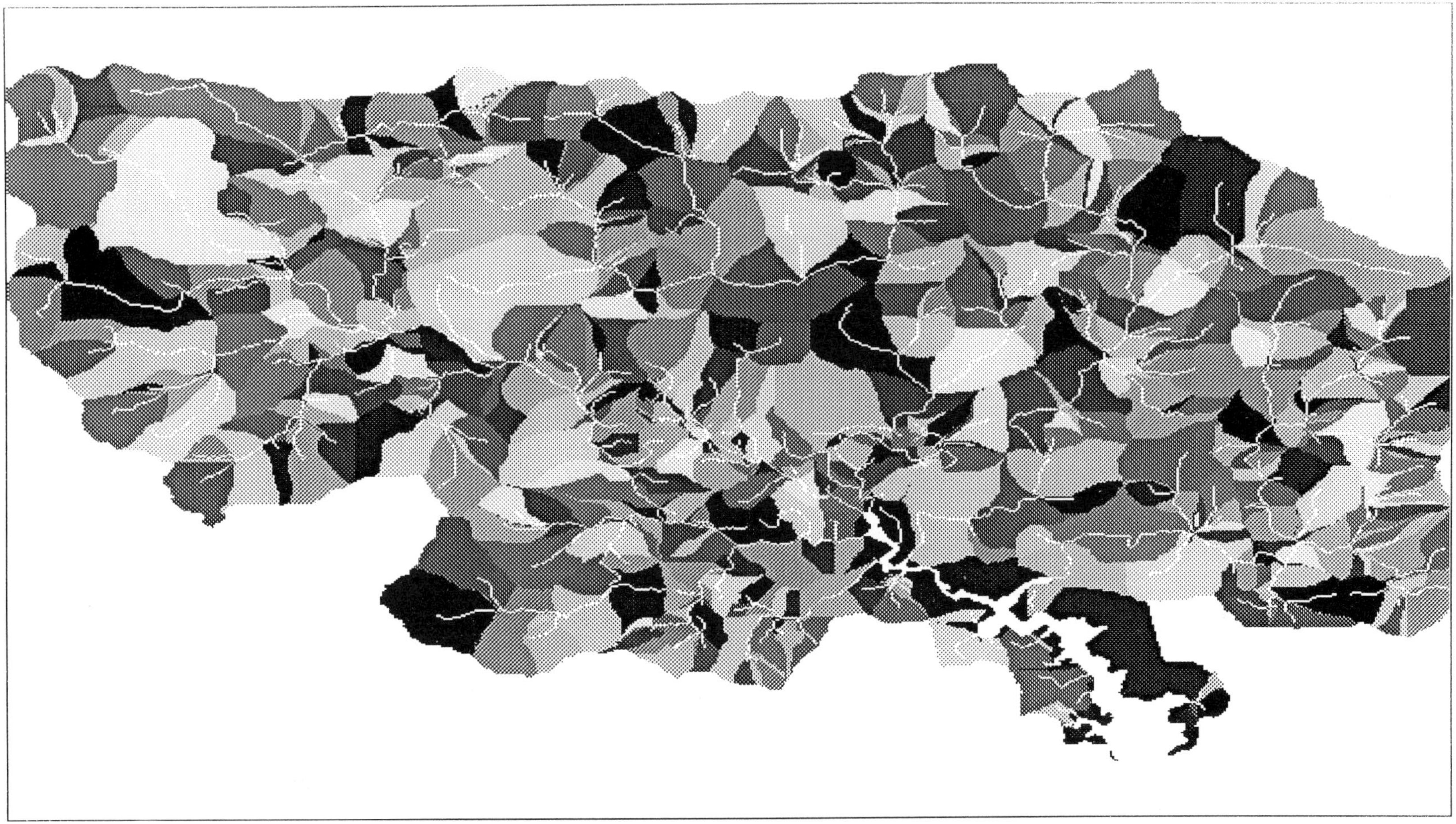

Figure 16.8 Subcatchment areas defined from stream links.

Applications in sediment yield modelling
The results in the previous two sections have clear uses for distributed sediment yield models both in pre- and post-processing. A data requirement of such models is an estimate of principal stream locations, and the definition of subcatchment boundaries enables model calibrations to be made alongside local field data. However, in tropical developing regions where data are sparse, the routing algorithm may be used more directly in the calculation itself. The absence of infiltration data, comprehensive rainfall records and locally calibrated friction indices makes distributed models difficult to apply. An alternative approach based on the older technique of erosion predictors uses spatially distributed estimates of soil loss coupled with distributed values of delivery ratio. Empirical predictors such as the Universal Soil Loss Equation (Wischmeier & Smith 1965) or the Soil Loss Equation for Southern Africa (Elwell 1978) can produce acceptable estimates of rill and sheet erosion at the small scale in the absence of more comprehensive input data. Pixel sizes may be chosen to be within the range for which these predictors were developed, and estimates of the annual soil loss for a given pixel constructed. The analyses of slope, land use, soil type and rainfall information required to make such predictions are well documented (Wischmeier 1976). In a large river basin, however, the redistribution of sediment between the outlet of the source pixel and the outlet of the basin is considerable. A simple model of redeposition can be implemented by estimating the delivery ratio, i.e. the proportion of sediment redeposited before reaching the basin outlet.

The use of computers allows individual delivery ratios to be constructed for each pixel, thus defining a delivery ratio surface that represents the redeposition processes of the whole basin. The calculation of the ratio for a given pixel may be divided into two components: overland redeposition and river and floodplain redeposition. The second component is calculated with reference to river hydraulics for which only channel dimensions and sediment and flow characteristics are needed. However, the routing algorithm has a central role in the calculation of the first component. The delivery ratio during the overland flow phase (i.e. until flow reaches a principal stream) is dependent on the length and slope of the flow path, and the overland discharge at each pixel in the flow path (see Foster & Meyer 1975). The algorithm presented may be used to provide estimates of all three parameters.

Conclusions

The aspect-driven routing algorithm outlined in this paper offers a simple and accurate procedure for the calculation of flow paths over digitally modelled terrain. The algorithm is easily applied to predict locations of principal streams within a river basin and may be further extended to provide estimates of sub-catchment boundaries. The quality of the results depends on the quality of the digital terrain model and on the memory constraints imposed by the computer.

Where data are sparse, the algorithm provides a simple substitute for a distributed hydrological model for estimating stream and catchment location and can form the basis of a simple sediment redeposition model to be used in conjunction with empirically developed soil loss estimators.

References

Abbot, M. B., J. C. Bathurst, J. A. Cunge, P. E. O'Connell, J. Rasmussen 1986. An introduction to the European Hydrologic System – Système Hydologique Européen. "SHE". 1: History and philosophy of a physically based, distributed modelling system. *Journal of Hydrology* **87**, 45–59.

Beven, K., A. Calver, E. M. Morris 1987. *The Institute of Hydrology Distributed Model.* IH report, No. 98. Wallingford, England: Institute of Hydrology.

Elwell, H. A. 1978. *Soil loss estimation – Compiled works of the Rhodesian Multidisciplinary Team on Soil Loss Estimation.* Salisbury, Rhodesia.

Foster, G. R. & L. D. Meyer 1975. Mathematical simulation of upland erosion by fundamental erosion mechanics. Present and prospective technology for predicting sediment yields and sources. In *Proceedings of the sediment yield workshop, USDA Sediment Laboratory, Mississippi, November 1972.*

Petrie, G. & T. J. M. Kennie 1986. Terrain modelling in surveying and civil engineering. *Conference on state of the Art in Stereo and Terrain Modelling, Cafe Royal 28 May, British Computer Society Displays Group.*

Wischmeier, W. H. 1976, Use and misuse of the USLE. *Journal of Soil and Water Conservation* **31**, 5–9.

Wischmeier, W. H. & D. D. Smith 1965. *Predicting rainfall erosion losses from cropland of the Rocky Mountains.* Agricultural Handbook 282, Agricultural Research Service, USDA, Washington DC, USA.

17
Modelling long-term soil loss and landform change

Jürgen Schmidt

Abstract

EROSION 2D is a physically based computer model for simulating sediment transport on slopes. The model calculates erosion and deposition per unit area, including the resulting changes in slope geometry. The input parameters of the EROSION 2D model are the altitude coordinates of the initial slope profile, the surface and soil properties and the vegetation cover of the slope. The characteristic variables can vary through space (changes in the long profile of the slope) and time. Sediment transport is always calculated on the basis of single precipitation events. These are characterized by the specific soil and canopy-cover conditions at the time of the event, by the duration of rainfall and the temporal variation of rainfall intensity. Several events can be linked to form a sequence representing a month or a year. These sequences can, in turn, be coupled repeatedly, thus extending the simulation period almost indefinitely.

Introduction

A review of landform change over the past millennium in central Europe (Bork 1988) has shown that, with increasing anthropogenic use of the soil, processes of wind and water erosion have increased in intensity and have substantially affected landform change during this period. Soil erosion research has thus become an important field of both historic-genetic and applied geomorphological research. Because of the many parameters involved, early efforts were made to develop mathematical models to describe the interaction of individual factors governing erosion. The first models were purely empirical, the best known being USLE (Wischmeier & Smith 1965). However, the application potential of purely empirical models is very limited. As well as the problem of non-transferability, it is mainly the low spatial and temporal resolution and the neglect of deposition processes that limit the use of empirical models, particularly USLE. For this reason, efforts have been made to develop physically based erosion models. One such physically based model is

EROSION 2D (Schmidt 1991a). This model describes the detachment, transport and deposition of soil particles, including the resulting changes in slope geometry. Possible applications of EROSION 2D range from purely practical questions – such as planning and calculating soil protection measures (Schmidt 1991b) – to simulating long-term landform change. In addition, the theoretical approach permits the user to view the influence of individual parameters in isolation (sensitivity analyses), thus opening up new possibilities of understanding the complex processes involved.

Description of the model

Because of the complexity of erosion processes it is advisable to split them up into various subprocesses and to represent each of these by an appropriate submodel. EROSION 2D distinguishes the following process components: the detachment of soil particles from the soil surface and the transport of the detached particles by runoff. As well as the complexity of the processes, the model should take into account the great temporal and spatial variations of the parameters governing erosion. To ensure an adequate consideration of parameter behaviour, the model's equations are applied to small spatial and temporal segments that are in themselves homogeneous. The following sections give an overview of the equations on which the model is based.

Detachment

For an erosional process to occur it is necessary that individual soil particles or small aggregates can be detached from the soil surface. This is only possible when the fluid forces generated by overland flow and raindrops overcome the particle's cohesion and gravity. Because of the heterogeneous conditions at the soil surface, direct measurement and theoretical description of the forces affecting the soil particles are practically impossible, especially when taking individual soil particles into consideration. For this reason some simplifications cannot be avoided. In particular, it is necessary to move from the microscopic scale of individual particles to a macroscopic view. An expression which summarizes the erosional effects of overland flow and raindrops in that way is the momentum flux exerted by the flow and falling droplets, respectively. The momentum flux φ_q exerted by the flow is defined as:

$$\varphi_q = w_q \, \Delta y \, v_q \tag{17.1}$$

where w_q is the mass rate of flow; Δy is the width of the slope segment; and

v_q is runoff velocity.

According to Equation 17.2:

$$w_q = q \, \rho_q \tag{17.2}$$

the mass rate of flow, w_q, is obtained from the volume rate of flow, q, and the fluid density, ρ_q. To obtain the volume rate of flow, q, the following relation is used:

$$q = (r_\alpha - i) \, \Delta x + q_{in} \tag{17.3}$$

where $r_\alpha = r \cos\alpha$ is the effective rainfall intensity related to the slope surface, i is the infiltration rate, Δx is the length of the slope segment and q_{in} is the inflow into the slope segment from the segment above.

For a sufficiently short time interval the flow velocity, v_q, contained in Equation 17.1 can be assumed as uniform. Under this condition the mean velocity of flow may be estimated according to Equation 17.4 from the coefficient of surface roughness, n, the slope, S, and the depth of flow, δ.

$$v_q = \frac{1}{n} \, \delta^{2/3} \, S^{1/2} \tag{17.4}$$

where

$$\delta = \left(\frac{q \, n}{S^{1/2}} \right)^{3/5} . \tag{17.5}$$

Equations 17.4 and 17.5 are based on the Manning equation, which was originally established for turbulent flow in channels. Various experiments have shown that the Manning equation is also valid for overland flow on slopes (Emmett 1970, Pearce 1976), as long as turbulent flow conditions can be assumed. This generally applies in the cases relevant for erosion (Bork 1988, 140). When calculating the flow velocity v_q according to Equation 17.4, it is assumed that runoff is distributed uniformly all over the slope segment. Local variations in flow velocity and flow depth within the slope segment are disregarded.

The momentum flux, $\varphi_{r,\alpha}$, exerted by the falling droplets is defined by analogy to Equation 17.1 as:

$$\varphi_{r,\alpha} = w_r \, A \, v_r \, \sin\alpha \qquad (17.6)$$

where w_r is the mass rate of rainfall, A is the area of the slope segment, v_r is the mean fall velocity of the droplets, and α is the slope angle. The effect of a canopy cover on rainfall impact is taken into account by the use of a ground-cover index, defined as $C_L = A_{leaf}/A$, where A_{leaf} is the segment area covered by vegetation or plant residues and A is the entire segment area. The ground-cover index may be combined with Equation 17.6 to give:

$$\varphi_{r,\alpha} = w_r \, A \, v_r \, \sin\alpha \left(1 - C_L\right). \qquad (17.7)$$

The mass rate of rainfall is determined by:

$$w_r = r_\alpha \, \rho_r \qquad (17.8)$$

where r_α is the effective rainfall intensity related to the slope surface and ρ_r is the fluid density of rainfall.

Substitution of this expression into Equation 17.7 yields:

$$\varphi_{r,\alpha} = r_\alpha \, \rho_r \, A \, v_r \, \sin\alpha \left(1 - C_L\right). \qquad (17.9)$$

The fall velocity of raindrops contained in Equation 17.9 is very difficult to measure under field conditions. Available data show that the size, and hence the velocity, of the droplets increase with rainfall intensity (Laws 1941, Laws & Parsons 1943). By making use of these data, we obtain the following empirical equation (Eq. 17.10), which provides a simple method of estimating the mean fall velocity of raindrops on the basis of rainfall intensity data:

$$v_r = 4.5 \, r^{0.12}. \qquad (17.10)$$

Here v_r is the mean fall velocity of raindrops in m s^{-1} and r the rainfall intensity in mm h^{-1}.

In order to characterize the erosional resistance of the soil, use is made of the fact that the occurrence of a measurable rate of erosion presupposes a minimum rate of overland flow, q_{crit} (Hjulström 1935). Substitution of q_{crit} in Equations 17.1 and 17.2 yields the critical momentum flux φ_{crit}, with which the specific erosional characteristics of the soil can be described, similar to the previously derived equations for the erosional effects of overland flow and raindrops:

$$\varphi_{crit} = q_{crit}\ \rho_q\ \Delta y\ v_q. \tag{17.11}$$

Here q_{crit} is the volume rate of flow at initial erosion (as a function of soil type, state of tillage, etc.). In addition, ρ_q is the fluid density, Δy is the width of the specified slope segment, and v_q is the flow velocity according to Equation 17.4. q_{crit} has to be determined experimentally for a given soil.

Because of their formal conformity, the model concepts describing the erosional effects of raindrops and overland flow can be linked with the concept describing the soil's resistance to give a dimensionless coefficient, E:

$$E = \frac{\varphi_q + \varphi_{r,\alpha}}{\varphi_{crit}}. \tag{17.12}$$

This coefficient characterizes the ability of a given flow ($q > 0$) to detach particles from the soil surface. Erosion occurs when $E > 1$, which means that the erosional effects of overland flow and raindrops (given by the momentum fluxes φ_q and $\varphi_{r,\alpha}$) exceeds the soil's resistance to erosion (given by the critical momentum flux φ_{crit}). $E \leq 1$ characterizes the erosion-free state of flow.

For quantitative results, the coefficient E is correlated with experimental data. Fifty experiments under simulated rainfall have been performed in a test flume filled with silty soil (Schmidt 1988). The data can be fitted by the following regression equation:

$$q_s = (1.75\ E - 1.75)\ 10^{-4} \tag{17.13}$$

where q_s is the sediment discharge of detached particles. Figure 17.1 shows the regression curve and the experimental data on which the curve is based. Because of the theoretical postulate that sediment cannot be eroded when $E \leq 1$ the regression curve must intersect the x-axis at $E = 1$. The mean relative deviation of the values calculated from Equation 17.13 from the measured values is $\pm 20\%$. Taking into account the inaccuracies in the measured values, we may presume that the theoretically derived erosion coefficient adequately describes the main factors governing detachment.

Transport

Due to gravity, the sediment particles suspended in a fluid sink to the bottom with a velocity largely dependent on particle size. This process can only be stopped or delayed when it is counteracted by a sufficiently large, vertical (turbulent) flow component. Hence, the size-dependent settling velocity of the particles and the vertical turbulence component within the flow are decisive for

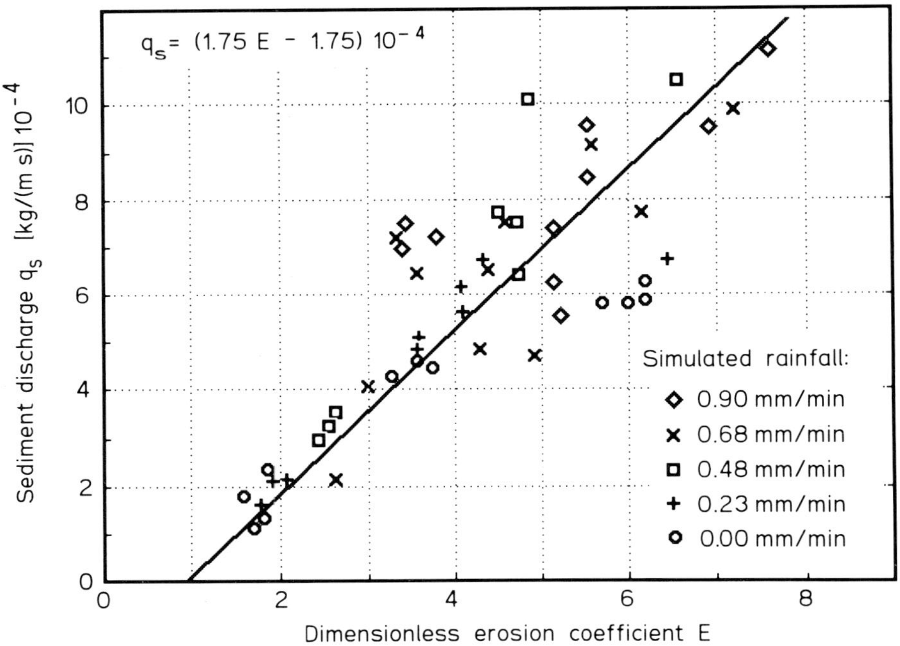

Figure 17.1 Measured sediment discharge, q_s, vs. erosion coefficient, E.

the suspended transport of particles. The settling velocity of a particle in a stagnant fluid is described by the Stokes equation:

$$v_p = \frac{1}{18} \frac{D^2 (\rho_p - \rho_q) g}{\eta} \tag{17.14}$$

in which D is particle size, ρ_p particle density, ρ_q fluid density, g acceleration of gravity, and η fluid viscosity.

If the settling velocity, $v_{p,}$ is multiplied by the mass rate of the settling particles, w_p, and by the segment area, A, we obtain the critical momentum flux of the suspended particles, $\varphi_{p,\text{crit}}$ (analogous to φ_{crit}), below which the particles are not maintained in suspension:

$$\varphi_{p,crit} = w_p \; A \; v_p \tag{17.15}$$

The mass rate of particles, $w_{p,}$ from Equation 17.15 can be expressed as:

$$w_p = c\,\rho_p\,v_p \tag{17.16}$$

where c is the concentration of suspended particles in the fluid, ρ_p is the particle density and v_p is the settling velocity of the particles according to Equation 17.14.

The critical momentum flux of the suspended particles, $\varphi_{p,crit}$, is counteracted by the vertical momentum flux component of the flow, $\varphi_{q,vert}$, which is assumed to be a fraction of the total momentum flux exerted by the flow and the falling droplets respectively.

$$\varphi_{q,vert} = \frac{1}{\kappa}\left(\varphi_q + \varphi_{r,\alpha}\right) \tag{17.17}$$

where κ is a factor.

When transport capacity has been reached, the vertical momentum flux component of flow equals the critical momentum flux of the suspended particles as defined in Equation 17.15:

$$\varphi_{p,crit} = \varphi_{q,vert}. \tag{17.18}$$

Substituting Equations 17.15, 17.16 & 17.17 into Equation 17.18 gives:

$$c_{\max}\,\rho_p\,A\,v_p^2 = \frac{1}{\kappa}\left(\varphi_q + \varphi_r\right) \tag{17.19}$$

where c_{max} is the concentration of particles at transport capacity.

Rearranging Equation 17.19 yields:

$$c_{\max} = \frac{1}{\kappa}\,\frac{\varphi_q + \varphi_r}{\rho_p\,A\,v_p^2}. \tag{17.20}$$

Transport capacity is then determined according to:

$$q_{s,\max} = c_{\max}\,\rho_p\,q. \tag{17.21}$$

According to Equation 17.21, it is possible to calculate the transport capacity for any particle size class separately. The transport capacities thus derived specify the maximum mass rate of particles that can be transported within this size class under the given flow conditions (assuming that transport is not limited by detachment). In order to determine the actual mass rate and the particle size distribution of the transported sediment, the following assumptions are made:

(a) The particle size distribution of the detached sediment is the same as in the original soil.

(b) Detachment occurs only if there is excess transport capacity.

This means that the particle size distribution of the transported sediment corresponds to that of the initial soil, as long as the mass rate of the detached particles does not exceed the transport capacity in any of the particle classes considered. If that is not the case, the mass rate of the particles and hence the size distribution of the transported sediment is controlled by transport capacity.

Erosion/deposition
In order to calculate the rate of erosion or deposition for each of the individual slope segments the following simple equation is used:

$$\gamma = \left[\frac{q_{s,in} - q_{s,out}}{\Delta x} \right] \qquad (17.22)$$

where γ is the rate of erosion ($\gamma < 0$) or deposition ($\gamma > 0$) per unit area, $q_{s,in}$ is the sediment discharge into the segment from the segment above, $q_{s,out}$ is the sediment discharge out of the segment and Δx is the length of the slope segment.

Program structure and operation

In order to make the model applicable for practical purposes the equations described above have been transferred to a computer program called EROSION 2D.

Input parameters and data management
EROSION 2D uses three main groups of input parameters: relief, soil and precipitation. These groups consist of the following individual parameters: The program also takes into account a number of other parameters, which are fixed and cannot be influenced by the user. They include fluid and particle density, fluid viscosity, and acceleration due to gravity.

Each group of parameters mentioned in Table 17.1 is organized in a separate file. The relief parameter file contains the geometry data of the initial slope

Table 17.1 Input parameters.

Relief	Soil	Precipitation
Slope length	Grain size	Precipitation intensity
Slope geometry (x-, y- coordinates)	Infiltration rate	
	Resistance to erosion (crit. momentum flux)	
	Surface roughness (Manning's n)	
	Canopy cover	

profile. The file comprises at least two pairs of coordinates, denoting the top and the bottom point of the slope profile. (However, more than two points are generally necessary to represent the slope geometry adequately.) The soil parameter file contains the soil- and canopy-specific data of any number of homogeneous slope segments ≥ 1 m, and the precipitation parameter file comprises the duration and the rainfall intensity data in 10 min intervals. Unlike the relief parameters, the soil and precipitation parameters refer to one actual event or to a time section of such an event ≥ 10 min.

Output parameters

Table 17.2 shows the output parameters supplied by the program.

Table 17.2 Output parameters.

Area-related parameters	Point-related parameters
Rate of erosion per unit area	Sediment discharge
	Sediment concentration
Rate of deposition per unit area	Clay, silt and sand fractions of the transported sediment

The area-related parameters each relate to the pre-set slope segments (Δx = 1 m, Δy = 1 m) or to their multiples (Δx = 1, 2, 3, ... n m, Δy = 1 m). The point-related parameters refer to a specific, user-selected slope position (e.g. the sediment concentration at the position x = 156 m). The time reference basis for all output parameters is the internally determined time interval (Δt = 10 min) or a corresponding multiple (Δt = 20, 30, 40, ... n min).

Long-term simulation

The long-term option of EROSION 2D is based on the coupling of any number of individual erosional events. Two complementary procedures are employed. The first procedure combines several events into a sequence, for example the erosional events of 1 year or of a particular season. The program automatically identifies and processes sequential files (containing the corresponding soil and precipitation data of each event). The sequential procedure is particularly suitable for simulating the seasonal or annual rate of erosion. If a longer period (several years to several hundred years) is to be simulated, an iterative procedure is employed, based on either a single event or on a sequence of events.

In both the sequential and the combined sequential–iterative procedures, the erosion/deposition-dependent changes in slope geometry are taken into account. After processing each of the successive events the slope profile is modified according to the calculated rates of erosion and deposition. At the end of simulation the initial and final slope profiles and the cumulated rates of erosion and deposition are displayed in graphical and numerical form.

Sensitivity analysis

In the following, some hypothetical examples are used to examine the dependence of predicted soil loss upon slope length and geometry, gradient, rainfall intensity and soil properties. First, the impact of slope length and geometry is investigated, using the following variants:

(a) straight slope;

(b) convex slope;

(c) concave slope;

(d) convex-concave slope; and

(e) irregularly shaped slope.

In all cases, total slope length is 130 m and the difference in height between the top point of the slope and the base is 14 m.

Figure 17.2 shows the results for the straight slope. The top half of Figure 17.2a describes the slope profile, the lower half the calculated soil loss

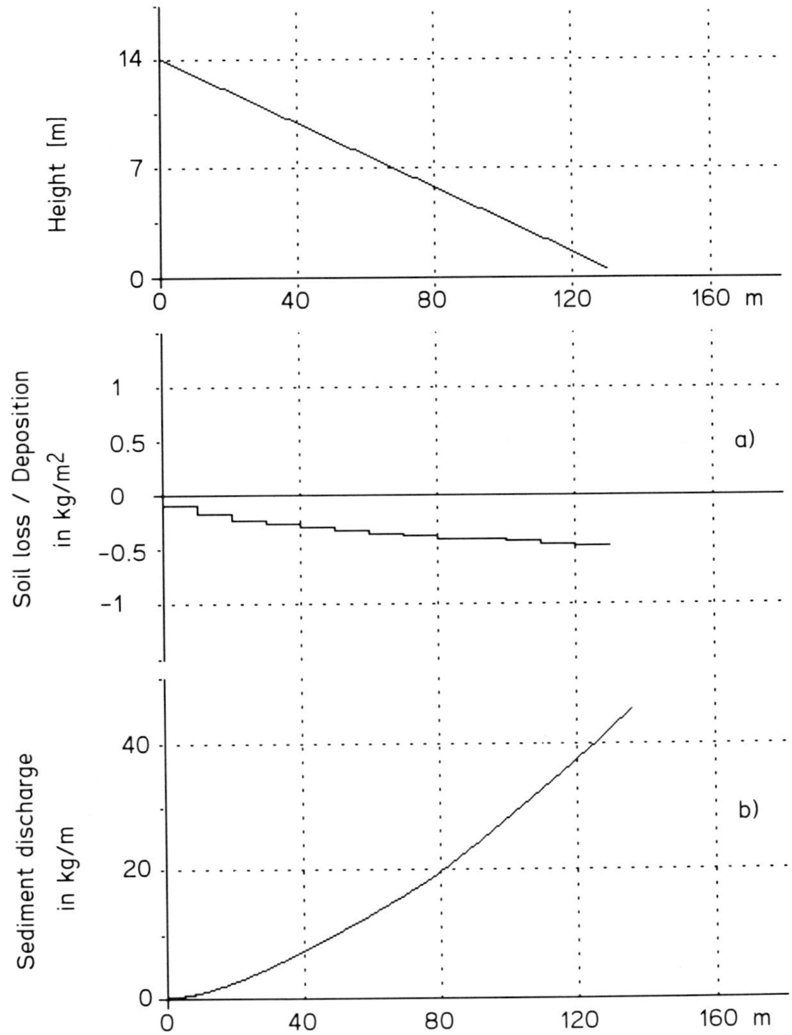

Figure 17.2 Calculated soil loss and sediment discharge on a straight slope.

(negative values) or deposition (positive values). We see that soil loss per unit area increases with slope length. However, the increase in soil loss diminishes as slope length increases. If, instead of soil loss, the sediment discharge is plotted against slope length (Fig. 17.2b), a comparison of both curves shows that the increase in soil loss on the upper slope is linked to an overproportional increase in sediment discharge. On the lower slope, however, sediment discharge increases more or less linearly, and soil loss per unit area is

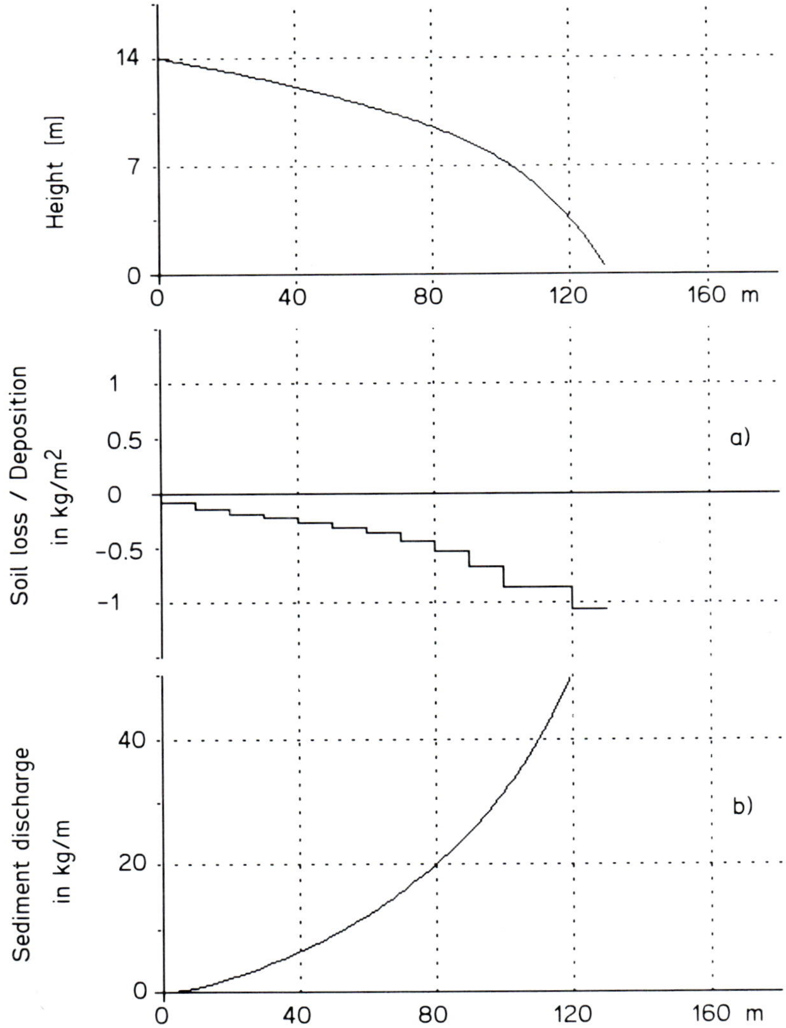

Figure 17.3 Calculated soil loss and sediment discharge on a convex slope.

approximately constant.

In the case of the convex slope also, soil loss increases with slope length (Fig. 17.3). In contrast to the straight slope, the increase in soil loss is larger with increasing slope length. The same is true of the sediment discharge.

The results are completely different in the case of the concave slope. Here soil loss per unit area is more or less constant throughout the entire slope (Fig. 17.4a). Accordingly, the increase in sediment discharge with slope length is aproximately linear (Fig. 17.4b).

Figure 17.5a shows soil loss at a convex–concave slope. As expected, soil

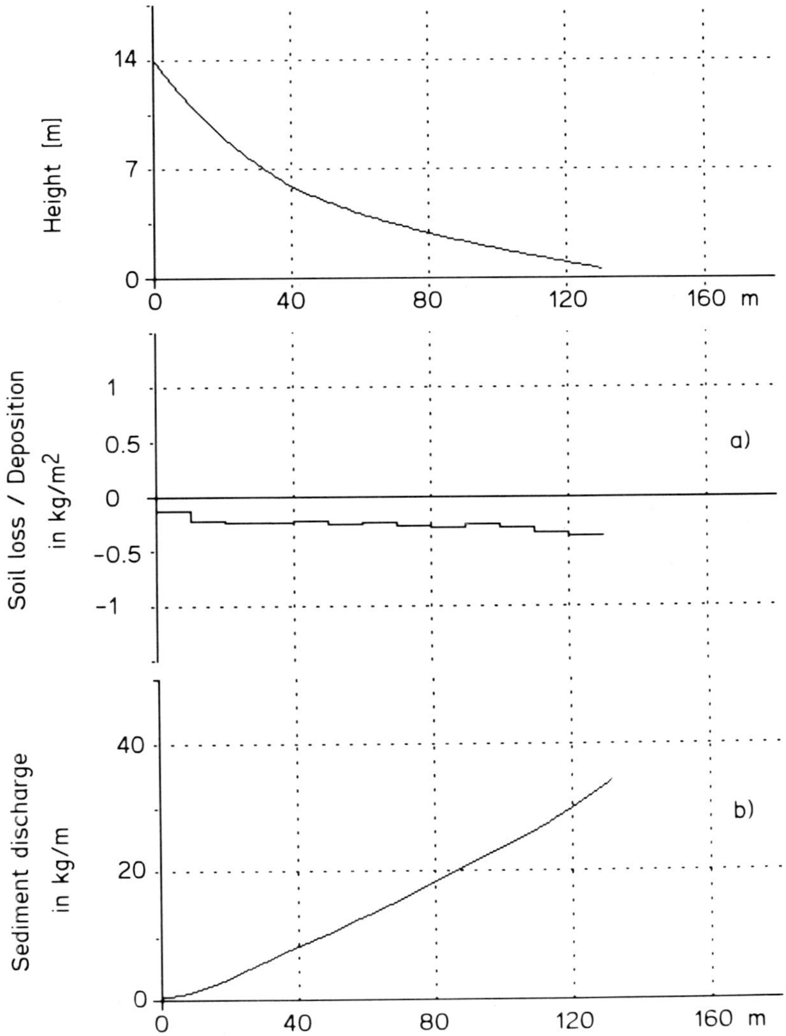

Figure 17.4 Calculated soil loss and sediment discharge on a concave slope.

loss increases with slope length on the upper, convex slope section, reaching a maximum in the middle third of the profile and rapidly decreasing at the transition to the lower, concave slope section. At the base of the slope, soil loss decreases to almost zero. Sediment discharge (Fig. 17.5b) first increases exponentially, then almost linearly with slope length, finally reaching an almost constant value. (As Fig. 17.5a shows, soil loss is then practically zero).

In the case of the irregular slope (Fig. 17.6a), soil loss peaks at the convex slope sections, while the concave part at mid-slope undergoes very little erosion. Sediment discharge increases exponentially in each of the convex

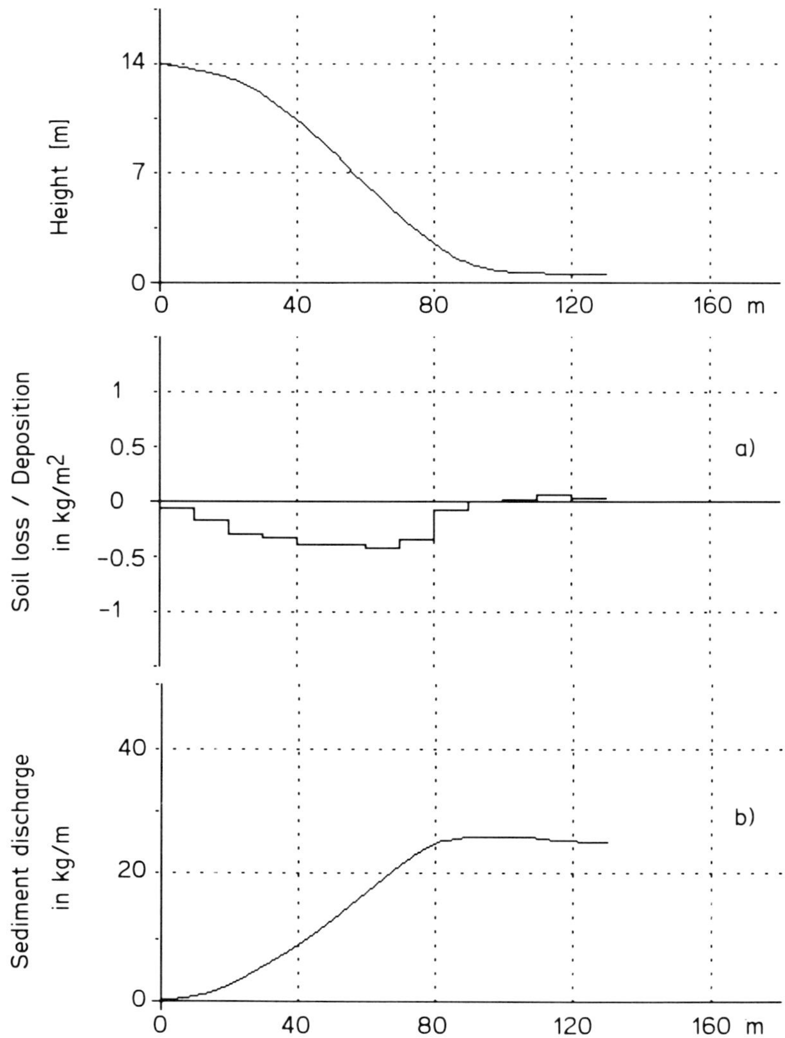

Figure 17.5 Calculated soil loss and sediment discharge on a convex–concave slope.

slope sections. By contrast, the concave sections have approximately constant values (Fig. 17.6b).

As the preceding figures have shown, the simulated variations in slope geometry affect both the relation of soil loss to slope length and the absolute amount of soil loss. Since slope length and difference in height have been kept constant, it is possible to compare the net amounts of soil loss calculated for each variant. Taking as reference value the soil loss on the straight slope (= 100%), we obtain the diagram shown in Figure 17.7. It is obvious that the

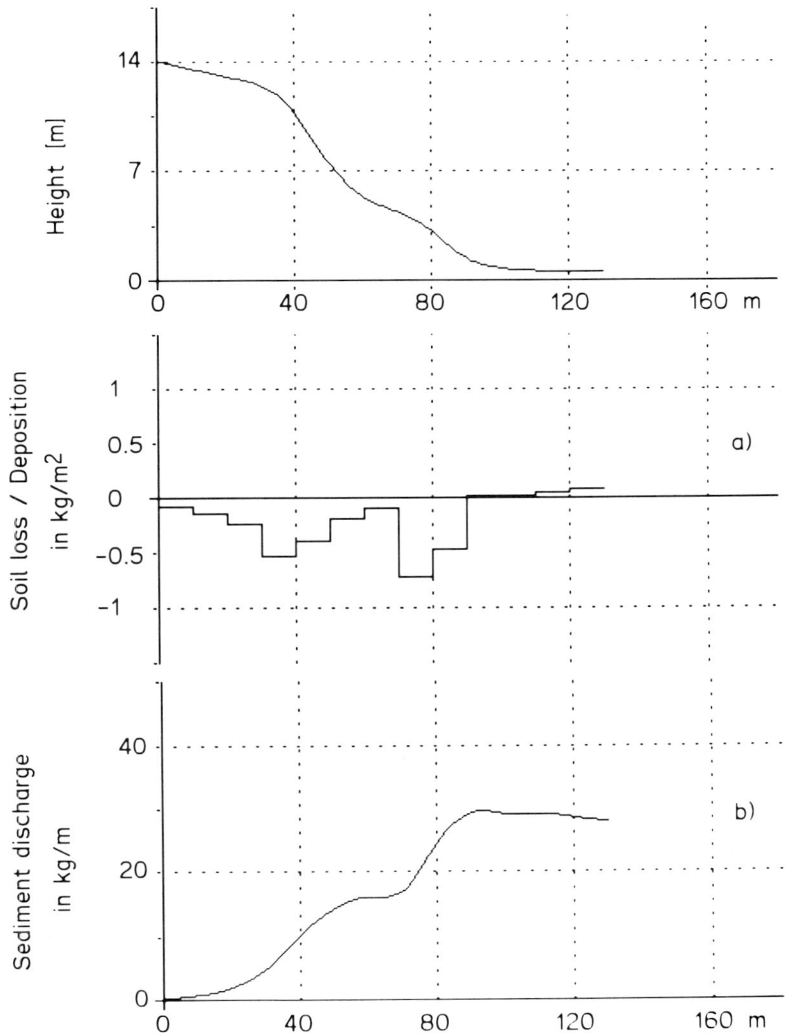

Figure 17.6 Calculated soil loss and sediment discharge on an irregular slope.

convex slope loses by far the most soil (139%), exceeding even the straight slope. The convex–concave slope loses least soil (57%). This is particularly interesting because most natural slopes belong in this category.

A further decisive factor governing soil loss is slope gradient. The straight slope demonstrates this most clearly (cf. Fig. 17.2). In a series of nine simulation runs, slope inclination was increased in steps of 5°. In Figure 17.8 the predicted soil loss (related to soil loss at 45°) is plotted against slope angle. We see that soil loss increases as the slope becomes steeper. However, the increase in soil loss decreases exponentially with increasing slope angle.

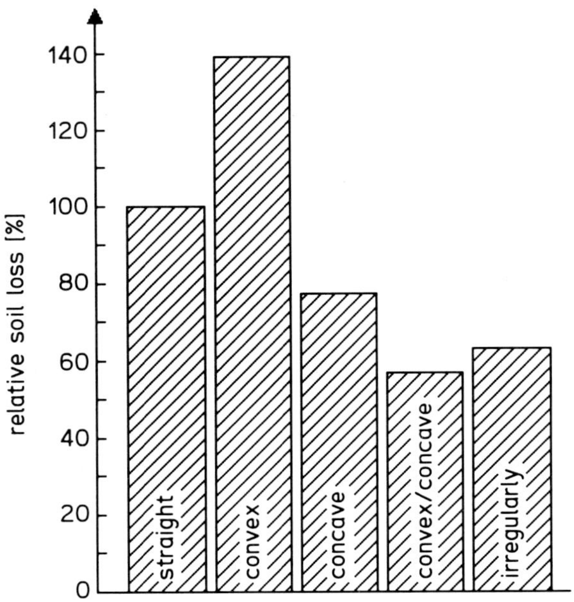

Figure 17.7 Dependence of soil loss upon slope geometry.

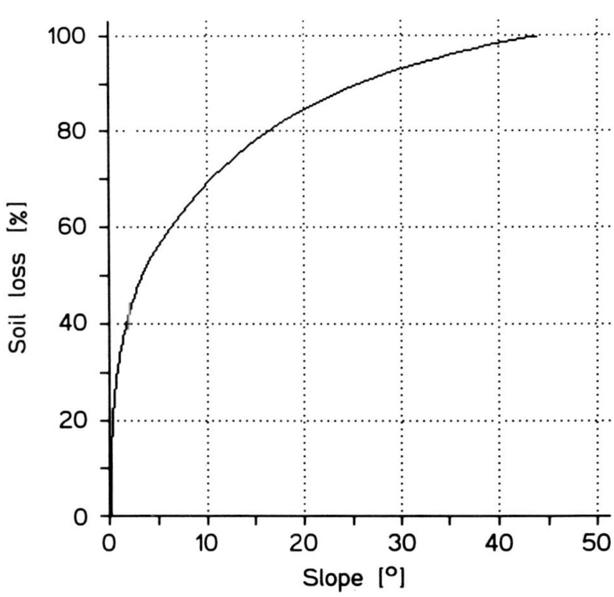

Figure 17.8 Soil loss as a function of slope gradient.

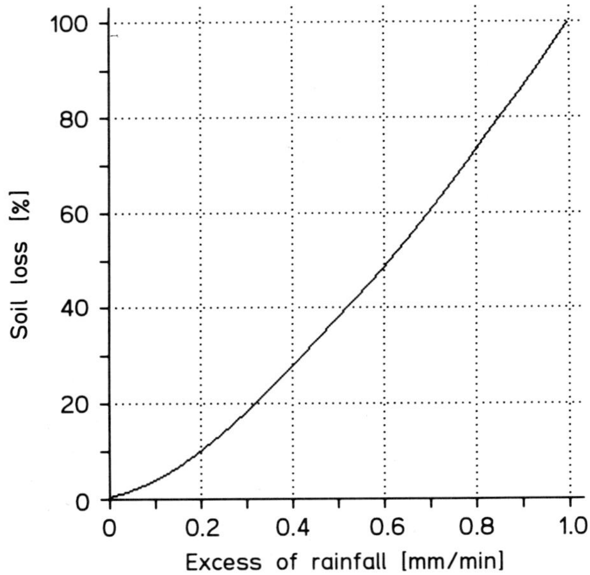

Figure 17.9 Soil loss as a function of rainfall intensity.

Figure 17.9 describes the relationship between soil loss and rainfall intensity. Again, the simulations are based on the straight slope shown in Figure 17.2. Rainfall intensity was varied in steps of 0.1 mm min^{-1}, rainfall duration remaining constant. Taking as reference value the soil loss at 1.0 mm min^{-1}, Figure 17.9 shows the predicted soil loss as a function of rainfall intensity. (Here, rainfall intensity relates to the excessive rainfall, i.e. the infiltration rate has already been subtracted.) We see that increasing intensity is always linked to increasing soil loss.

In the model, the soil and surface properties are described by means of the following parameters: grain size, resistance to erosion, surface roughness and canopy cover. Resistance to erosion and surface roughness are, in turn, dependent on grain size and canopy cover. The canopy cover, in particular, is subject to considerable spatial and temporal fluctuations, depending on land use. Because of the interdependence of the various parameters, it is not practical to vary one of the above parameters separately while the others remain constant. Since the simulation examples described in the following section give an idea of the interaction of the various individual parameters, further analyses can be dispensed with here.

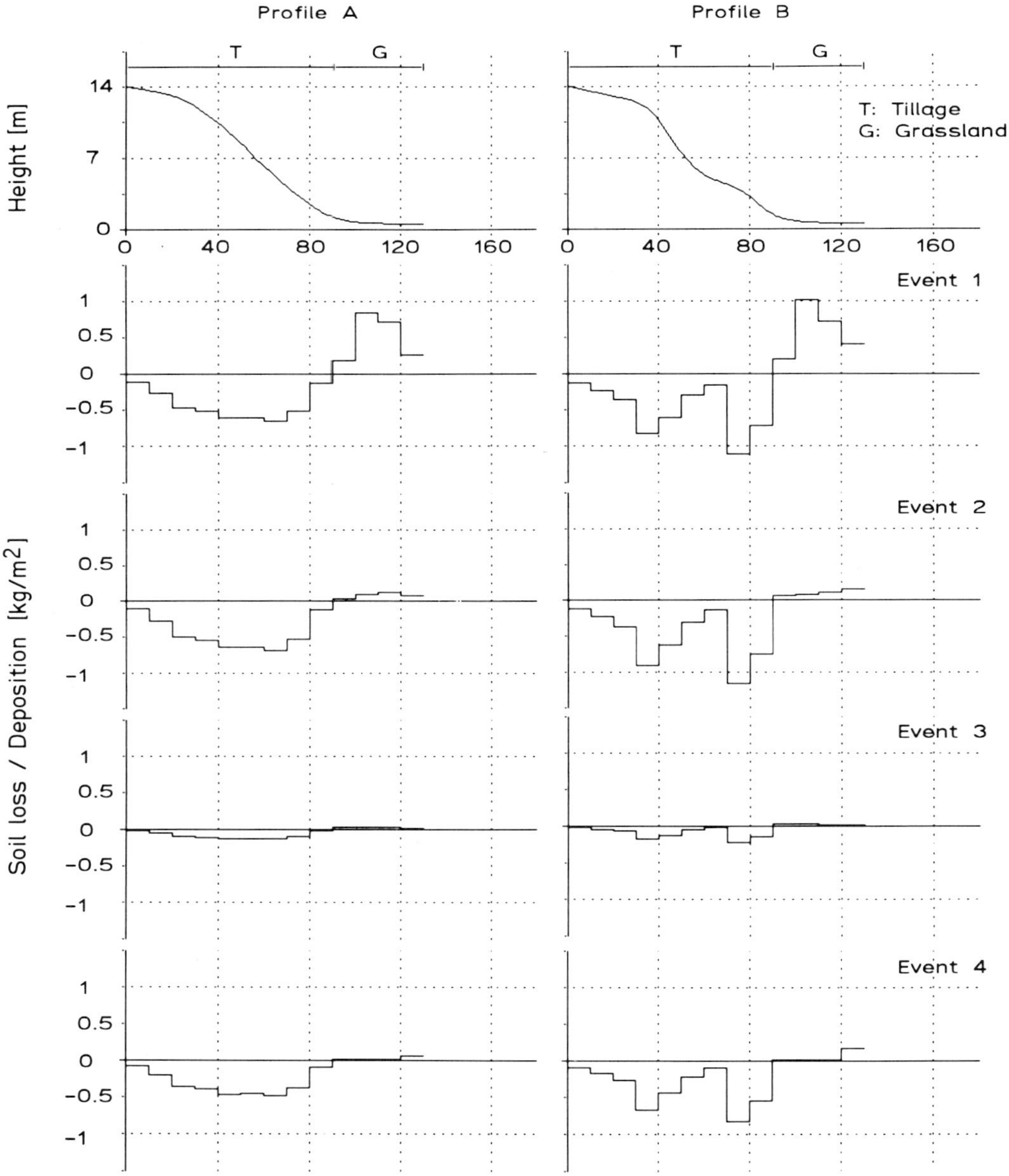

Figure 17.10 Single events 1–4 (Profile A).

Figure 17.11 Single events 1–4 (Profile B).

Long-term simulations

Data

Simulations are based on four, sequentially linked events representing different seasonal conditions. Event 1 applies to a bare slope and represents the start of the vegetation period. The events shown in Table 17.3 characterize conditions of increasing cover. The rainfall duration of each individual event is 40 min. Rainfall intensity varies according to Table 17.4. Beyond that, the following variations in agricultural use are taken into account: I, tillage over the entire slope; II, tillage interrupted by strips of permanent vegetation (green strips); III, tillage followed by permanent vegetation (e.g. grassland) at the bottom of the slope.

Table 17.3 Soil and canopy data.

	Event			
	1	2	3	4
Infiltration rate (mm min^{-1})	0.1	0.2	0.2	0.2
Resistance to erosion ([kg m] s^{-2})	0.0005	0.0010	0.0015	0.0020
Surface roughness (s $m^{-1/3}$)	0.01	0.02	0.03	0.04
Canopy cover (%)	0	20	50	80

Table 17.4 Rainfall intensity data (in mm min^{-1}).

	Event			
Time	1	2	3	4
0 – 10 min	0.2	0.2	0.2	0.4
10 – 20 min	0.3	0.7	0.4	0.8
20 – 30 min	0.3	0.8	0.5	1.2
30 – 40 min	0.2	0.2	0.3	0.6

Single events

Looking at each of the single events (Figs 17.10 & 17.11), it is easy to see how they differ. Precipitation was clearly less heavy, but more material was redeposited during the first event than during the following ones. In particular, deposition at the slope base was considerably greater than in the following events. The main reasons for this are the lack of plant cover during event 1 (start of the vegetation period) and the decreased resistance to erosion offered by the tilled soil.

Long-term simulations

The following long-term simulations are based on the iteration of the sequence described above. Figures 17.12–17.16 each show the initial slope profile and the superimposed curve of the slope profile as calculated after 250 iterations. In addition, the cumulated soil loss and deposition is plotted in the lower half of the diagrams.

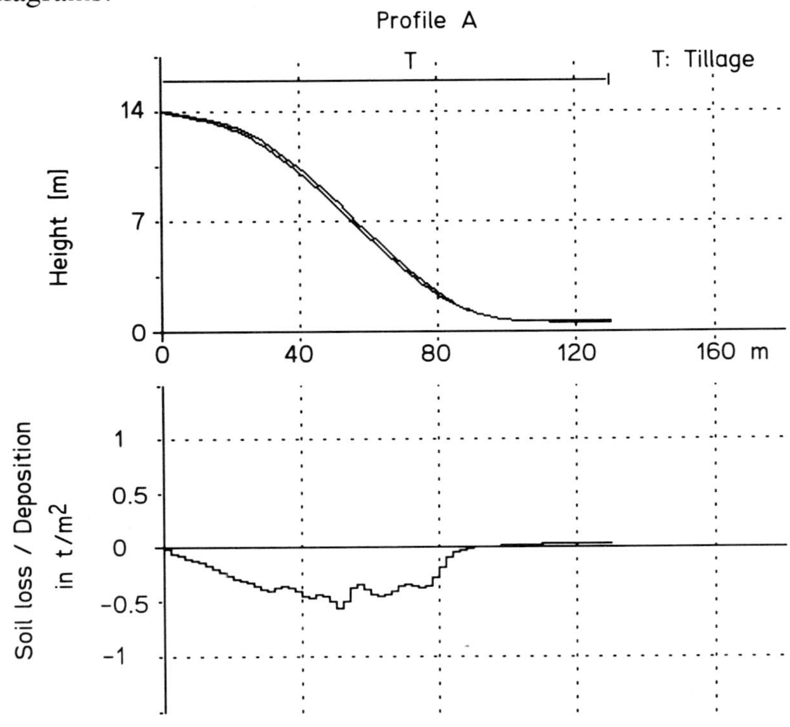

Figure 17.12 Long-term simulation (Profile A/I).

Figures 17.12 & 17.13 show conditions when the slope is used for tillage only. As Figure 17.12 shows, a more or less smooth, convex–concave slope (Profile A) undergoes almost parallel backward displacement with no substantial change in geometry. This is not the case when slope shape is more complex (Profile B, Fig. 17.13). Here erosion tends to smooth down the convex parts of the slope overlying the general slope profile. This is particularly noticeable on the lower parts of the slope, owing to the increase of runoff with slope length. In comparison, the concave section mid-slope is subject to considerably less erosion.

The examples below show the geometrical changes of the slope profile when slope use is not homogeneous. Figure 17.14 describes conditions when the

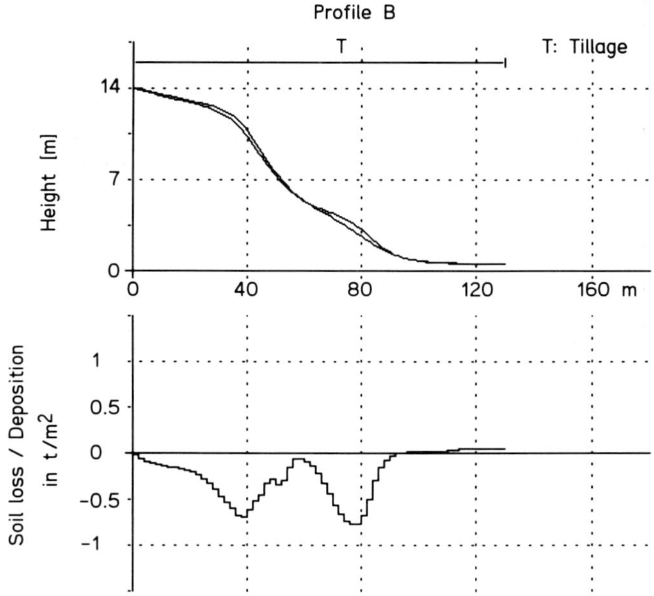

Figure 17.13 Long-term simulation (Profile B/I).

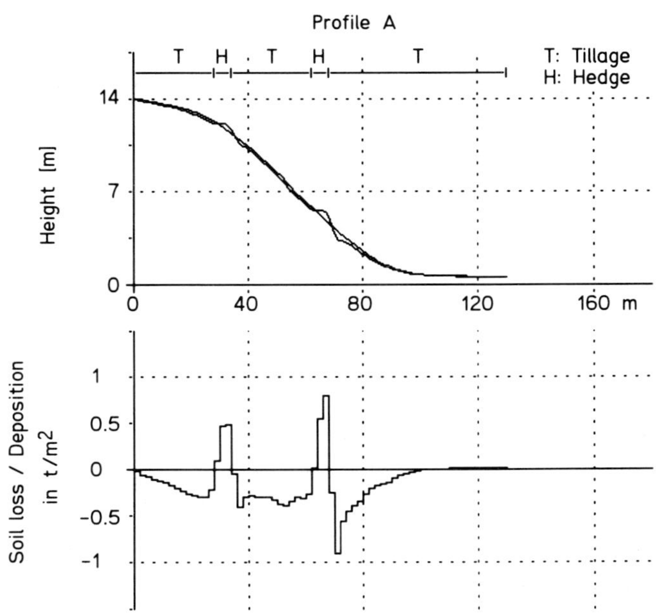

Figure 17.14 Long-term simulation (Profile A/II).

slope is partitioned into three sections, divided from each other by strips of permanent vegetation. The green strips decelerate surface runoff and thus lead

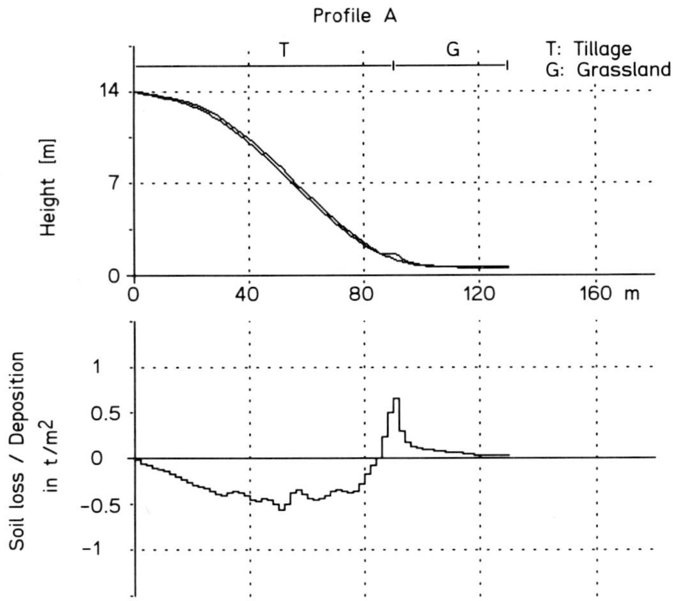

Figure 17.15 Long-term simulation (Profile A/III).

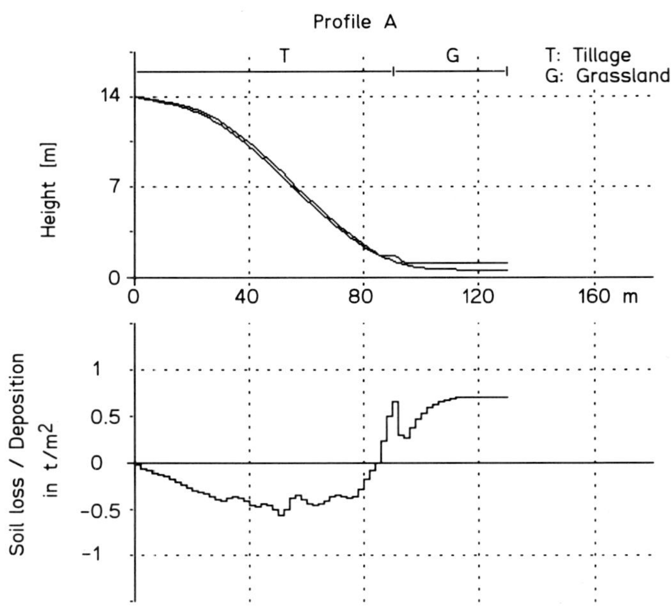

Figure 17.16 Long-term simulation (Profile A/III) taking deposition of flood sediments into account.

to partial or complete deposition of the soil eroded further up slope. Runoff is therefore almost devoid of sediment when leaving the green strips, and erosion starts again below these strips with increased intensity. The combination of both these processes – accumulation within the green strips, increased erosion below them – leads to the formation of small steps. As Figure 17.14 shows, these steps are more pronounced in the lower part of the slope than in the upper because runoff increases with slope length.

In the examples depicted in Figures 17.15 & 17.16 the upper slope is tilled and the lower part is covered by grassland. As in the case of the green strips, soil eroded up slope is deposited in the grassland down slope. The curve of sediment deposition shows that sedimentation quickly reaches a maximum when overland flow enters the grassland and then continuously decreases towards the bottom of the slope. The steady accumulation of soil in the transitional area between tilled soil and grassland finally also leads to the formation of a small step; however, unlike the steps described above, this is purely colluvial in origin.

In the last example (Fig. 17.16) it is assumed that the base level of the slope rises by 0.5 mm per iteration step, owing to the deposition of flood sediments. In contrast to the previous examples, the slope foot merges into a completely flat area of sedimentation. In other respects the profile does not differ substantially from the curve in Figure 17.15.

Assessment of results

The results yielded by simulation seem plausible. Hence we may assume that the model accurately describes the processes of erosion by water. To validate the model further it is necessary to compare its results with field data. Such studies are currently underway. Initial results are encouraging: an average deviation of 18% was noted. This is within the desired accuracy limits. However, since comparative data are scarce, it is not yet possible to give a final assessment of the model's accuracy.

Symbols and units

A	area of slope segment	m^2
C_L	ground cover	
c_{max}	concentration of particles at transport capacity	$m^3\ m^{-3}$
D	particle size	m
E	erosion coefficient	
g	acceleration due to gravity	$m\ s^{-2}$
i	infiltration rate	$m\ s^{-1}$

n	roughness coefficient	$s\ m^{-1/3}$
q	volume rate of flow	$m^3\ (m\ s)^{-1}$
q_{crit}	volume rate of flow at initial erosion	$m^3\ (m\ s)^{-1}$
q_s	sediment discharge	$kg\ (m\ s)^{-1}$
r	rainfall intensity	$m\ s^{-1}$
S	slope gradient	$m\ m^{-1}$
Δt	time interval	s
v_p	settling velocity of soil particles	$m\ s^{-1}$
v_q	mean flow velocity	$m\ s^{-1}$
v_r	mean fall velocity of raindrops	$m\ s^{-1}$
w_p	mass rate of particles	$kg\ (m^2\ s)^{-1}$
w_q	mass rate of flow	$kg\ (m\ s)^{-1}$
w_r	mass rate of rainfall	$kg\ (m^2\ s)^{-1}$
Δx	length of slope segment	m
Δy	width of slope segment	m
α	slope angle	
γ	erosion ($\gamma < 0$) or deposition ($\gamma > 0$)	$kg\ (m^2\ s)^{-1}$
δ	mean flow depth	m
η	fluid viscosity	$kg\ (m\ s)^{-1}$
κ	factor	
ρ_p	particle density	$kg\ m^{-3}$
$\rho_{q,r}$	fluid density	$kg\ m^{-3}$
φ_{crit}	critical momentum flux	$kg\ m\ s^{-2}$
$\varphi_{p,crit}$	critical momentum flux of suspended particles	$kg\ m\ s^{-2}$
φ_q	momentum flux exerted by flow	$kg\ m\ s^{-2}$
$\varphi_{q,vert}$	vertical momentum flux component of flow	$kg\ m\ s^{-2}$
φ_r	momentum flux exerted by raindrops	$kg\ m\ s^{-2}$

Acknowledgements

The research project on which this paper is based was financially supported by the Bundesminister für Forschung und Technologie, registration no. 0339233A.

References

Bork, H.-R. 1988. Bodenerosion und Umwelt. *Landschaftsgenese und Landschaftsökologie* **13**

Emmett, W. W. 1970. *The hydraulics of overland flow on hillslopes.* US Geological Survey Paper, 662A. Washington DC: US Government Printing Office.

Hjulström, F. 1935. Studies of the morphological activity of rivers as illustrated by the River

Fyris. *Bulletin of the Universty of Upsala.*
Laws, J. D. 1941. Measurements of the fall-velocity of water-drops and raindrops. *Transactions of the American Geophysical Union* **21**, 709–21.
Laws, J. D. & D. A. Parsons 1943. The relation of raindrop size to intensity. *Transactions of the American Geophysical Union* **24**, 452–60
Pearce, A. J. 1976. Magnitude and frequency of erosion by Hortonian overland flow. *Journal of Geology*, **84** 65–80
Schmidt, J. 1988. Wasserhaushalt und Feststofftransport an geneigten, landwirtschaftlich bearbeiteten Nutzflächen. Unpublished dissertation. Free University of Berlin.
Schmidt, J. 1991a. A mathematical model to simulate rainfall erosion. *Catena Supplement* **19**, 101–9.
Schmidt, J. 1991b. Predicting the sediment yield from agricultural land using a new soil erosion model. In *Proceedings of the 5th International Symposium on River Sedimentation (ISRS)*, Karlsruhe (in press).
Wischmeier, W. H. & D. D. Smith 1965. *Predicting rainfall-erosion losses from cropland east of the Rocky Mountains.* US Department of Agriculture Handbook 282, Washington DC.

Index

References to figures are given in *italics*; those to tables are given in **bold** type.